Advances in Intelligent Systems and Computing

Volume 202

For further volumes:
http://www.springer.com/series/11156

Jagdish Chand Bansal ·
Pramod Kumar Singh · Kusum Deep
Millie Pant · Atulya K. Nagar
Editors

Proceedings of Seventh International Conference on Bio-Inspired Computing: Theories and Applications (BIC-TA 2012)

Volume 2

 Springer

Editors

Jagdish Chand Bansal
South Asian University
Chankya Puri, New Delhi
India

Pramod Kumar Singh
ABV-IIITM, Gwalior
Gwalior, Madhya Pradesh
India

Kusum Deep
Department of Mathematics
Indian Institute of Technology Roorkee
Roorkee
India

Millie Pant
Department of Applied Science
    and Engineering
Indian Institute of Technology Roorkee
Roorkee
India

Atulya K. Nagar
Department of Mathematics
    and Computer Science
Liverpool Hope University
Liverpool
UK

ISSN 2194-5357   ISSN 2194-5365 (electronic)
ISBN 978-81-322-1040-5 ISBN 978-81-322-1041-2 (eBook)
DOI 10.1007/978-81-322-1041-2
Springer New Delhi Heidelberg New York Dordrecht London

Library of Congress Control Number: 2012954374

Preface

Human beings have always been fascinated by nature and especially by biological diversity and their evolutionary process. This has resulted into inspirations drawn from natural or biological systems, and phenomenon, for problem solving and has seen an emergence of a new paradigm of computation known as Natural Computing with Bio-inspired Computing as its subset. The widely popular methods, e.g., evolutionary computation, swarm intelligence, artificial neural networks, artificial immune systems, are just some examples in the area. Such approaches are of much use when we need an imprecise, inaccurate but feasible solution in a reasonable time as many real-world problems are too complex to be dealt using traditional methods of finding exact solutions in a reasonable time. Therefore, bio-inspired approaches are gaining popularity as the size and complexity of the real-world problems require the development of methods which can give the solution within a reasonable amount of time rather than an ability to guarantee the exact solution. Bio-inspired Computing can provide such a rich tool-chest of approaches as it tends to be, just like its natural system counterpart, decentralized, adaptive and environmentally aware, and as a result have survivability, scalability and flexibility features necessary to deal with complex and intractable situations.

Bio-Inspired Computing: Theories and Applications (BIC-TA) is one of the flagship conferences on Bio-Computing bringing together the world's leading scientists from different branches of Natural Computing. Since 2006 the conferences have taken place at Wuhan (2006), Zhengzhou (2007), Adelaide (2008), Beijing (2009), Liverpool and Changsha (2010), Penang (2011). BIC-TA has attracted wide ranging interest amongst researchers with different backgrounds resulting in a seventh edition in 2012 at Gwalior. It is our privilege to have been part of this seventh edition of the BIC-TA series which is being hosted for the first time in India.

This volume in the AISC series contains papers presented at the Seventh International Conference on Bio-Inspired Computing: Theories and Applications (BIC-TA 2012) held during December 14–16, 2012 at ABV-Indian Institute of Information Technology and Management Gwalior (ABV-IIITM Gwalior), Madhya Pradesh, India. The BIC-TA 2012 provides a unique forum to researchers and practitioners working in the ever growing area of bio-inspired computing methods and their applications to solve various real-world problems.

BIC-TA 2012 attracted attention of researchers from all over the globe and we received 188 papers related to various aspects of bio-inspired computing with umpteen applications, theories, and techniques. After a thorough peer-review process a total of 91 thought-provoking research papers are selected for publication in the Proceedings, which is in two volumes (Volume 1 and 2). This thus corresponds to an acceptance rate of 48% and is intended to maintain a high standard in the conference proceedings. We hope that the papers contained in this proceeding will serve the purpose of inspiring more and more researchers to work in the area of bio-inspired computing and its application.

The editors would like to express their sincere gratitude to the authors, plenary speakers, invited speakers, reviewers, and members of international advisory committee, programme committee and local organizing committee. It would not have been possible to come out with the high quality and standard of the conference as well as this edited Proceeding without their active participation and whole hearted support. It would not be fair on our part if we forget to mention special thanks to the ABV – Indian Institute of Information Technology and Management Gwalior (ABV-IIITM Gwalior) and its Director Prof. S. G. Deshmukh for providing us all the possible help and support including excellent infrastructure of the Institute to make this conference a big success. We express our gratitude to the Department of Mathematics and Computer Science, Liverpool Hope University, Liverpool, UK headed by Prof. Atulya K. Nagar for providing us much valued and needed support and guidance. Finally, we would like to thank all the volunteers; their untiring efforts in meeting the deadlines and managerial skills in managing the resources effectively and efficiently which has ensured a smooth running of the conference.

It is envisaged that the BIC-TA conference series will continue to grow and include relevant future research and development challenges in this exciting field of Computing.

Jagdish Chand Bansal, South Asian University, New Delhi, India
Pramod Kumar Singh, ABV-IIITM, Gwalior, India
Kusum Deep, Indian Institute of Technology, Roorkee, India
Millie Pant, Indian Institute of Technology, Roorkee, India
Atulya K. Nagar, Liverpool Hope University, Liverpool, UK

Editors

Jagdish Chand Bansal
Pramod Kumar Singh
Kusum Deep
Millie Pant
Atulya K. Nagar

About Editors

Dr. Jagdish Chand Bansal is an Assistant Professor with the South Asian University New Delhi, India. Holding an excellent academic record, he is a budding researcher in the field of Swarm Intelligence at the International Level.

Dr. Pramod Kumar Singh is an Associate Professor with the ABV-Indian Institute of Information Technology and Management, Gwalior, India. He is an active researcher and has earned a reputation in the areas of Nature-Inspired Computing, Multi-/Many-Objective Optimization, and Data Mining.

Dr. Kusum Deep is a Professor with the Department of Mathematics, Indian Institute of Technology Roorkee, Roorkee, India. Over the last 25 years, her research is increasingly well-cited making her a central International figure in the area of Bio-Inspired Optimization Techniques, Genetic Algorithms and Particle Swarm Optimization.

Dr. Millie Pant is an Associate Professor with the Department of Applied Science and Engineering, Indian Institute of Technology, Roorkee, Roorkee, India. At this age, she has earned a remarkable International reputation in the area of Genetic Algorithms, Differential Algorithms and Swarm Intelligence.

Prof. Atulya K. Nagar is the Professor and Head of Department of Mathematics and Computer Science at Liverpool Hope University, Liverpool, UK. Prof. Nagar is an internationally recognized scholar working at the cutting edge of theoretical computer science, natural computing, applied mathematical analysis, operations research, and systems engineering and his work is underpinned by strong complexity-theoretic foundations.

Organizing Committees

BIC-TA 2012 was held at ABV- Indian Institute of Information Technology and Management Gwalior, India. Details of the various organizing committees are as follows:

Patron: S. G. Deshmukh, ABV-IIITM Gwalior, India

General Chairs: Atulya Nagar, Liverpool Hope Uuniversity Liverpool, UK
Kusum Deep, IIT Roorkee, India

Conference Chairs: Jagdish Chand Bansal, South Asian University New Delhi, India
Pramod Kumar Singh, ABV-IIITM Gwalior, India

Program Committee Chairs: Millie Pant, IIT Roorkee, India
T. Robinson, MCC Chennai, India

Special Session Chair: Millie Pant, IIT Roorkee, India

Publicity Chairs: Manoj Thakur, IIT Mandi, India
Kedar Nath Das, NIT Silchar, India

Best Paper Chair: Kalyanmoy Deb, IIT Kanpur, India
(Technically Sponsored by KanGAL, IIT Kanpur, India)

Conference Secretaries: Harish Sharma, ABV-IIITM Gwalior, India
Jay Prakash, ABV-IIITM Gwalior, India
Shimpi Singh Jadon, ABV-IIITM Gwalior, India
Kusum Kumari Bharti, ABV-IIITM Gwalior, India

Local Arrangement Committee: Jai Prakash Sharma
(ABV-IIITM Gwalior, India) Narendra Singh Tomar
Alok Singh Jadon
Rampal Singh Kushwaha
Mahesh Dhakad
Balkishan Gupta

International Advisory Committee: Atulya K. Nagar, UK
Gheorghe Paun, Romania
Giancarlo Mauri, Italy
Guangzhao Cui, China
Hao Yan, USA
Jin Xu, China
Jiuyong Li, Australia
Joshua Knowles, UK
K G Subramanian, Malaysia
Kalyanmoy Deb, India
Kenli Li, China
Linqiang Pan, China
Mario J. Perez-Jimenez, Spain
Miki Hirabayashi, Japan
PierLuigi Frisco, UK
Robinson Thamburaj, India
Thom LaBean, USA
Yongli Mi, Hong Kong

Special Sessions:

Session 1: Computational Intelligence in Power and Energy Systems, Amit Jain, IIIT Hyderabad, India

Session 2: Bio-Inspired VLSI and Embedded System, Balwinder Raj, NIT Jalandhar, India

Session 3: Recommender System: Design Using Evolutionary & Natural Algorithms, Soumya Banerjee Birla Institute of Technology Mesra, India & Shengbo Guo, Xerox Research Centre Europe, France

Session 4: Image Analysis and Pattern Recognition, K. V. Arya, ABV-IIITM Gwalior, India

Session 5: Applications of Bio-inspired Techniques to Social Computing, Vaskar Raychoudhury, IIT Roorkee, India

Keynote Speakers:

Title: Spiking Neural P Systems
Speaker: Pan Linqiang

Title: Advancements in Memetic Computation
Speaker: Yew-Soon Ong

Title: Machine Intelligence, Generalized Rough Sets and Granular Mining: Concepts, Features ans Applications
Speaker: Sankar Kumar Pal

Title: Of Glowworms and Robots: A New Paradigm in Swarm Intelligence
Speaker: Debasish Ghose

Title: Advances in Immunological Computation
Speaker: Dipankar Dasgupta

Title: Selection of Machinery Health Monitoring Strategies using Soft Computing
Speaker: Ajit Kumar Verma

Title: Can Fuzzy logic Formalism via Computing with Words Bring Complex Environmental Issues into Focus?
Speaker: Ashok Deshpande

Technical Program Committee:

Abdulqader Mohsen, Malaysia
Abhishek Choubey, India
Adel Al-Jumaily, Australia
Aitor Rodriguez-Alsina, Spain
Akila Muthuramalingam, India
Alessandro Campi, Italy
Amit Dutta, India
Amit Jain, India
Amit Pandit, India
Amreek Singh, India
Anand Sharma, India
Andre Aquino, Brazil
Andre Carvalho, Brazil
Andrei Paun, USA
Andres Muñoz, Spain
Anil K Saini, India
Anil Parihar, India
Anjana Jain, India
Antonio J. Jara, Spain
Anupam Singh, India
Anuradha Fukane, India
Anurag Dixit, India
Apurva Shah, India
Aradhana Saxena, India
Arnab Nandi, India
Arshin Rezazadeh, Iran
Arun Khosla, India
Ashish Siwach, India
Ashraf Darwish, Egypt
Ashwani Kush, India
Atulya K. Nagar, UK
B.S. Bhattacharya, UK
Bahareh Asadi, Iran
Bala Krishna Maddali, India
Balaji Venkatraman, India
Balasubramanian Raman, India
Banani Basu, India
Bharanidharan Shanmugam, Malaysia
Carlos Coello Coello, Mexico
Carlos Fernandez-Llatas, Spain
Chang Wook Ahn, Korea
Chi Kin Chow, Hong Kong
Chu-Hsing Lin, Taiwan
Chun-Wei Lin, Taiwan
Ciprian Dobre, Romania
D.G. Thomas, India
Dakshina Ranjan Kisku, India
Dana Petcu, Romania
Dante Tapia, Spain
Deb Kalyanmoy, India
Debnath Bhattacharyya, India
Desmond Lobo, Thailand
Devshri Roy, India

Dipti Singh, India
Djerou Leila, Algeria
Asoke Nath, India
K K Shukla, India
Kavita Burse, India
Mrutyunjaya Panda, India
Shirshu Varma, India
Raveendranathan K.C., India
Shailendra Singh, India
Eduard Babulak, Canada
Eric Gregoire, France
Erkan Bostanci, UK
F N Arshad, UK
Farhad Nematy, Iran
Francesco Marcelloni, Italy
G.R.S. Murthy, India
Gauri S. Mittal, Canada
Ghanshyamsingh Thakur, India
Gheorghe Paun, Romania
Guoli Ji, China
Gurvinder Singh-Baicher, UK
Hasimah Hj. Mohamed, Malaysia
Hemant Mehta, India
Holger Morgenstern, Germany
Hongwei Mo, China
Hugo Proença, Portugal
Ivica Boticki, Croatia
Jaikaran Singh, India
Javier Bajo, Spain
Jer Lang Hong, Malaysia
Jitendra Kumar Rai, India
Joanna Kolodziej, Poland
Jose Pazos-Arias, Spain
Juan Mauricio, Brazil
K K Shukla, India
K V Arya, India
K.G. Subramanian, Malaysia
Kadian Davis, Jamaica
Kamal Kant, India
Kannammal Sampathkumar, India
Katheej Parveen, India
Kazumi Nakamatsu, Japan
Kedar Nath Das, India
Khaled Abdullah, India
Khelil Naceur, Algeria
Khushboo Hemnani, India
Kittipong Tripetch, Thailand
Kunal Patel, USA
Kusum Deep, India
Lalit Awasthi, India
Lam Thu Bui, Australia
Li-Pei Wong, Malaysia
Lin Gao, China

Linqiang Pan, China
M.Ayoub Khan, India
Madhusudan Singh, Korea
Manjaree Pandit, India
Manoj Saxena, India
Manoj Shukla, India
Marian Gheorghe, UK
Mario Koeppen, Japan
Martin Middendorf, Germany
Mehdi Bahrami, Iran
Mehul Raval, India
Michael Chen, China
Ming Chen, China
Mohammad A. Hoque, United States
Mohammad Reza Nouri Rad, Iran
Mohammed Abdulqadeer, India
Mohammed Rokibul Alam Kotwal, Bangladesh
Mohd Abdul Hameed, India
Monica Mehrotra, India
Monowar T, India
Mourad Abbas, Algeria
Mps Chawla, India
Muhammad Abulaish, Saudi Arabia
N.Ch.Sriman Narayana Iyengar, India
Nand Kishor, India
Narendra Chaudhari, India
Natarajamani S, India
Navneet Agrawal, India
Neha Deshpande, India
Nikolaos Thomaidis, Greece
Ninan Sajeeth Philip, India
O. P. Verma, India
P. G. Sapna, India
P. N. Suganthan, Singapore
Philip Moore, U.K
Pierluigi Frisco, UK
Ponnuthurai Suganthan, Singapore
Pramod Kumar Singh, India
Vidya Dhamdhere, India
Kishan Rao Kalitkar, India
Punam Bedi, India
Qiang Zhang, China
R. K. Singh, India
R. N. Yadav, India
R. K. Pateriya, India
Rahmat Budiarto, Malaysia
Rajeev Srivastava, India
Rajesh Sanghvi, India
Ram Ratan, India
Ramesh Babu, India
Ravi Sankar Vadali, India
Rawya Rizk, Egypt
Razib Hayat Khan, Norway
Reda Alhajj, Canada

Ronaldo Menezes, USA
S. M. Sameer, India
S. R. Thangiah, USA
Sami Habib, Kuwait
Samrat Sabat, India
Sanjeev Singh, India
Satvir Singh, India
Shan He, UK
Shanti Swarup, India
Shaojing Fu, China
Shashi Bhushan Kotwal, India
Shyam Lal, India
Siby Abraham, India
Smn Arosha Senanayake, Brunei
Darussalam Sonia Schulenburg, UK
Sotirios Ziavras, United States
Soumya Banerjee, India
Steven Gustafson, USA
Sudhir Warier, India
Sumithra Devi K A, India
Sung-Bae Cho, Korea
Sunil Kumar Jha, India
Suresh Jain, India
Surya Prakash, India
Susan George, Australia
Sushil Kulkarni, India
Swagatam Das, India
Thambi Durai, India
Thamburaj Robinson, India
Thang N. Bui, USA
Tom Hendtlass, Australia
Trilochan Panigrahi, India
Tsung-Che Chiang, Taiwan
Tzung-Pei Hong, Taiwan
Umesh Chandra Pati, India
Uzay Kaymak, Netherlands
V. Rajkumar Dare, India
Vassiliki Andronikou, Greece
Vinay Kumar Srivastava, India
Vinay Rishiwal, India
Vittorio Maniezzo, Italy
Vivek Tiwari, India
Wahidah Husain, Malaysia
Wei-Chiang Samuelson Hong, China
Weisen Guo, Japan
Wenjian Luo, China
Yigang He, China
Yogesh Trivedi, India
Yoseba Penya, Spain
Yoshihiko Ichikawa, Japan
Yuan Haibin, China
Yunong Zhang, China
Yuzhe Liu, US

Contents

Wavelet-ANN Model for River Sedimentation Predictions

Raj Mohan Singh[1]

[1] Associate Professor, Department of Civil Engineering, MNNIT Allahabad, India
E-mail: rajm@mnnit.ac.in; rajm.mnnit@gmail.com

Abstract. The observation of peak flows into river or stream system is not straight forward but complex function of hydrology and geology. Accurate suspended sediment prediction in rivers is an integral component of sustainable water resources and environmental systems modeling. Agricultural fields' fertility decays, rivers capacity decreases and reservoirs are filled due to sedimentation. The observation of suspended sediment flows into river or stream system is not straight forward but complex function of hydrology and geology of the region. There are statistical approaches to predict the suspended sediments in rivers. Development of models based on temporal observations may improve understanding the underlying hydrological processes complex phenomena of river sedimentation. Present work utilized temporal patterns extracted from temporal observations of annual peak series using wavelet theory. These patterns are then utilized by an artificial neural network (ANN). The wavelet-ANN conjunction model is then able to predict the daily sediment load. The application of the proposed methodology is illustrated with real data.

Keywords: Wavelet analysis, ANN, Wavelet-ANN, Time series modeling, Suspended sediment event prediction.

1 Introduction

Sediment amount which is carried by a river is complex functions of river's flow rate, and the characteristics of the catchment. Though discharge (flow rate) can be measured at a site, determination of catchment characteristics accurately is not always an easy task. The correlation between rivers' flow and sediment observation results and the basin's characteristics should be determined for well planning studies on soil and water resources development [1]. In rivers, a major part of the sediment is transported in suspension. Recently, the importance of correct sediment prediction, especially in flood-prone areas, has increased significantly in water resources and environmental engineering. A great deal of research has been devoted to the simulation and prediction of river sediment yield and its dynamics

J. C. Bansal et al. (eds.), *Proceedings of Seventh International Conference on Bio-Inspired Computing: Theories and Applications (BIC-TA 2012)*, Advances in Intelligent Systems and Computing 202, DOI: 10.1007/978-81-322-1041-2_1, © Springer India 2013

[2][3]. The daily suspended sediment load (S) process is among one of the most complex nonlinear hydrological and environmental phenomena to comprehend, because it usually involves a number of interconnected elements [4]. Classical models based on statistical approach such as multilinear regression (MLR) and sediment rating curve (SRC) are widely used for suspended sediment modeling [5].

Hydrologic time series are generally autocorrelated. Autocorrelation in time series such as streamflow usually arises from the effects of surface, soil, and groundwater storages Sometimes significant autocorrelation may be the result of trends and/or shifts in the series [6]. The application of ANN to suspended sediment estimation and prediction has been recently used [7][8]. An ANN model was also employed to estimate suspended sediment concentration (SSC) in rivers, achieved by training the ANN model to extrapolate stream data collected from reliable sources [9]. Bhattacharya et al. (2005) devised an algorithm for developing a data-driven method to forecast total sediment transport rates using ANN [10]. Raghuwanshi et al. (2006) proposed an ANN model for runoff and sediment yield modeling in the Nagwan watershed in India [11]. The ANN models performed better than the linear regression models in predicting both runoff and sediment yield on daily and weekly simulation scales.

Present work utilized temporal patterns extracted from temporal observations of daily discharge and suspended sediment observed series using wavelet theory. These patterns are then utilized by an artificial neural network (ANN). The wavelet-ANN conjunction model is then utilized to predict the yearly sediment flows in a stream. The application of the proposed methodology is illustrated with real data.

2 Artificial Neural Network

ANN is a broad term covering a large variety of network architecture, the most common of which is a multilayer perceptron feedforwrd network (Fig. 1) with backpropogation algorithm [12]. There is no definite formula that can be used to calculate the number hidden layer(s) and number of nodes in the hidden layer(s) before the training starts, and usually determined by trial-and-error experimentation.

The back propagation algorithm is used for training of the feed forward multilayer perceptron using gradient descent, applied to sum-of-squares error function. This algorithm involves an iterative procedure for minimization of error function, with adjustments to weights being in series of sequence of steps. There are two distinct stages at each step. In the first stage errors are propagated backwards in order to evaluate the derivatives of the error function with respect to weights. In the second stage, the derivatives are used to compute the adjustments to be made with weights [13] [14]. (Bishop, 1995; Singh et al., 2004). Present paper utilized Levenberg- Marquardt (LM) algorithm to optimize the weights and biases in the network. LM algorithm is more powerful and faster than the conventional gradient descent technique [15] [16]. Basics and details of ANN are available in literature [17].

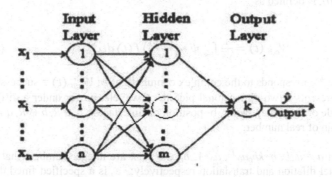

Fig.1. Three layered ANN model architecture

3 Wavelet Analysis

Wavelets are mathematical functions that give a time-scale representation of the time series and their relationships to analyze time series that contain nonstationarities. Wavelet analysis allows the use of long-time intervals for low frequency information and shorter intervals for high frequency information. Wavelet analysis is capable of revealing aspects of original data like trends, breakdown points, and discontinuities that other signal analysis techniques might miss. Furthermore, it can often compress or denoise a signal. Basics of wavelets are available in literature [18] [19].

Discrete wavelet transform (DWT) operates two sets of function (scaling and wavelets) viewed as high-pass and low-pass filters. The original time series are passed through high-pass and low-pass filters and separated at different scales. The time series is decomposed into one comprising its trend the approximation and one comprising the high frequencies and the fast events (the detail). In the present study, the detail coefficients (D) and approximation (*A*) subtime series are obtained using MATLAB wavelet tool box [20]. Wavelet-ANN model is assumed to perform satisfactory when its performance evaluation statistics is not improving or constant. Developed methodology is implemented in MATLAB 7.0 platform (MATLAB, 2004). Model evaluations criteria are utilized to judge the predictive capability of the best performing ANN models. The background information about the wavelet transform is presented below.

Wavelet function $\psi(t)$ is called the mother wavelet, can be defined as:

$$\int_{-\infty}^{\infty} \psi(t)\,dt = 0 \tag{1}$$

Mathematically, the time-scale wavelet transform of a continuous time signal, $f(t)$, is defined as

$$W_{a,b}(t) = \frac{1}{\sqrt{a}} \int_{-\infty}^{\infty} \psi * \left(\frac{(t-b)}{a}\right) f(t)\, dt \qquad (2)$$

where * corresponds to the complex conjugate of ψ; $W_{a,b}(t)$ = successive wavelet and presents a two-dimensional picture of wavelet power under a different scale; a=scale or frequency factor, b=position or time factor; $a \in R, b \in R, a \neq 0$ and R= domain of real number.

Let $a = a_0{}^j$, $b = k b_0 a_0{}^j$, $a_0 > 1$, $b_0 \in R$; j, k are integer numbers that control the wavelet dilation and translation respectively; a_0 is a specified fined dilation step greater then 1; and b_0 is the location parameter and must be greater than zero. The discrete wavelet transform (DWT) of $f(t)$ can be written as:

$$W(a,b) = a_0{}^{-j/2} \int_{-\infty}^{\infty} \psi^*(a_0{}^{-j/2} - k b_0) f(t)\, dt \qquad (3)$$

The most common and simplest choice for the parameters a_0 and b_0 is two and one time steps, respectively. This power of two logarithmic scaling of the time and scale is known as dyadic grid arrangement and is the simplest and most efficient case for practical purposes [21]. Putting $a_0=2$, and $b_0=1$ in the above equation, the dyadic wavelet can be written in more compact notation as:

$$W(a,b) = 2^{-j/2} \int_{-\infty}^{\infty} \psi^*(2^{-j/2} - k) f(t)\, dt \qquad (4)$$

For a discrete time series $f(t)$, which occurs at different time t i.e., here integer time steps are used, the DWT for wavelet of scale $a=2^j$, $b=2^j k$., can be defined as:

$$W(a,b)_D = 2^{-j/2} \sum_{j=0}^{j=J} \psi^*(2^{-j/2} - k) f(t) \qquad (5)$$

Now, the original time series may be represented (reconstructed) as:

$$f_i = \bar{C} + \sum_{j=1}^{J} \sum_{k=0}^{2^{J-k}-1} W(a,b)_D \qquad (6)$$

which may be further simplified as:

$$f_i = \bar{C} + \sum_{j=1}^{J} W_j(t) \qquad (7)$$

where first term, $\bar{C}$, is called approximation sub-signal at level J and second term, $W_j(t)$, are details subsignals (low scale, high frequency) at levels j = 1, 2, . . .,J.

4 Wavelet-ANN Conjunction Model for Flood Events Prediction

Wavelet-ANN conjunction model utilized wavelet decomposed coefficients obtained from wavelet analysis to the ANN technique for daily sediment load prediction as shown in Fig.2. First, the measured daily time series, Q (discharge) (m^3/s) and/or S(suspended sediment) (mg/l) were decomposed into several multi-frequency time series comprising of details (low scale, high frequency) - $Q_{D1}(t)$; $Q_{D2}(t)$;...; $Q_{Di}(t)$ for discharge and $S_{D1}(t)$; $S_{D2}(t)$;...; $S_{Di}(t)$; and approximate (high scale, low frequency) – $Q_a(t)$ for discharge and $S_a(t)$ by DWT. The representation Di presents the level 'i' decomposed details time series and 'a' denotes approximation time series. The decomposed Q(t) and S(t) time series were utilized as inputs to the ANN model and the original time series of observed suspended sediment load at the next step S(t+1) is output to the Wavelet-ANN model. The schematic representation of methodology is presented in Fig. 2.

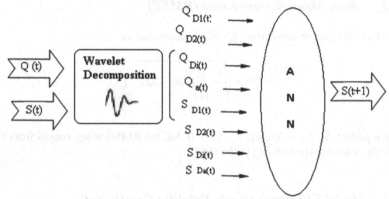

Fig.2. Time Series Wavelet-ANN conjunction Model

5 Model Evaluation Criteria

The performances of the developed models are evaluated based on some performance indices in both training and testing set. Varieties of performance evaluation criteria are available [22] [23], which could be used for evaluation and inter comparison of different models. Following performance indices are selected in this study based on relevance to the evaluation process. There can be other criteria for evaluation of performance.

5.1   Correlation Coefficient (R)

The correlation coefficient measures the statistical correlation between the predicted and actual values. It is computed as:

$$R = \frac{\sum\limits_{i=1}^{n}(Xai - \overline{Xai})(Xpi - \overline{Xpi})}{\sqrt{\sum\limits_{i=1}^{n}(Xai - \overline{Xai})^2 \sum\limits_{i=1}^{n}(Xpi - \overline{Xpi})^2}} \tag{8}$$

where Xai and Xpi are measured and computed values of diffuse pollution concentration values in streams; $\overline{Xai}$ and $\overline{Xpi}$ are average values of Xai and Xpi values respectively; i represents index number and n is the total number of concentration observations.

5.2   Root Mean Square Error (RMSE)

The root mean squared error (RMSE) is computed as:

$$RMSE = \sqrt{\frac{1}{n}(\sum\limits_{i=1}^{n}(Xai - Xpi)^2)} \tag{9}$$

For a perfect fit, $Xa_i = Xp_i$ and RMSE = 0. So, the RMSE index ranges from 0 to infinity, with 0 corresponding to the ideal.

5.3   Model Efficiency (Nash–Sutcliffe Coefficient)

The model efficiency (ME_{Nash}), an evaluation criterion proposed by Nash and Sutcliffe (1970) [22], is employed to evaluate the performance of each of the developed model. It is defined as:

$$ME_{Nash} = 1.0 - \frac{\sum\limits_{i=1}^{n}(Xa_i - X_{pi})^2}{\sum\limits_{i=1}^{n}(X_{ai} - \overline{X}_{ai})^2} \tag{10}$$

5.4 Index of Agreement (IOA)

It seeks to modify the Nash–Sutcliffe Coefficient by penalizing the differences in the mean of predicted and observed values. However, due to more squaring terms, this index is overly sensitive to outliers in the data set. It is defined as:

$$IOA = 1.0 - \frac{\sum\limits_{i=1}^{n} (Xa_i - X_{pi})^2}{\sum\limits_{i=1}^{n} \left[\left| (X_{ai} - \overline{Xai}) \right| + \left| (X_{pi} - \overline{Xpi}) \right| \right]^2} \tag{11}$$

5.5 Mean Absolute Error (MAE)

It is defined as:

$$MAE = \frac{1}{n} (\sum\limits_{i=1}^{n} (Xai - Xpi)) \tag{12}$$

6 Application Wavelet-ANN Conjunction Model

The developed models require uninterrupted time series data pertaining to Q and S at a gauging station for calibration and verification periods. The data derived from the IMISSISSIPPI RIVER AT TARBERT LANDING, MS (USGS Station Number 7295100; latitude: 310030; longitude: 0913725) were employed to train and test all the models developed in this study. The daily time series of Q and S for this station were downloaded from the USGS web server (http://co.water.usgs.gov/sediment/ seddatabase.cfm). Data from October 1, 1979 to September 30, 1986 (seven years) and the data from October 1, 1986 to September 30, 1989 (three years) were used as training and testing sets, respectively. Fig. 3 and Fig. 4 show the time series of data related to daily Q and S respectively.

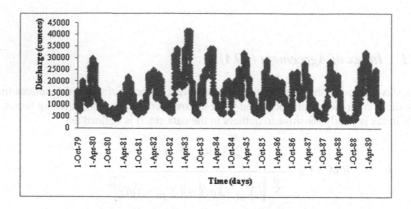

Fig. 3. Time series of discharge data

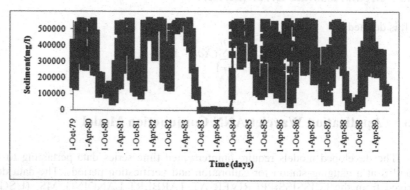

Fig. 4. Time series of sediment data

Original data series is subjected to 1-level of wavelet transform comprising of its approximation and details. Fig. 5 and Fig.6 shows approximation and detail sub-signal of original discharge measurement data and Fig. 7 and Fig. 8 show approximation and detail sub-signal of original suspended sediment measurement data of the original series obtained using DWT by Haar wavelet (level 1) for the discharge and sediment series respectively.

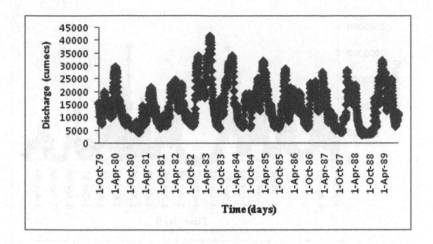

Fig. 5. Approximation sub signal of daily discharge measurement values by DWT (Haar wavelet level 1)

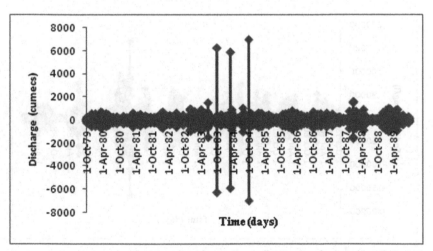

Fig. 6. Details sub signal of daily discharge measurement values by DWT (Haar wavelet level 1)

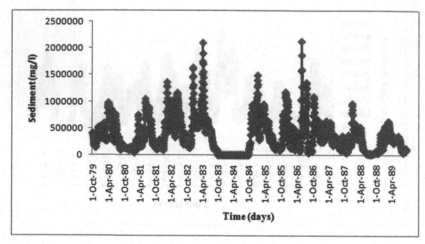

Fig. 7. Approximation sub signal of daily suspended sediment concentration measurement values by DWT (Haar wavelet level 1)

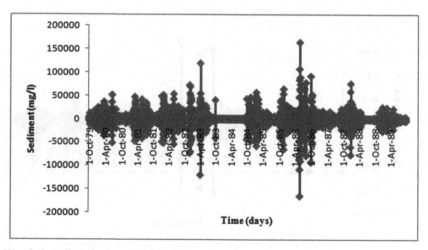

Fig. 8. Details sub signal of daily suspended sediment concentration measurement values by DWT (Haar wavelet level 1)

7 Results and Discussion

Wavelet-ANN conjunction model is implemented on MATLAB platform. Back propagation ANN training algorithm is implemented for obtaining the optimal architecture with the internal parameters as: number of epoch=1000; momentum coefficient=0.8. In addition, optimum combinations of transfer functions in the hidden and output layer are obtained with 'trainlm' function. Performance of Wavelet–ANN (W-ANN) conjunction model is compared with actual time series (without wavelet decomposition) ANN model (T-ANN). Two W-ANN models, W-ANN 1 and W-ANN 2 are developed. W-ANN-1 model has two inputs (wavelet decomposition-approximation and details coefficients of discharge data at time t by wavelet transform DWT Haar wavelet level 1). The output for W-ANN 1 model is suspended sediment concentration the river at time t+1. The W-ANN-2 model has four inputs (wavelet decomposition-approximation and details coefficients of both discharge and sedimentation data at time t by wavelet transform DWT Haar wavelet level 1). The output is same as WANN-1 model. Two T-ANN models, T-ANN 1 and T-ANN 2, are also developed. T-ANN 1 model has one input i.e. actual discharge at time t and one output (sediment load at time t+1. The T-WANN 2 model has two inputs i.e. both actual discharge and sediment at time t. The output is same as T-WANN 1 model. Thus, output of all the models is same.

Experimentation with varying number of hidden nodes and training algorithm are performed. The error statistics of best performing W-ANN and T-ANN models in training and testing are shown in Table 1. When only discharge data is used for sedimentation prediction, results are not promising in case of both W-ANN 1 and T-ANN 1 models. When both discharge and sediment data are employed in sedimentation prediction, results improve considerably for both W-ANN 2 and T-ANN 2 models. W-ANN 2 model (4-3-1) performs considerably better than T-ANN 2 (2-1-1) as can be seen by error statistics presented in Table 1. Use of multi levels of wavelet decomposition may further improve the results as more details components will be incorporated as inputs. Results obtained with this work may also be compared with wavelet Neuro-fuzzy fuzzy model.

Table 1. Training and testing errors for W-ANN conjunction model and T-ANN

Models (Trainlm)	Errors Training/ Testing	R	RMSE	E	IOA	MAE
W-ANN 1 2-1-1	Training	0.578	2.8E+05	0.335	0.706	1.8E+05
	Testing	0.823	1.5E+05	0.464	0.873	1.1E+05
W-ANN 2 4-3-1	Training	0.996	2.9E+04	0.993	0.998	1.7E+04
	Testing	0.996	1.7E+04	0.993	0.998	1.2E+04
T-ANN 1 1-1-1	Training	0.578	2.8E+05	0.334	0.705	1.8E+05
	Testing	0.822	1.5E+05	0.466	0.873	1.1E+05
T-ANN 2 2-1-1	Training	0.990	3.3 E+04	0.991	0.997	1.8E+04
	Testing	0.996	1.9 E+04	0.991	0.997	1.2E+04

8   Conclusions

The study presents the general framework for evaluating sedimentation in a river system. Methodology for wavelet-ANN conjunction model for suspended sediment prediction in rivers is demonstrated through illustrative real daily discharge and suspended sediment data. The values of statistical performance evaluation criteria indicate the W-ANN time series model is able to simulate the complex sedimentation event in rivers. Wavelet decomposition improved the results considerably. Multi levels of wavelet decomposition may further improve the testing results.

References

[1]   Yenigun K, Gumus V, Bulut H (2008). Trends in streamflow of Euphrates Basin, Turkey, ICE Water Management 161(4): 189-198.

[2]   Yang, C. T. (1996). Sediment transport, theory and practice, McGraw-Hill, New York.

[3]   Verstraeten, G., and Poesen, J. (2001). Factors controlling sediment yield from small intensively cultivated catchments in a temperate humid climate. Geomorphology, 40(1-2), 123-144.

[4]   Rajaee, T., Nourani, V., Mirbagheri, S. A., Zounemat-Kermani, M., and Kisi, O. (2011). River suspended sediment load prediction: Application of ANN and wavelet conjunction model. Journal of Hydrologic Engineering, ASCE, 16(8), 613-627.

[5]   Kisi, O. (2005). Suspended sediment estimation using neuro-fuzzy and neural network approaches. Hydrol. Sci. J., 50(4), 683-696.

[6]   Salas, J.D., 1993. Analysis and modeling of hydrologic time series. Chapter 19 in the McGraw Hill Handbook of Hydrology, D. Maidment, Editor.

[7]   Cigizoglu, H. K. (2004). Estimation and forecasting of daily suspended sediment data by multi layer perceptrons. Adv. Water Resour., 27(2), 185–195.

[8]   Cigizoglu, H. K., and Kisi, O. (2006). Methods to improve the neural network performance in suspended sediment estimation. J. Hydrol., 317(3-4), 221–238.

[9]   Nagy, H. M., Watanabe, K., and Hirano, M. (2002). Prediction of load concentration in rivers using artificial neural network model. J. Hydraul. Eng., 128(6), 588–595.

[10] Bhattacharya, B., Price, R. K., and Solomatine, D. P. (2005). Data-driven modelling in the context of sediment transport. Phys. Chem. Earth, 30 (4-5), 297–302.

[11] Raghuwanshi, N., Singh, R., and Reddy, L. (2006). Runoff and sediment yield modeling using artificial neural networks: Upper Siwane River, India. J. Hydrol. Eng., 11(1), 71–79.

[12] Rumelhart, D.E., Hinton, G.E., Williams, R.J. (1986). Learning internal representation by error propagation. Parallel Distributed Processing, 1, 318-362, MIT Press, Cambridge, Mass.

[13] Bishop, C.M. (1995). Neural Networks For Pattern Recognition, Oxford University Press, India

[14] Singh, R.M, Datta, B., Jain, A. (2004). Identification of unknown groundwater pollution sources using artificial neural networks, Journal of Water Resources Planning and Management, ASCE, 130(6), 506-514.

[15] Hagan, M. T., and Menhaj, M. B. (1994). Training feed forward networks with the Marquaradt algorithm, IEEE Trans. Neural Netw., 6, 861–867.
[16] Kisi, O. (2007). Streamflow forecasting using different artificial neural network algorithms, J. Hydrol. Eng., 12 (5), 532–539.
[17] Haykin, S. (1994). *Neural networks: A comprehensive foundation*. Mac- Millan, New York, 696.
[18] Daubechies, I. (1988). Orthonormal bases of compactly supported wavelets, commun, Pure and Appllied Mathmatics XLI, 901–996.
[19] Kang, S., and Lin, H. (2007). Wavelet analysis of hydrological and water quality signals in an agricultural watershed, Journal of Hydrology, 338, 1-14.
[20] MATLAB (2004). The Language of Technical Computing, The MathWorks Inc., Natick, Mass., USA.
[21] Mallat, SG (1998). A wavelet tour of signal processing. Academic, San Diego.

[22] Nash, J. E., Sutcliffe, J. V. (1970). River flow forecasting through conceptual models. Part 1-A: Discussion principles, Journal of Hydrology, 10, 282–290.
[23] ASCE Task Committee on Definition of Criteria for Evaluation of Watershed Models (1993). Criteria for evaluation of watershed models, Journal of Irrigation and Drainage Engineering, ASCE, 119(3), 429–442.

Fuzzy Modeling and Similarity based Short Term Load Forecasting using Swarm Intelligence-A step towards Smart Grid

Amit Jain[1], M Babita Jain[2]

[1] Lead Consultant, Utilities, Infotech Enterprisese Limited, Hyderabad, Andhra Pradesh, India

[2] Professor, Rungta Engineering College, Raipur, Chattisgarh, India

{Amit.Jain@infotech-enterprises.com; jain.babita@gmail.com}

Abstract. There are a lot of uncertainties in planning and operation of electric power system, which is a complex, nonlinear, and non-stationary system. Advanced computational methods are required for planning and optimization, fast control, processing of field data, and coordination across the power system for it to achieve the goal to operate as an intelligent smart power grid and maintain its operation under steady state condition without significant deviations. State-of-the-art Smart Grid design needs innovation in a number of dimensions: distributed and dynamic network with two-way information and energy transmission, seamless integration of renewable energy sources, management of intermittent power supplies, real time demand response, and energy pricing strategy. One of the important aspects for the power system to operate in such a manner is accurate and consistent short term load forecasting (STLF). This paper presents a methodology for the STLF using the similar day concept combined with fuzzy logic approach and swarm intelligence technique. A Euclidean distance norm with weight factors considering the weather variables and day type is used for finding the similar days. Fuzzy logic is used to modify the load curves of the selected similar days of the forecast by generating the correction factors for them. The input parameters for the fuzzy system are the average load, average temperature and average humidity differences of the forecasted previous day and its similar days. These correction factors are applied to the similar days of the forecast day. The tuning of the fuzzy input parameters is done using the Particle Swarm Optimization (PSO) and Evolutionary Particle Swarm Optimization (EPSO) technique on the training data set of the considered data and tested. The results of load forecasting show that the application of swarm intelligence for load forecasting gives very good forecasting accuracy. Both the variants of Swarm Intelligence PSO and EPSO perform very well with EPSO an edge over the PSO with respect to forecast accuracies.

Keywords: Euclidean norm, Evolutionary particle swarm optimization, Fuzzy logic approach, Particle swarm optimization, Short term load forecasting, Similar day method.

J. C. Bansal et al. (eds.), *Proceedings of Seventh International Conference on Bio-Inspired Computing: Theories and Applications (BIC-TA 2012)*, Advances in Intelligent Systems and Computing 202, DOI: 10.1007/978-81-322-1041-2_2, © Springer India 2013

1 Introduction

Short term load forecasting (STLF) is a time series prediction problem that ana-
lyzes the patterns of electrical loads. Basic operating functions such as unit

Fig. 1 Overview of the methodology followed

comitment, economic dispatch, fuel scheduling and maintenance can be performed
efficiently with an accurate load forecast [1]-[3]. STLF is also very important for
electricity trading. Therefore, establishing high accuracy models of the STLF is
very important and this faces many difficulties. Firstly, because the load series is
complex and exhibits several levels of seasonality. Secondly, the load at a given
hour is dependent not only on the load at the previous hour, but also on the load at
the same hour on the previous day and because there are many important exogenous variables that must be considered, specially the weather-related variables [4].

Traditional STLF methods include classical multiply linear regression, automatic regressive moving average (ARMA), data mining models, time-series models and exponential smoothing models [5]-[13]. Similar-day approach and various
artificial intelligence (AI) based methods have also been applied [4, 5, 7 and 14].
Evolutionary and behavioural random search algorithms such as genetic algorithm
(GA) [15]-[17]-[20, 21], particle swarm optimization (PSO) [18, 19], etc. have
been previously implemented for different problems.

There also exist large forecast errors using ANN method when there are rapid
fluctuations in load and temperatures [4, 23]. In such cases, forecasting methods
using fuzzy logic approach have been employed. S. J. Kiartzis et al [22, 24], V.
Miranda et al [25], and S. E. Skarman et al [26] described applications of fuzzy
logic to electric load forecasting as well as many others [27]-[29].

The above discussed literature aims at making an accurate STLF for helping the
grid work efficiently. For making the distribution grid smarter it is required to
deploy communications and leverage advanced controls that are commonplace in
substation automation, remedial action schemes, power management systems, and
industrial closed-loop power automation [38]-[40].

In this paper, we propose an approach for the short term load forecasting using
similarity and the fuzzy parameters tuned by the PSO and EPSO algorithms for
better power generation and distribution management aiming to make the power
system a smart grid. In this method, the similar days to the forecast day are selected from the set of previous days using a Euclidean norm based on weather variables and day type [30]. There may be a substantial discrepancy between the
load on the forecast day and that on similar days, even though the selected days
are very similar to the forecast day with regard to weather and day type. To rectify
this problem load curves on the similar days are corrected to take them nearer to
the load curve of the forecast day using correction factors generated by a fuzzy inference system which is tuned with two techniques PSO and EPSO. This tuned

fuzzy inference system (FIS) is developed using the history data. The suitability of the proposed approach is verified by applying it to a real time data set. This paper contributes to the short term load forecasting by developing a PSO and EPSO tuned FIS for reducing the forecasting error and finally coming out with the best suitable technique for STLF. The overview of the methodology followed is shown in the Fig. 1.

The paper is organized as follows: Section II deals with the PSO and EPSO for STLF and data analysis; Section III gives the overview of the proposed forecasting methodology; Section IV presents the tuning of fuzzy parameters using PSO; Section V presents the tuning of fuzzy parameters using EPSO Section VI presents comparison of simulation results of the proposed forecasting methodology i.e. PSO and EPSO tuned fuzzy parameters results followed by conclusions in Section VII.

2 PSO and EPSO for STLF and Variables Impacting Load Pattern

EPSO is a general-purpose algorithm, whose roots are in Evolutions Strategies (ES) [31]-[33] and in Particle Swarm Optimization (PSO) [34] concepts. The PSO is an optimization algorithm that was introduced in 1995 and some researchers have tried its application in the power systems field with reported success [35, 36]. The EPSO technique, a new variant in the meta-heuristic set of tools, is capable of dealing with complex, dynamic and poorly defined problems that AI has problem with, has an advantage of dealing with the nonlinear parts of the forecasted load curves, and also has the ability to deal with the abrupt change in the weather variables such as temperature, humidity and also including the impact of the day type. PSO has recently found application in STLF where PSO has been applied to identify the autoregressive moving average with exogenous variable (ARMAX) model of the load [37]. According to a thorough literature survey performed by authors, any application of EPSO to STLF has not been reported in literature as of today.

The analysis on the monthly load and weather data helps in understanding the variables which affect load forecasting. The data analysis is carried out on data containing hourly values of load, temperature, and humidity of 3 years. In the analysis phase, the load curves are drawn and the relationship between the load and weather variables is established [38].

2.1 Variation of Load with Day type

The load curves for a winter test week (12th – 18th Jan, 1997) and summer test week (13th – 19th July, 1997) are shown in Fig 2. The observations from the load curves show that there exists weekly seasonality but the value of load scales up and down and the load curves on week days show similar trend and the load curves on the weekends show similar trend. It can also be seen that this weekly seasonality feature holds good for all the seasons of the year. The only variation is in the load which is more in summer than in winter due the increased temperatures of summer and this correlation can be seen in the Fig.2.

Based on the above observations in the present study, days are classified as four categories. First: normal week days (Tuesday - Friday), second: Monday, third: Sunday and the fourth category being Saturday.   Monday is accounted to be different to weekdays so as to take care for the difference in the load because its previous day is a weekend.

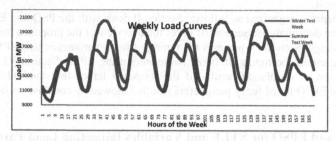

Fig. 2 Weekly Load curves of winter and summer test weeks

2.2 *Variation of Load with Temperature and Humidity*

The variation of the temperature and humidity variables results in a significant variation in the load. Fig 3 shows a plot between the maximum temperatures versus average demand and average humidity. The graph shows a positive correlation between the load and temperature and load and humidity i.e. demand increases as the temperature and humidity increases.

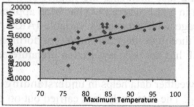 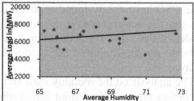

Fig. 3 Maximum Temperature Vs Average Load Curve for the month of July'97 and Average Humidity Vs Average Load Curve for the month of July'97

3 Short Term Load Forecasting using Fuzzy Logic

This section presents in detail the architecture details and implementation procedure of the fuzzy inference system for the proposed STLF. A very important observation is made from the Fig. 4 which shows the annual load curves of 1997, 1998 and 1999 generated using their daily average loads. It can be seen that the similar months of different years follow a similar load curve pattern. Hence for the selection of similar days the previous year's similar months will also have considerable effect.The load forecasting at any given hour not only depends on the load at the previous hour but also on the load at the given hour on the previous day and also on the load of the previous day of previous years'. Assuming same trends of relationships between the previous forecast day and previous similar days as that of the forecast day and its similar days, the similar days can thus be evaluated by analyzing the previous forecast day and its previous similar days. Also the Euclidean Norm alone is not sufficient to obtain the similar days; hence the evaluation of similarity between the load on the forecast day and that on the similar days is done using the adaptive fuzzy inference system.

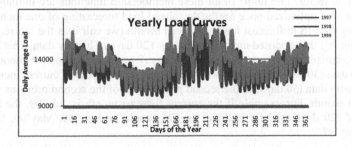

Fig. 4 Daily Average Load curves of Year 1997, 1998, 1999

In the fuzzy inference system (FIS), difference of the previous forecast day and its similar days' load, temperature and humidity are fed as input, resulting in correction factors, which are used to correct the similar days of the forecast day and then averaged to obtain the load forecast. The parameters of the fuzzy inference system used for the forecast of the current month are already optimized using the data of previous month and its history using PSO and EPSO.

3.1 Calculation of Weights and selection of Similar Days

The first task in building the FIS is to identify the similar days of the forecast previous day and the similarity is judged on the basis of the Euclidean Distance Norm given by the formula:

$$EN = \sqrt{W_1(\Delta T_{max})^2 + W_2(\Delta H_{avg})^2 + W_3(\Delta D)^2} \quad (1)$$

Where

$\Delta T_{max} = T_{max} - T^p_{max}, \Delta H_{avg} = H_{avg} - H^p_{avg} \text{ and } \Delta D = D - D^p$

Where, T_{max} and H_{avg} are the forecast day maximum temperature and average humidity respectively. Also, T^p_{max} and H^p_{avg} are the maximum temperature and average humidity of the searched previous days , D and D^p are the day type values of the forecast day and the searched previous days and w_1, w_2, w_3 are the weight factors determined by least squares method based on the regression model constructed using historical data.

The data available is 38 months data. For weight calculation first 26 months data is used. The equations when formulated using matrix algebra form:$L = AY$, where A is a [790][5] matrix, L is a [790][1] and Y is a [5][1] matrix. Y is the weight matrix.

3.2 Formulation of Fuzzy System

The formulation of the developed Fuzzy Inference System comprises of three input membership functions:ΔE_L, ΔE_T, ΔE_H, average load difference, average temperature difference and average humidity difference respectively of the forecast day and its selected similar days and one output membership function i.e. the correction factor. The limits of all these membership functions are initially fixed and are later optimized once for the day ahead load forecasting of one month. For building the FIS to forecast load of a given month (we call it as the 'current forecast month') the proposed methodology uses 120 days of history data. This history data comprises of two months data (60 days) prior to the current month, one month data (30 days) of second prior month of previous year of current month and one month data (30 days) of the second prior month of the second previous year of current month. For example if the current forecast month is July'99, the history data of 120 days used for building the FIS would be June'99, May'99, May'98 and May'97.

Table 1. Fuzzy Rules of the Inference System

Rule No	E_L	E_T	E_H	Output Value
R1	H	H	H	PVB (Positive Very Big)
R7	M	M	H	PB2 (Positive Big 2)
R14	M	M	M	ZE (Zero Error)
R23	L	L	H	NB1 (Negative Big 1)

4   Optimization of Fuzzy parameters using Particle Swarm Optimization

4.1 Insight into Particle Swarm Optimization

PSO is initialized with a group of random particles (solutions) and then searches for optima by updating generations. In each iteration, each particle is updated by following two "best" values. The first one is the best solution (fitness) it has achieved so far. (The fitness value is also stored.) This value is called pbest. Another "best" value that is tracked by the particle swarm optimizer is the best value, obtained so far by any particle in the population. This best value is a global best and called gbest. When a particle takes part of the population as its topological neighbours, the best value is a local best and it is called lbest.

After finding the two best values, the particle updates its velocity and positions with equations (2) and (3).

$$v[\ ] = v[\ ] + c1 * rand(\ ) * (pbest[\ ] - present[\ ]) + c2 * rand(\ )$$
$$* (gbest[\ ] - present[\ ]) \tag{2}$$
$$present[\ ] = present[\ ] + v[\ ] \tag{3}$$

$v[\ ]$ is the particle velocity, $present[\ ]$ is the current particle (solution). $pbest[\ ]$ and $gbest[\ ]$ are defined as stated before. $rand(\ )$ is a random number between $(0,1)$. $c1$, $c2$ are learning factors. usually $c1 = c2 = 2$.

Particles' velocities on each dimension are clamped to a maximum velocity Vmax. If the sum of accelerations would cause the velocity on that dimension to exceed Vmax, which is a parameter specified by the user then the velocity on that dimension is limited to Vmax.

4.2 PSO Implementation for FIS optimization

Optimization of the fuzzy parameters a1......a6 is done using the particle swarm optimization technique. For the data set considered the fuzzy inference system has been optimized for six parameters (maxima and minima of each of the input fuzzy variable E_L, E_T, E_H), considering 49 particles. Hence each particle is a six dimensional one. The initial values of the fuzzy inference system are obtained by using the 120 days data as discussed in Section III. These values are incorporated into the fuzzy inference system to obtain the forecast errors of forecast previous month (in the example case forecast previous month is June '99).

The particle swarm optimization function accepts the training data i.e. 120 days, and the objective is to reduce the RMS MAPE error of the 30 forecast days (June '99) using the 90 days history data (details given in Section III-C). The MAPE is taken as the fitness function and the particle swarm optimizer function is run for 100 iterations (by then the RMS MAPE is more or less fixed and comes less than 3%). After each iteration, the particle swarm optimizer updates the latest particle position using the optimizer equations based on the PBest and Gbest of the previous iteration if the fitness function value is better than the previous one. The parameters thus obtained after the PSO optimization are the final input parameters of the designed FIS. These fuzzy parameter values are set as the input parameter limits of the fuzzy inference system and this FIS is used to forecast the load of the current forecast month (in the example case current forecast month is July '99).

4.3 Forecast of current forecast month load

The data of the current forecast month (Example: July1 to July30) is taken as the testing dataset for the problem at hand. The short term load forecasting for the month of July is now done using the FIS optimized by the PSO i.e PSO-FIS. The five similar days are selected from the history 90 days (for July1 the history 90 days are June'99, June'98, June'97) of the forecast day and the hourly correction factors to these similar days are obtained by the five similar days of the forecast previous day (for June 30 the previous 90 days are May31 to June 29 of 99, 98 and 97) and the PSO-FIS. These five correction factors are then applied to the five similar days of the forecast day and the average of the corrected five values is considered as the load forecasting for each hour. The same procedure is done for all the 24 hours of the day. The same procedure is followed for all days of July i.e. Jul 1 to Jul 30. The MAPE is calculated for the each day of the 30 days of forecast of the Jul data (using the actual hourly values and the forecast hourly values). The FIS is formulated and optimized for every month of 1999 using the same methodology and is then implemented for the load forecasting of all the months of 1999 year. The results obtained for the STLF using PSO-FIS have been quite satisfactory and further analysis and study of the performance of the PSO-FIS for STLF is done in Section VII.

5 Optimization of Fuzzy parameters using Evolutionary Particle Swarm Optimization

5.1 Insight into Evolutionary Particle Swarm Optimization

The *particle movement* rule for EPSO is that given a particle x_i, a new particle x_i^{new} results from:

$$x_i^{new} = x_i + v_i^{new} \tag{4}$$

$$v_i^{new} = w_{i0} * v_i + w_{i1} * (b_i - x_i) + w_{i2} * (b_g^* - x_g) \tag{5}$$

This formulation is very similar to classical PSO – the movement rule keeps its terms of inertia, memory and cooperation. However, the weights, taken as object parameters, undergo mutation which is not the case with PSO:

$$w_{ik}^* = w_{ik} + \mu N(0,1) \tag{6}$$

Where N (0, 1) is a random variable with Gaussian distribution, 0 mean and variance 1.

The global best b_g is randomly disturbed to give:

$$b_g^* = b_g + \mu' N(0,1) \tag{7}$$

The logic behind this modification from PSO is the following: a) if the current global best is already the global optimum, this is irrelevant; but b) if the optimum hasn't yet been found, it may nevertheless be in the neighbourhood and it makes all sense not to aim *exactly* at the current global best – especially when the search is already focused in a certain region, at the latter stages of the process.

The μ, μ' are learning parameters (either fixed or treated also as strategic parameters and therefore subject to mutation-fixed in the present case).

5.2 EPSO implementation for FIS optimization

Same as in the case of PSO-FIS the fuzzy inference system has been optimized for six parameters (maxima and minima of each of the input fuzzy variable E_L, E_T, E_H), considereing 49 particles. Hence each particle is a six dimensional one. The initial values of the fuzzy inference system are obtained by using the 120 days data as discussed in Section III. These values are incorporated into the fuzzy inference system to obtain the forecast errors of forecast previous month.

5.3 Forecast of July month load

The procedure here is same as in the case of PSO-FIS given in Section IV-C with the only difference that the EPSO-FIS is used instead of PSO-FIS. The MAPE is less than 3% for maximum days of forecast of the whole year of 1999. The results obtained for the STLF using EPSO-FIS have been quite satisfactory and further analysis and study of the performance of the EPSO-FIS for STLF is done in Section VII.

6 Simulation Results

The performance of the proposed PSO optimized FIS and EPSO optimized FIS for the STLF is tested by using the 38 months data, Nov'96 to Dec'99 of a real data set. The PSO, and EPSO implementation has been done using the MATLAB coding and the Fuzzy Inference System has been developed using fuzzy logic toolbox available in MATLAB and load forecasting is done for the all days of all months of the year 1999.

The parameters of the PSO and EPSO algorithms used for the tuning of fuzzy input variables are given in Table 2. The forecasted results of one winter week and one summer week are presented. These two weeks include four categories of classified days of week in the present methodology namely Saturday, Sunday, Monday, and Tuesday and also the effectiveness of the technique for all seasons.

Table 2. Parameters of the PSO and EPSO algorithms

Parameters	PSO	EPSO
Population Size	49	49
Number of Iterations	100	50
$C1/w_{i0}^*$ (initial)	2.0	0.6
$C2/w_{i1}^*$ (initial)	2.0	0.1
$V(0)/w_{i2}^*$ (initial)	1.0	0.3
$\mu=\mu^*$	NA	1.5

The figure 5 shows the graphical representation of the comparative load forecasted of a winter test week for all day types by the two proposed methodologies in comparison with the actual load.

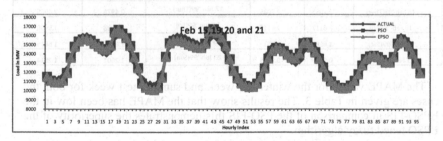

Fig. 5 Winter Week load forecast of PSO-FIS and EPSO-FIS

The forecast results deviation from the actual values are represented in the form of MAPE, which is defined as in the equation 8 and the MAPE plots of the actual hourly load, forecasted hourly load with PSO-FIS and EPSO-FIS for the 4 representative days of the summer test week of June '99 representing four categories of classified days of week for all the three cases are given figure 6.

$$MAPE = \frac{1}{N} \sum_{i=1}^{N} \frac{\left| P_A^i - P_F^i \right|}{P_A^i} \times 100 \tag{8}$$

P_A, P_F are the actual and forecast values of the load. N is the number of the hours of the day i.e. 24 and i = 1, 2,.....,.24.

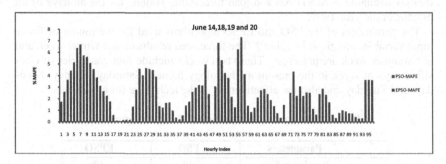

Fig. 6 June 14, 18, 19 and 20 '99 hourly MAPE comparison of PSO-FIS, EPSO-FIS

Table 3. Comparative MAPE of Winter and summer test weeks

Forecast Day	PSO-FIS	EPSO-FIS	Forecast Day	PSO-FIS	EPSO-FIS
15 Feb '99(Mon)	1.858	1.8559	14 Jun '99(Mon)	2.8487	2.8476
16 Feb '99(Tue)	2.4176	2.4125	15 Jun '99(Tue)	2.8723	2.8001
17 Feb '99(Wed)	0.9752	0.9747	16 Jun '99(Wed)	1.7475	1.7423
18 Feb '99(Thur)	2.7089	2.7085	17 Jun '99(Thur)	2.4978	2.4965
19 Feb '99(Fri)	2.6117	2.6113	18 Jun '99(Fri)	1.9928	1.9908
20 Feb '99(Sat)	1.6188	1.61	19 Jun '99(Sat)	1.6905	1.6815
21 Feb '99(Sun)	0.9379	0.9295	20 Jun '99(Sun)	1.2601	1.2514

The MAPE values for the winter test week and summer test week for both the cases are given in Table 3. The results show that the MAPE has been low in the EPSO-FIS in comparison of the PSO-FIS this demonstrates the superiority of the EPSO tuned fuzzy algorithm.

7   Conclusions

This paper proposes a novel method for comparative short term load forecasting using two different variants of particle swarm technique which are PSO and EPSO optimized fuzzy inference system. As the State-of-the-art Smart Grid design needs innovation in a number of dimensions: distributed and dynamic network with two-way information and energy transmission, seamless integration of renewable energy sources, management of intermittent power supplies, real time demand response, and energy pricing strategy the proposed architecture is a step towards efficiently managing the real time demand and managing the intermittent power supplies by making a very accurate STLF and hence helping the grid work smarter. Also, a new Euclidean norm including temperature and humidity and day type is proposed, which is used for the selection of similar days. For the first time the distance based fuzzy system has been optimized using the swarm intelligence and applied for the short term load forecasting. All the two proposed systems are used to evaluate the correction factor of the selected similar days to the forecast day using the information of the previous forecast day and its similar days. The results clearly indicate that all the proposed two systems are very robust and effective for all day's types and all seasons. Still, the fuzzy inference system with EPSO algorithm is proved to be the better compared to PSO-FIS as we observed during our simulation study where weather variables, temperature as well as humidity, are used, as it gives load forecasting results with very good accuracy. The reason analyzed for the excellent performance of EPSO-FIS is the ability to update the object parameters, which are the particles to be optimized and also it updates its strategic parameters which helps in faster convergence and better accuracy as can be seen in the results shown in Table 3. The EPSO-FIS is able to produce very accurate load forecast in lesser number of iterations in comparison to the PSO-FIS. The use of three years of historical data is also greatly responsible for the very good quality results indeed for both of the techniques with almost all the MAPE values very much less than 3%. The selection of the similar days from 90 days of history data comprising of the forecast previous month of the same year, of the previous year and also of the two years' previous year is a novel concept which helps in getting the similar most load curves with respect to temperature, humidity, day type to be selected for optimized fuzzy correction.Authors hope that the proposed methodology will further propagate research for short term load forecasting using swarm intelligence and new optimization techniques to get even more improvement in forecasting results.

8 References

[1]   O.A. Alsayegh, "Short-Term Load Forecasting Using Seasonal   Artificial   Neural   Networks", *International Journal of Power and Energy   Systems*, vol. 23, 2003.

[2]   Tomonobu Senjyu, Hitoshi Takara, Katsumi Uezato, and Toshihisa Funabashi, "One Hour-Ahead Load Forecasting Using Neural Network", *IEEE Transactions on Power Systems*, vol. 17, Feb. 2002.

[3]   A.G.Baklrtzis, V.Petrldis, S.J.Klartzis, M.C.Alexiadls and A.H.Malssis, "A Neural Network Short Term Load Forecasting Model for the Greek Power System", *IEEE Transactions on Power Systems*, vol.11, May 1996.

[4]   Henrique Steinherz Hippert, Carlos Eduardo Pedreira, and Reinaldo Castro Souza, "Neural Networks for Short-Term Load Forecasting: A Review and Evaluation", *IEEE Transactions on Power Systems*, vol. 16, Feb. 2001.

[5]   A. D. Papalexopoulos and T. C. Hesterberg, "A regression-based approach to short-term load forecasting," *IEEE Transactions on Power Systems*, vol. 5, pp. 1535-1550, 1990.

[6]   T. Haida and S. Muto, "Regression based peak load forecasting using a transformation technique," *IEEE Transactions on Power Systems*, vol. 9, pp. 1788-1794, 1994.

[7]   S. Rahman and O. Hazim, "A generalized knowledge-based short term load-forecasting technique," *IEEE Transactions on Power Systems*, vol. 8, pp. 508-514, 1993

[8]   S. J. Huang and K. R. Shih, "Short-term load forecasting via ARMA model   identifica-
      tion including nongaussian process considerations," *IEEE Transactions on Power Sys-
      tems*, vol. 18, pp. 673-679, 2003.
[9]   H.Wu and C. Lu, "A data mining approach for spatial modeling in small area load fore-
      cast," *IEEE Transactions on Power Systems*, vol. 17, pp. 516-521, 2003.
[10] S. Rahman and G. Shrestha, "A priority vector based technique for load forecasting," *IEEE
      Transactions on Power Systems*, vol. 6, pp. 1459-1464, 1993.
[11] Ibrahim Moghram, Saifure Rahman, "Analysis Evaluation of Five Short Term Load Fore-
      casting Techniques", *IEEE Transactions on Power Systems*, vol. 4, 1989.
[12] K.Lru, S.Subbarayan, R.R.Shoults, M.T.Manry, C.Kwan, F.L.lewis and J.Naccarino,
      "Comparison of very short term load forecasting techniques", *IEEE Transactions on Power
      Systems*, vol. 11, 1996.
[13] Eugene A. Feinberg and Dora Genethliou,"Applied Mathematics for Power Systems: Load
      Forecasting".
[14] Kun-Long Ho, Yuan-Yih Hsu, Chih-Chien Liang and Tsau-Shin Lai, "Short-Term Load
      Forecasting of Taiwan Power System Using A Knowledge-Based Expert Systems", *IEEE
      Transactions on Power Systems*, vol. 5, Nov. 1990.
[15] J. H. Holland, "Adaptation in Natural and Artificial Systems", Ann Arbor, MI: Univ.
      Michigan Press, 1975.
[16] D. T. Pham and D. Karaboga, "Intelligent Optimization Techniques, Genetic Algorithms,
      Tabu Search, Simulated Annealing and Neural Networks", New York: Springer-Verlag,
      2000.
[17] L. Davis, "Handbook of Genetic Algorithms", New York: Van Nostrand Reinhold, 1991.
[18] R. C. Eberhert and J. Kennedy, "A new optimizer using particle swarm theory", Proceeding
      of the *Sixth International Symposium on Micro Machine and Human Science*, pp. 39-43,
      Nagoya, Japan, 1995.
[19] J.B. Park, K.S. Lee, J.R. Shin, K.Y. Lee, "A particle swarm optimization for economic dis-
      patch with non-smooth cost function", *IEEE Transactions on Power Systems*, vol. 20, pp.
      34–42, 2005.
[20] D.B. Fogel, "Evolutionary Computation: Toward a New Philosophy of Machine Intelli-
      gence", second ed., *IEEE* Press, Piscataway, NJ, 2000.
[21] R.C. Eberhart, Y. Shi, "Comparison between genetic algorithms and particle swarm optimi-
      zation", in: Proc. *IEEE International Conf. Evolutionary Computing*, May, 1998, pp. 611–
      616.
[22] S. J. Kiartzis and A. G. Bakirtzis, "A Fuzzy Expert System for Peak Load Forecasting: Ap-
      plication to the Greek Power System", Proceedings of the *10th Mediterranean Electro
      technical Conf.*, 2000, pp. 1097-1100.
[23] V. Miranda and C. Monteiro, "Fuzzy Inference in Spatial Load Forecasting", Proceedings
      of *IEEE Power Engineering Winter Meeting*, 2, 2000, pp. 1063-1068.
[24] S. E. Skarman and M. Georgiopoulous, "Short-Term Electrical Load   Forecasting
      using a Fuzzy ARTMAP Neural Network", Proceedings of *SPIE*, 1998, pp. 181-191.
[25] T. Senjyu, Mandal. P, Uezato. K, Funabashi. T, "Next Day Load Curve Forecasting using
      Hybrid Correction Method," *IEEE Transactions on Power Systems*, vol. 20, pp. 102-109,
      2005.
[26] P. A. Mastorocostas, J. B. Theocharis, and A. G. Bakirtzis, "Fuzzy modeling for short
      term load forecasting using the orthogonal least squares method," *IEEE Transactions on
      Power Systems*, vol. 14, pp. 29-36, 1999.
[27] M. Chow and H. Tram, "Application of fuzzy logic technology for spatial load forecast-
      ing," *IEEE Transactions on Power Systems*, vol. 12, pp. 1360-1366, 1997.
[28] T. Senjyu, Uezato. T, Higa. P, "Future Load Curve Shaping based on similarity   using
      Fuzzy Logic Approach," *IEE Proceedings of Generation, Transmission, Distribu-
      tion*, vol. 145, pp. 375-380, 1998.
[29] Rechenberg, I., "Evolutionsstrategie – Optimierung technischer Systeme nach Prinzipen der
      biologischen Evolution", Frommann-Holzboog, Stuttgart, 1973.
[30] Schwefel, H.-P., "Evolution and Optimum Seeking", Ed. Wiley, New York NY, 1995.
[31] Fogel, D.B., "Evolving Artificial Intelligence", Ph.D. Thesis, University of California, San
      Diego, 1992 .

[32] Kennedy, J., R.C. Eberhart, "Particle Swarm Optimization", *IEEE International Conference on Neural Networks*, Pert, Australia, *IEEE* Service Center, Piscataway, NJ., 1995

[33] Fukuyama, Y. and Yoshida, H., "A particle swarm optimization for reactive power and voltage control in electric power systems", *IEEE Proc. of Evolutionary Computation* 2001 , vol.1 , pp. 87 -93, 2001.

[34] Yoshida, H., Fukuyama, Y., Takayama, S. and Nakanishi, Y., "A particle swarm optimization for reactive power and voltage control in electric power systems considering voltage security assessment", *IEEE* Proc. of *SMC '99*, vol. 6, pp.497 -502, 1999.

[35] C. Huang, C. J. Huang, and M. Wang, "A particle swarm optimization to identifying the ARMAXmodel for short-term load forecasting", *IEEE Transactions on Power Systems*, vol. 20, pp. 1126–1133, May 2005.

[36] S. Rahman and R. Bhatnagar, "An expert system based algorithm for short term load forecast", *IEEE Transactions on Power Systems*, vol. 3, pp. 392-399, 1988.

[37] Miranda, V., Fonseca, N., "New Evolutionary Particle Swarm Algorithm (*EPSO*) Applied to Voltage/Var Control", Proceedings of *PSCC'02 – Power System Computation Conference*, Spain, June 24-28, 2002.

[38] *Markushevich*, Cross-cutting aspects of Smart Distribution Grid applications, *IEEE Power and Energy Society General Meeting, 2011*

[39] *Meliopoulos, S.; Cokkinides, G.; Huang, R.; Farantatos, E.; Sungyun Choi; Yonghee Lee; Xuebei Yu*, Smart Grid Infrastructure for Distribution Systems and Applications, 2011 44th Hawaii International Conference on System Sciences (HICSS)

[40] *Dolezilek, David*, Case study of practical applications of smart grid technologies, 2011 2nd IEEE PES Innovative Smart Grid Technologies (ISGT Europe)

Universal Web Authentication System Using Unique Web Identity

M. Mallikarjuna[1], Bharathi M A[2], Rahul Jayarama Reddy[1] and Pramod Chandra Prakash[1]

[1] Department of ISE, Reva Institute of Technology and Management, Bengaluru, India
{mathada.mass@gmail.com}, {rahul.jayaram@hotmail.com}, {pramodcp15@gmail.com}

[2] Department of CSE, Reva Institute of Technology and Management, Bengaluru, India
{bmalakreddy@yahoo.com}

Abstract. Number of people accessing the Internet daily is increasing at an exponential rate. The Web has become a common place for Rich Interactive Applications and with such improvements to the Web, authenticating users is very important. Having this understanding that Web is the future, this paper is focused on creating a safe future for the web. The model is a Universal Web Authentication System, which is used to authenticate people on the web with a Universal Web Identity (U-WI) that is generated upon genuine registration, which will eliminate anonymous users from the Web and also eliminate the database overhead of various social networking sites and other Web applications that require authorized access.

Keywords: Web, authentication, UID, validation, identity, registration

1 Introduction

Universal Web Authentication System (UWAS) is mainly based on the Unique Web Identity (U-WI). The U-WI is used to uniquely identify users on the web. UWAS can authorize users and grant access to various websites. Websites can make use of UWAS by embedding simple APIs on their webpages. UWAS will have its own database that stores all the necessary information about the user. Websites using UWAS need not maintain their own database for authenticating users, hence saving additional resource overhead. With UWAS anonymous users and impersonating of people or profiles can be eliminated from social networking sites.

J. C. Bansal et al. (eds.), *Proceedings of Seventh International Conference on Bio-Inspired Computing: Theories and Applications (BIC-TA 2012)*, Advances in Intelligent Systems and Computing 202, DOI: 10.1007/978-81-322-1041-2_3, © Springer India 2013

1.1 Related Work

Yu Sheng and Zhu Lu [2] discuss the scheme that combines the password entered by the user, the password associated with private key protected by trusted platform module, and user certificate provided by trusted computing platform, thieving only the password on the web will not have an effect on user's security. Ayu Tiwari and Sudip Sanyal [1] have proposed in their work a new protocol using multifactor authentication system that is both secure and highly usable. It uses a novel approach based on Transaction Identification Code and SMS to enforce extra security level with the traditional Login/password system, which is simple to use and deploy, that does not require any change in infrastructure or protocol of wireless networks.

Subhash Chander and Ashwani Kush [9] have outlined about Aadhaar card that provides an identity to each individual in India, which can be used to avail services provided at the various government centers.

Aadhaar is a 12-digit unique number [3] which the Unique Identification Authority of India (UIDAI) will issue for all residents in India. The number will be stored in a centralized database and linked to the basic demographic and biometric information – photograph, ten fingerprints and iris – of each individual. It is unique and robust enough to eliminate the large number of duplicate and fake identities in government and private databases. The unique number generated will be devoid of any classification based on caste, creed, religion and geography.

It is believed that Unique National IDs will help address the rigged state elections and widespread embezzlement that affects subsidies. Addressing illegal immigration into India and terrorist threats is another goal of the program.

Government distributed benefits are fragmented by purpose and region in India, which results in widespread bribery, denial of public services and loss of income, especially afflicting poor citizens. As the unique identity database comes into existence, the various identity databases (voter ID, passports, ration cards, licenses, fishing permits, border area ID cards) that already exist in India are planned to be linked to it. Instead, it will enroll the entire population using its multi-registrar enrollment model using verification processes prescribed by the UIDAI. This will ensure that the data collected is clean right from the beginning of the program. However, much of the poor and underserved population lack identity documents and the UID may be the first form of identification they will have access to.

Single Sign-On (SSO) which is worked out on similar lines for the proliferation of web applications force users to remember multiple authentication credentials (usernames and passwords) for each application. Faced with the impractical task of remembering multiple credentials, users reuse the same passwords, pick weak passwords, or keep a list of all usernames and passwords. Managing multiple authentication credentials is annoying for users and weakens security for the authentication system. Web Single Sign-On (Web SSO) systems [7] allow a single username and password to be used for different web applications. For the user, Web SSO systems help to create what is called a federated identity.

Federated identity management benefits both the user and the application provider. Users only remember one username and password, so they do not have to suffer from password-amnesia. Application providers also reduce their user management cost. They neither need to support a redundant registration process nor deal with one-time users creating many orphan accounts.

2 Architecture of the System

There are various components that make up the architecture of UWAS. The components are as follows.

2.1 Unique Web Identity

The Unique Web Identity (U-WI), which identifies a person, will give individuals the means to clearly establish their identity in the web across the globe. U-WI is provided during the registration process where a person's demographic and biometric information are collected and verified by the Government authority. After successful verification U-WI is issued to the person.

A U-WI is merely a string assigned to an entity that identifies the entity uniquely. Biometric identification system and checks would be used to ensure that each individual is assigned one and only U-WI and the process of generating a new U-WI would ensure that duplicates are not issued as valid U-WI numbers.

As shown in Figure 1, U-WI is a 14 character string. The first two characters i.e., U0 and U1 of the U-WI indicate the 2-Character ISO code [4] for the country where the person was born. The remaining part of the string (i.e., U2 - U13) is a 12 digit random number, which is independent of the person's demographic or personal information.

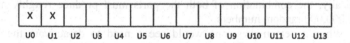

Fig. 1. Structure of U-WI

The random number is generated using the built-in function Random() of the System.Random [10] class available on the .NET platform.

The 12-digit random number is split into 2 parts of 6-digits each; Part1 and Part2. Initially the numbers 111111 and 999999 are passed as the seeds to Random() function of each part respectively. Using these seeds the corresponding Random() functions of both the parts are instantiated. The function Random.Next() generates a random number based on the seed value passed. These random numbers from both the parts are together concatenated along with the 2-Character ISO country code of the user to generate the final and complete U-WI.

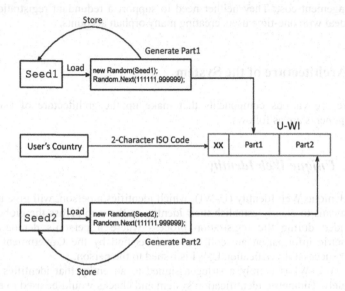

Fig. 2. Process for generating U-WI

The random numbers generated as Part1 and Part2 are stored and used as the seeds for generating the next set of random numbers. This cycle of generating a random number and reusing it as a seed for generating the next random number, keeps repeating. The range for generating the random numbers of each part is from 111111 to 999999, which can yield up to 888888 random numbers. Theoretically, the combination of both the parts must produce roughly around 790,000,000,000 random numbers.

When a person dies, one would see a need to de-activate the U-WI associated with the person. One simple way to deal with that is to flag U-WI record as inactive once one confirms the death. In a world of more than 7 billion people, updating U-WI records based on the death register is not easy, especially since a large number of cases of death are not re-ported making it difficult to update them at a central UWAS.

One way to ensure that U-WI's are not misused by others after a person's death is to inactivate the U-WI if it has not been used say in a year (timeout can be changed). In the case that a U-WI is inactivated of a person who has simply not authenticated him-self/herself in a long time, he/she can simply activate their U-WI by a simple re-activation procedure

2.2   User Credentials

The registration process requires the user's fingerprints, to check for duplication of records, the same fingerprints will also be stored on the database. These fingerprints will act as the basic and default credentials for validation of the user on the web.

Avoiding the use of passwords seemed to be quite impossible with the current technology and the downside of Biometric devices, specified later in this document. Hence UWAS also includes the Password for validating users on the web. Once the user has obtained the U-WI he/she can set a password for use with the websites and also set up a few security questions in order to retrieve the password, in case they forget it.

2.3   Authentication System

Universal Web Authentication is the process wherein a person's U-WI, along with other attributes, including biometrics, are submitted for validation through the web. Once the person's credentials are verified then a positive response is sent back to the browser.

An initial request for a protected page causes UWAS API to redirect the user's browser to a central authentication server. The authentication server and the user interact to establish the user's identity. This normally involves the user providing the U-WI and credentials over a secure connection. Then the authentication server may set a session cookie [2] so that it can respond to future authentication requests without needing to ask for the credentials again.

Fig. 3 shows how the database is organized within the server. It mainly consists of three tables; GeneralData, AuthenticationData and Address. GeneralData consists of attributes like FirstName, LastName, Gender etc. AuthenticationData consists of binary attributes FP_L5, FP_L4... for storing the ten fingerprints of the user. The Address table consists of attributes like City, ZipCode, State etc. to store the user's address.

On a global scale the authentication system follows a distributed approach. There are two types of servers; Local and Central. Each country will have its own local server along with the database. First the central server checks the U-WI and determines to which locale the U-WI belongs, and then it redirects the request to the local server of the corresponding locale. The local server then verifies the credentials and then sends a response to the client browser along with basic information like name, date of birth and gender of the user.

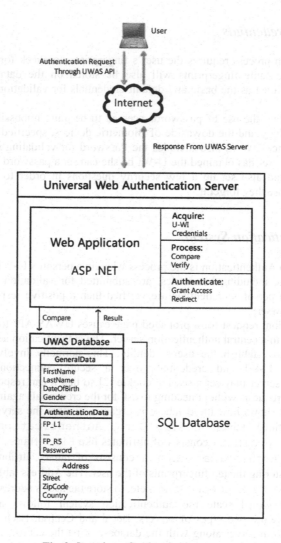

Fig. 3. Overview of authentication process

The role of the central server is to just process the U-WI, determine the first two characters (2-Character ISO code for the country) and transfer the request to the corresponding local server of that country. Then it is the job of the local server to compare the credentials and send a response back to the user's browser.

Fig.4 shows the global distributed approach of the authentication system. For the UWAS system to be implemented globally, all the countries must agree upon certain guidelines and must propose the geographical location for placing the Central Servers.

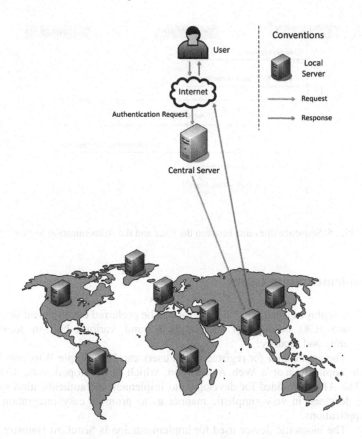

Fig. 4. Global distributed approach of UWAS

The Central servers can further be distributed across multiple locations in order to manage the high volume of authentication requests form the users. This will also ensure that the authentications systems are still working even if any one of the server fails. The same concept can also be used for safeguarding the local servers within each country.

Fig. 5 shows the sequence of events that happen when the user and/or the service provider request for the authentication of the user.

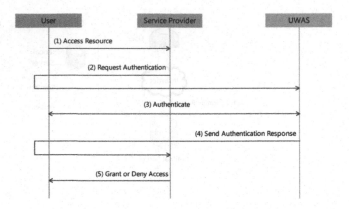

Fig. 5. Sequence of events between the User and the Authentication System

3 Implementation Issues

UWAS is deployed using the .NET platform. The preferred development language is C-Sharp (C#) because of its simplicity and various built-in tools for cryptography and security.

The application for registering the users can be a simple Windows Form Console application or a Web Application, which is developed using C-Sharp (C#). The API's provided for developers to implement the authentication system can be designed in very simplistic manner as to promote easy integration with their applications.

The biometric device used for implementation is SecuGen Hamster Plus. The fingerprints are obtained using the SDK tools provided by SecuGen. The extracted finger print templates are based on the SecuGen 400 [8] template, but can also be extracted using standard ANSI [8] template.

The entire UWAS model is simulated as a local scenario using just two computers, one acting as an authentication server and the other a client machine. The implementation results are provided in the later section of this document.

3.1 Biometric Devices

One major issue related to the use of biometric devices is, each device follows its own algorithm for feature extraction and creation of templates. Also it is not possible for all users to use the same biometric devices all the time. Hence it is difficult to perform feature extraction using a common template.

Until a standard toolkit is available for all biometric devices to obtain the templates using a standard format [6], it is difficult to force users to use biometric credentials over the dominant passwords. Also there may be scenarios where a person (user) may not be physically enabled to use the biometric devices, in case of birth defect or a severe accident the person might have lost either arms or eyes, where the biometric devices are of no use. Hence the use of passwords cannot be eliminated completely.

3.2 Application Compatibility

Another issue for big applications is to migrate from the current authentication systems on which the entire core of the application is based. In some cases it might be required to change the entire architecture of the application to use UWAS, which might not be feasible.

3.3 Geographical Discrepancies

Each country follows its own set of Rules and Regulations. For the entire world to promote the use of UWAS, all the countries will have to agree upon some guidelines on which the UWAS might function.

4 Security Analysis

The proposed model poses a huge security risk. Security is a major issue in web based authentication system. There are various internet threats [1] which affect the security of the system and increase the risk of storing the user's data on a single server.

Since not all users will be at the reach of biometric devices, passwords will remain predominant form of authentication. Passwords are known to have many problems. Passwords are vulnerable to dictionary attacks and can be easily phished [5] using a spoofed web site. Moreover, since users use a single U-WI and password for all the websites, a single server compromise can result in account takeover at many other sites.

Privacy is another matter. If a user can login from any location to access data and applications, it's possible the user's privacy could be compromised. One way to overcome this issue is to use security techniques such as SSL [7] and Certificates.

5 Results and Discussions

Based on the implementation of the local scenario, we can now successfully simulate the use of UWAS API to authenticate users.

The user registration console has a very simple user interface. It has various fields for obtaining information like First Name, Middle Name, Last Name, Date of Birth, Gender, Country, Street, City, State and Zip Code.

The console also has tools to capture the ten fingerprints and extract the minutiae data from the captured fingerprints. The U-WI is generated only after all the information about the user is provided. A function checks the country field and determines the 2-Character ISO code for that country and then generates the U-WI.

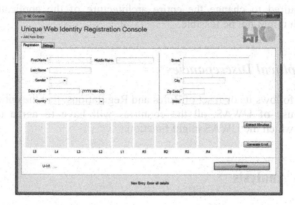

Fig. 6. An empty form in the User Registration Console

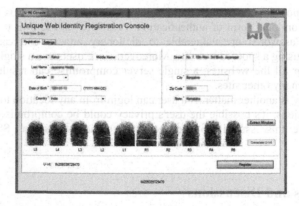

Fig. 7. Fully filled form in the User Registration Console

Fig. 8. API for authenticating users

The Authentication application has three fields; the first field is the U-WI, the second field allows the user to choose the finger which he wishes to use for scanning and the third field to select the fingerprint scanning device.

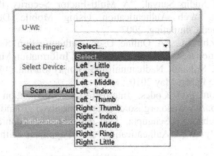

Fig. 9. API for authenticating users

As shown in Fig. 9, the second field provides the user with the option to select finger which he/she wishes to use. When the user clicks the "Scan and authenticate" button (Fig. 8), the scanner scans the fingerprint and the server compares the scanned fingerprint with the one stored in the database. If there is a match then it sends a positive response to the user's browser, else sends a negative response.

6 Conclusion

In the presented version of the model we have focused only on how to streamline the authentication process and how to pass the user's information to the service provider.

Overall we have a proposed a model for authenticating users based on a single Unique ID, which is very effective in stopping users from providing false information on the websites and also avoid impersonating people on the social networking sites, which was the sole aim of the project.

7 Acknowledgement

This work wouldn't have been made possible without the help of our beloved Principal Dr. Rana Pratap Reddy, Dean R&D Dr. Sunil Kumar Manvi. We thank them for their continuous guidance, support and motivation.

References

[1]. Ayu Tiwari and Sudip Sanyal, "A Multi-Factor Security Protocol For Wireless Payment- Secure Web Authentication Using Mobile Devices", International Conference Applied Computing, 2007

[2]. Yu Sheng and Zhu Lu, "A Online User Authentication Scheme for Web-based services", International Seminar on Business and Information Management, 2008

[3]. Hemant Kanakia, Srikanth Nadhamuni and Sanjay Sarma, "A UID NUMBERING SCHEME", White Paper, May 2010

[4]. Documentation on ISO Codes, International Organization for Standardization. (Available at: http://www.iso.org/iso/country_codes/background_on_iso_3166.htm)

[5]. Ben Dodson, Debangsu Sengupta, Dan Boneh, and Monica S. Lam, "Secure, Consumer-Friendly Web Authentication and Payments with a Phone", Stanford University

[6]. Planning Commision, UIDAI, "UIDAI BIOMETRIC CAPTURE API Draft", White Paper, 2010

[7]. Shakir James, "Web Single Sign-On Systems", Computer Science Department, WUSTL, Dec 2007

[8]. SecuGen, "SecuGen .NET Programming Manual", 2011

[9]. Subhash Chander and Ashwani Kush, "Impact of Aadhaar (Unique ID) in Governance" published in proceedings of "National Conference Organized by Computer Society of India, Dec 2011

[10]. Documentation on Random Classes in .NET, MSDN, Microsoft (Available at: http://msdn.microsoft.com/en-us/library/system.random.aspx)

An Overview of the Applications of Particle Swarm in Water Resources Optimization

Rosemary Cyriac[1], A. K. Rastogi[2]

[1]M.Tech Student, Department of Civil Engineering, Indian Institute of Technology Bombay, Mumbai 400 076, India
[2] Professor, Department of Civil Engineering, Indian Institute of Technology Bombay, Mumbai 400 076, India

{rosemaryc89@gmail.com, akr@civil.iib.ac.in}

Abstract: Optimization methods have evolved over the years to solve many water resources engineering problems of varying complexity. Today researchers are working on soft computing based meta heuristics for optimization as these are able to overcome several limitations of conventional optimization methods. Particle Swarm is one such swarm intelligence based optimization algorithm which has shown a great potential to solve practical water resources management problems. This paper examines the basic concepts of Particle Swarm Optimization (PSO) and its successful application in the different areas of water resources optimization.

Keywords: Water Resources Engineering, Particle Swarm Optimization, Swarm Intelligence

1 Introduction

Planning, development and management of water resources falls within the domain of water resources engineering. Freshwater demand for domestic, irrigational, industrial and recreational purposes already exceeds supply in many parts of the world and continues to rise due to rapid urbanization and population growth. Proper management of the available ground water and surface water resources in all user sectors is of utmost importance for any nation for the best utilization of the available sources of water.

One of the areas where this is more important than others is irrigation sector since over 80% of water in India is diverted towards agriculture. An entire spectrum of activities involving reservoir releases, groundwater withdrawals, use of new irrigation techniques call for optimal solutions to obtain maximum benefits from the available water while also meeting all the demands timely. Similarly, to minimize floods and droughts, to ensure water quality considerations and for well field installations water resources management is necessary to meet the competing demands. In this context, the importance of optimization in certain specific areas of water resources is considered in this paper.

2 Optimization

Optimization tools are utilized to facilitate optimal decision making in the planning, design and operation of especially large water resources systems. The entire gamut of operations involved with large water resources projects are complex and directly influence the people. The application of optimization techniques is therefore necessary and also challenging in water projects, due to the large number of decision variables involved. This is further demanded by the stochastic nature of the inputs and multiple objectives such as irrigation, hydropower generation, flood control, industrial and drinking water demands which a project has to meet simultaneously. Presently certain specific cases where optimization practices have been used successfully are considered as follows:

- Reservoir planning, design and operation

- River water pollution control using optimal operation policy

- Regional scale groundwater pollution and utilization management

- Identification of unknown groundwater pollution sources

- Estimation of unknown aquifer parameters in groundwater flow through inverse modelling

- Optimal design of water distribution and waste water systems

Fundamentally, optimization involves systematically choosing solutions from an allowed set of decision variables for maximizing the benefits and minimizing the losses. The conventional numerical optimization methods (viz. linear, nonlinear and dynamic programming) which were used in the past have limited scope in problems of water resources management where objective functions are often non convex, nonlinear, not continuous and non-differentiable with respect to the decision variables. Nonlinear programming methods have rather slow rate of convergence and often result in local optimal solutions since they depend upon initial estimations of variables, whereas the dynamic programming approach suffers from the curse of dimensionality [13]. Thus the conventional methods which utilize gradients or higher order derivatives of objective functions are not suitable for many real world problems in water resources management. For the last two decades non-conventional, metaheuristic techniques have been used successfully for obtaining optimal solutions. Although metaheuristic techniques do not have a rigorous mathematical proof like the conventional numerical methods, they follow a certain logical procedure that allows them to deliver a near global optimum solution.

3 Particle Swarm Optimization

Evolutionary Computation is the general term for several computational techniques which are based to some degree on the evolution of biological life in the natural world. Particle swarm optimization (PSO) is an evolutionary computation technique based upon the behaviour of a flock of birds or a school of fish [24]. When a swarm looks for food, the individuals will spread in the environment and move around independently. Each individual has a degree of freedom and randomness in its movements which enables it to find food deposits. Sooner or later, one of them will find something digestible and being social, announce this to its neighbours. These can then approach the source of food too.

Like the other evolutionary computation techniques, PSO is a population-based search algorithm and is initialized with a population of random solutions, called particles. Unlike in the other evolutionary computation techniques, each particle in PSO is also associated with a velocity. This velocity connotes an improvement in the solution which gets added to the initially assumed solution to make it move towards the optimum solution. Particles fly through the search space with velocities which are dynamically adjusted according to their historical behaviours. Therefore, the particles have a tendency to fly towards the better and better search area over the course of search process. Since its introduction by Eberhart and Kennedy [4], PSO has attracted considerable attention from the researchers around the world and seen gradual improvements with the passage of time.

3.1 Original PSO Algorithm

The basic concept of the PSO can be technically summarized in the following steps:

1. Initialize a population of random solutions on D dimensions in the search space. In the D dimensional search space the ith individual (assumed solution or a particle having a position equal to the assumed solution) of the population can be represented by a D dimensional vector

$$X_i = (x_{i1}, x_{i2} \ldots \ldots x_{id})^T \tag{1}$$

2. Each of the above elements of the assumed solution set is modified in each iteration in a probabilistic manner. The improvement made to each of them in each iteration is referred to as velocity. Thus the velocity (position change or change in solution) of the particle can be represented by another D dimensional vector which is also initialized with some random values.

$$Vi = (v_{i1}, v_{i2} \ldots v_{id}) \tag{2}$$

3. For each particle (position or assumed solution) evaluate the desired optimization fitness function in D variables.

4. The best previously obtained fitness value of each particle and the corresponding value of the particle is noted. They are stored in a D dimensional vector

$$p_{id} = (p_{i1}, p_{i2} \ldots p_{id})^T \tag{3}$$

5. The best fitness value obtained so far by any particle in the population space is noted and the value of the particle is stored as p_{gd}

6. Each of the initially assumed solutions (particles) is improved upon in each iteration through the following equation. The improvement in solution is denoted by v_{id} (velocity).

$$v_{id}^{m+1} = v_{id}^m + c_1 \, rand1(p_{id} - x_{id}^m) + c_2 rand2(p_{gd} - x_{id}^m) \tag{4}$$

$$x_{id}^{m+1} = x_{id}^m + v_{id}^{m+1} \tag{5}$$

Where c_1 and c_2 are positive constants, and rand1 and rand2 are two random functions in the range [0, 1], m is the number of iterations;

7. Loop to step (2) until a criterion is met, which is either a sufficiently good fitness or depends upon maximum number of iterations. At the end of n iterations the modified x_{id} for which the best fitness value has been obtained in all these iterations is denoted by p_{gd} (global best) and in the nth iteration the value of x_{id} in the solution set which gives the best fitness value is denoted by $p_{id.}$ Thus in each iteration initially assumed solution is updated with respect to the best fitness value obtained among all the other members of the population set and its own previous best. Like other evolutionary algorithms, PSO algorithms is a population based search algorithm with random initialization, and interactions among population members. However, unlike the other evolutionary algorithms, in PSO, each particle flies through the solution space, and has the ability to remember its previous best position, and survives from generation to generation [8].

3.2 Parameters of PSO

The first new parameter added into the original PSO algorithm is the inertia weight (Eberhart and Shi 1998a, 1998b).They modified dynamic equation (4) of PSO as:

$$v_{id}^{m+1} = \omega v_{id}^m + c_1 rand1(p_{id} - x_{id}^m) + c_2 rand2(p_{gd} - x_{id}^m) \qquad (6)$$

where a new parameter, inertia weight ω is introduced. Equation (5), however remains unchanged. The inertia weight is introduced to balance between the global and local search abilities. The large inertia weight facilitates global search while the small inertia weight facilitates local search. A value of $0.1 - 0.9$ is recommended in many of the research papers. The introduction of the inertia weight also eliminates the requirement of carefully setting the maximum velocity V_{max}each time the PSO algorithm is used. The V_{max}can be simply set to the value of the dynamic range of each variable and the PSO algorithm still performs satisfactorily.

Another parameter - constriction coefficient was introduced to accelerate PSO convergence [1][2]. A simplified method of incorporating it appears in Equation (7), where k is a function of c_1 and c_2 as seen in Equation (8).

$$v_{id}^{m+1} = k[v_{id}^m + c_1 rand1(p_{id} - x_{id}^m) + c_2 rand2(p_{gd} - x_{id}^m)] \qquad (7)$$

$$k = \frac{2}{|2 - \varphi - \sqrt{\varphi^2 - 4\varphi}|} \qquad (8)$$

where $\varphi = c_1 + c_2, \varphi > 4$

Mathematically, Equation (6) and (7) are equivalent by setting inertia weight ω to be k, and c_1 and c_2 meet the condition $\varphi = c_1 + c_2, \varphi > 4$. The PSO algorithm with the constriction factor can be considered as a special case of the PSO algorithm with inertia weight while the three parameters are connected through Equation (8). As a rule of thumb a better approach is to utilize the PSO with constriction factor while limiting $Vmax$to $Xmax$, the dynamic range of each variable on each dimension, or utilize the PSO with inertia weight while selecting ω, $c1$ and $c2$ according to Equation (8)[6].

When Clerc's constriction method is used, φ is commonly set equal to 4.1 and the constant multiplier kis approximately 0.729. This is equivalent to the PSO with inertia weight when $\omega \approx 0.729$ and $c_1 = c_2 = 1.49445$. Since the search process of a PSO algorithm is nonlinear and complicated, a PSO with well-selected parameter set can have good performance, but much better performance could be obtained if a dynamically changing parameter is well designed. Intuitively, the PSO should favour global search ability at the beginning of PSO while it should favour local search ability at the end of PSO.

Shi and Eberhart [5] first introduced a linearly decreasing inertia weight to the PSO over the course of PSO, then they further designed fuzzy systems to nonlinearly change the inertia weight [7][8]. The fuzzy systems have some measurements of the PSO performance as the input and the new inertia weight as the output of the fuzzy systems. In a more recent study, an inertia weight with a random component [0.5 + (rand/2.0)] rather than time decreasing is utilized. This produces a randomly varying number between 0.5 and 1.0, with a mean of 0.75 which is similar to Clerc's constriction factor described above [8].

4 Applications of PSO in Water Resources Engineering

Researchers have attempted a wide range of problems in water resources engineering using PSO. Certain problems where particle swarm techniques have been successfully applied in water resources are examined as follows:

4.1 Reservoir Planning Design and Operation

Reservoir Operation optimization involves determining the optimum amount of water that should be released for flood control, irrigation, hydropower generation, navigation and municipal water supply. Being a complex problem it involves many decision variables, multiple objectives as well as considerable risk and uncertainty [14].

Kumar and Reddy [13] discussed the implementation of Particle Swarm Optimization in multipurpose reservoir operation. They considered Bhadra reservoir system in India which serves irrigation and hydropower generation. It was required to obtain the optimum releases to the left and right bank canals (utilized for irrigation and hydropower generation) and to the river bed turbine (for hydropower generation). To handle multiple objectives of the problem, a weighted approach was adopted. The objective function dealt with minimizing the annual irrigation deficits and maximizing the annual hydropower generation with greater weightage for minimizing irrigation deficits. The decision variables were the monthly releases that should be made to the left and right bank canal and the river bed turbine in a year. The optimization was carried out under a set of constraints which included mass balance, storage, canal capacity, power production and water quality requirements.

The performance of the standard PSO algorithm was improved by incorporating an Elitist Mutated PSO (EMPSO) in which a certain number of the best performing solutions (elites) were retained with mutation during each successive iteration to increase population diversity and enhance the quality of the population. The results obtained demonstrated that EMPSO consistently performed better than the standard PSO and genetic algorithm techniques. They concluded that EMPSO is yielding better quality solutions with less number of function evaluations.

4.2 Groundwater utilization

Gaur et.al.[9] used Analytic Element Method and Particle Swarm Optimization based simulation optimization model for the solution of a groundwater management problem. The AEM-PSO model developed was applied to the Dore river basin, France to solve two groundwater hydraulic management problems: (1) maximum pumping from an aquifer, and (2) minimize the cost to develop the new pumping well system. Discharge as well as location of the pumping wells were taken as the decision variables. The influence of the piping length was examined in the total development cost for new wells. The optimal number of wells was also calculated by applying the model to different sets of wells. The constraints of the problem were identified with the help of water authority, stakeholders and officials which included maximum and minimum discharge limits for the well pumping, minimum allowable groundwater drawdown and water demand.

The AEM flow model was developed to facilitate the management model in particular, as in each iteration optimization model calls a simulation model to calculate the values of groundwater heads. The AEM-PSO model was found to be efficient in identifying the optimal location and discharge of the pumping wells. A penalty function approach was used to penalize constraint violations and this was found to be valuable in PSO and also acceptable for groundwater hydraulic management problems.

4.3 Groundwater Pollution Control

In many parts of our country and in the world ground water is excessively contaminated due to various anthropogenic and industry related activities. Pollution of groundwater happens due to the leachate from animal and human waste dumped on the land, fertilizer application, industrial effluents and municipal waste dumped into surface water bodies. Mategaonkar and Eldho[15] presented a simulation optimization (SO) model for the remediation of contaminated groundwater using a PAT system. They developed a simulation model using Mesh Free Point Collocation Method (PCM) for unconfined groundwater flow and contaminant transport and an optimization model based upon PSO. These models are coupled to get an effective SO model for the groundwater remediation design using pump and treat mechanism. In groundwater pollution remediation using PAT, optimization is aimed at identification of cost-effective remediation designs, while satisfying the constraints on total dissolved solids concentration and hydraulic head values at all nodal points. Also, pumping rates at the pumping wells should not be more than a given specified rate. Only minimization of the remediation cost is considered as the objective function in this remediation design. The decision variables were the pumping or injection rates for the wells considered and the purpose of the design process is to identify the best combination of those decision variables. The cost function includes both the capital and operational costs of extraction and treatment. The PCM PSO model is tested for a field unconfined aquifer near Vadodara, Gujarat, India.

4.4 Estimation of unknown aquifer parameters in groundwater flow through inverse modeling

Jianqing and Hongfei[11] applied the PSO algorithm to the function optimization problem of analyzing pumping test data to estimate aquifer parameters of transmissivity and storage coefficient. The objective function was to minimize the difference between simulated and observed groundwater head values with transmissivity and storage coefficient as the decision variables. The results showed that 1) PSO algorithm may be effectively applied to solve the function optimization problem of analyzing pumping test data in aquifer to estimate transmissivity and storage coefficient, 2) the convergence of PSO algorithm and the computation time are influenced by the number of particles and that fewer iterations are needed in computation with the larger number of particles and 3) the ranges of initial guessed values of transmissivity may also bring some effect on the convergence of PSO algorithm and the computation time. They found that larger the ranges are, the more number of iterations and longer computation time are needed for a guaranteed convergence of PSO algorithm.

A few other applications of PSO are listed below in Table1.

Table 1. Applications of PSO

Author/s	Application	Decision variable	Empirical constants Chosen
Mattot et.al [17]	PSO is used for the cost minimization of a pump and treat optimization problem.	Extraction and Injection Rates of the wells and Number of wells required	
Zhou et.al. [25]	Training of Artificial Networks by PSO to classify and predict water quality	Weights of the input and hidden layers of ANN	1) $\omega = .9-.4$ 2) No of particles - 80 3) c1 = c2 = 2 4) k is not used5) Termination - 1000 iterations
Gill et.al [10]	Multi Objective PSO (MOPSO) to calibrate the (i) Sacramento soil moisture accounting model model and (ii) a support vector machine model for soil moisture prediction	Parameters of both the models (16+3)	1) ω(linearly varying) - 0.9 - 0.01 2) No of particles - (i) 100 (ii) 50 3) c1 = c2 = 0.5
Izquierdo et.al [12]	design of (i) 2 water distribution networks, the Hanoi new water distribution network and the New York tunnel water supply system (ii) the design of a waste water network and (iii) the calibration and identification of leaks in a water distribution network	pipe diameters and slopes	1). $\omega = 0.5 + 1/(2(\ln(k) +1))$;k -iteration no 2). No of particles - (i) 100 (ii) 100 (iii) 300 3). c1 = 3 ; c2 = 2 4). Termination after no of iterations - (i) 200 (ii) 800 (iii) 200
Mathur et al [16]	optimal schedule of irrigation from lateral canals	no of minor canals(21) and no of days (120)	1)ω(0.9-0.4) 2) particles - 200 3) c1=c2=1.5 4)Stop after 200steps

5 Conclusions and further Scope

Particle Swarm Optimization has been successfully used in various complex water resources engineering problems to decide water management policies.
Some of the advantages of PSO are as follows:
1. In comparison to other evolutionary algorithms PSO is simpler to understand and implement.
2. The method does not depend on the nature of the function it maximizes or minimizes. Thus approximations made in conventional techniques are avoided.
3. It uses objective function information to guide the search in problem space. Therefore it can easily deal with non differentiable and non convex objective functions.
4. Non Linear Programming solutions are dependent upon the initial estimation of solutions. Therefore different initial estimates of parameters give different suboptimal solutions. PSO method is not affected by the initial searching points, thus ensuring a quality solution with high probability of obtaining the global optimum for any initial solution.
5. In PSO particle movement uses randomness in its search. Hence, it is a kind of stochastic optimization algorithm that can search a complicated and uncertain area. Thus it is more flexible and robust than conventional methods.
6. The convergence is not affected by the inclusion of more constraints.
7. It also has the flexibility to control the balance between the global and local exploration in search space. This property enhances the search capabilities of the PSO technique and yields better quality solutions with fewer function evaluations.
8. The algorithm of PSO, demands fewer adjusted key parameters of the algorithm and its arithmetic process is convenient and programmable. It can be easily implemented, and is computationally inexpensive, since memory and CPU speed requirements are low.
PSO has been highly successful and within little more than a decade hundreds of papers have reported successful applications of PSO. As it is a technique of recent origin, the number of applications of PSO in water resources engineering is relatively less and there is still a lot of scope for a wider application of PSO to solve water related problems. Therefore there is a possibility that it may emerge as a powerful optimization tool in water resources research. Some of the possible areas in water resources where further research may be done is as follows

➢ Ground water – utilization management, detection of unknown groundwater pollution sources, contaminant remediation, estimation of unknown aquifer parameters, estimation of water table by geo physical methods, optimization for design of multi layered sorptive systems, management of salt water intrusion in coastal aquifers.

The decision variables are specific to the problem under study. It can include the location, number and discharge of pumping wells, unknown aquifer parameters, depth of water table etc

➤ Reservoir – planning, design and operation.
The decision variables may include the optimum discharge values for each time period such that the all the demands are met.

➤ Hydrology – Calibration of hydrological and ecological models , Time Series Modelling, stream flow forecasting,
The calibration of various models involve the estimation of the various parameters associated with them. It may not be possible to obtain them from physical observations. Hence optimization methods have a definite advantage.

➤ Irrigation – scheduling of irrigation canals, Canal design
➤ River Stage forecasting , River Water Quality Control and Prediction
➤ Design of Water Distribution Networks, Calibration and improvement of urban drainage systems, Detection of leaks and its rectification
➤ Climate Variability and Change, Calibration of climate models

There are efforts by many researchers to develop better variations of PSO to increase population diversity and ensure global convergence of the algorithm. These researches may make it more suitable for large scale complex combinatorial optimization problems.

REFERENCES

[1] Clerc, M., The swarm and the queen: towards a deterministic and adaptive particle swarm optimization. *Proc. Congress on Evolutionary Computation, 1999 Washington, DC, pp 1951 - 1957*. Piscataway, NJ: IEEE Service Centre (1999)

[2] Clerc, M., and Kennedy, J., The particle swarm explosion, stability, and convergence in a multi-dimensional complex space. *IEEE Transactions on Evolutionary Computation, vol. 6, p. 58-73* (2002)

[4] Eberhart, R.C., and Kennedy, J., A new optimizer using particle swarm theory. *Proceedings of the Sixth International Symposium on Micro Machine and Human Science, Nagoya, Japan, 39-43*. Piscataway, N J: IEEE Service Centre (1995)

[5] Eberhart, R. C., and Shi, Y.,.Evolving artificial neural networks.*Proc. Conference on Neural Networks and Brain, 1998, Beijing, P.R.C., PL5-PL3.* (1998)(a)

[6]Eberhart, R. C., and Shi, Y., Comparing inertia weights and constriction factors in particle swarm optimization. *Proc. Congress on Evolutionary Computation 2000, San Diego, CA, pp 84-88*. Piscataway, NJ: IEEE Service Centre (2000)

[7]Eberhart, R.C., and Shi, Y., Tracking and optimizing dynamic systems with particle swarms. *Proc. Congress on Evolutionary Computation 2001, Seoul, Korea*. Piscataway, NJ: IEEE Service Centre (2001a)

[8]Eberhart, R.C., and Shi, Y., Particle swarm optimization: developments, applications and resources. *Proc. Congress on Evolutionary Computation 2001, Seoul, Korea*. Piscataway, NJ: IEEE Service Centre (2001b)

[9] Gaur, S., Chahar, B.R., and Graillot, D., Analytic Element Method and particle swarm optimization based simulation-optimization model for groundwater management. *Journal of Hydrology 402, 217 -227* (2011)

[10] Gill, K.M., Kaheil, Y.H., Khalil, A., McKee, M., and Bastidas,L. Multiobjective particle swarm optimization for parameter estimation in hydrology. *Journal of Water Resources Research, 42, W07417, 14 PP* (2006)

[11]Hongfei, Z., and Jianqing, G., The Application of Particle Swarm Optimization Algorithms to Estimation of Aquifer Parameters from Data of Pumping Test. *Proc. 5th International Conference on Computer Sciences and Convergence information Technology(ICCIT) 2010,Seoul,Korea* (2010)

[12]Izquierdo J., Montalvo. I, Perez R., and Tavero M., Optimization in Water Systems: a PSO approach. *Proc. Spring Simulation Multi conference 2008, Ottawa, Canada* (2008)

[13] Kumar, D.N and Reddy, M.J., Multipurpose Reservoir Operation using Particle Swarm Optimization. *Journal of Water Resources Planning and Management 133:3,192-201* (2007)

[14]Loucks P. D., and Oliveira R. Operating rules for multi reservoir systems.*Water Resources Research, vol. 33, no. 4, p. 839* (1997)

[15]Mategaonkar M., and Eldho T. I., Groundwater remediation optimization using a point collocation method and particle swarm optimization. *Environmental Modelling& Software 32,37-38* (2012)

[16]Mathur, Y. P., Kumar.R. , and Pawde, A., A binary particle swarm optimisation for generating optimal schedule of lateral canals.*The IES Journal Part A: Civil & Structural Engineering, 3:2, 111-118* (2010)

[17]Mattot S. L., Rabideau J.A., and Craig R. J., Pump-and-treat optimization using analytic element method flow models. *Advances in Water Resources 29, 760–775* (2006)

[18]Moradi, J.M., Marino, M.A., Afshar, A., Optimal design and operation of irrigation pumping station. *Journal of Irrigation and Drainage Engineering129(3), 149–154* (2003)

[19]Poli R., An Analysis of Publications on Particle Swarm Optimisation Applications *Department of Computer Science University of Essex Technical Report CSM-469*(2007)

[20]Poli, R., Kennedy, J., and Blackwell, T., Particle swarm optimization. An overview.*Swarm Intelligence, 1(1), 33-57* (2007)

[21] Samuel P.M. and Jha K.M., Estimation of Aquifer Parameters from Pumping Test Data by Genetic Algorithm Optimization Technique. *Journal of Irrigation and DrainageEngineering, 129(5): 348-359* (2003)

[22]Theis, C. V., The relation between the lowering of the piezometric surface and the rate and duration of discharge of a well using ground-water storage. *Trans. of American Geophysical Union 16:519-52* (1935)

[23] Sharma, A.K., and Swamee, P.K., Cost considerations and general principles in the optimal design of water distribution systems. In: *ASCE Conf. Proc., vol. 247, p.85* (2006)

[24]Yuhui Shi., Particle Swarm Optimization. Electronic Data Systems, Inc.*IEEE Neural Network Society Magazine* (2004)

[25] Zhou C.,Gao L., GaoHaibing , and Peng C., Pattern Classification and Prediction of Water Quality by Neural Network with Particle Swarm Optimization. *Proceedings of the 6th World Congress on Control and Automation, June 21 - 23, 2006, Dalian, China* (2006)

Quasi-based hierarchical clustering for land cover mapping using satellite images

J. Senthilnath[a1], **Ankur raj**[b2], **S.N. Omkar**[a3], **V. Mani**[a4], **Deepak kumar**[c5]

[a]Department of Aerospace Engineering, Indian Institute of Science, Bangalore, India

[b]Department of Information technology, National Institute of Technology, Karnataka, India

[c]Department of Electrical Engineering, Indian Institute of Science, Bangalore, India

{[1]snrj@aero.iisc.ernet.in;[2]ankurrj7@gmail.com; [3]omkar@aero.iisc.ernet.in; [4]mani@aero.iisc.ernet.in; [5]deepak@ee.iisc.ernet.in}

Abstract. This paper presents an improved hierarchical clustering algorithm for land cover mapping problem using quasi-random distribution. Initially, Niche Particle Swarm Optimization (NPSO) with pseudo/quasi-random distribution is used for splitting the data into number of cluster centers by satisfying Bayesian Information Criteria (BIC).The main objective is to search and locate the best possible number of cluster and its centers. NPSO which highly depends on the initial distribution of particles in search space is not been exploited to its full potential. In this study, we have compared more uniformly distributed quasi-random with pseudo-random distribution with NPSO for splitting data set. Here to generate quasi-random distribution, Faure method has been used. Performance of previously proposed methods namely K-means, Mean Shift Clustering (MSC) and NPSO with pseudo-random is compared with the proposed approach - NPSO with quasi distribution(Faure).These algorithms are used on synthetic data set and multispectral satellite image (Landsat 7 thematic mapper). From the result obtained we conclude that use of quasi-random sequence with NPSO for hierarchical clustering algorithm results in a more accurate data classification.

Keywords: Niche Particle Swarm Optimization, Faure sequence, Hierarchical clustering.

1 Introduction

Nature has given a lot to mankind and land is one such resource. We need actual information regarding the features of land to make good use of it. Using satellite images, we can accurately plan and use land efficiently. Satellite images offer a method of extracting this temporal data that can be used in gaining knowledge

J. C. Bansal et al. (eds.), *Proceedings of Seventh International Conference on Bio-Inspired Computing: Theories and Applications (BIC-TA 2012),* Advances in Intelligent Systems and Computing 202, DOI: 10.1007/978-81-322-1041-2_5, © Springer India 2013

regarding land use. Recent advances in the realm of computer science has allowed us perform "intelligent" jobs. This has established a vast research area in solving the automatic image clustering problem. The image clustering using satellite image for land cover mapping problem is useful for auditing the land-usage and city planning [1].

The main objective of clustering problem is to minimize the intra-cluster distance and maximize the inter-cluster distance [2]. One of the main task of clustering problem is to locate the cluster centres for a given data set; it is basically a problem of locating maxima of a mapped function from a discrete data set. Recently researchers are interested in capturing multiple local optima of a given multi-modal function for this purpose nature inspired algorithms are used. Brits *et.al* [3] developed Niche PSO (NPSO) for optimization of standard benchmark functions, later Senthilnath *et.al* [2] applied the same for locating multiple centres of a data set for hierarchical clustering problem.

NPSO is a population based algorithm its performance has shown high dependency on initial distribution of population in search space it has been observed in literatures that performance of particle swarm optimization has improved by using more uniform distribution of particle in search space [4]. Kimura *et.al* [5] have used Halton sequence for initializing the population for Genetic Algorithms (GA) and have shown that a real coded GA performs much better when initialized with a quasi-random sequence in comparison to a GA which is initialized with a population having uniform probability distribution (i.e. pseudo-random distribution). Instances where quasi-random sequences have been used for initializing the swarm in PSO can be found in [4, 5, 6, 7]. Nguyen *et.al* [7] has given a detailed comparison of Halton, Faure and Sobol sequences for initializing the swarm. It has been observed that performance of Faure sequence takes over the performance of Halton sequence in terms of uniformity in space.

In this paper, the comparison is done between pseudo and quasi based distribution for initializing NPSO to capture multiple local maxima for a given data set. In our study the data set used are synthetic data set and Landsat satellite image for hierarchical clustering. In earlier studies [4, 5, 6, 7] for optimization problem, it has been observed that use of quasi sequence for initializing population in PSO has given a better performance. The same approach has been applied in this study using the quasi sequence with NPSO for hierarchical clustering algorithm. NPSO is used to split complex large data set into number of cluster satisfying Bayesian Information criteria (BIC) which is commonly used in model selection [8]. These cluster centres are used for merging the data set to their respective group. The challenge is how to get better classification efficiency using quasi distribution in NPSO with hierarchical clustering algorithm.

2 Random Sequences

The clustering using population based methods require initial random distribution of points to extract optimal cluster centres. To generate truly random numbers there is a requirement of precise, accurate, and repeatable system measurements of absolutely non-deterministic processes. Computers normally cannot generate truly random numbers, but frequently are used to generate sequences of pseudo-random numbers. There are two principal methods used to generate random numbers. One measures some physical phenomenon that is expected to be random and then compensates for possible biases in the measurement process. The other uses computational algorithms that produce long sequences of apparently random numbers called pseudo-random. It may be possible to find a more uniform distribution using low-discrepancy sequence known as quasi-random numbers. Such sequences have a definite pattern that fills in gaps evenly, whereas pseudo-random sequence unevenly distributes the sequence, this leads to larger gaps in search space.

2.1   Pseudo-random sequences

A pseudo-random process is a process that appears to be random but is not. Pseudo-random sequences typically exhibit statistically randomness while being generated by an entirely deterministic casual process. These are generated by some algorithm, but appear for all practical purposes to be random. Random numbers are used in many applications, including population based method involving distribution of initial points using random numbers (pseudo number). A common pseudo-random number generation technique is called the linear *congruential* method [9]. The pseudo-random numbers are generated using following equation.

$$A_{n+1} = (Z * A_n + I) \, mod \, M \qquad 1$$

where A_n is the previous pseudo number generated, Z is a constant multiplier, I is a constant increment, and M is a constant modulus. For example, suppose Z is 7, I is 5, and M is 12 if the first random number (usually called the *seed*) A_0 is 4, then next pseudo number $A_1 = (7*4+5) mod \, 12 = 9$. In this way we can generate the pseudo-random sequence.

2.2   Quasi random sequence

The quasi-random numbers have the low-discrepancy (LD) property that is a measure of uniformity for the distribution of the point mainly for the multi-dimensional case. The main advantage of quasi-random sequence in comparison to pseudo-random sequence is it distributes evenly hence there is no larger gaps and no cluster formation, this leads to spread the number over the entire region. The concept of LD is associated with the property that the successive numbers are added in a position as away as possible from the other numbers that is, avoiding clustering (grouping of numbers close to each other). The sequence is constructed based on some pattern such that each point is separated from the others, this leads to maximal separation between the points. This process takes care of evenly distribution random numbers in the entire search space [10, 11].

The most fundamental LD sequence for one dimension is generated by Van der corput method, further to continue random sequence in higher dimension Faure and Halton methods are used.

2.2.1 Faure sequence

Faure sequence is a method to generate LD sequence; it extends the idea of Van der corput sequence in higher dimension. The most basic way to generate quasi-random sequence is Van der corput method, this method uses two basic equations eq.2 and eq.3 to transform a number n in 1-dimensional space with base b.

$$n = \sum_{j=0}^{m} a_j(n) b^j \qquad\qquad 2$$

$$\phi(n) = \sum_{j=0}^{m} a_j(n) b^{-j-1} \qquad\qquad 3$$

where m is the lowest integer that makes $a_j(n)$ as 0 for all $j > m$. Above equations are used at base b, which is a prime number in respective dimension. The Van der corput sequence, for the number n and base b, is generated by a three step procedure:

Step-1: The decimal base number n is expanded in the base b using eq.2
$$3 = 0 * 2^0 + 0 * 2^1 + 1 * 2^2 = 011$$
Step-2: The number in base b is reflected. In this example it is 011 is reflected 110
Step-3: Now the reflected number is written as fraction less than one using eq.3

writing 011 gives $\phi(3) = 1 * 2^{-0-1} + 1 * 2^{-1-1} + 0 * 2^{-2-1} = \dfrac{3}{4}$

Now let us consider $n=4$ length of sequence to be generated, and let $n_1=1$, $n_2=2, n_3=3$ and $n_4=4$ then quasi-random sequence in 1-dimensional space will be generated as follow.

For $n_1=1$, using eq.2
$$n_1 = 1 * 2^0 + 0 * 2^1 + 0 * 2^2 \qquad \text{here} a_0 = 1, a_1 = 0, a_2 = 0$$

Now using eq.3

$$\phi(n_1) = a_0 * 2^{-0-1} + a_1 * 2^{-1-1} + a_2 * 2^{-2-1} = \dfrac{1}{2}$$

Similarly calculating for 2 and 4 gives $\dfrac{1}{4}$ and $\dfrac{1}{8}$ respectively, hence the first 4

numbers of Van der corput sequence are $\dfrac{1}{2}, \dfrac{1}{4}, \dfrac{3}{4}, \dfrac{1}{8}$

This is basic LD sequence in one dimension, for higher dimension LD sequences are generated using Halton and Faure method. In Halton method the sequence numbers are generated using different prime base for each k-dimension. For k^{th}-dimension the Nth number of the sequence is obtained by

$\phi(N, b_1), \phi(N, b_2) \ldots \phi(N, b_k)$ where $i=1 \ldots k$, b_i is the prime number greater than or equal to i.

Faure sequence is similar to Halton sequence with two major differences: Faure uses same base for all dimension and vector elements are permutated in higher dimension. For dimension one it uses Van der corput sequence to generate sequence for higher dimensions vector permutation is carried out using eq.4.

$$a_j^d(n) = \left(\sum_{j \geq i}^{m} \frac{j!}{i!(j-i)!} a^{d-1}(n) \right) \mathrm{mod}\ b \qquad\qquad 4$$

where $c_i = \frac{j!}{j!(j-1)!}$

The base of a Faure sequence is the smallest prime number which is greater than or equal to the number of dimensions in the problem, say for one dimensional problem base 2 is taken. The sequence number is selected between [0,1), the quantity of number generated to complete a cycle increases as the number of dimension increases. For e.g. in base two, for a cycle two numbers are picked within an interval [0,1) i.e. for the first cycle (0,1/2) and for second cycle (1/4,3/4) are selected, similarly for base 3 in the first cycle (0,1/3,2/3) are picked and for second cycle (1/9,4/9,7/9) are selected, hence long cycles have the problem of higher computational time. As Halton sequence uses different bases in each dimension so it has a problem of long cycle length, but in case of Faure this problem has been reduced by taking same base for each dimension. By reordering the sequence within each dimension, a Faure sequence prevents some problems of correlation for high-dimensions, whereas Halton sequence fails to minimize the correlation [12].

2.2.2 Illustration of Faure sequence

Let us consider the same example as discussed in section 2.2.1. The Faure sequence in 1st dimension corresponding to first 4 numbers (n_1=1, n_2 =2, n_3=3 and n_4=4) will be same as that of Van der corput sequence i.e. 1/2, 1/4, 3/4 and 1/8 now numbers for second dimension using Faure Method will be calculated as follow:

For n_1=1 representing at base b=2, $a_2 a_1 a_0$=001, now using eq.4 for vector permutation

$for\ a_0^2(1)\ i = 0,\quad a_0 = \left({}^0c_0 a_0 + {}^1 c_0 a_1 + {}^2 c_0 a_2 \right) \mathrm{mod}\ b = (1 + 0 + 0) \mathrm{mod}\ 2 = 1$

$for\ a_1^2(1)\ i = 1\quad a_1 = \left({}^1 c_1 a_1 + {}^2 c_1 a_2 \right) \mathrm{mod}\ b = (0 + 0) \mathrm{mod}\ 2 = 0$

$for\ a_2^2(1)\ i = 2\quad a_2 = \left({}^2 c_2 a_2 \right) \mathrm{mod}\ b = 0\ \mathrm{mod}\ 2 = 0$

$for\ n_2 = 2\ at\ base\ b\ a_2 a_1 a_0 = 010\ now\ applying\ equation\ 4$

$a_0^2(2) = \left({}^0 c_0 a_0 + {}^1 c_0 a_1 + {}^2 c_0 a_2 \right) \mathrm{mod}\ b = (0 + 1 + 0) \mathrm{mod}\ 2 = 1$

$a_1^2(2) = \left({}^1 c_1 a_1 + {}^2 c_1 a_2 \right) \mathrm{mod}\ b = (1 + 0)\ \mathrm{mod}\ 2 = 1$

$a_2^2(2) = \left({}^2 c_2 a_2 \right) \mathrm{mod}\ 2 = 0\ \mathrm{mod}\ 2 = 0$

Now applying eq.3we get

$\phi(1) = a_0^2(1) \times 2^{(-0-1)} + a_1^2(1) \times 2^{(-1-1)} + a_2^2(1) \times 2^{(-2-1)} = \frac{1}{2}$ and

$$\phi(2) = a_0^2(2) \times 2^{(-0-1)} + a_1^2(2) \times 2^{(-1-1)} + a_2^2(2) \times 2^{(-2-1)} = \frac{3}{4}$$

Similarly other numbers are generated. The first four numbers of Faure sequence in 2-dimension are (1/2, 1/2), (1/4,3/4), (3/4,1/4), (1/8,5/8).

2.2.3 Quasi and Pseudo distribution

Fig-1 shows the distribution of 100 particle in the search space of [-2,2]. Two dimensional Faure sequence has been taken for quasi-random number. It can be seen that in Fig-1a quasi sequence is very uniformly distributed in space (each grid has at-least one point) whereas pseudo sequence as shown in Fig-1b which is generated by matlab *random number generator (rand() function)* is not very uniform.

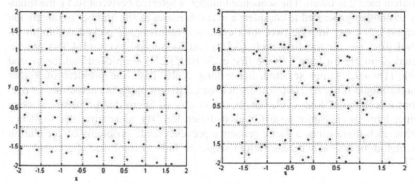

Fig-1a:Quasi-random distribution   Fig-1b:Pseudo-random distribution

3   Cluster splitting and merging

The cluster analysis forms the assignment of data set into cluster based on some similarity measures. In this study an attempt to improve the performance of previously proposed hierarchical clustering is compared with quasi-random distribution with NPSO. The hierarchical splitting technique uses Kernel function for mapping discrete data set to an objective function. This is done by using a Gaussian Kernel, based on Euclidian distance between two data points *(r)* which is given by [13]

$$K(r) = e^{-\|r\|^2}$$

5

It is very difficult to predict how number of clusters is optimal for a given data set, as this is dependent on data distribution. A platform is provided using Bayesian Information criteria (BIC) which is a model fitting approach that provides an optimal number of clusters. The splitting of data set using BIC into number of cluster is given by [8, 15].

$$BIC \approx L(\theta) - \frac{1}{2} \times k_j \times \log(n)$$

6

where $L(\theta)$ is log-like hood measure, k_j is number of free parameters for specific number of cluster and n is no of data point for a given data set.

Niching techniques are modelled after a phenomenon in nature where animal species specialize in exploration and exploitation of different kinds of resources. The introduction of this specialization, or Niching, in a search algorithm allows it to divide the space in different areas and search them in parallel. The technique has proven useful when the problem domain includes multiple global and local optimal solutions. Brits et. al [3] implemented Niche particle swarm optimization (NPSO) which is a variant of PSO [14], based on flock of birds aimed to capture multiple optima in a multi-modal function.

The objective function of all the particles is calculated using Kernel function, using eq.5, if the variance in objective function value of the particle for some fixed number of iteration is less than some threshold value ε then it is named as subswarm leader.

The swarm is divided into several overlapping sub-swarms in order to detect multiple peaks. Sub-swarms are created with all particles around the local centre within the swarm radius. These particles are made to converge towards the local best position i.e. sub-swarm leaders

$$v_{i,j}(t+1) = wv_{i,j}(t) + \rho(t)\left(\hat{y}(t) - x_{i,j}(t)\right) \qquad 7$$

$$x_{i,j}(t+1) = -x_{i,j}(t) + v_{i,j}(t+1) \qquad 8$$

where $-x_{i,j}(t)$ resets the particle position towards the local best position, $\hat{y}_{i,j}(t)$ within sub-swarm radius, $w*v_{i,j}$ is the search direction, and $\rho(t)$ is the region for the better solution. The personal best position of particle is updated using eq.10 where f denotes the objective function.

$$y_i(t+1) = \begin{cases} x_i(t+1) & \text{if } f(x_i(t+1)) \le f(y_i(t)) \\ y_i(t) & \text{if } f(x_i(t+1)) \ge f(y_i(t)) \end{cases} \qquad 9$$

The cluster centres generated using NPSO is grouped using agglomerative approach. These cluster centres are used for initializing K-means to perform agglomerative clustering [16, 17, 18]. Here parametric method is used to group the data points to the closest centres using similarity metric.

Merging data set algorithm:

Step-1: Results obtained as cluster centres from NPSO is given to K-means clustering.

Step-2: Merge data points to closest centres.

Step-3: Use voting method for each data points in the cluster.

Step-4: Cluster is grouped agglomerative using labels.

Step-5: Assign each data points to one of the class.

4 Results and discussion

In this section, we discuss the cluster splitting and merging by comparing pseudo and quasi-random distribution. This distribution is assigned initially for n particles in NPSO. We evaluate the performance of NPSO on synthetic and satellite data sets using the classification matrix of size $n \times n$, where n is the number of classes. A value $A_{i,j}$ in this matrix indicates the number of samples of class i which have been classified into class j. For an ideal classifier, the classification matrix is diagonal. However, we get off-diagonal elements due to misclassification. Accuracy assessment of these classification techniques is done using individual (η_i), average (η_a) and overall efficiency (η_o) [15, 19].

4.1 Synthetic data set

The above algorithm is been applied for classification of a synthetic data set, the original data set consists of two classes, in each class there are 500 samples as shown in Fig-2a. The BIC analysis is carried out as shown in Fig-2b, the optimal clusters for this data set is 8. The hierarchical clustering technique using NPSO is used to generate the cluster centres by initializing the population based on pseudo and quasi-random distribution, in quasi-random Faure method is used.

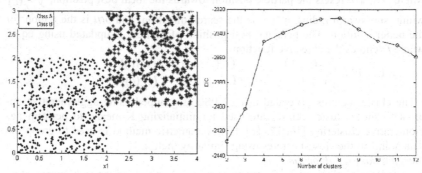

Fig-2a: Synthetic data set   Fig-2b: BIC for synthetic data set

In NPSO, to set the parameter value for sub-swarm radius, inertia weight and weight of leader follower (ρ) different runs are carried out. From Fig-3 we can observe that the optimal parameter value for weight of leader follower (ρ) is 0.4. As it can be observed from Fig-4 that using quasi-random distribution more uniform variation of number of clusters with weight of leader follower (ρ), hence it is easy to predict the parameter value. In contrast for pseudo-random distribution variation is very high which makes difficult to initialize the parameter. The other parameter values assigned are: number of population is 300, sub-swarm radius is 0.1 and inertia weight is adaptively varied in interval [0.6, 0.7]. The 8 cluster centres obtained from NPSO is merged to obtain exact number of classes using merging technique as discussed above. The classification matrix obtained after merging using NPSO based on quasi-random is as shown in table 1.

Also the same experiment is repeated using NPSO based on pseudo-random distribution by keeping all the parameter to be same. The classification matrix obtained by NPSO pseudo-random to split the clusters and merging the data set to their class labels is as shown in table 2.

From table 1 and table 2 we can compare that NPSO quasi-random distribution performed better for all the performance measures to that of NPSO pseudo-random distribution.

Table 1: NPSO-quasi based classification

DATA Class	Class-1	Class-2	Individual efficiency
Class-1 (η_1)	500	0	100%
Class-2 (η_2)	27	473	94.6%
OE (η_o)	97.3	AE (η_a)	97.3

Table 2: NPSO-pseudo based classification

Data class	Class-1	Class-2	Individual Efficiency
class-1 (η_1)	492	8	98%
class-2 (η_2)	50	450	90%
OE (η_o)	94%	AE (n_a)	94.2%

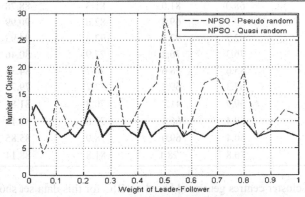

Fig-3: Effect of weight of leader follower in NPSO with quasi and pseudo distribution respectively

4.2 Landsat image

In this study, the Landsat image used is 15 X 15.75 Km^2 (500 X 525 pixels) and has 30m spatial resolution. The aim is to classify 9 land cover region using Landsat image. Senthilnath *et. al*[15] provides a detail description of data set. There are 9 level-2 land cover region for this image which include deciduous(C_1), deciduous pine(C_2), pine(C_3), water(C_4), agriculture(C_5), bareground(C_6), grass(C_7), urban (C_8) and shadow(C_9).

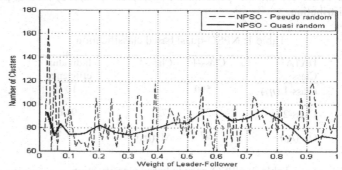

Fig-4 :Effect of weight of leader follower in NPSO with FAURE distribution In LANDSAT IMAGE

Table 3: Performance measure for K-means, MSC, and NPSO using Landsat data

Classification Efficiency	K-means[2]	MSC[2]	NPSO[2] (Pseudo)	NPSO (Faure)
η_1	82.1	85.9	85.0	90.39
η_2	68.0	81.0	81.3	88.78
η_3	53.9	69.3	82.6	90.99
η_4	92.3	92.4	92.4	94.93
η_5	76.3	76.2	77.9	80.46
η_6	35.6	38.6	70.7	84.35
η_7	67.3	69.2	72.2	79.76
η_8	39.9	41.7	70.8	81.84
η_9	28.4	66.8	77.8	81.16
η_a	60.4	69.0	78.9	85.85
η_o	70.8	78.1	81.8	88.14

Maximum cluster centres generated based on BIC for this data set should be 80 [2]. For this data set the NPSO parameter value assigned are: number of population is 500, sub-swarm radius is 0.1, inertia weight is adaptively varied in the interval [0.7,0.6], and weight of leader follower (ρ) is equal to 0.4. Among these parameter weight of leader follower plays an important role to generate the 80 cluster centres. The weight of leader follower for NPSO (ρ) was observed as most dominant factor, Fig-4 shows the variation of number of cluster centres generated with (ρ), using pseudo-random distribution abrupt variation is observed due to high degree of randomness, whereas when Faure sequence is used as initial distribution as expected a more smooth curve is obtained. These cluster centres obtained from NPSO is merged to obtain exact number of classes using merging technique as discussed in section 3. The classification matrix obtained after merging is as shown in table 3.

From Table 3 we can observe that performance measure in all aspect using NPSO with Faure distribution based hierarchical clustering and classification is better in comparison to NPSO with pseudo based clustering for Landsat data. Fig-5 shows the classification results obtained for Landsat image using NPSO with Faure distribution.

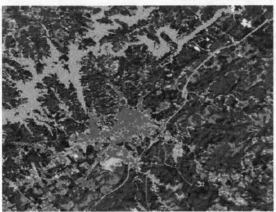

Fig-5: Classification using NPSO with Faure distribution

5 Conclusions and discussion

In this paper, we have presented an improved hierarchical clustering algorithm based on quasi-random distribution using NPSO. Initially NPSO with pseudo and quasi-random distribution is used to initialize the search space to split the data set into cluster centres by satisfying BIC. Here to generate the quasi-random sequence Faure method is used. Since by using the quasi-random sequence particles are distributed in the search space more uniformly, resulting in the accurate convergence of the particle to the centres.

An effect of weight of leader follower parameter has been analysed, it is observed that using quasi-random sequence to initialize weight of random component parameter of NPSO minimizes the random behaviour of the algorithm. This is useful to select a weight of leader follower value more accurately. The performance is measured using classification efficiency - individual, average and overall of the proposed algorithm. We observed that use of quasi-random sequence as initial distribution with NPSO results in better efficiency of classification for synthetic and Landsat data set.

References

[1] David, L.: Hyperspectral image data analysis as a high dimensional signal processing problem. IEEE Signal processing Mag. 19 (1), 17–28 (2002)

[2] Senthilnath, J., Omkar, S.N., Mani, V., Tejovanth, N., Diwakar, P.G., Shenoy, A.B.: Hierarchical clustering algorithm for land cover mapping using satellite images. IEEE journal of selected topics in applied earth observations and remote sensing. 5 (3), 762-768 (2012)

[3] Brits, R., Engelbrecht, A.P., van den Bergh, F.: A niching Particle Swarm Optimizer. In proceedings of the fourth Asia Pacific Conference on Simulated Evolution and learning. 692 –696 (2002)

[4] Parsopoulos, K.E., Vrahatis, M.N.: Particle swarm optimization in noisy and continuously changing environments. in Proceedings of International Conference on Artificial Intelligence and soft computing. 289-294 (2002)

[5] Kimura, S., Matsumura, K.: Genetic Algorithms using low discrepancy sequences. in proc of GEECO. 1341 –1346 (2005)

[6] Brits, R., Engelbrecht, A.P., van den Bergh, F.: Solving systems of unconstrained equations using particle swarm optimization. in proceedings of the IEEE Conference on Systems. Man and Cybernetics. 3, 102 – 107 (2002)

[7] Nguyen, X.H., Mckay, R.I., Tuan, P.M.: Initializing PSO with Randomized Low-Discrepancy Sequences: The Comparative Results, In Proc. of IEEE Congress on Evolutionary Algorithms. 1985 – 1992 (2007)

[8] Schwarz, G.: Estimating the dimension of a model. the Annals of statistics. 6 (2), 461-464 (1978)

[9] Donald, K.: Chapter 3 – Random Numbers". The Art of Computer Programming. Seminumerical algorithms (3 ed.) (1997)

[10] Niederreiter, H.: Quasi-Monte Carlo Methods and Pseudo Random Numbers. Bulletin of American Mathematical Society. 84(6) 957-1041 (1978)

[11] Marco A.G.D.: Quasi-Monte Carlo Simulation. http://www.puc-rio.br/marco.ind/quasi_mc2.html

[12] Galanti, S., Jung, A.: Low-Discrepancy Sequences: Monte Carlo Simulation of Option Prices. Journal of Derivatives. 63-83 (1997)

[13] Comaniciu, D., Meer, P.: Mean shift :a robust approach towards feature space analysis. IEEE Trans .pattern Anal .machIntell. 24 (5), 603-619 (2002)

[14] Kennedy, J., Eberhart, R.C.: Particle swarm optimization. Proceedings of the IEEE International Conference on Neural Networks, IV (Piscataway, NJ), IEEE Service Center. 1942–1948 (1995)

[15] Senthilnath, J., Omkar, S.N., Mani, V., Tejovanth, N., Diwakar, P.G., Archana, S.B.: Multispectral satellite image classification using glowwarm swarm optimization. in proc. IEEE int. Geoscience and Remote Sensing Symp (IGARSS).47-50 (2011)

[16] Li, H., Zang, K., Jiang ,T.: The regularized EM algorithm. in proc.20thNat.conf.Artificial Intelligence. 807-8 (2005)

[17] MacQueen ,J.: Some methods for classification and analysis of multi-variate observations. in proc .5th BerkeleySymp. 281-297 (1967)

[18] Senthilnath, J., Omkar, S. N., Mani, V.: Clustering using firefly algorithm – Performance study. Swarm and Evolutionary Computation. 1 (3), 164-171 (2011)

[19] Suresh, S., Sundararajan, N., Saratchandran, P.: A sequential multi-category classifier using radial basis function networks. Neurocomputing. 71, 1345-1358 (2008)

Clustering using Levy Flight Cuckoo Search

J. Senthilnath[a1], Vipul Das[b2], S.N. Omkar[a3], V. Mani[a4]

[a] Department of Aerospace Engineering, Indian Institute of Science, Bangalore, India

[b] Department of Information technology, National Institute of Technology, Karnataka, India

{[1]snrj@aero.iisc.ernet.in; [2]vipulramdas@gmail.com; [3]omkar@aero.iisc.ernet.in; [4]mani@aero.iisc.ernet.in}

Abstract. In this paper, a comparative study is carried using three nature-inspired algorithms namely Genetic Algorithm (GA), Particle Swarm Optimization (PSO) and Cuckoo Search (CS) on clustering problem. Cuckoo search is used with levy flight. The heavy-tail property of levy flight is exploited here. These algorithms are used on three standard benchmark datasets and one real-time multi-spectral satellite dataset. The results are tabulated and analysed using various techniques. Finally we conclude that under the given set of parameters, cuckoo search works efficiently for majority of the dataset and levy flight plays an important role.

Keywords: Genetic algorithm, Particle swarm optimization, Cuckoo search, Levy flight, Clustering.

1 Introduction

Clustering is an unsupervised learning method where objects with closer resemblance are grouped together to form a cluster based on a similarity measure. The objective of clustering is to minimize intra-cluster distance while inter-cluster distance is maximized [1]. Clustering has various applications which include data analysis, machine learning, image analysis and other engineering applications.

Clustering can be classified into two types: hierarchical and partition. In hierarchical clustering, objects belong to more than one cluster forming a hierarchical pattern. Hierarchical clustering is carried out by splitting and merging the dataset. In splitting the number of cluster centres generated would be greater than the number of classes while merging is to group the dataset to exact number of classes. In partition clustering, objects are clustered into disjoint groups without forming a hierarchy. In both methods, similarity measure is used to generate cluster centres.

Previously, the most popularly used and tested partition based algorithm is k-means clustering. The main disadvantage of k-means clustering is convergence to

J. C. Bansal et al. (eds.), *Proceedings of Seventh International Conference on Bio-Inspired Computing: Theories and Applications (BIC-TA 2012)*, Advances in Intelligent Systems and Computing 202, DOI: 10.1007/978-81-322-1041-2_6, © Springer India 2013

the local minima [2]. In literature, nature inspired algorithms are used effectively in clustering problem as it converges to global minima [2, 3]. These algorithms are based on the exploration and exploitation behaviour observed in nature and is effectively used in optimization problems.

In this paper, a comparative performance study is carried out based on the results obtained using three nature inspired algorithms namely genetic algorithm (GA), particle swarm optimization (PSO) and cuckoo search algorithm (CS) on clustering problem. The standard benchmark clustering data used in our study are the same that is available in the (UCI machine learning repository) literature [4] and a real-time multi-spectral satellite image for crop type classification. Xin-She et.al [5] has implemented and analyzed CS algorithm by comparing with GA and PSO using standard benchmark functions. In their study, CS algorithm is used with levy flight and is found to be performing better compared to the other two methods. In literature, CS has been used without levy distribution for clustering problem on satellite image [3]. In our study, we use CS with levy flight as used in [5], on clustering data set by comparing with GA and PSO. The important property of levy flight is it makes sure that the whole search space is covered, which is due to the heavy-tailed property of levy distribution [6-10]. In our study, we split the data into training and testing samples. The cluster centres are determined using the algorithms on the training dataset and the testing dataset is used to determine the classification error percentage (CEP).

The remaining sections are in the following order: in section 2 the problem formulation for clustering is discussed, in section 3 a brief discussion of the algorithms is presented, in section 4 and section 5 we discuss analysis of the results obtained and conclusion respectively..

2 Problem Formulation

The clustering is done based on unsupervised learning. Here the data is divided into training set and testing set. The training set data is used to generate the cluster centres. The aim of clustering is to minimize the objective function [2].

$$f(k) = \sum_{k=1}^{K} \sum_{i=1}^{n_k} (x_i - c_k)^2 \qquad\qquad 1$$

where $k=1,2,...K$ is the number of clusters, x_i , $i=1,2,...n_k$ are the patterns in the kth cluster, c_k is centre of the kth cluster. Here the cluster centres are represented by

$$c_k = \frac{1}{n_k} \sum_{i=1}^{n_k} x_i \qquad\qquad 2$$

In this study, the nature-inspired algorithms are used to find the cluster centers from the training data set. This is done by placing each object to their respective cluster centers using the distance measure. The testing data set is used to calculate percentage error using classification matrix.

3 Methodology

This section gives brief introduction about the algorithms used in our study, the way it has been applied for clustering problem and also the pseudo-code for the algorithms are discussed.

3.1 Genetic algorithm

This algorithm is based on the natural selection process seen in nature [11, 12]. The best fit organism of the current generation carries on the genes to the next generation. The concept of genetic operators (cross-over and mutation) is included in the algorithm wherein a change in the gene structure is introduced that produces an entirely different trait. The main idea behind genetic algorithm is the operators used namely reproduction, crossover and mutation.

This algorithm takes a predetermined number of random solutions (population) in the search space called chromosomes. Here the convergence criterion is used to terminate the algorithm. At each iteration the chromosomes are made to crossover using single point crossover and the fitness of each chromosomes is calculated using

$$f_i = f(x_i) \quad i=1,2,...,n \qquad\qquad 3$$

where $f(x_i)$ is the fitness function given by Eq. 1 considering the clusters individually and n is the population size.

The fittest chromosomes (solutions) among the entire population are considered for the next generation (iteration). At any random point the chromosomes undergo mutation based on the mutation rate. The fitness is calculated and the best solutions carryon till termination criteria is reached. Thus the cluster centres are generated using the training data set.

Pseudo-code
1. Initialize population of n chromosomes
2. Repeat till stopping criteria
   a) Calculate fitness using Eq. 3
   b) Apply elitism by sorting the fitness value of the population
   c) Retain the best fit solutions (reproduction)
   d) Crossover the adjacent chromosomes at a random position using single point crossover
   e) Mutate randomly selected point within a chromosome
3. Cluster centre will be the best fit solution from the population

3.2 Particle Swarm Optimization

This is a population based method which iteratively improves the solution by moving the solutions closer to the optimal solution. Here each particle moves towards the optimal solution with a velocity v_i at each iteration. Eventually all particles converge to an optimal position [13].

Initially n particles are created and randomly distributed in the search space. The fitness of each particle is evaluated using Eq.3 and Eq.1, considering the classes individually. All the particles are made to move one step towards the fittest particle (global best solution) as well as towards its personal best position with a velocity v_i given by

$$v_i(t+1)=w*v_i(t)+b_p*rand*(p_i-c_i)+b_g*rand*(g-c_i) \qquad 4$$

where p_i is the personal best position of the particle, c_i is the current position of the particle, g is the global best of the entire particle, w is the inertial constant, b_p is the personal best constant and b_g is the global best constant, $i=1, 2,..., n$. Each particle moves using

$$c_i(t+1)=c_i(t)+v_i \qquad 5$$

The fitness of each particle is calculated and the personal best position and the global best are determined. This process is repeated until stopping criteria is met. The global best position will be the cluster centre to the given data set.

Pseudo-code

1. Initialize n particles
2. Repeat till stopping criteria met
    a) Calculate fitness of each particle using Eq.3
    b) global best position is the best fit particle
    c) move all the particles towards the global best position using Eq.4 and Eq.5
    d) for each particle if (fitness of current position < fitness of personal best) then personalbest = current position
    e) update personal best position for each particle
    f) global best fitness value is retained
3. Cluster centre is the global best position

3.3 Cuckoo Search

This algorithm is based on the breeding pattern of parasitic cuckoos [3, 5, 14]. Some species of cuckoo namely ani and Guira lay their eggs in the nest of other birds. The possibility of occurrence of such act leads to i) the host birds' eggs being destroyed by the cuckoo itself or the cuckoo chick upon hatching; ii) the host birds may realise the presence of a foreign egg in its nest and may throw away these eggs or abandon the nest altogether and build a new nest elsewhere [5].

These are the processes in nature that this algorithm inculcates. The basic assumptions made are: 1) At a time each cuckoo lays one egg and dumps it into randomly chosen nest; 2) The best nest with high quality eggs will carry over to the next generation; 3) Each nest contains only one egg and the number of host nests are fixed and; 4) The probability that the host bird discovers the cuckoo egg is p_a. This implies that the fraction p_a of n nests is replaced by new nests (with new random solutions) [5].

Each nest represents a solution and a cuckoo egg represents a new solution. The aim is to use the new and potentially better solutions (cuckoo eggs). An initial population of host nest is generated randomly. The algorithm runs till the convergence is reached. At each iteration a cuckoo is selected at random using levy flight as given [5]

$$x_i(t+1)=x_i(t) + \alpha*L \qquad\qquad 6$$

where α is the step-size, L is a value from the Levy distribution, $i=1,2,...,n$, n is the number of nests considered. The fitness of the cuckoo is calculated using Eq.3 and Eq.1, considering the classes individually.

Choose a random nest from the given population of nests and evaluate its fitness from Eq.6. If the fitness of the new solution is better than the older one then replace the older one with the new one. A fraction p_a of the total number of nests is replaced by new nests with new random solution. The best nests with the fittest egg (solution) are carried-on to the next generation.

This is continued till the termination criteria is reached and the best nest with fittest egg is taken as the optimal value. Thus the cluster centres can be generated using this optimal value.

Pseudo-code
1.   Initialise n nests
2.   Repeat till stopping criteria is met
     a)   Randomly select a cuckoo using levy flight using Eq.6
     b)   Calculate its fitness using Eq.3 (F_c)
     c)   Randomly select a nest
     d)   Calculate its fitness using Eq.3 (F_n)
     e)   If ($F_c < F_n$) then Replace the nest with the cuckoo
     f)   A fraction p_a of nest are replaced by new nests
     g)   Calculate fitness and keep best nests
     h)   Store the best nest as optimal fitness value
3.   Cluster centre will be the best nest position

4   Results and discussion

In this section the results and the performance evaluation are discussed. The specifications of the clustering data used in this study are given in Table 1. The training data are randomly picked from the dataset for vehicle dataset and glass dataset. The training data for image segmentation dataset are as in the UCI repository.

Table 1. Specifications of the clustering dataset used

Dataset	Total data	Training data	Test data	Attributes	Classes
Image segmentation	2310	210	2100	19	7
Vehicle	846	635	211	18	4
Glass	214	162	52	9	6*
Crop Type	5416	2601	2815	4	6

*Glass dataset has 7 classes. The data for the fourth class is unavailable.

The performance measures used in this paper are classification error percentage [2], Statistical significance test [4], Receiver operating characteristic [15, 16] and time complexity analyses.

4.1. Classification error percentage

The result of application of the algorithms on clustering data is given in terms of classification error percentage. This is the measure of misclassification of the given dataset using the particular algorithm. Let n be the total number of elements in the dataset and m be the number of elements misclassified after finding out the cluster centre using the above algorithms, then classification error percentage is given by

$$CEP = \frac{m}{n} *100 \qquad\qquad 7$$

Table 2. Classification error percentage

Dataset \ Algorithms	GA	PSO	CS
Image segmentation	32.6857	32.45716	30.56188
Vehicle	61.61138	60.18956	58.76636
Glass	61.15386	55.76924	45.76926
Crop type	19.3677	20.0710	20.0355

The algorithms are run five times and the average of the results is as shown in Table 2. The values are obtained using the testing dataset. The parameters such as the maximum generation and the number of initial random solution are kept the same for all the algorithms. Each algorithm is run till it converged to a point with a tolerance of 0.01. In GA, the best 40% of the parent generation and the best 60% of the offspring generation are carried on to the next generation. In PSO, the inertial constant (w), the personal best constant (b_p) and the global best constant (b_g) are all set to 1. In CS algorithm, the probability factor p_a is set to 0.25.

4.2. Statistical significance test

Statistical significance is done to ascertain that the results are obtained consistently. No matter where the initial random solutions are picked up from, they would always converge to the global optimum position (cluster centre). This would imply that an algorithm which performed better than the other algorithms will always perform better when run under similar initial conditions. In this study a binomial test is conducted [3] between CS and GA and also CS and PSO based on the result obtained on image segmentation dataset.

Assume the test is carried between CS and GA. Here the total number of test-runs is N, i.e., the result of CS and GA differ in N places. Let S (success) is the number of times CS gave correct result and F (failure) is the number times GA gave correct result. Now, calculating the *p-value* (probability of S successes out of N trials) using the binomial distribution as

$$P = \sum_{j=S}^{N} {}^{N}C_j * p^j * q^{N-j}$$ 8

Here p and q are the probability that the algorithms CS and GA will succeed. Let p and q value be set to 0.5, assuming each algorithm to behave the same. The results of comparison of CS with GA and CS with PSO are as shown in Table 3. With a low value of P, we can say that cuckoo search gives better result than GA and PSO, the chance has nothing to do with the better performance of CS algorithm.

Table 3. Binomial Test on image segmentation dataset

	N	S	F	P
GA	255	153	102	$8.44e^{-04}$
PSO	62	35	27	0.1871
CS	-	-	-	-

4.3. Receiver Operating Characteristics

Receiver operating characteristics [15, 16] are used to evaluate the performance of a binary classifier. An experiment will have actual values and prediction values. If the prediction value is P and the actual value is also P, then it is called true positive (TP). If prediction value is P and the actual value is N, then it is called false positive (FP). Likewise, true negative (TN) if prediction value is N and actual value is N and false negative (FN) when the prediction value is N and actual value is P. The above can be shown using a 2×2 contingency matrix given in Table 4.

With the above statements, we define three parameters namely – sensitivity or true positive rate (TPR), false positive rate (FPR) and accuracy (ACC) given by

$$TPR = \frac{TP}{TP + FN}$$ 9

$$FPR \;=\; \frac{FP}{FP \;+\; TN} \qquad\qquad 10$$

$$ACC \;=\; \frac{TP \;+\; TN}{TP \;+\; FN \;+\; FP \;+\; TN} \qquad\qquad 11$$

Sensitivity defines how many correct positive results occur among all the positive samples available during the test i.e., in our case the number of elements that have been correctly clustered amongst all the elements that belonged to the particular class. FPR defines how many incorrect positive results occur among all negative samples available during the test i.e., the number of misclassified elements amongst all the other elements that does not belong to the particular class. Accuracy defines how many samples have been correctly classified to their respective classes.

Table 4. ROC Contingency matrix

		Predicted value	
		True	False
Actual Value	True	True Positive	False Negative
	False	False Positive	True Negative

In our case, we analyse on the image segmentation dataset. We give an example of the analyses using cuckoo search. The classification matrix obtained after applying cuckoo search algorithm on image segmentation data is given in Table 5.

Table 5. Classification matrix of image segmentation dataset using CS algorithm

	Class1	Class 2	Class 3	Class 4	Class 5	Class 6	Class 7
Class 1	126	0	91	14	69	0	0
Class 2	0	294	0	6	0	0	0
Class 3	60	1	171	14	54	0	0
Class 4	69	12	10	183	9	14	3
Class 5	30	0	50	21	195	0	4
Class 6	9	0	9	0	0	249	33
Class 7	0	0	17	0	41	4	238

In the above representation, row indicates the class the element belongs to and the column indicates the class the elements are classified into after using the cluster centre based on the CS algorithm. The principal diagonal elements represent correctly classified elements. Consider class 1, from Table 4, we have TP=126, FN= 174, FP=168, TN=1632. From this data, we calculate the true positive rate, false positive rate and the accuracy of the given algorithm on class 1 of the given clustering dataset. From Eq. 9, Eq. 10 and Eq. 11, TPR is 0.4200, FPR is 0.0933 and ACC is 0.8371. This implies that 42% of what actually belonged to class 1 was correctly classified and 9% of the data which did not belong to class 1 were added to class 1. The overall efficiency of the algorithm with respect to class 1 is 83%. Similarly the ROC analyses for all the classes of image segmentation dataset for the above three algorithms are given in Table 6.

Table 6. ROC analyses for image segmentation data using CS, GA and PSO

Class	CS			GA			PSO		
	TPR	FPR	ACC	TPR	FPR	ACC	TPR	FPR	ACC
1	42%	9.3%	83%	73%	21%	77%	46%	21%	77%
2	98%	0.7%	99%	99%	0.8%	99%	98%	0.8%	99%
3	57%	9.8%	85%	6%	0.3%	86%	46%	0.3%	86%
4	61%	3.0%	92%	56%	2.8%	91%	61%	2.8%	91%
5	65%	9.6%	86%	63%	9.7%	86%	67%	9.7%	86%
6	83%	1.0%	96%	87%	2.1%	96%	84%	2.1%	96%
7	79%	2.2%	95%	82%	1.2%	96%	78%	1.2%	96%

4.4. Time complexity analysis

The pseudo-codes of the algorithms are discussed in section 3. The time complexity of each algorithm can be derived from the pseudo-code. The time complexity analysis gives us an insight into the complexity of calculation involved in the algorithm, in order to know the time taken to produce the output. The time complexities of the algorithms are given in Table 7.

Table 7. Time complexity

Algorithm	Time complexity
GA	$O(clnum*gen*(comp\_fit + sort\_inb + m))$
PSO	$O(clnum*gen*(comp\_fit * m))$
CS	$O(clnum*gen*(comp\_fit * m))$

The algorithm is run till the stopping condition is met which in this case is till the solutions converge to a point with a tolerance of 0.01. Let the total number of iterations be *gen* and the number of clusters is *clnum*. Thus the total number of outer iterations is *clnum* gen*. Let *m* be the population size and *n* be the number of fitness evaluation to generate each cluster center. Thus in each iteration, fitness is calculated with a time complexity of $O(n)$. Let this $O(n)$ be called *comp_fit*. In GA, additional operation is performend by sorting the population using m fitness values. This is done using a Matlab inbuilt function. Let the complexity of this function be *sort_inb*. Crossover and mutation takes *(m/2)* and m run respectively. Thus in each iteration, the overall operations executed will be of the order *(comp_fit + sort_inb + m/2 + m + C)*. Thus the overall order is *(clnum*gen*(comp_fit + sort_inb + m/2 + m + C))*. Thus the time complexity of GA used in this paper is $O(clnum*gen*(comp\_fit + sort\_inb + m))$. Similarly for PSO and CS, *clnum, gen, m* and *comp_fit* implies the same as in GA.

The algorithms are run on a system with core i-5 processor, 4 GB memory on Matlab version 7.12.0.635. The execution time in secs taken by these algorithms to converge to the solution on glass dataset is given in Table 8.

Table 8. Time taken by the algorithms on glass dataset (in seconds)

Algorithms	Trial 1	Trial 2	Trial 3	Trial 4	Trial 5
GA	47.8299	62.7020	58.2859	33.6453	41.2319
PSO	760.5055	661.8087	1051.3	676.1566	695.6928
CS	163.95	147.8284	141.5073	159.0653	141.6662

5 Conclusion and Discussions

In this paper, we have implemented and analyzed three nature inspired techniques for clustering problem. Here we observe that the average classification error percentage of clustering dataset using cuckoo search with levy flight algorithm is less than GA and PSO for the benchmark problems and is at par with GA and PSO for crop type dataset. The statistical significance test proves that the cuckoo search was not better by chance. The obtained p-value being very small implies that the cuckoo search is better than GA and PSO with a high confidence level. The ROC analyses further gives us an insight into the efficiency of cuckoo search.

In cuckoo search, the levy flight factor plays a major role here. The fact that levy flights are heavy-tailed is used here. This helps in covering the output domain efficiently. Looking into the time complexity measure, we see that GA has one additional computation compared to the other two i.e., sorting of the population (solutions) according to the fitness values. But this takes negligible time as the number of agents or the population size is only 20. Thus GA takes less time as expected but CS takes a far lesser time compared to PSO. This can be attributed to the fact that CS algorithm uses levy flight. Thus we can clearly observe that the heavy-tailed property of levy flights helps to converge to the solution fast thereby increasing the efficiency.

References

[1]   Anitha, E.S., Akilandeswar, J., Sathiyabhama, B. : A survey on partition clustering al-
      gorithms. International Journal of enterprise and computing and business systems. 1
      (1), 1 – 14 (2011)
[2]   Senthilnath, J., Omkar, S. N., Mani, V.: Clustering using firefly algorithm – Perfor-
      mance study. Swarm and Evolutionary Computation. 1 (3), 164-171 (2011)
[3]   Suresh, S., Sundararajan, N., Saratchandran, P.: A sequential multi-category classifier
      using radial basis function networks. Neurocomputing. 71, 1345-1358 (2008)
[4]   Samiksha, G., Arpitha, S., Punam, B.: Cuckoo search clustering algorithm: a novel
      strategy of biomimicry. World Congress on Information and Communication Tech-
      nologies. IEEE proceedings (2011)
[5]   Xin-She, Y., Suash, D.: Cuckoo search via levy flight. World Congress on Nature and
      Biologically Inspired Algorithms. IEEE publication. 210-214 (2009)
[6]   Viswanathan, G.M., Afanasyev, V., Sergey, V.B., Shlomo, H., Da Luz, M.G.E., Rapo-
      so, E.P., Eugene, S.H.: Levy flight in random searches. Physica A 282, 1-12 (2000)
[7]   Viswanathan, G.M., Bartumeus, F., Sergey V.B., Catalan, J., Fulco, U.L., Shlomo, H.,
      Da Luz, M.G.E., Lyra, M.L., Raposo, E.P., Eugene, S.H.: Levy flights in Biological
      systems. Physica A 314, 208-213 (2002)
[8]   Peter, I., Ilya, P.: Levy flights: transitions and meta-stability. Journal of Physics A:
      Mathematical and General. J. Phys. A: Math. Gen. 39 L237–L246 (2006). doi:
      10.1088/0305-4470/39/15/L01
[9]   Ilya P.: Cooling down Levy flight. Journal of Physics A: Mathematical and Theoreti-
      cal. J. Phys. A: Math. Theor. 40 12299–12313 (2007). doi: 10.1088/1751-
      8113/40/41/003
[10] John P.N.: Stable Distributions – Models for Heavy Tailed Data. chapter 1. Processed
      (2009)
[11] Yannis, M., Magdalene, M., Michael, D., Nikolaos, M., Constantin, Z.: A hybrid sto-
      chastic genetic–GRASP algorithm for clustering analysis. Oper Res Int J 8:33–46
      (2008). doi: 10.1007/s12351-008-0004-8
[12] Whitley, L. D.: A genetic algorithm tutorial, Statist. Comput. 4:65–85 (1994)
[13] De Falcao, I., Dello Ciappo, A., Tarantino, E.: Facing classification problems with
      Particle swarm optimization – Applied Soft Computing 7. 652-658 (2007)
[14] Walton, S., Hassan, O., Morgan, K., Brown, M.R.: Modified cuckoo search: A new
      gradient free optimization algorithm- Chaos, Solition and Fractals 44. 710-718 (2011)
[15] Christopher D.B., Herbert T.D.: Receiver operating characteristics curves and related
      decision measures: A tutorial. Chemometrics and Intelligent Laboratory Systems 80.
      24-38 (2006)
[16] Tom F.: ROC Graphs: Notes and Practical Considerations for researchers. HP Labora-
      tories, MS 1143, 1501 Page Mill Road, Palo Alto, CA 94304 (2004)

VLSI Architecture of Reduced Rule Base Inference for Run-Time Configurable Fuzzy Logic Controllers

Bhaskara Rao Jammu[1], Sarat Kumar Patra[2], Kamala Kanta Mahapatra[3]

[1] Senior Research Fellow , Deptartment Of ECE, NIT Rourkela, Rourkela,India.

[2] Professor, SMIEE, Department Of ECE, NIT Rourkela, Rourkela, India.

[3] Professor, Department Of ECE, NIT Rourkela, Rourkela,India.

{ j.bhaskararao@gmail.com; skpatra@nitrkl.ac.in;kkm@nitrkl.ac.in}

Abstract. In this paper, a new VLSI architecture is provided for the application of quad-input and dual-output Fuzzy Logic Controller (FLC) with maximum seven fuzzy membership functions. Our approach is based on classical three stage implementation process – fuzzification, rule inference and defuzzification cores. An innovative design methodology is proposed by splitting the process between DSP processor and FPGA to implement run time configurable FLC. Since the target application takes a maximum of 4 inputs and 7 membership functions, the rule base comprises of 2401 (7^4) rules. It increases the complexity of the overall system. To minimize this effect, rule reduction VLSI architecture is suggested to bind the no of rules to 16 (2^4). The Rule inference is designed for seeking the maximum frequency of operation for targeted Virtex 5 LX110T FPGA. The simulation results obtained with Modelsim 6.3g show satisfactory results for all test vectors.

Keywords: FPGA, FLC, FSM, VERILOG, VLSI, hardware implementation.

1 Introduction

In recent times fuzzy logic is addressing complex control problems such as robotic arm movement, chemical or manufacturing control process and automatic transmission control with more precision than conventional control techniques. The important principles inside the FLC have been broadly covered in the literature [1] [2]. Fuzzy logic is a methodology for expressing operation laws of a system in linguistic terms instead of mathematical operations. Fuzzy logic linguistic terms provide a useful method to define the operation characteristic of a system which is too complex to model accurately even with complex mathematical equations. The field of fuzzy systems and control has been making rapid progress in recent years. Due to practical success in consumer-product and industrial process control there has been rigorous research and development and theoretical studies.

J. C. Bansal et al. (eds.), *Proceedings of Seventh International Conference on Bio-Inspired Computing: Theories and Applications (BIC-TA 2012),* Advances in Intelligent Systems and Computing 202, DOI: 10.1007/978-81-322-1041-2_7, © Springer India 2013

This has led to a tremendous increase in the amount of work in the field of fuzzy systems and fuzzy control. Due to the increased complexity level of the plant, the demand for controllers in the market is increasing day by day. In order to meet the market demand, controllers have to be designed according to the market needs. Some of the market needs highlighted are increasing in computational speed, decrease in computational complexity, ease of know-how of the product, easy working with the product and less turnaround time in terms of design. One of the best solutions to meet the above market demand is to switch over to a digital domain. The lookout for such a device, where thousands of gates can be incorporated ended up in Field Programmable Gate Arrays (FPGA). The application specific FPGA based architectures for FLC [3] [4] [5] show the maximum implementation efficiency in terms of silicon utilization and processing speed.

In general, fuzzy logic is implemented in 3 phases. They are Fuzzification (Crisp input to fuzzy set mapping), Inference (fuzzy rule generation) and Defuzzification (fuzzy to crisp out transformation). This paper gives Very Large Scale Integration (VLSI) Architecture for general purpose Inference module suitable for all applications where as fuzzification and defuzzification are performed in DSP processor. The architecture includes the design for rule base reduction and the interface between the DSP and the FPGA to read and write the fuzzified and inference output data. This paper provides the design of all modules with its module level verification and FPGA implementation. The processor interface register information is provided for software programming.

This paper is organized as follows: in section 2 provides a brief introduction to the specification of the generalized FLC. Section 3 describes the VLSI architecture of the rule base reduction module or inference module. Section 4 lists design choices we have made. Section 5 outlines the test bench and test vector generation. The developed model is simulated and synthesized as sketched in section 6. Finally, conclusions are presented in section 7.

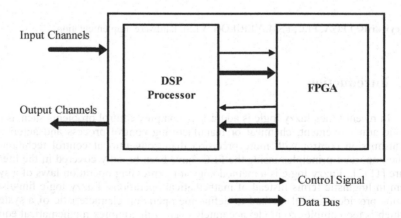

Fig. 1. System Platform to implement general purpose fuzzy logic controller.

2 Specification Of Fuzzy Logic Controller

The fuzzy logic controller proposed in this paper is standalone, configurable and generalized for any control application or requirement. Here the idea is to implement a run time configurable algorithm which can be configured according to requirement and can tune the parameters at any point of time. The main parameters and their limitations are given as follows:

1) No of inputs : maximum 4 (configurable)
2) No of outputs: maximum 2 (configurable)
3) Shape of membership function : Triangular
4) No of membership function for each input and output : maximum 7 (configurable)
5) Implication model: Mamdani
6) Aggregation model: Mamdani
7) Inference rules should be field programmable
8) For each input variable, the overlapping degree of its membership function is a maximum of two.

Since the standalone configurable FLC hardware takes maximum four inputs, each with maximum seven membership functions. The rule base comprises of 2401 (7^4) rules and the fuzzification and defuzzification algorithms need more complex mathematical operations. Hence the complexity of the algorithm increases if the FLC is implemented in FPGA alone [6] [7] [8]. It is proposed in this paper that the data acquiring, fuzzification and defuzzification can accommodate by the DSP processor because of its proficiency in handling complex mathematical functions. The FPGA is selected to generate Rule Base and inference. The suggested platform for this model is shown in Fig. 1.

3 VLSI Architecture of reduced rule base module

Once membership functions are defined for input and output variables a control rule base can be developed to relate the output actions of the controller to the observed inputs. This phase is known as the inference or a rule definition portion of the fuzzy logic. There are N^m (where N= No of Inputs and m= No of Membership functions) rules can be created to define the actions of fuzzy logic controller. This section gives VLSI Architecture of the rule base module for a generalized fuzzy logic controller with the specifications defined in the earlier section.

The architecture also includes the interface between DSP and FPGA to program or to tune FLC parameters. Inside the inference we can observe that the design of rule base consists of 2401 (with Maximum of 4 inputs and 7 membership functions) rules and consuming much gate count in the FPGA, to reduce the device utilization one way is to reduce the no of rules in the rule base.

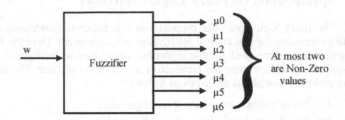

Fig. 2. The Fuzzification Unit

From limitation number 8 in specification mentioned in the previous section the fuzzifier gives two non-zero fuzzy term sets at most as seen in figure 2. Hence the no of rules will be reduced to 4^2 (for 4 inputs and 2 membership functions), Results in a reduction in logic utilization and improves the overall system performance, in our design we have achieved maximum frequency of operation as 175.809MHz. The rule selector unit outputs these two fuzzified values with its Index number and its fuzzy value. The index number is used by the address generator to generate activated rule with respect to linguistic variables. Fuzzy value is used by the inference engine with Mamdani max-min inference rule.

The proposed VLSI architecture (see Fig. 3) for a rule base module with rule reduction includes five principal units:

1) Rule selector unit to select non zero fuzzy term set;

2) An address generator unit which generates an address to read the appropriate active rule;

3) Rule base memory unit to store the user defined rules, provision is provided to program at any interval of time.

4) An inference engine unit which performs approximate reasoning by associating input variables with fuzzy rules.

5) The CPU registers unit is memory with 32 bit data width is used to store fuzzified values for four input variables, inference output values and Register to program no of inputs and no of membership functions (Tuning parameters).

6) Here the inference engine involves in another sub-block which performs Mamdani min-max implication operation and calculates the degree of applicability of all active rules selected from the rule base memory by address generator. The results are stored in the CPU registers for defuzzification process. Typically the DSP processor and FPGA runs in different clock frequencies interrupt handler is used here as a status flag to tell internal state machine that new data is available. Subsequently a Finite State Machine (FSM) within the FPGA can use the interrupt handler to generate an interrupt to DSP Processor. The architecture specifies 32 bit memory

space is in Table I. The rule reduction algorithm used in this architecture is given in Fig. 4.

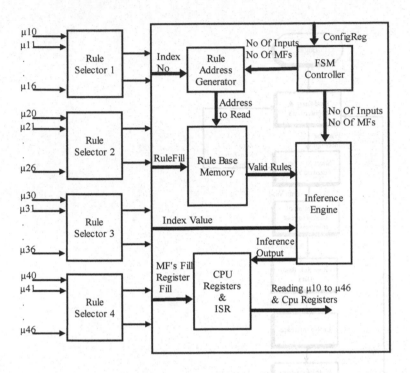

Fig. 3. VLSI Architecture of the Reduced Rule Base

4   Design Choices of Internal Modules

As we have discussed from previous section the rule base is filled with 2401, the software fills this data with address from 0 to 2400. To read the appropriate rule for non-zero fuzzified values it needs to generate matching addresses with its index numbers. For example if non-zero value of input 1 is with index numbers 1 and 2, input 2 is 2 and 3, input 3 is 4 and 5, input 4 is 5 and 6, it needs to read the rule addresses as shown in Fig. 5.

The module rule address generator from Fig. 3 arranges index numbers to address shown in Fig. 5 with the programming value of no of inputs, if no of inputs are 4 the rules are 16 and to 3 inputs rules become 8 and to 2 inputs rules become 4. The address generator module does the mapping of the rule index with the rule base address (filled by the software). Part of the VERILOG code to evaluate rule address to match with rule base memory addresses is provided in Fig. 6.

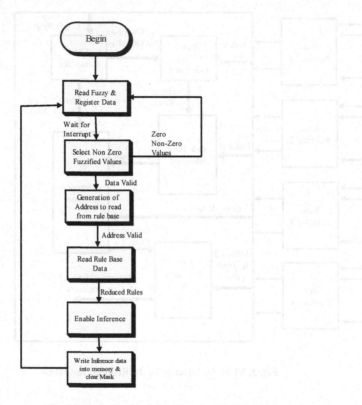

Fig. 4. Rule Reduction Algorithm for VLSI Implementation

001010100101	1 3 4 5	2 3 4 5
1 2 4 5	1 3 4 6	2 3 4 6
001010100110	1 3 5 5	2 3 5 5
1 2 4 6	1 3 5 6	2 3 5 6
001010101101	2 2 4 5	
1 2 5 5	2 2 4 6	
001010101110	2 2 5 5	
1 2 5 6	2 2 5 6	

Fig. 5. Addresses of rules

The module rule address generator from Fig. 3 arranges index numbers to address shown in Fig. 5 with the programming value of no of inputs, if no of inputs are 4 the rules are 16 and to 3 inputs rules become 8 and to 2 inputs rules become 4. The address generator module does the mapping of the rule index with the rule base address (filled by the software). Part of the VERILOG code to evaluate rule address to match with rule base memory addresses is provided in Fig. 6.

Table 1. Memory Space

SI. No	Memory Information			
	Address	Access	Name	Description
1	0X00H to 012D	Read	NRULE	2401 Rules are filled in this Memory
2	012E to 0137	Read/ Write	FUZZYV	28 fuzzified values, inference output values
3	0138	Read	CONTROL REG	No of Inputs, No Of Member-ship functions for input and output
4	0139	Read	ISR	Interrupt status register
5	013A	Read/ Write	IMR	Interrupt Mask Register

The block model of the rule evaluator along with index values is shown in the Fig. 7. Based on the inputs the decision of which output is to be chosen is decided here. This module consists of the two memories one memory is used to read the reduced rules continuously to find maximum values. Another memory is stored with minimum values of corresponding Index values of rule selector. In this design memories have been chosen to utilize the memory blocks in the FPGA and to reduce the logic count and corresponding delay and power dissipation. Provision is provided for all parameters in the design unit using `define and ifdef compiler directives provided by VERILOG [7].

```
if (RuleIndex[11:3] == 0)
        RuleAddress <= RuleIndex;
else if (RuleIndex[11:6] == 0)
        RuleAddress <= RuleIndex - {3'h0,RuleIndex[11:3]};
else if (RuleIndex[11:9] == 0)
        RuleAddress <= RuleIndex - (RuleIndex[8:6] * 7 + RuleIndex[5:3]
                + RuleIndex[8:6]*8 );
else
        RuleAddress <= RuleIndex - (RuleIndex[11:9] * 169 + RuleIndex[8:6] * 15
                + RuleIndex[5:3] );
```

Fig.6. Part of VERILOG Description for Address generation to read reduced Rules.

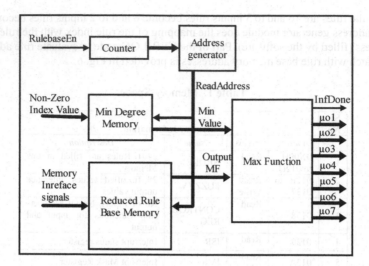

Fig.7. The Block Model of Mamdani Inference engine

5 Test Bench and Test Vector Generation

The fundamental verification principle is an implementation of Register Transfer Level (RTL) code. RTL code must follow the completion of specification to avoid unnecessary complex and unverifiable designs. Testbench usually refers to simulation code used to create a predetermined input sequence to test the response of the output. Fig. 8 shows the interaction of test bench with its DUT (Design under test). The verification of the designed modules has gone through three stages:

    1)   Unit level verification

    2)   Block and core verification

    3)   FPGA verification

The verification plan (Test Cases) in Table II included features that are to be verified from the specification. Some of the important test cases that are covered with this paper are described here:

- Different range of values

- Sequence of transactions

- Relevant interactions between one module to another module

- Synchronization of all module table type styles

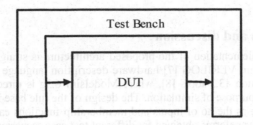

Fig.8. Generic model of a testbench and design under test.

Table 2. The Verification Plan

Sl . No	Test Plan		
	Name	*Description*	*Result*
1	Fuz4In7Mf8 bit.v	Test case with no inputs =4, No of membership functions =7, data width =8	Passed
2	Fuz3In7Mf8 bit.v	Test case with no inputs =3, No of membership functions =7, data width =8	Passed
3	Fuz3In6Mf8 bit.v	Test case with no inputs =3, No of membership functions =6, data width =8	Passed
4	Fuz2In7Mf8 bit.v	Test case with no inputs =2, No of membership functions =7, data width =8	Passed
5	Fuz2In6Mf8 bit.v	Test case with no inputs =2, No of membership functions =6, data width =8	Passed
6	Fuz4In7Mf16bit.v	Test case with no inputs =4, No of membership functions =7, data width =16	Passed
7	Fuz4In5Mf16bit.v	Test case with no inputs =4, No of membership functions =5, data width =16	Passed
8	Fuz4In4Mf16bit.v	Test case with no inputs =4, No of membership functions =5, data width =16	Passed
9	Fuz4In3Mf16bit.v	Test case with no inputs =4, No of membership functions =3, data width =16	Passed
10	Fuz3In7Mf16bit.v	Test case with no inputs =3, No of membership functions =7, data width =16	Passed
11	Fuz2In7Mf16bit.v	Test case with no inputs =2, No of membership functions =7, data width =16	Passed
12	Fuz2In5Mf16bit.v	Test case with no inputs =2, No of membership functions =5, data width =16	Passed

6 Results and discussion

The implementation of the proposed architecture is straightforward by coding all modules in VERILOG [7] hardware description language implemented by Xilinx foundation 13.3 tools [8], where Modelsim 6.3g is directed as an integrated tool for the purpose of simulation. The design of the rule base is highly flexible and configurable as the no of inputs and membership functions can be easily changed. Simulation waveforms obtained for different test case scenarios with varied number of system inputs and membership functions are presented in Fig.9 to Fig11.

Fig.9. CPU Register Writing and Reading

Fig.10. Index values, index numbers of rule selectors and their inference output for no of inputs 2.

Fig.11. Index values, index numbers of rule selectors and their inference output for no of inputs 4.

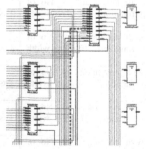

Fig.12. RTL Schematic of the interface between Rule Selectors and Rule base Module.

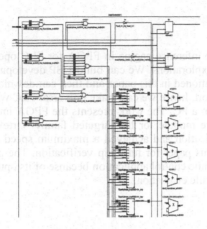

Fig.13. RTL View of the Address Generator.

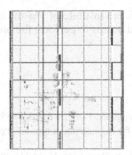

Fig.14. Implemented Physical design in FPGA (device xc5vlx110t).

Table 3. Device Utilization Summary

Selected Device	xc5vlx110tff1136-1
Number of Slices registers	1039 out of 69120 1%
Number of Slice LUTs	1456 out of 69120 2%
Number of Bonded IOBs	51 out of 640 7%
Number of Block RAMs	1 out of 148 0%
Number of GCLKs	1 out of 32 3%
Maximum Frequency	175.809MHz

This implemented FPGA chip is largely efficient to all the constituents of FLC addressed in this paper. This was possible since the chip contains 2495 slices and 4990 logic cells as well three 3×8 multipliers and etc. The RTL view of the reduced rule base module is illustrated in Fig 12 and Fig. 13. Table III shows the FPGA logic resources used to develop the same. Implemented physical design with logic power consumption of 0.17 mw is illustrated in Fig 14.

7 Conclusion

The development of the inference engine in FPGA for FLC opens up with a line of approach to several explorations. We can build DSP development software to define the parameters mentioned in this paper and then can download them into the FPGA through parallel port, hence we can make universal FLC where parameters can be programmed from a PC. The work presents the FPGA implementation of the reduced rule inference module for the targeted family Virtex 5 LX110T. Which shows less logic utilization and achieved a maximum speed of operation. Test cases are provided in this paper for full chip verification. The proposed method can be applied for real time control application because of its options run time configurablilty and speed of rule evaluation.

Acknowledgements

The Authors would like to thank the Institute of Plasma research (IPR), Ahmedabad for providing the material and boards to perform the present work.

References

[1]   L.A. Zadeh, "Fuzzy Logic" IEEE Computer, 1 (4), pp 83-93, 1988.
[2]   Mendel, J.M, "Fuzzy logic systems for engineering: a tutorial" Proc. IEEE, 83 (3), pp 345-377, March 1995.
[3]   A. Mesa, A. Mallet, A Massi Pavan, A. Guessoum, H. Mekki, "FPGA-based implementation of a fuzzy controller (MPPT) for photovoltaic module," Energy Conversion and Management, vol. 52, pp 2695-2907, March 2011.
[4]   Nasri Sulaiman, Zeyad Assi Obaid, M. H. Marhaban, M. M. Hamidon, "FPGA- Based Fuzzy Logic: Design and Applications – a Review" IACSIT International Journal of Engineering and Technology, vol. 1, pp 491-503, December 2009.
[5]   Dajin Kim, "An Implementation of Fuzzy Logic Controller on the reconfigurable FPA system" IEEE Trans. Industrial Electronics., vol. 47, pp 703-715, June 2000.
[6]   K. M. Deliparaschos, F. I. Nenedakis, S. G. Tzafestas, "Design and implementation of fast digital fuzzy logic controller using FPGA Technology" Intelligent and robotic systems, Springer, vol. 45, pp 77-96, 2006.
[7]   Shabiul islam, Nowshad amin, M. S. Bhuyan, Mukter Zaman, Bakri Madon and Masuri Othman, "FPGA Realization of Fuzzy Temperature Controller for Industrial Application", Wseas Transactions On Systems And Control, ISSN: 1991-8763, Issue 10, vol. 2, p. 484-490, Sep. 17, 2007.
[8]   Fazel Taeed, Zainal Salam, Shahrin M. Ayob, "FPGA Implementation of a single input fuzzy logic controller for a boost converter with the absence of an external Analog-to-Digital Converter" Industrial Electronics, IEEE, vol. 59, No. 2, pp 1208-1217, February 2012.
[9]   Peng Bao, Fan Ting Ting, Man Jianquo, "Application of VERILOG HDL language in FPGA development" Control & automation, n 10A, pp 2-88,2004.
[10] Xilinx ISE 13.3 Software Manuals: http://www.xilinx.com/support/sw_manuals.

Analyzing Different Mode FinFET Based Memory cell at different power supply for Leakage Reduction

Sushil Bhushan[1], Saurabh Khandelwal[2], Dr. Balwinder Raj[3]

[1]Research Scholar, I.T.M. University Gwalior, M.P.

[2]Research Scholar, I.T.M. University Gwalior, M.P.

[3]Assistant Professor, N.I.T. Jalandhar, Punjab

{ er.sushil.bhushan@gmail.com; saurabhkhandelwal52@yahoo.com; balwinderraj@gmail.com }

Abstract. FinFET are more versatile than traditional single-gate field effect transistors because it has two gates that can be controlled independently. Usually, the second gate of FinFET is used to dynamically control the threshold voltage of the first gate in order to improve circuit performance and reduce leakage power [1]. A self-controllable-voltage-level (SVL) circuit which can supply a maximum DC voltage to an active-load circuit on request or can decrease the DC voltage supplied to a load circuit in standby mode was developed. This SVL circuit can drastically reduce standby leakage power of CMOS logic circuits with minimal overheads in terms of chip area and speed [2]. In this paper we propose new leakage power reduction techniques namely series LSVL (lower self controlled voltage level) and after using it, leakage power reduces 20% for every increment of series transistor in lower ground connection. Leakage is found to contribute more amount of total power consumption in power-optimized FinFET logic circuits. This paper mainly deal with the various logic design styles to obtain the Leakage power savings through the judicious use of FinFET logic styles using NOR based design at 45 nm technolgy [3]. FinFET circuits are superior in performance and produce less static power when compared to 32nm circuits [4]. FinFET can be designed at 32nm. Finally, implementation of the schematics in CMOS NOR MODE, SG MODE, IG MODE, IG/LP MODE, LP MODE of NOR based FINFET is simulated by cadence virtuoso tools version 6.1 to obtain Leakage Power and Power Dissipation. By applying this we obtain 88% Leakage power savings through the judicious use of FinFET logic styles having NOR based design at 45 nm technology.

Keywords: CMOS scaling, low power, FinFET, DG devices, Series LSVL.

J. C. Bansal et al. (eds.), *Proceedings of Seventh International Conference on Bio-Inspired Computing: Theories and Applications (BIC-TA 2012),* Advances in Intelligent Systems and Computing 202, DOI: 10.1007/978-81-322-1041-2_8, © Springer India 2013

1. Introduction

Steady miniaturization of transistors with each new generation of bulk CMOS technology has yielded continual improvement in the performance of digital circuits. The scaling of bulk CMOS, however, faces significant challenges in the future due to fundamental material and process technology limits [5]. Primary obstacles to the scaling of bulk CMOS to sub-45nm gate lengths include short channel effects, sub-threshold leakage, gate-dielectric leakage and device-to-device variations [6]. It is expected that the use of FinFETs, which provide better control of short-channel effects, lower leakage and better yield in aggressively scaled CMOS process, will be required to overcome these obstacles to scaling [7, 8]. It was estimated that active-mode leakage power might account for as much as 40% of the total power consumption in CMOS circuits at the 70nm technology node [9] and by using 45nm technology node 62.56% of the total power consumption in CMOS circuits. The widespread use of Fin-FETs will somewhat mitigate this problem. In order to avoid this problem, we are using independent control of FinFETs (i.e.) DG devices. Here independent control of front and back gate in DG devices (FinFET) can be effectively used to improve performance and reduce power consumption. Independent gate control can be used to merge parallel transistors (source and drain terminals tied together) in non-critical paths [10]. FinFETs have been shown to provide much lower sub-threshold leakage currents than bulk CMOS transistors at the same gate length [11].

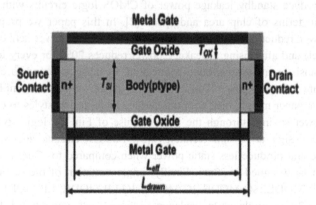

Fig. 1. Cross section of a 15nm FinFET model designed in Taurus.

Table 1. Device parameters of the Taurus FinFET model

DEVICE PARAMETERS	VALUES
Drawn Channel Length L drawn	17nm
Effective Channel Length L_{eff}	15nm
Oxide Thickness T_{ox}	1nm
Body Thickness T_{Si}	2nm
Device Height H	22nm
Vdd	0.7V
VT	0.12V

2. FinFET Transistor with double Gate

Double-gate devices have been used in a variety of innovative ways in digital and analog circuit designs. DG devices with independent gates (separate contacts to back and front gates) have been recently developed. In the context of digital logic design, the ability to independently control the two gates of a DG-FET has been utilized chiefly in two ways: by merging pairs of parallel transistors to reduce circuit area and capacitance, and the next way through the use of a back-gate voltage bias to modulate transistor threshold voltage. A parallel transistor pair consists of two transistors with their source and drain terminals tied together. In Double-gate (DG) FinFETs, the second gate [12] is added opposite the traditional (first) gate, which have been recognized for their potential to better control short-channel effects (SCEs) and as well as to control leakage current. The structure of the FinFET is shown in the Fig.1. The two gates for FinFET provide effective control of the short-channel effects without aggressively scaling down the gate-oxide thickness and increasing the channel doping density. The separate biasing in DG device easily provides multiple threshold voltages. It can also be exploited to reduce the number of transistors for implementing logic functions [13].

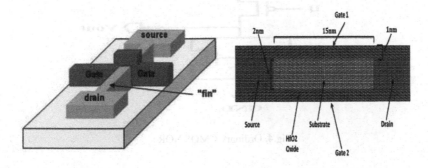

Fig.2. FinFET structure [12]   **Fig. 3.** Schematic of a double gate device [1].

The goal of this paper is to explore FinFET logic design styles (layout) and study their implications for low-power design. It was estimated that leakage power might account for as much as half of the of the total power consumption in CMOS circuits. Leakage power consumption was observed to remain around more amount of the total power [8] on an average we explore methods to efficiently overcome this challenge through a combination of circuit design techniques and logic-level optimization. It considers the use of IDDG-FETs in digital CMOS design, focusing on the use of independent-gate FinFET [12].

3. Methodology:

In this paper, four modes of FinFET operation are identified, such as the shorted-gate (SG) mode with transistor gates tied together, the independent gate (IG) mode where independent digital signals are used to drive the two device gates, the low-power (LP) mode where the back-gate is tied to a reverse-bias voltage to reduce leakage power and the hybrid (IG/LP) mode, which employs a combination of LP and IG modes [10] . This paper has considered four design styles for digital logic structures using FinFETs. In the interest of brevity, this section presented data only for two-input NOR gates. To evaluate the utility of the different FinFET modes, this paper has constructed layout design for the four modes of NOR gates using **CADENCE VIRTUOSO TOOLS version IC 6.1**. The circuit diagram of different FinFET-based NOR gate designs along with the ordinary CMOS is shown in the Fig 4-8.

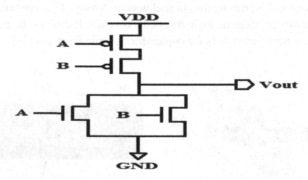

Fig.4. Ordinary CMOS NOR

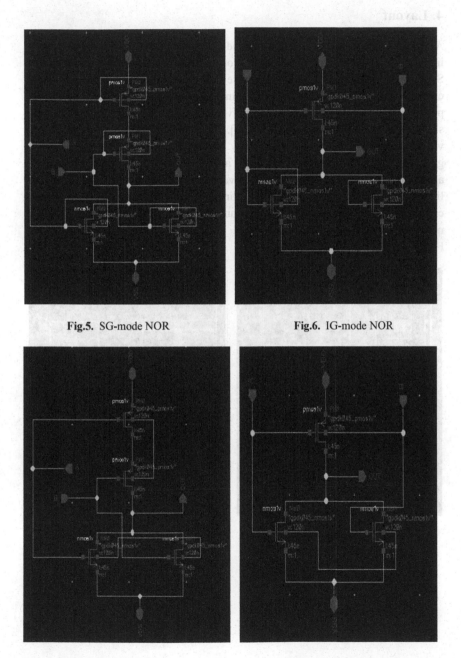

Fig.5. SG-mode NOR   **Fig.6.** IG-mode NOR

Fig.7. LP-mode NOR   **Fig.8.** IG/LP-mode NOR

4. Layout

As layout dimensions continue to be reduced, lithographic considerations will impose additional constraints on the layout of future nanoscale SRAM layout. Sources of mismatch in dense nanoscale SRAM devices due to variations in channel doping (both random and systematic) may be attributed to the use of pushed design rules and alignment sensitive doping variation sources such as halo shadowing, lateral implant straggle [14]. The general subject of non-random variation in dense SRAM devices may be further expanded to include the geometric sources of mismatch. These arise from the non-ideal environment associated with pushed design rules, variation in alignment and additional lithography effects such as corner rounding and line end foreshortening. These effects are layout topology dependent and can also contribute to the overall mismatch in the dense bit cell devices. Fig 9,10,11,12 shows different mode FinFET

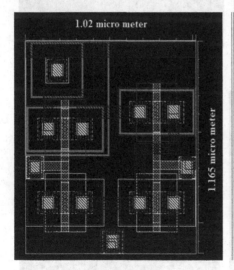 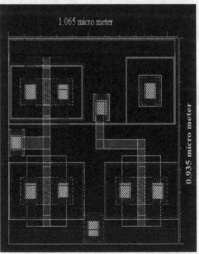

Fig.9. SG-mode NOR   Fig.10. IG-MODE NOR

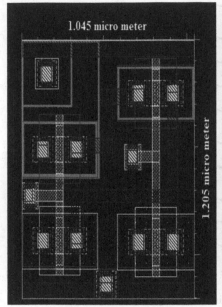

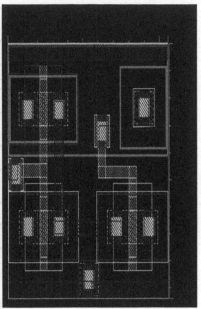

Fig.11. LP-mode NOR   **Fig.12.** IG/LP-mode NOR

5. Proposed Work

There are two well-known techniques for reducing stand-by power (Pst). One is to use a multi-threshold-voltage CMOS (MTCMOS). It has serious drawbacks such as the need for additional fabrication processes for higher Vth and the fact that storage circuits based on this technique cannot retain data. The other technique involves using a variable threshold-voltage CMOS (VTCMOS). Which reduces leakage current by increasing substrate-bias (Vsub).This technique also faces some serious problems, such as very slow substrate-bias controlling operation, large area penalty, and large power penalty due to substrate-bias supply circuits. To solve the above-mentioned drawbacks, a self-controllable-voltage-level (SVL) circuit, which can significantly decrease Pst while maintaining high-speed performance, has been developed. While the load circuits are in the active mode, the developed SVL circuit supplies the maximum DC voltages [2].

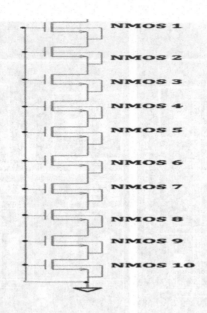

Fig .13. Series LSVL Techniques for leakage reduction [12]

Now new proposed techniques namely series LSVL improved leakage reduction drastically with respect to SVL technique. It reduces leakage power approximately 20% at every increment of transistor in lower ground and after the implementation in circuit at different mode approximately 80% reduces the leakage power after using 10 NMOS in series.

The four different modes of FinFET based NOR gate logic styles such as SG, IG, LP and hybrid (IG/LP) modes have been analyzed using the **CADENCE VIRTUOSO IC 6.1 TOOLS**. The table 1 shows the performance evaluation in which power dissipated values are obtained for the supply voltage from 0 to 0.7 volt respectively for various FinFET based NOR gate logic styles along with that of ordinary CMOS NOR gate at the time when 10 NMOS transistor using in series at the lower ground terminal.

With the growing use of portable and wireless electronic systems, reduction in power consumption has become one of the main concerns in today's VLSI circuit and system design. For a CMOS digital circuit, power dissipation includes three components [15]: switching power dissipation (Pswitching) ,short-circuit power dissipation (P Short Circuit), and static leakage power dissipation (P leakage) .The average power dissipation can be expressed by,

$$P \text{ average} = P \text{ switching} + P \text{ short-circuit} + P \text{ leakage}$$

$$= \alpha CL \, V^2 DD \, fclk + Isc \, VDD + I \text{ leakage } VDD$$

where CY is the switching activity (average number of switching per clock period) , C L is the load capacitance, fclk is the clock frequency, Is, is the directpath short circuit current, $I_{leakage}$ is the leakage current, and VDD is the supply voltage. Lowering supply voltage is obviously the most effective way to reduce the power consumption. With the scaling of the supply voltage, the transistor threshold voltages should also be scaled in order to satisfy the performance requirements Unfortunately, such scaling leads to the increase of the leakage current through a transistor[15]. Therefore, the leakage power cannot be ignored for low voltage low power circuit designs.

The obtained results are shown in table 2.

Table 2. Comparisons of normal CMOS NOR with various modes of FinFET

VOLTAGE	CMOS NOR		SG MODE		IG MODE		IG/LP MODE		LP MODE	
Vdd (v)	Leakage Power(µw)		Leakage Power(µw)		Leakage Power(µw)		Leakage Power(µw)		Leakage Power(µw)	
	NS	S	NS	S	NS	S	NS	S	NS	S
	LSVL	LSVL	LSVL	LSVL	LSVL	LSVL	LSVL	LSVL	LSVL	LSVL
0.25	22.4	17.92	22.2	17.76	22.5	18.00	22.6	18.08	22.8	18.24
0.30	25.1	20.08	25.6	20.48	25.4	20.32	25.7	20.56	25.2	20.16
0.35	26.8	21.44	26.3	21.04	26.4	21.12	26.9	21.52	26.1	20.88
0.40	31.6	25.28	31.1	24.88	31.4	25.12	31.8	25.44	31.9	25.52
0.45	32.4	25.92	32.2	25.76	32.7	26.16	32.9	26.32	32.1	25.68
0.50	35.7	28.56	35.2	28.16	35.5	28.40	35.3	28.24	35.9	28.72
0.55	36.2	28.96	36.5	29.20	36.8	29.44	36.7	29.36	36.9	29.52
0.60	37.1	29.68	37.3	29.84	37.5	30.00	37.7	30.16	37.9	30.32
0.65	38.8	31.04	38.2	30.56	38.4	30.72	38.6	30.88	38.1	30.48
0.70	39.2	31.36	39.3	31.44	39.4	31.52	39.5	31.60	39.7	31.76

6. Different modes of FinFET- based NOR gate design:

The I-V characteristics are also obtained for the four different modes of FinFET based NOR gate logic styles along with ordinary NOR gate. From the obtained I-V characteristics graph we can calculate the ON current for the four different modes. The ION current obtained is suited best for IG mode. Similarly the off current is low in IG mode since the ION is increased. Thus the power dissipation is decreased. The ION current has been calculated by setting the parameters (i.e) threshold voltage as 0.12V, body thickness as 0.2 nm, mobility of electrons as 0.03 eV respectively. Hence power consumption is found to be low in IG mode. The ION current for the four different modes of FinFET based NOR gate logic styles along with ordinary NOR gate is shown below in terms of a tabular column in table 4. By placing a second gate on the opposite side of the device, the gate capacitance of the channel is doubled and the channel potential is better controlled by the gate electrode, thus limiting I_{off}. Reducing the body thickness further decreases I_{off}. As a result the leakage current is further reduced, this in turn causes reduction in power dissipation. The Percentage of Power Dissipation reduced at different modes is shown in Table 3.

Table 3. Comparisons of Different modes of FinFET- based NOR gate design

MODES	POWER DISSIPATION
CMOS NOR	45%
SG MODE	76%
IG MODE	36%
IG/LP MODE	80%
LP MODE	58%

Table 4. Different modes of FinFET- based NOR gate design and its I_{ON} Current at different power supply

MODES	I_{ON} CURRENT(μA)									
	0.25 V	0.30 V	0.35 V	0.40 V	0.45 V	0.50 V	0.55 V	0.60 V	0.65 V	0.70 V
CMOS NOR	89.66	83.67	76.50	79.00	72.0	71.4	65.81	61.83	59.69	56.00
SG MODE	88.83	85.34	75.15	77.75	71.5	70.4	66.36	62.16	58.76	56.14
IG MODE	90.06	84.62	75.40	78.56	72.6	71.0	66.90	62.50	59.07	56.28
IG/LP MODE	90.46	85.65	76.80	79.51	73.1	70.6	66.72	62.83	59.38	56.42
LP MODE	91.20	84.07	74.55	79.73	71.3	71.8	67.09	63.16	58.61	56.71

7. Conclusion

In conclusion, we have discussed various logic styles for low-power FinFET circuits. We demonstrated that the rich diversity of design styles, made possible by independent control of FinFET gates, can be used effectively to reduce total active power consumption in digital circuits. Four modes are investigated using CADENCE VIRTUOSO IC 6.1 TOOLS. Our results indicate that on an average, 76% of the total active power dissipation is obtained in IG mode, 36% in SG mode,80% in LP mode and 58% in hybrid (IG/LP) mode circuits. From the above results obtained we would like to conclude that IG mode dissipates less power compared to all other modes. On comparing with ordinary CMOS NOR gate logic, the leakage power will be less in IG mode. Hence the power consumption in IG mode will be less compared to that of all other modes. As a result independent control of double gate transistors based on NOR gate logic styles, can be used effectively to reduce total power consumption in digital circuits. The comparison of different modes of FinFET based NOR gate design is shown in table 2.

References

[1]   Michael C. Wang. "Independent-Gate FinFET Circuit Design Methodology". IAENG International Journal of Computer Science, 37:1, IJCS_37_1_06.

[2]   Tadayoshi Enomoto, Yoshinori Oka, Hiroaki Shikano, and Tomochika Harada, "A Self Controllable-Voltage-Level (SVL) Circuit for Low-Power, High-Speed CMOS Circuits" ESSCIRC 2002.

[3]   Nirmal, Vijaya Kumar , Sam Jabaraj. "NAND GATE USING FINFET FOR NANOSCALE TECHNOLOGY". International Journal of Engineering Science and Technology Vol. 2(5), 2010, 1351-1358.

[4]   Brian Swahn and Soha Hassoun, "Gate Sizing: FinFETs vs 32nm Bulk MOSFETs". DAC 2006,July 24–28, 2006, San Francisco, California, USA. Copyright 2006 ACM 1-59593-381-6/06/0007 ...$5.00.

[5]   E. J. Frank, R. H. Dennard, E. Nowak, P. M. Solomon, Y. Taur, and H.-S. P. Wong. "Device scaling limits of Si MOSFETs and their application dependencies". Proc. IEEE, 89(3):259–288, (2001).

[6]   Etienne Sicard, Sonia Delmas, "Basics of CMOS cell design" book, (2006).

[7]   T.-J. King, "FinFETs for nanoscale CMOS digital integrated circuits". In Proc. Int. Conf. Computer-Aided Design, pages 207–210, (2005).

[8]   L. Wei, Z. Chen, and K. Roy, "Double gate dynamic threshold voltage (DGDT) SOI MOSFETs for low power high performance designs." In Proc. IEEE Int. SOI Conf., pages 82–83, (1997).

[9]   W. Zhang, J. G. Fossum, L. Mathew, and Y. Du, "Physical insights regarding design and performance of independent-gate FinFETs". IEEE Electronic Device Lett, 52(10):2189–2206, (2005).

[10] Anish Muttreja, Niket Agarwal and Niraj K. Jha, "CMOS logic design with independent-gate FinFETs" ©2007 IEEE

[11] E. J. Nowak, I. Aller, T. Ludwig, K. Kim, R. V. Joshi, C.-T. Chuang, K. Bre, and R. Puri. "Turning silicon on its edge." IEEE Circuits and Devices Magazine, 20(1):20–31, (2004).

[12] I. Aller. "The double-gate FinFET: Device impact on circuit design." In Proc. Int. Solid-State Circuits Conf., pages 14–15 (and visual supplements, pp. 655–657), (2003).

[13] P. Beckett, "A fine-grained reconfigurable logic array based on double gate transistors." In Proc. IEEE Int. Field-Programmable Technology Conf., pages 260–267, (2002).

[14] Randy W. Mann and Benton H. Calhoun, "New category of ultra-thin notchless 6T SRAM cell layout topologies for sub-22nm" 2010, IEEE proceeding on 11th Int'l Symposium on Quality Electronic Design

[15] Zhanping Chen, Liqiong Wei and Kaushik Roy , "REDUCING GLITCHING AND LEAKAGE POWER IN LOW VOLTAGE CMOS CIRCUITS" , (1997). ECE Technical Reports. Paper 85. http://docs.lib.purdue.edu/ecetr/85.

ANN Based Distinguishing Attack on RC4 Stream Cipher

Ashok K Bhateja and Maiya Din

SAG, DRDO, Metcalfe House, Delhi-110054

{akbhateja@gmail.com; anuragimd@gmail.com}

Abstract. RC4 is the most widely used stream cipher in many applications. Many Distinguishing Attacks on RC4 system have been published which are based on statistical approaches. These statistical Distinguishing Attacks exploit distribution of bytes, diagraphs & trigraphs in RC4 generated output key stream with respect to random key stream. This paper presents an Artificial Neural Network (ANN) based approach to distinguish RC4 key stream from random key stream. The Joint Mutual Information (JMI) criterion has been used in effective feature selection. The prominent features are used in Back-propagation learning of Multilayer Perceptron (MLP) network to distinguish RC4 system.

Keywords: RC4 Stream Cipher, Distinguishing Attack, Joint Mutual Information, Multilayer Perceptron Network, Back-Propagation Learning.

1 Introduction

In cryptography, RC4 is the most widely-used software based stream cipher and is used in popular protocols such as Secure Sockets Layer (SSL) to protect Internet traffic and WEP to secure wireless networks. RC4 was created by Rivest for RSA Securities Inc in 1987. Its key size varies from 40 to 256 bits. It has two parts namely Key Scheduling Algorithm (KSA) and a Pseudo-Random Generator Algorithm (PRGA). KSA turns a random key into an initial permutation S of $\{0,1,\ldots, N\text{-}1\}$, where N is the size of the RC4 permutation. PRGA uses this permutation to generate a pseudo-random output sequence.

RC4 [2, 10] has weaknesses that argue against its use in new systems although it is remarkable for its simplicity and speed in software. It is especially vulnerable when the beginning of the output key-stream is not discarded, nonrandom or related keys are used, or a single key stream is used twice. Some ways of using RC4 can lead to very insecure cryptosystems such as WEP. Mantin (2005) mentioned

Diagraph Repetition Bias Attack on this cipher. According to this attack for small strings T (<16), the pattern ABTAB occurs with the approximate probability $(1/N^2+1/N^3)$. For this 2^{29} sample are required for a success probability 0.9 when N is 256.

The rest of the paper has been organized in the following way: In section 2 we describe RC4. In section 3 we present previous distinguishing attacks on RC4. The details of data preparation and prominent feature selection based on Joint Mutual Information (JMI) criterion is discussed in section 4. The Back-propagation learning of MLP network is described briefly in section 5. The simulation details of ANN based distinguishing attack along with the results obtained is mentioned in section 6. The results are analyzed and concluded in section 7.

2 RC4 Stream Cipher

RC4 Stream Cipher [10] is a variable key size stream cipher with byte oriented operations. In the RC4 algorithm, there are two stages process during encryption as well as decryption. The algorithm is dividing into the two parts KSA (Key scheduling Algorithm) and PRGA (Pseudo Random Generator Algorithm). KSA as the first stage of algorithm also knows as initialization of permutation vector S and PRGA known as stream generation in the RC4.

2.1 The key-scheduling algorithm (KSA)

The key scheduling algorithm generates initial permutation S of $\{0, \cdots, N-1\}$ from a (random) key of length l bytes. Typically l lies in the range between 5 and 32. The key length N may have maximum value 256 bits. The array S of size N, is initialized to the identity permutation and then mixes bytes of the key within it.

```
for i from 0 to N - 1
    S[i] = i
end
j = 0
for i from 0 to N - 1
    j = (j + S[i] + key[i mod keylength]) mod N
    swap(S[i], S[j])
end
```

2.2 The pseudo-random generation algorithm (PRGA)

PRGA uses the permutation S generated by KSA, to generate a pseudo-random output sequence. The output byte is selected by looking up the values of $S(i)$ and $S(j)$, adding them together modulo N, and then looking up the sum in S; $S(S(i) + S(j))$ is used as a byte of the key stream.

$$i = j = 0$$
while loop Generating_Output key stream
$$i = (i + 1) \bmod N$$
$$j = (j + S[i]) \bmod N$$
Swap (S[i], S[j])
Output = S[(S[i] + S[j]) \bmod N]
end while loop

Many stream ciphers are based on linear feedback shift registers (LFSRs), which are efficient in hardware but less efficient in software. The design of RC4 avoids the use of LFSRs, and is ideal for software implementation, as it requires only byte manipulations. It uses 256 bytes (for $N = 256$) of memory for the array, S[0] through S[255], l bytes of memory for the key, key[0] through key[l-1], and integer variables, i, j. Performing a *modulus* 256 can be done with a bitwise AND with 255 (or on most platforms, simple addition of bytes ignoring overflow).

3 Distinguishing Attacks

The idea behind the Distinguishing Attack [10] on a Stream Cipher is to distinguish its key-stream from a random bit stream. Similarly in case of Block Cipher, output of the Block Cipher is distinguished with respect to random permutation. Distinguishing Attacks employ techniques from tests of randomness on a specific event of the concerned cipher.

Cryptanalysis of RC4 is divided into two main parts [10], analysis of the initialization of RC4 and analysis of the key-stream generation. The first part focuses on the KSA, the PRGA initialization and the integration of both, whereas the last focuses on the internal state and the round operation of the PRGA. The simplicity of the initialization part and the key-stream generation part attracted a lot of attention in the cryptographic community and indeed various significant weakness discovered like classes of weak keys, patterns that appears twice and thrice the expected probability, propagation of key patterns through the KSA to the initial permutation and through the PRGA initialization to the prefix of the stream and modes of operation that allow related key attacks.

In 1995, A. Roos [1] observed that the first byte of the key-stream is correlated to the first three bytes of the key and the first few bytes of the permutation after the KSA are correlated to some linear combination of the key bytes. The key-stream generated by the RC4 is biased in varying degrees towards certain sequences. The best such attack is due to Itsik Mantin and Adi Shamir [2] who showed that the second output byte of the cipher was biased toward zero with probability 1/128 (instead of 1/256). This is because if the third byte of the original state is zero, and the second byte is not equal to 2, then the second output byte is always zero.

In 2005, Andreas Klein [5] presented an analysis of the RC4 Stream Cipher showing more correlations between the RC4 key-stream and the key. S. Paul and Preneel [3] mentioned in their attack that the probability of the first two bytes is same and equal to $1/N\,(1-1/N)$, Where N is number of states in S-Permutation used in RC4 Stream Cipher. The number of outputs required to reliably distinguish RC4 outputs from random strings using this bias is 2^{26} bytes. After the first N bytes are thrown, the bias reduces to $1/N\,(1-1/N^2)$. Similar kind of Distinguishing Attack based on two consecutive bytes of RC4 key-stream was developed by Basu, Ganguly, Maitra & G. Paul (2008), but it require 2^{42} bytes for a success probability 0.9772 and $N = 256$.

We have proposed Artificial Neural Network (Bio-Inspired) based Distinguishing Attack on RC4 by selecting only the prominent features to reduce required number of key-stream bytes and to reduce the computing time.

4   Data Preparation and Feature Selection

The output key-stream was generated for both RC4 crypto system and Perfect Random System. Total 600 frames (each frame of 10, 00,000 bytes) were generated based on different key seeds. The 300 frames were generated from each system. Then 60 normalized features (20 high frequent monograms, 20 high frequent digraphs & 20 high frequent differences of digraphs and corresponding reversed digraphs of bytes) were computed for each frame. These selected features were further reduced to 10 features based on Joint Mutual Information (JMI) criterion. The achieved prominent features were used in training and testing of ANN, which also save computational time. The network was trained on 400 feature vectors using Back-propagation Learning Technique. Remaining 200 feature vectors were used for testing the network.

4.1   Effective Feature Selection

High dimensional data sets pose significant challenges for machine learning. In some of the most difficult problems, such as Crypto System Classification, selection of discriminating features is also a challenge. Input feature selection [4] is the most important part of classification modeling process, because it interprets the data modeling problem by specifying those explanatory features most relevant to the target variables. There are various methods of prominent feature selection according to their distinguishing/classifying power.

Feature selection techniques [7] can be broadly grouped into approaches that are classifier-dependent ('Wrapper' and 'embedded' techniques), and classifier-independent ('Filter' techniques). Wrapper techniques search the space of feature subsets using training accuracy of a particular classifier. In contrast, filter techniques (Duch, 2006) separate the classification and feature selection components. In general, filters are faster than embedded techniques. A primary advantage of filters is that they are relatively cheap in terms of computational expense. The defining component of Filter technique is the relevance index (selection score), quantifying the utility of including a particular feature in the set.

4.2 Entropy and Mutual Information

The fundamental unit of information is the 'entropy' of a random variable; it is denoted by $H(X)$ for variable 'X'. The 'entropy' is low when there is a little uncertainty over the outcome. If all the events are equally likely, that is maximum uncertainty over the outcome, then 'entropy' is maximal. It is defined as:

$$H(X) = -\sum_{x \in X} p(x) \log p(x)$$

Here, x denotes a possible value that the variable X can adopt. The entropy can be conditioned on other events. The conditional entropy of X given Y is denoted, as

$$H(X/Y) = -\sum_{y \in Y} p(y) \sum_{x \in X} p(x/y) \log p(x/y)$$

This can be thought of as the amount of uncertainty remaining in X after we learn the outcome of Y. Mutual Information [Shanon, 1948] between X and Y, that is the amount of information shared by X and Y, as follows:

$$I(X;Y) = H(X) - H(X/Y)$$

$$= \sum_{x \in X} \sum_{y \in Y} p(xy) \log \frac{p(xy)}{p(x)p(y)}$$

This is the difference of two entropies- the uncertainty before Y is known, $H(X)$, and the uncertainty after Y is known, $H(X/Y)$. This can also be interpreted as the amount of uncertainty in X which is removed by knowing Y. So 'Mutual

Information' is the amount of information that one variable provides about another. The 'Mutual Information' is symmetric, that is, $I(X;Y) = I(Y;X)$. It is zero if and only if the variables are statistically independent i.e. $p(xy) = p(x)p(y)$.

The 'Mutual Information' can also be conditioned; the 'Conditional Information' [7] is defined as,

$$I(X;Y/Z) = H(X/Z) - H(X/Y.Z)$$

$$= \sum_{z \in Z} p(z) \sum_{x \in X} \sum_{y \in Y} p(xy/z) log \frac{p(xy/z)}{p(x/z)p(y/z)}$$

This can be thought of as the information still shared between X and Y after the value of a third variable, Z is revealed.

4.3 Joint Mutual Information

Filter techniques [7] are defined by a criterion 'Relevance Index' denoted by J which is intended to measure how potentially useful a feature may be when used in a classifier. An intuitive J would be some measure of correlation between the feature and the class label i.e. the intuition being that a stronger correlation between these should imply a greater predictive ability when using the feature. For a class label Y, the 'Mutual Information' index for a feature X_k is as follows:

$$J_{mim}(X_k) = I(X_k;Y)$$

Here, 'mim' stands for *Mutual Information Maximization*. An important limitation is that this assumes that each feature is independent of all other features and effectively ranks the features in descending order of their individual mutual information content. However, where features may be interdependent, this is known to be suboptimal. In general, it is widely accepted that a useful and parsimonious set of features should not only be individually *relevant*, but also should not be *redundant* with respect to each other features should not be highly correlated.

Battiti (1994) presented the 'Mutual Information Feature selection (MIFS) criterion:

$$J_{mifs}(X_k) = I(X_k, Y) - \beta \sum_{X_j \in S} I(X_k, X_j)$$

Here, S is the set of currently selected features. This includes the $I(X_k;Y)$ term to ensure feature *relevance*, but introduces a penalty to enforce low correlations with features already selected in S. The β in the MIFS criterion is a configurable parameter, which must be set experimentally. Using $\beta = 0$ would be equivalent to $J_{mim}(X_k)$, selecting features independently, while a larger value will place more emphasis on reducing inter-feature dependencies. In experiments, Battiti found that $\beta = 1$ is often optimal, though with no strong theory to explain why. The MIFS criterion focuses on reducing redundancy; an alternative approach was proposed by Yang and Moody (1999), and also later by Meyer et al. (2008) using the Joint Mutual Information (JMI), to focus on increasing complementary information between features.

The JMI index for feature X_k is defined as:

$$J_{jmi}(X_k) = \sum_{X_j \in S} I(X_k X_j; Y)$$

This is the information between the targets and a joint random variable $X_k\, X_j$, defined by paring the candidate X_k with each previously selected feature. The idea is that if the candidate feature is complementary with existing features, it should be included. The JMI criterion is used to reduce initial 60 features to 10 prominent features applied for training of the MLP network.

5 Back-Propagation Learning based ANN

An artificial neural network (ANN), usually called neural network (NN), is a mathematical model or computational model that is inspired by the structure and/or functional aspects of biological neural networks. A neural network consists of an interconnected group of artificial neurons, and it processes information using a connectionist approach to computation. In most cases an ANN is an adaptive system that changes its structure based on external or internal information that flows through the network during the learning phase. Modern neural networks are non-linear statistical data modeling tools. They are usually used to model complex relationships between inputs and outputs or to find patterns in data.

A Multilayer Perceptron (MLP) Network [8] is a feed-forward artificial neural network model that maps sets of input data onto a set of appropriate output. An MLP consists of multiple layers of nodes in a directed graph, with each layer fully connected to the next one. Except for the input nodes, each node is a neuron (or processing element) with a nonlinear activation function. MLP utilizes a supervised learning technique called Back-propagation for training the network. MLP is a modification of the standard linear perceptron, which can distinguish data that is not linearly separable.

An ANN having multi layers (Input layer, Hidden layers and Output layers) is called Multilayer ANN [9] as shown in the Fig. 1.

Back-propagation algorithm [8] can be used to train the network. It uses a gradient search technique to minimize cost function equal to least mean square error between desired and actual net outputs.

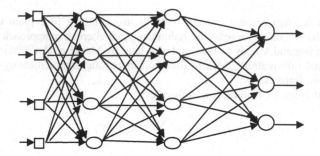

Fig. 1: Multilayer Artificial Neural Network

6 ANN based Distinguishing Attack

An ANN based Distinguishing Attack has been developed to distinguish RC4 key-stream with respect to Random key-stream. In this attack Joint Mutual Information (JMI) criterion is applied in effective feature selection. The reduced 10 prominent features are used in Back-propagation learning of Multilayer Perceptron (MLP) network having 3 layers. The no. of neurons used in the network layers were 10, 5, 2 (Input layer to output layer respectively) based on Sigmoid Activation Function.

The network was trained on 400 feature vectors (200 of RC4 and 200 of Random key-streams) each having 10 effective features. The variable learning rate used was in the range 1.5 to 0.75 for achieving required error threshold value. The achieved distinguishing score was 78% on training data. The network was tested on 100 feature vectors of each class and achieved distinguishing score was 69%.

7 Conclusion

In this paper, an Artificial Neural Network (ANN) based approach to distinguish RC4 key stream from Random Key Stream is presented. This ANN based attack requires 2^{20} bytes. The Joint Mutual Information (JMI) criterion is applied in effective feature selection. The reduced 10 prominent features are used in Back-propagation learning of Multilayer Perceptron (MLP) network to distinguish RC4 system. The achieved distinguishing score was 69%.

Acknowledgement

Authors would like to thank Dr P. K. Saxena, Director SAG for his interest and encouragement for this work. We would like to thank Dr S. S. Bedi, Associate

Director for valuable suggestions to carry out this work. Authors also thank to Mr Ram Ratan, Scientist 'F' for valuable suggestions in preparing and submitting the paper to BICTA 2012 conference.

References

[1] Roos, A.: Class of Weak Keys in the RC4 Stream Cipher. Post in sci.crypt, (1995)

[2] Fluhrer, S., Mantin I. and Shamir A.: Weakness in the Key Scheduling Algorithm of RC4, LNCS 2259, 1-24, Springer Verlag (2001)

[3] Paul, Souradyuti and Preneel, Bart: A New Weakness in the RC4 Key-stream Generator and an Approach to Improve the Security of the Cipher, FSE-2001, 245-259, Springer Verlag (2004)

[4] Mantin, I.: Predicting and Distinguishing Attacks on RC4 Key-stream Generator 491-505, Springer Verlag (2005)

[5] Klein, Andreas: Attacks on the RC4 Stream Cipher, Journal of Design, Code and Cryptography, vol 48, issue 3, 269-286, (2008)

[6] Yang, H. H. and Moody, J.: Feature Selection based on Joint Mutual Information, Journal of Computational Intelligence Methods and Applications, Int. Computer Science Convention, Vol.13, 1-8, (1999)

[7] Brown, G., Pocock, A. and Jhao, M.J.: Conditional Likelihood Maximization: A Unifying Framework for Information Theoretic Feature Selection. Journal of Machine Learning Research, Vol.13, 27-66, (2012)

[8] Haykin, S.: Neural Networks- A Comprehensive Foundation, Macmillan, New York, (2001)

[9] Katagiri, S.: Hand Book of Neural Networks for Speech Processing, Artech House, London, 1st edition,(2000)

[10] www.security.iitk.ac.in/hack.in/2009/bimal_keynote_hack.in.pdf

Optimal Municipal Solid Wastes Management with Uncertainty Characterization Using Fuzzy Theory

Nekram Rawal[1] and Raj Mohan Singh[1]

[1]Department of Civil Engineering, Motilal Nehru National Institute of Technology, Allahabad-211002, India

{E-mail1: nrrawal@mnnit.ac.in; E-mail2: rajm@mnnit.ac.in }

Abstract. There is an increasing concern by environmental managers and planners to follow a sustainable approach to municipal solid waste management (SWM) and to integrate strategies that will produce the comprehensive optimal practicable option. The selection of a suitable SWM process is driven by the type of waste source and quality of waste produced. This study demonstrates application of optimization based methodology that would facilitate optimal management of collection and transportation of solid waste in urban cities. Uncertainty in solid waste management due to uncertain amount of solid wastes is characterized using fuzzy logic. Uncertainty analysis result shows that uncertainty in the optimal waste management (in terms of total optimal cost) is approximately 1.5 times uncertainty in waste amount.

Keywords: Municipal solid waste (MSW), optimization formulation, landfills, fuzzy logic and uncertainty characterization.

1   Introduction

Solid wastes are the result of urbanization and development, and have emerged a serious threat to environment. Improper disposal of solid wastes may adversely affect environment and human health. Solid waste management (SWM) is needed for the solution of the problem of solid waste concerning protection of the environment and conservation of natural resources. Solid waste essentially managed by municipal authorities in India to keep net and clean a city [1]. The consideration of complex interactions among collection and transportation systems and the facilities for waste management (waste reduction and disposal) is required for solid waste management.

Land use patterns, and urban growth and development patterns must also be taken into account in response to local waste management needs. Municipal solid waste management (MSWM) is one of the major environmental problems of Indian cities [2]. The order of preference in terms of solid waste management strategies may be source reduction (selection of optimized waste source centers), reuse, or recycling, incineration with energy recovery or without energy recovery, and landfill disposal [3]. Many factors must be evaluated in the planning of an integrated waste management, the system is generally structured into the four phase of collection, transportation, processing and disposal [4; 5; 6]. Uncertainty plays an important role in most solid waste management problems. Fuzziness is one type of random character, which is linguistic in nature and generally cannot be described by traditional probability distributions. Such impreciseness refers to the absence of sharp boundaries in information and frequently exists in the decision marking process. Koo's [7] proposed a frame work using Waste Resources Allocation Program (WRAP) and fuzzy set theory to address the trade-off among the objectives of economic efficiency, environmental quality, and administrative efficiency, such that the optimal site for a hazardous waste treatment facility could be determined [8;9;10]

In recent years, many works have been presented with the aim of providing useful and comprehensive decision models [11;12; 13; 14; 15] which should be both significantly close to reality and computationally tractable in order to help planners in managing solid waste disposal and treatment in urban areas, taking into account multidisciplinary aspects involving economic, technical, normative, and environmental sustainability issues. Specifically, considerable efforts have been directed towards the development of economic-based optimization models for MSW flow allocation. [16; 17; 18 and 19]. Recent advances in solid waste management can be seen in Minciardi's [20]; and Papachristou's [21]. Rawal's [22] presents comprehensive MSW application using integer linear model.

This paper presents a detailed of the mixed integer MSW model that can be used as tools for decision makers of a municipality in the day to day planning and management of comprehensive programs of solid waste collection, incineration, recycling, treatment, and disposal. The main focus of this study is performance evaluation results of optimization models for the MSW. The mathematical formulation of the optimization model was solved using commercially available optimization software Lingo 10.0 [23]. Uncertainty in solid waste management due to uncertain amount of solid wastes is characterized using fuzzy logic. Uncertain waste amount is represented as fuzzy numbers. Optimization formulation with fuzzified waste amount is then solved to characterize uncertainty is cost. The results in this work establish potential application of the methodology for solid waste management and uncertainty characterization in urban areas.

2 Mixed Integer Linear Optimization Model for Municipal Solid Waste Management

The methodologies first address the issue of collection of municipal solid waste by minimization of the vehicle routes. The optimized collection points are further utilized in the development optimization model of MSW. Thus, the present work proposed methodology to address the issues of MSW in a comprehensive manner. In addition the model proposed in this paper has been formulated taking conditions of typical city in Asian countries.

The physical components of the MSW optimization model are shown in Fig. 1. The optimization model essential consists of total cost owing to investment and management costs, transportation costs, and operational costs from the use of repairing of trucks. The benefits from energy generation, RDF production, compost, and recycling is to be subtracted from the overall cost.

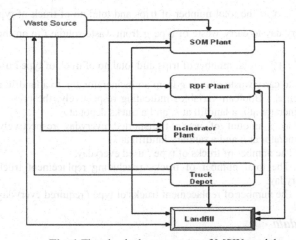

Fig. 1 The physical component of MSW model

The constraints include waste flow constraints; facility availability; capacity of available facilities; site selection; environmental; landfill saturation constraints etc. It is being assumed that waste source are located at one point of an individual areas; and MSW separation is done at the waste source locations (collection points). MSW operations proposed in the model are on daily basis. The optimization model is presented as a mixed integer linear program where the transportation cost is the actual amount of waste transported by each truck. The description of new variables and the models are presented below.

Following are the indices used in this paper

$i = 1, 2, \ldots$, :location of waste sources (collection points); $j = 1, 2, \ldots$, : location of incinerators; $k = 1, 2. \ldots$, : location of sanitary landfills; $r = 1, 2, \ldots$, :location of replacement trucks depots; $l = 1, 2, \ldots$, : truck type; and $g = 1, 2, \ldots$,: waste type.

Here are variables of the problems

β_{lijg}, β_{likg}, : respectively amount of waste (in tons) of type g collected everyday by trucks of type l from a waste source i to an incinerator at j, and a landfill at k.

λ_{ljkg}, λ_{lhkg} : respectively amount of waste (in tons) of type g collected everyday by trucks of type l from an incinerator at j, to a landfill at k.

X_{ij}^{lg}, x_{ij}^{lg} : the total number of trips and total no of truck of type l used every day to carry waste of type g from waste source i to an incinerator at j.

Y_{ij}^{lg}, y_{ij}^{lg} : total number of trips and total no of truck of type l used every day to carry waste of type g from an incinerator at j to a landfill at k.

z_j, z_k, z_r : Boolean variables indicating respectively, the presence of an incinerator at j, a landfill at k, and a trucks depot at r.

w_j, w_k : amount of waste transported everyday respectively, to an incinerator at j, and a sanitary landfill at k.

T_l : The number of trucks of type l used everyday.

T : The total number of trucks (excluding replacement trucks) used everyday.

Z_l : The number of replacement trucks of type l required everyday.

2.1   Input data

a_{ij}^{l}, a_{ik}^{l} : The expected number of trips a truck of type l can make respectively, per day between waste source at i and an incinerator at j, and a landfill at k.

b_{ik}^{l} : The expected number of trips a truck of type l can make respectively, per day between an incinerator at j, and a landfill at k.

α_l: capacity (in tonnes) of a truck of type l.

ρ_l: probability that a truck of type l breaks down in a day.

v_{ij}^l, v_{ik}^l : the transportation cost per unit of waste carried by a truck of type l from a waste source at i to an incinerator at j, and a landfill at k.

e_{lrk}, e_{lrj}, e_{lri}: respectively the cost of moving a truck of type l from a replacement trucks depot at r to a landfill at k, an incinerator at j, and a waste source at i.

d_{jk}^l : the transportation cost per unit of waste carried by a truck of type l from an incinerator at j, to a landfill at k.

v_j: revenue generated per unit of waste at an incinerator at j.

f_{cl} : the cost of repairing of truck of type l, $l = 1 \ldots , L$.

W_i : amount of waste at source i.

γ_j : fraction (%) of unrecovered waste respectively, at an incinerator at j that requires disposal to a landfill.

Q_j, Q_k, Q_r : capacity per day respectively, for an incinerator at j, a landfill at k, and a replacement trucks depot at r.

C_j, C_k, C_r : respectively fixed cost incurred in opening an incinerator at j, an a landfill at k, and a replacement trucks depot at r.

c_j, c_k : respectively variable cost incurred in handling a unit of waste at an incinerator at j, , and a landfill at k.

2.2   Objective Function

Minimized $I_{MI} = (I_1 + I_2 + I_3)_{MI} - A_{MI}$ $\hspace{2cm}$ (1)

I_1 $(z, w, X, Y) = \{$ Part I (Investment and management expenses) + part II (Transportation cost)

$$Part - I = [\sum_j (C_j z_j + c_j w_j) + \sum_k (C_k z_k + c_k w_k)] \hspace{1cm} (2)$$

$$Part - II = [\sum_{glij} v_{lij} \beta_{lijg} + \sum_{glik} v_{lik} \beta_{likg} + \sum_{gljk} w_{ljk} \lambda_{ljkg}] \hspace{1cm} (3)$$

The component I_2 gives expenses owing to the use of replacement trucks,

$$I_2 (n, z) = \sum_{rkl} e_{lrk} n_{lrk} + \sum_{rjl} e_{lrj} n_{lrj} + \sum_{ril} e_{lri} n_{lri} + \sum_r C_r z_r \hspace{1cm} (4)$$

Component I_3 gives the total cost for repairing of all trucks required in the daily management of waste.

$$I_3 (x, y, n) = \sum_l fc_l (T_l + Z_l) \hspace{1cm} (5)$$

Component A gives the benefits at the plants owing to the production of electric energy, compost and refuse derived fuel.

$$A(w) = \sum_j v_j (1 - \gamma_j) w_j \tag{6}$$

2.3 Constraints

In general, the constraints are the capacity, site selection, facility availability, environmental, and landfill saturation constraints. In constraint, we make sure that the total waste moved from each waste collection point i is at least be equal to the amount of waste found at that point.

$$\sum_{glj} \beta_{lijg} + \sum_{glk} \beta_{likg} \geq d_i \tag{7}$$

In this constraints, we guarantee that the amount of waste carried away from every plant to a landfill, is at least be equal to the amount of waste found at that plant.

$$\gamma_j w_j \leq \sum_{gkl} \lambda_{ljkg} \tag{8}$$

Next constraints, the maximum capacities for the processing plants are accounted for. These constraints mean that the amount of waste taken to these plants should not exceed the plant capacities.

$$w_j \leq Q_j z_j, \tag{9}$$

In constraint the same thing is done for sanitary landfills

$$t_k \leq Q_k z_k, \tag{10}$$

This Constraint means that the total number of replacement trucks of type l cannot be less than the expected number of daily truck breakdowns of the type l. With constraint, we ensure that there is at least one depot for the replacement trucks.

$$\sum_{rk} n_{lrk} + \sum_{rj} n_{lrj} + \sum_{ri} n_{lri} \geq \gamma_l T_l \tag{11}$$

$$\sum_r^R z_r \geq 1 \tag{12}$$

In constraint, we codify that the number of trucks in a depot cannot exceed its capacity and other constraint means that the total number of replacement trucks is not too big compared to the total number of trucks used per day.

$$\sum_{lk} n_{lrk} + \sum_{lj} n_{lrj} + \sum_{li} n_{lri} \leq Q_r z_r \tag{13}$$

$$\sum_r Q_r z_r \leq T \tag{14}$$

Constraints mean that once the flow to either plant or sanitary landfill is positive, that plant or landfill must actually exist.

$$\beta_{lijg} \leq Q_j z_j, \qquad \beta_{likg} \leq Q_k z_k, \qquad \lambda_{ljkg} \leq Q_k z_k, \tag{15}$$

Constraints can be referred to as waste flow fixing constraints. The reason is that when there are benefits at some node there is a tendency to move as much waste as possible to that node as long as there is space on the truck. In such a case, what is "carried" on the truck, that includes false waste, may go beyond the amount at a waste source; this is undesirable because the interest is in the precise amount of waste picked from the source.

$$\beta_{lijg} \leq W_i, \qquad \beta_{likg} \leq W_i, \qquad \lambda_{ljkg} \leq \gamma_j w_j, \tag{16}$$

2.4   Variables Conditions

The variables in constraints are defined as non-negative; these give the amount of waste that flows between various nodes.

$$\beta_{lijg} \geq 0, \qquad \beta_{likg} \geq 0, \qquad \lambda_{ljkg} \geq 0, \tag{17}$$

The variables in constraints are defined as non-negative integers. These give the number of trucks used between two nodes in the model per day, excluding replacement trucks.

$$x_{lijg}, \text{integer} \geq 0, \qquad x_{likg}, \text{integer} \geq 0, \qquad y_{ljkg}, \text{integer} \geq 0, \tag{18}$$

The variables in constraints are defined as non-negative integers. These give the number of replacement trucks required everyday in the waste management program. We note that the breakdown of a truck can occur anywhere in the road network followed by the trucks. For purposes of locating the truck depots, it is assumed that these breakdowns occur at either a waste collection point or at a plant or at a landfill.

$$n_{lrk}, \text{integer} \geq 0, \qquad n_{lrj}, \text{integer} \geq 0, \qquad n_{lri}, \text{integer} \geq 0, \tag{19}$$

The variables in are defined as boolean. These are used to determine the existence of either a plant or a landfill.

$$z_j \in [0,1], \qquad z_k \in [0,1], \qquad z_r \in [0,1], \tag{20}$$

In this constraints, we guarantee that the amount of waste carried away from every plant to a landfill, is at least be equal to the amount of waste found at that plant.

3   Illustrative Application of the Optimization Model

The usefulness of the optimization model represented by equations (1) to (20) in real field situation, data for the performance evaluation are taken according to the situation in Allahabad city in India shown in Figure 2. Allahabad, one of the holy cities of India lies at the confluence of Ganga, Yamuna and hidden Saraswati River. It is located at 25°25'N latitude and 80°58'E longitude and 81°58'E longitude at the height of 98.0 metres above the Mean Sea Level.. As per information provided by Allahabad Nagar Nigam (ANN)   [24], the entire city is divided into 80 municipal wards within 20 sanitary wards and generated 430 gram per capita per day. For the purpose of solid waste management, the Allahabad city is divided into 5 zones as shown in Figure 2. In this study, the major Zone 3 is considered the schematic representation of is shown in Figure 3 which comprises of 1 Source, 1 Incinerator, 1 Landfill and 1 Depot and covers 16 wards out of 80 wards. At the moment city does not have existing waste management component, so some of the parameters value is subjective.

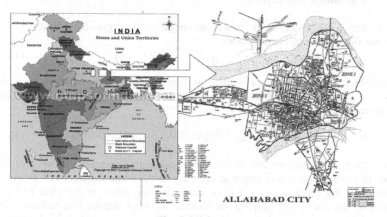

Fig: 2 Study Area

It is assumed that there is number of waste source (wards) in a zone. The waste specified generated in a particular ward is as termed as waste sources. The capacity of vehicle starts at a first waste source as a common collection point and visit to nearest waste source and pick up the waste, when a vehicle not full, it needs to go to the closest available waste source. When a vehicle full, it needs to return to the first waste source.

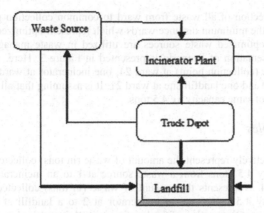

Fig.3 Schematic representation of scenario

4   Uncertainty   Characterization   in   the   Solid   Waste Management Model

In the presence of limited, inaccurate or imprecise information, simulation with fuzzy numbers represents an alternative to statistical tool to handle parametric uncertainty. Fuzzy sets offer an alternate and simple way to address uncertainties even for limited exploration data sets. In the present work, the optimal design is first obtained assuming a deterministic value of total waste amount in optimization model. Uncertainty in optimization is then characterized using fuzzy numbers for waste amount in the waste management model. Uncertainty characterization is based on Zadeh's extension principle [25]. In this study only total waste amount produced is considered to be imprecise. Waste amount as imprecise parameter, is represented by triangular fuzzy numbers with different α-cuts. The reduced TM [26] is used in the present study. The measure of uncertainty used is the ratio of the 0.1-level support to the value of which the membership function is equal to 1 [27].

5   Results and Discussion

The Vehicle routing model is able to solve optimality modest-sized zone 3 consisting of 16 wards for location of waste source. Rawal et al (2010) gives the total distance travelled for collection of all waste from ward to common collection point and w24 (66302.07 m) is the minimum distance wards which represent optimized waste source location. These optimized waste sources are utilized in waste management model. Schematic representation of Scenario is presented in Figure 3. Here, we considered one waste source (collection point) at ward 24, one incinerator at ward 53, one truck depot at ward 10 and one landfill site at ward 21. It is assuming that all the truck used to carry waste is of same capacity i.e 4.5 tons.

5.1   Variables

u12, u14 : respectively represent the amount of waste (in tons) collected everyday by trucks of capacity 4.5 tons from a waste source at 1 to an incinerator at 2, and a landfill at 4; v24 : represents the amount of waste (in tons) collected everyday by trucks of capacity 4.5 tons from an incinerator at 2 to a landfill at 4; x12, x14 : respectively represent the number of trucks of capacity 4.5 tons used everyday to carry waste from a waste source at 1 to an incinerator at 2, and to a landfill at 4; y24 : number of trucks of capacity 4.5 tons used everyday to carry waste from an incinerator at 2 to a landfill at 4; n31, n32, n34 : respectively represent the number of trucks of capacity 4.5 tons used everyday from a replacement trucks bank at 3 to a waste source at 1, an incinerator at 2, and a landfill at 4; w2 (= u12), t4 : respectively represent the amount of waste transported everyday to an incinerator at 2, and a landfill at 4.

5.2   Input data/Parameters used

10, 6 respectively are the expected number of trips (single trips) a truck of capacity 4.5 tons can make everyday from a waste source at 1 to an incinerator at 2, and a landfill at 4; 6 is the expected number of trips a truck of capacity 4.5 tons can make everyday between and incinerator at 2, and a landfill at 4; 0.28, 5.39 respectively are the transportation costs per ton of waste transported from a waste source at 1 to an incinerator at 2, and a landfill at 4; 5.36 is the transportation per ton of waste moved from an incinerator at 2 to a landfill at 4; 2.53, 2.83 and 5.56 respectively are the costs of moving a replacement truck of capacity 4.5 tons from a replacement trucks depot at 3 to a waste source at 1, an incinerator at 2, and a landfill at 4; 1500 is the revenue per unit of waste from an incinerator at 2; 108.04 is the amount of waste (in tons) at a waste source at 1; 0.30 is the fraction (%) of unrecovered waste at an incinerator at 2;75, 3, and 200 are the respective capacities for an incinerator at 2, a repairing of trucks depot at 3, and a landfill at 4; 800 and 44.56 are the respective costs of handling a ton of waste at an incinerator at 2, and a landfill at 4; 10000 is cost of repairing truck; 0.13 : probability that a truck breaks down in a day.

The result obtained by model runs for 0.3 fraction (%) of un-recovered waste is treated as base case model runs for both models. The objective function value of minimum cost is 43987.59 for MIL model with above input. The feasible solution obtained by MIL model is. $x12$ is 2 numbers, $x14$ is 2 numbers and $y24$ is 1 numbers respectively. Difference is only the solution obtained by MIL model $u12$ is 75 tons, $u14$ is 31.4 tons and of $v24$ is 23 tons of waste respectively.

5.3 Sensitivity Analysis

In order to demonstrate sensitivity of total cost with respect to fraction (%) of un-recovered waste (γ) at an incinerator at 2. Detail sensitivity analysis is performed by varying un-recovered waste γ varies over the different intervals from (0.0, 0.8) as shows in Figure 4. It is observed that the total cost falls with lower values of γ; this is because the lower the value of γ, the more efficient the plant is, and consequently the more benefits will be obtained. In other word increase in cost as fraction of un-recovered waste were found to increase in both the models. After the value of γ = 0.4 a constant cost of Rs. 55314.68 is obtained by mixed integer linear model. As the % of un-recovered waste increases, benefits from incinerator decreases and transportation cost from incinerator to landfill also increases. This cause overall increase in the cost.

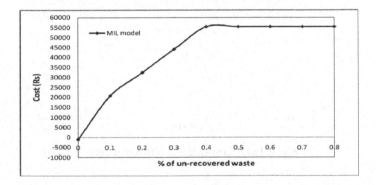

Fig. 4. Results of sensitivity analysis

5.4 Uncertainty Results

Fuzzy representation of waste amount is shown in Figure 5 with different α-cuts. The central value of triangular representation is 108 ton. The resulting output i.e. minimum cost obtained by the solution of optimization model is also fuzzy numbers characterized by their membership functions (Figure 6). It is found that for 15 percent uncertainty in waste amount, uncertainty is cost (for 0.3 % of unrecovered waste) is 26 percent.

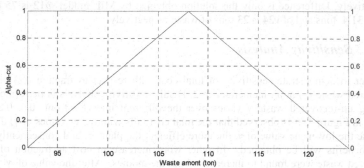

Fig.ure 5. Fuzzy representation of waste amount

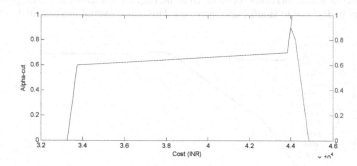

Fig. 6. Fuzzy cost values at different α-cuts

6   Conclusions

The mixed integer linear programming model presented in this work is suitable for MSW management for Indian cities like Allahabad in India. The route optimization is embedded in MSW management. In this way, the methodology presented here, demonstrate a comprehensive framework of MSW management. The sensitivity analysis indicated that the total cost lowered with high quality of incinerators (with lesser unrecovered wastes) in given case. The work certainly has advantages over previous approaches which uses a deterministic value of waste amount. Results of uncertainty analysis show that uncertainty in the optimal waste management (in terms of total optimal cost) is more than (approximately 1.5 times) uncertainty in waste amount. The application of the methodology to urban city like Allahabd shows the applicability of the methodology to other urban cities.

References

[1]   Ministry of Environment and Forests (MoEF),.The Gazette of India. Municipal Solid Waste (Management and Handling) Rules, New Delhi, India (2000).

[2]   Sharholy M., Ahmad K., Gauhar M., Trivedi R.C., , Municipal solid waste management in Indian cities – A review, Waste Management (2008). 28(2), 459-467.

[3]   Peavey, H.S., Donald, R.R. and Gorge, G.. Environmental Engineering. McGraw-Hill Book Co, Singapore (1985).

[4]   Caruso, C., Colorni, A., Paruccini, M., , The regional urban solid waste management system: A modelling approach, European Journal of Operational Research, (1993). 70, 16-30.

[5]   Ahluwalia, P. K., and Nema, A. K.. "Methodology for Estimating Optimum Reuse Time Span: Application to Reusable Computer Waste ." Journal of Practice Periodical of Hazardous, Toxic, and Radioactive Waste Management ASCE (2010), 14 (3), 178–184.

[6]   Tatarakis, A., Minis, I., , Stochastic single vehicle routing with a predefined customer sequence and multiple depot returns, European Journal of Operational Research (2009). 197(2), 557-571.

[7]   Koo, J.K., Shin, H.S., and Yoo, H.C.. "Multi-objective siting planning for a regional hazardous waste treatment center", Waste management and Research (1991). 9, pp 205-218.

[8]   Lee, Y.W., Bogardi, I., and Stansbury, J.. "Fuzzy decision making in dredged-material management", Journal of Environmental Engineering ASCE (1991). 117(5), pp 614-628.

[9]   Chang, N.B., and Wang, S.F.. "Managerial fuzzy optimal planning for solid waste management systems", Journal of Environmental Engineering ASCE (1996). 122(7), pp 649-657.

[10]   Chang, N.B., and Wang, S.F.. "A fuzzy goal programming approach for the optimal planning of metropolitan solid waste management systems", European Journal of Operation Research (1997). 99(2), pp 287-303.

[11]   Shukla, S.K. and Chandra, S. (1994). "A Generalized Mechanical Model for Geosynthetic-Reinforced Foundation Soil." Geo-textiles and Geo-membranes 13, pp 813-825.

[12]   Mahar, P.S. and Datta, B.. "Optimal identification of ground-water pollution sources and parameter estimation." Journal of Water Resources Planning and Management ASCE (2001). 127(1) pp 20– 29.

[13]   Mishra, A., Singh, R. and Raghuwanshi, N.S.. "Development and Application of an Integrated Optimization-Simulation Model for Major Irrigation Projects". Journal of Irrigation and Drainage Engineering (2005). 131(6), pp 504-513.

[14]   Zhang, K., Li., H., and Achari, G. "Incorporating Uncertain Site Information into Groundwater Modeling", The 12th International Conference of International Association for Computer Methods and Advances in Geomechanics (IACMAG),1-6 October, Goa, India, pp 2513-2520 (2008).

[15]   Babu, G.L. S., Reddy, K.R. and Chouksey, S.K.. "Constitutive model for municipal solid waste incorporating mechanical creep and biodegradation-induced compression" Waste Management (2010). 30 pp 11–22.

[16]   Fiorucci, P., Minciardi, R., Robba, M., Sacile, R., , Solid waste management in urban areas development and application of a decision support system, Resources, Conservation and Recycling (2003). 37, 301-328.

[17]   Shih L., Lin, Y., , Multi-criteria optimization for infectious medical waste collection system planning, Practice Periodical of Hazardous, Toxic, and Radioactive Waste Management System (2003). 7(2), 78-85.

[18]   Nema, A.K., Gupta, S.K., , Multi-objective risk analysis and optimization of regional Hazardous waste management system, Practice Periodical of Hazardous, Toxic, and Radioactive Waste Management System ASCE (2003). 7(2), 69-77.

[19]   Costi, P., Minciardi, R., Robba, M., Rovatti, M., Sacile, R., An environmentally sustainable decision model for urban solid waste management, Waste Management (2004). 24, 277-295.

[20]   Minciardi R., Paolucci M., Robba M., Sacile R., , Multi-objective optimization of solid waste flows: Environmentally sustainable strategies for municipalities, Waste Management (2008). 28(11), 2202-2212.

[21]   Papachristou, E., Hadjianghelou, H., Darakas, E., Alivanis, K., Belou, A., Ioannidou, D., Paraskevopoulou, E., Poulios, K., Koukourikou, A., Kosmidou, N., Sortikos, K., ,

Perspectives for integrated municipal solid waste management in Thessaloniki, Greece, Waste Management (2009). 29(3), 158-1162.

[22]   Rawal, N; Singh, R.M; and Vaishya R.C.. "Mathematical Models for Optimal Management of Solid Wastes in Urban Areas" In 3rd International Perspective on Current & Future State of Water Resources & the Environment International conference (ASCE) will be held on the Indian Institute of Technology (IIT) Madras on Jan,5-7, 2010.

[23]   Lingo 10.0, , User's Guide, Lindo Systems Inc, 1415 North Dayton Street, Chicago, Illinois, 60622. (2006)

[24]   Allahabad Nagar Nigam ANN, "A DPR report on Integrated Plan of MSW Management for the City of Allahabad" by M/s Tetra Tech India Ltd. Delhi (2007).

[25]   Zadeh, L. A.: Fuzzy algorithms. Information and Control 12, 94-102 (1968)

[26]   Hanss, M., Willner, K.: On using fuzzy arithmetic to solve problems with uncertain model parameters. In Proc. of the Euromech 405 Colloquium, Valenciennes, France, 85-92 (1999)

[27]   Abebe, A.J., Guinot, V., Solomatine, D.P.: Fuzzy alpha-cut vs. Monte Carlo techniques in assessing uncertainty in model parameters. 4th Int. Conf. Hydroinformatics, Iowa, USA. (2000)

Artificial Bee Colony Based Feature Selection for Motor Imagery EEG Data

Pratyusha Rakshit[1], Saugat Bhattacharyya[2], Amit Konar[1], Anwesha Khasnobish[2], D.N.Tibarewala[2], R. Janarthanan[1]

[1] Dept. of Electronics & Telecommunication Engg., Jadavpur University, Kolkata-700032

[2] School of Bioscience & Engg., Jadavpur University, Kolkata-700032

{ pratyushar1@gmail.com; saugatbhattacharyya@gmail.com; konaramit@yahoo.co.in; anweshakhasno@gmail.com; biomed.ju@gmail.com; srmjana_73@yahoo.com }

Abstract. Brain-computer Interface (BCI) has widespread use in Neuro-rehabilitation engineering. Electroencephalograph (EEG) based BCI research aims to decode the various movement related data generated from the motor areas of the brain. One of the issues in BCI research is the presence of redundant data in the features of a given dataset, which not only increases the dimensions but also reduces the accuracy of the classifiers. In this paper, we aim to reduce the redundant features of a dataset to improve the accuracy of classification. For this, we have employed Artificial Bee Colony (ABC) cluster algorithm to reduce the features and have acquired their corresponding accuracy. It is seen that for a reduced features of 200, the highest accuracy of 64.29%. The results in this paper validate our claim.

Keywords: Brain-computer Interface, Electroencephalography, Motor Imagery, Feature Selection, Power Spectral Density, Artificial Bee Colony.

1 Introduction

Brain-computer interface (BCI) is the fastest emerging trend in neuro-rehabilitation [1]. Their functions are to decode the bio-potential signals obtained from the different region of the brain for various applications, like in robotics, communication, and gaming [2, 3]. It finds its greatest use in the rehabilitation of persons suffering from paralysis, Amyotropic Lateral Sceloris (ALS), loss of limb and like [4], [5]. These bio-potential signals are extracted, decoded and studied with the help of various brain measures like Magnetoencephaography (MEG), functional Magnetic Resonance Imaging (fMRI), Electro-oculography (ECoG) and Electroencephalography (EEG) [6].

J. C. Bansal et al. (eds.), *Proceedings of Seventh International Conference on Bio-Inspired Computing: Theories and Applications (BIC-TA 2012)*, Advances in Intelligent Systems and Computing 202, DOI: 10.1007/978-81-322-1041-2_11, © Springer India 2013

The brain signals recorded using EEG are non-linear, non-stationary, complex and non-Gaussian. EEG recording is preferred to other modalities as it is portable, easy to use, inexpensive, and has a higher temporal resolution. EEG acquired during motor imagery from the motor cortex area of the brain can be used to drive assistive devices. Motor imagery signals are obtained when a person has an intention for any sort of action in form of movement, which is found in the alpha (8-12 Hz) and central Beta (16-25 Hz) band [7]. Thus, motor imagery data are obtained from the C3 and C4 electrodes whose locations are directly above the motor cortex area of the brain [8]. The basic BCI module consists of the following steps: Preprocessing of the signal, Feature Extraction and Classification. The classified results lead to the generation of the control signals required to drive an assistive device. The main concern in BCI research has been the high dimensionality of the features and the selection of relevant features, such that they have the highest discriminability[9-11]. Often it is observed that due to the presence of a large number of redundant features in the feature set, the accuracy of the classifier is greatly reduced. In this regard, another module is added along with the Feature Extraction module, which is known as *Feature Selection*. Commonly, used Feature Selection techniques are Principal Component Analysis [12], Singular Value Decomposition [13], and Independent Component Analysis [14].

In this study, we aim to reduce the size of the features from their original sizes with an aim to improve the accuracy of the features to correctly differentiate among the various data-points of the complete dataset to their respective classes. For this task we have employed the use of Artificial Bee Colony Algorithm. It is shown that Artificial Bee Colony (ABC) [15], inspired from stochastic behavior of foraging in bees, with a modification of food source representation scheme, can give very promising results if applied to the clustering problem. We here apply the algorithm to the clustering problem. Although any stochastic optimization algorithm, such as genetic algorithm (GA), particle swarm optimization technique (PSO), differential evolution (DE) and the like could have been used for the problem, we have selected ABC because of its faster convergence and qualitative time-optimal solution [16].

In the proposed evolutionary learning framework, a number of trial solutions come up with different pre-defined number of features as well as cluster center coordinates for the same data set. Correctness of each possible grouping is quantitatively evaluated with a global validity index (e.g., the CS measure [17]). Then, through a mechanism of mutation and natural selection, eventually, the best solutions start dominating the population, whereas the bad ones are eliminated. Ultimately, the evolution of solutions comes to a halt (i.e., converges) when the fittest solution represents a near-optimal partitioning of the data set with respect to the employed validity index. In this way, the optimal pre-defined number of features along with the accurate cluster center coordinates can be located in one run of the evolutionary optimization algorithm.

The rest of this paper is organized as follows. Section 2 defines the clustering problem in a formal language. Section 3 outlines the proposed ABC-based clustering algorithm. Section 4 describes the real data sets used for experiments and gives a detail of the filtering and feature extraction technique applied. Results of clustering over real-life data sets are presented in Section 5. Conclusions are provided in Section 6.

2 Formulation of the Problem

2.1 Problem Definition

A *pattern* is a physical or abstract structure of objects. It is distinguished from others by a collective set of attributes called *features*, which together represent a pattern [18, 19]. Let $X_{N \times D} = \left\{ \vec{X}_1, \vec{X}_2, ..., \vec{X}_N \right\}$ be a set of N patterns or data points, each having D features. Given such $X_{N \times D}$ matrix, a partitional clustering algorithm tries to find out a partition $C = \left\{ C_1, C_2, ..., C_K \right\}$ of K classes, such that the similarity of the patterns in the same cluster is maximum and patterns from different clusters differ as far as possible. The partitions should maintain three properties

- Each cluster should have at least one pattern assigned, i.e.,
  $C_i \neq \Phi, \forall i \in \left\{ 1, 2, ..., K \right\}$.

- Two different clusters should have no pattern in common, i.e.,
  $C_i \cap C_j \neq \Phi, \forall i \neq j$ and $i.j \in \left\{ 1, 2, ..., K \right\}$.

- Each pattern should definitely be attached to a cluster i.e., $\bigcup_{i=1}^{K} C_i = X$.

2.2 Similarity Measure

Clustering is the process of recognizing clusters in multidimensional data based on some similarity measures [18, 19]. The most popular way to evaluate similarity between two patters amounts to the use of a *distance measure*. The most widely used distance measure is the Euclidean distance, which between any two d-dimensional patterns $\vec{X}_i$ and $\vec{X}_j$ is given by

$$d(\vec{X}_i, \vec{X}_j) = \left\| \vec{X}_i - \vec{X}_j \right\| = \sqrt{\sum_{p=1}^{D} \left(X_{i,p} - X_{j,p} \right)^2} \tag{1}$$

2.3 Clustering Validity Index: CS Measure

Clustering validity index corresponds to the statistical-mathematical functions used to evaluate the results of a clustering-algorithm on a quantitative basis [18]. It should take care of two aspects of partitioning.

- Cohesion: The patterns in one cluster should be as similar to each other as possible.
- Separation: Clusters should be well-separated.

Recently, Chou et al. have proposed the CS measure [17] for evaluating the validity of a clustering scheme. Let $\bar{m}_i$ be the centroid of i-th data cluster. The CS measure is then defined as

$$CS(K) = \frac{\sum\limits_{i=1}^{K}\left[\dfrac{1}{N_i}\sum\limits_{\vec{X}_i \in C_i}\max\limits_{\vec{X}_q \in C_i}\left\{d(\vec{X}_i,\vec{X}_q)\right\}\right]}{\sum\limits_{i=1}^{K}\left[\min\limits_{j \in K, j \neq i}\left\{d(\vec{m}_i,\vec{m}_q)\right\}\right]} \tag{2}$$

As can be easily be perceived, this measure is a function of the ratio of the sum of within-cluster-scatter to between-cluster separation.

3 Artificial Bee Colony (ABC)-Based Clustering

3.1 Artificial Bee Colony Algorithm

In ABC algorithm, the colony of artificial bees contains three groups of bees:
- A bee waiting on a dance area for making decision to choose a food source is called an *onlooker* bee.
- A bee going to the food source visited by it previously is named as *employed* bee.
- A bee carrying out random search is called a *scout* bee.

In ABC algorithm, the position of a food source represents a possible solution of the optimization problem and the nectar amount of a food source corresponds to the fitness of the associated solution. The number of employed bees and onlooker bees is equal to the number of solutions in the population.

ABC consists of following steps.

A. Initialization

ABC generates a randomly distributed initial population P ($g=0$) of NP solutions (food source position) where NP denotes the size of population. Each solution $\vec{Z}_i$ ($i= 0, 1, 2\ldots$ NP-1) is a "dim" dimensional vector.

B. Placement of employed bees on the food sources in memory

An employed bee produces a modification on the position (solution) in her memory depending on the local information (visual information) as stated by equation (4) and tests the nectar amount of the new source. Provided that the nectar amount of the new one is higher than that of the previous one, the bee memorizes the new position and forgets the old one. Otherwise, she keeps the position of the previous one in her memory.

C. Placement of onlooker bees on the food sources in memory

After all employed bees complete the search process; an onlooker bee evaluates the nectar information from all employed bees and chooses a food source depending on the probability value associated with that food source, p_i, calculated by the following expression:

$$p_i = \frac{fit_i}{\sum_{j=0}^{NP-1} fit_j} \tag{3}$$

Here fit_i is the fitness value of the solution i evaluated by its employed bee. After that, onlooker bee produces a modification on the position in her memory and checks the nectar amount of the candidate source. Providing that its fitness is better than that of the previous one, bee memorizes the new position.

A new food source $\vec{Z}_i' = \left(z_{i0}, z_{i1}, ..., z_{i(j-1)}, z_{ij}', z_{i(j+1)}, ..., z_{i(dim-1)}\right)$ in the neighborhood of food source $\vec{Z}_i = \left(z_{i0}, z_{i1}, ..., z_{i(dim-1)}\right)$ has being generated by altering the value of one randomly chosen solution parameter j and keeping other parameters unchanged. The value of z_{ij}' parameter in $\vec{Z}_i'$ solution is computed using the following expression:

$$z_{ij}' = z_{ij} + u \times (z_{ij} - z_{kj}) \tag{4}$$

Here u is a uniform variable in $[-1, 1]$ and k is any number between 0 to NP-1 but not equal to i. If a parameter produced by this operation exceeds its predetermined limit, the parameter can be set to an acceptable value.

D. Send scout bee to search food source in memory

In the ABC algorithm, if a position cannot be improved further through a predefined number of cycles called 'limit', the food source is abandoned. This abandoned food source is replaced by the scouts by randomly producing a position.

After that again steps (B), (C) and (D) will be repeated until the stopping criteria is met.

3.2 Food Source Representation

In the proposed method, for N data points, each D dimensional, and user-specified maximum number of features $d \in [1, D]$, a food source is a vector of real numbers of dimension $d + d \times K$ for K number of clusters. The first d entries of

$\vec{Z}_i$ are positive integers numbers in [1, D]. The value of $Z_{i,j}=p$ ($p \in [1, D]$) indicates that the p-th feature is activated i.e., to be really used for classifying the data. The remaining entries are reserved for K cluster centers, each d dimensional. For example, the food source $\vec{Z}_i$ is shown in the following equation.

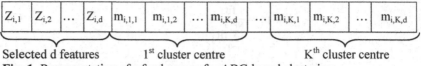

Selected d features 1st cluster centre K^{th} cluster centre

Fig. 1. Representation of a food source for ABC-based clustering

3.3 Fitness Function Evaluation

In the proposed clustering method using ABC, CS clustering validity measure is used for fitness function evaluation. The CS-measure based fitness function for i-th food source with K clusters can be described as

$$\text{fit}(\vec{Z}_i) = \frac{1}{CS_i(K) + eps} \tag{5}$$

Here, *eps* is very small positive integer constant.

3.4 Pseudo Code

Input: A set of N patterns each with D features $X_{N \times D}$, maximum number of activated features d, number of clusters K and algorithm parameter "limit".

Output: A set of d features selected from given D number of features and the class level of N patterns belong_to_cluster.

Procedure ABC_cluster ($X_{N \times D}$)

For i=1to NP

    Initialize $\vec{Z}_i$ to contain d number of randomly selected features and K randomly chosen cluster centroids as shown in Fig. 1.

    Set trial$_i$=0.

    fit ($\vec{Z}_i$)= Evaluate_cost ($\vec{Z}_i$).

End For.

For g=1 to Gmax

    For each i-th **employed bee**

        Produce a new food source $\vec{Z}_i'$ as n (4).

        fit ($\vec{Z}_i'$)= Evaluate_cost ($\vec{Z}_i'$).

        If fit ($\vec{Z}_i'$)>fit ($\vec{Z}_i$)

$\vec{Z}_i = \vec{Z}_i'$.

fit ($\vec{Z}_i$)= fit ($\vec{Z}_i'$).

Set $trial_i=0$.

Else

Set $trial_i= trial_i+1$.

End If.

End For.

For each t-th **onlooker bee**

Select a food source $\vec{Z}_i$ to be modified based on its probability as given in (3).

Produce a new food source $\vec{Z}_i'$ as n (4).

fit ($\vec{Z}_i'$)= Evaluate_cost ($\vec{Z}_i'$).

If fit ($\vec{Z}_i'$)>fit ($\vec{Z}_i$)

$\vec{Z}_i = \vec{Z}_i'$.

fit ($\vec{Z}_i$)= fit ($\vec{Z}_i'$).

Set $trial_i=0$.

Else

Set $trial_i= trial_i+1$.

End If.

End For.

$\vec{Z}_{best} = arg(arg(ma x(fit(\vec{Z}_1), fit(\vec{Z}_2),..., fit(\vec{Z}_{NP}))))$.

Set maxtrial=arg(max($trial_1$, $trail_2$, …, $trial_{NP}$)).

Set maxtrial_index=arg(arg(max($trial_1$, $trail_2$, …, $trial_{NP}$))).

If maxtrial>limit

Reinitialize the food source $\vec{Z}_{max\ trial\_index}$ by scout bee.

End If.

End For.

For each data vector $\vec{X}_p$ calculate its distance metric $d(\vec{X}_p, \vec{m}_{best,j})$, $\forall j \in [1, K]$

from all K cluster centres represented by food source $\vec{Z}_{best}$ using the selected d features.

Set

belong_to_cluster(p) = k if $d(\vec{X}_p, \vec{m}_{best,k}) = \min_{\forall j\in[1,K]} \{d(\vec{X}_p, \vec{m}_{best,j})\}$ $\forall p \in [1, NP]$.

Return belong_to_cluster.

Procedure Evaluate_cost ($\vec{Z}_i$)

For each data vector $\vec{X}_p$ calculate its distance metric $d(\vec{X}_p, \vec{m}_{i,j})$, $\forall j \in [1, K]$

from all K cluster centres represented by food source $\vec{Z}_i$ using the selected d features.

Set

$$\text{belong\_to\_cluster}(p) = k \quad \text{if } d(\vec{X}_p, \vec{m}_{i,k}) = \min_{\forall j \in [1,K]} \left\{ d(\vec{X}_p, \vec{m}_{i,j}) \right\} \quad \forall p \in [1, NP].$$

Evaluate the value of CS measure of the clusters represented by food source $\vec{Z}_i$ as in (2).

Determine $\text{fit}(\vec{Z}_i)$ as in (5).

Return $\text{fit}(\vec{Z}_i)$.

4 Data Analysis

4.1 The Dataset

This dataset was provided by Fraunhofer FIRST, Intelligent Data Analysis Group and Campus Benjamin Franklin of the Charité-Univeristy Medicine Berlin, Department of Neurology, and Neurophysics Group [20]. We have used the data given from subject *aa* for our study. This dataset contains only data from the 4 initial sessions without feedback. Visual Cues indicated for 3.5 seconds which of the two motor imageries the subject should perform: Right hand (Class 1) and Right Foot (Class 2). The presentation of target cues was followed by periods of relaxation of random length, 1.75 to 2.25 seconds.

4.2 Experimental Setup

The recording of the EEG signal has been done using BrainAmp Amplifiers and a 128 Ag/AgCl electrode cap from ECI. 118 channels were used for the measurement of the EEG signals at positions of the international 10/20 electrode system. Signals were bandpass filtered between 0.05 and 200Hz and digitized at 1000Hz with 16bit accuracy, which was further down sampled to 100Hz to obtain the final dataset [20].

4.3 Preprocessing

For our study, C3 and C4 electrodes are taken. Before the signals are fed to the feature extraction algorithms, data from C3 and C4 were filtered using Laplace Filtering, as shown in (6). In this technique, the average of the neighborhood electrodes are subtracted from each individual channels.

$$C3 = C3 - \frac{1}{8}(FCC\,5 + FC\,3 + C1 + CCP\,3 + CP\,3 + CCP\,5 + C5 + FCC\,5) \tag{6}$$

$$C4 = C4 - \frac{1}{8}(FCC4 + FC4 + FCC6 + C6 + CCP6 + CP4 + CCP4 + C2 + FCC4)$$

4.4 Power Spectral Density

In this study, we have used Welch based Power Spectrum Density Technique [21], [22], to prepare the initial feature set. Spectrum estimation describes the power distribution contained in a signal over frequency based on a finite set of data. For our study, we have used the Welch's Periodogram for the spectral estimation of the EEG data. Here, the data segments are overlapped and windowed prior to the calculation of the periodogram.
Let the j-th data segment be denoted as

$$y_j(t) = y((j-1)k + 1) : t = 1, \ldots, M \text{ and } j = 1, \ldots, S.$$

(j-1)k is the starting point for the j-th sequence of data.

The windowed periodogram corresponding to $y_j(t)$ given by

$$\hat{\phi}_j(\omega) = \frac{1}{MP} \left| \sum_{t=1}^{M} v(t) y_j(t) e^{-i\omega t} \right|^2 \tag{7}$$

Here P denotes the power of the temporal window $\{v(t)\}$

$$P = \frac{1}{M} \sum_{t=1}^{M} |v(t)|^2 \tag{8}$$

The Welch estimate of PSD is determined by averaging the windowed periodogram in (8)

$$\hat{\phi}_W(\omega) = \frac{1}{S} \sum_{j=1}^{S} \hat{\phi}_j(\omega) \tag{9}$$

Welch method allows the overlap between data segments gets more periodograms to be arranged in (9); thus decreases the variance of the estimated PSD. More control over the bias/resolution properties of the estimated PSD is obtained by introducing the window in the periodogram computation. Welch method is more effective in reduction of variance via averaging in (9). The windowed periodograms in Welch method offer more flexibility in controlling the bias properties of the estimated spectrum. These characteristics make this method highly suitable for analysis of a non-stationary signal.

Here, the frequency range was taken from 8Hz to 25Hz, to include the mu-rhythm and central beta rhythm. The PSD was obtained separately for both the electrodes C3 and C4. The total number of estimates obtained for each electrode was 350. Thus, the total size of the feature set was 350 × 2=700.

5 Results

The features extracted in the previous section are used as inputs to the ABC clustering algorithm. The target of the clustering algorithm is to differentiate among the various data-points correctly to their respective classes, that is, Right hand (Class 1) and Right Foot (Class 2). As per the requirements of our problem, we have reduced the feature set from its original dimension of 700 to various dimensions smaller than 700 (as shown in Table 1). Table 1 shows that the average accuracy for d=50 to d=300 is more than d=700, but the accuracy from d= 350 to d=650 is lower than d=700 in most cases. It clearly signifies that from d=350 onwards, the feature set contains more redundant features which affects the accuracy of the dataset. Thus, it is observed that there is an improvement of accuracy for d < N/2, where N is the original size of the feature set, where it can be stated that ABC has increased the number of differentiable (relevant) features and reduced the number of redundant features in the complete feature set. It is observed that for d=200, the algorithm gives the best average accuracy of 64.29% and has obtained the highest accuracy of 68.67% in its 20 run.

We have also compared our result with Harmonic Search (HS) based clustering and the comparative results are shown in Table 2. It is observed that ABC has a significantly higher accuracy compared to that of HS. HS has obtained a highest accuracy of 55.17% while ABC has obtained a 64.29% in this regard.

Table 1. Accuracy level obtained using ABC-based clustering

Dimension d	Best Accuracy (in %)	Average Accuracy over 20 Runs (in %)
50	64.71	63.22
100	63.86	62.14
150	65.50	60.17
200	68.67	64.29
250	64.64	62.11
300	62.36	61.82
350	59.29	55.95
400	58.57	56.39
450	60.00	53.18
500	61.79	60.22
550	59.64	58.25
600	58.93	55.43
650	62.07	57.33
700	62.07	57.58

Table 2.1. Accuracy level obtained using ABC and HS-based clustering for d=50-150

Dimension d	Best Accuracy (in %)	Best Accuracy (in %)
50	64.71	44.28
100	63.86	55.71
150	65.50	55.71

Table 2.2. Accuracy level obtained using ABC and HS-based clustering for d=200-650

Dimension d	Best Accuracy (in %)	Best Accuracy (in %)
200	68.67	44.28
250	64.64	44.28
300	62.36	44.28
350	59.29	44.28
400	58.57	44.28
450	60.00	44.28
500	61.79	44.28
550	59.64	55.71
600	58.93	55.71
650	62.07	44.28

We have also compared our result with Harmonic Search (HS) based clustering and the comparative results are shown in Tables 2.1 and 2.2. It is observed that ABC has a significantly higher accuracy compared to that of HS. HS has obtained a highest accuracy of 55.17% while ABC has obtained a 64.29% in this regard.

6 Conclusion

This paper proposes a novel technique for feature selection technique based on a clustering algorithm. Our proposed approach is validated on a dataset using Power Spectral Density as the feature and Artificial Bee Colony as the clustering algorithm. The results thus obtained have justified our claim that an improvement of accuracy is observed when the dataset is reduced to half of its original size, containing mostly the relevant features. Simultaneously the computational complexity has also been reduced. Further study in this direction will aim to optimize the feature selection, extraction and classification techniques to be implemented in the online classification of the EEG data for BCI research, and thus to ultimately develop a complete stand-alone system for an EEG driven neuro-prosthetic control for rehabilitation purpose.

References

[1] Nijholt Anton, Tan Desney, "Brain computer interfacing for intelligent systems," IEEE intelligent systems, 2008

[2] Vaughan T.M., Heetderks W.J., Trejo L.J., Rymer W.Z., Weinrich M., Moore M.M., Kubler A., Dobkin B.H., Birbaumer N., Donchin E., Wolpaw E.W., Wolpaw J.R. " Brain computer interface technology: A review of the second international meeting", IEEE Trans. Neural Syst. Rehab. Eng. 11(2), June 2003, 94-109

[3] Wolpaw J.R., Birbaumer N., Heetderks W.J., McFarland D.J., Peckham P.H., Schalk G., Donchin E., Quatrano L.A., Robinson C.J., Vaughan T.M. " Brain computer inter-

face : A review of the first international meeting", IEEE Trans. Rehabilitation Eng. 8(2), June 2000, 164-173

[4] Daly Janis J, Wolpaw Jonathan R , "Brain–computer interfaces in neurological rehabilitation", Lancet Neurol 2008; 7: 1032–43

[5] Schwartz A.B., Cui X.T., Weber D.J., Moran D.W. "Brain Controlled Interfaces: Movement Restoration using Neural Prosthetics." Neuron 52, October 2006, 205-220

[6] Anderson R.A., Musallam S., Pesaran B., "Selecting the signals for a brain-machine interface", Curr Opin Neurobiol 14(6), December 2004, 720-726

[7] Pfurscheller G., Neuper C., "Motor imagery activates primary sensorimotor area in humans", Neuroscience Letters 239, December 1997, 65-68

[8] Pfurscheller G., Neuper C., Schlogl A., Lugger K. "Separability of EEG signals recorded during right and left motor imagery using Adaptive Autoregressive Parameters." IEEE transaction on rehabilitation engineering 6 (3), September 1998, 316-325.

[9] S. Theodoridis and K. Koutroumbas, Pattern Recognition, 3rd ed. Academic Press, 2006.

[10] A. Rakotomamonjy, V. Guigue, G. Mallet, and V. Alvarado, "Ensemble of svms for improving brain computer interface", International Conference on Artificial Neural Networks, 2005, pp 300.

[11] F Lotte, M Congedo, A Lecuyer, F Lamarche and B. Arnaldi, "A Review of Classification Algorithms for EEG-based Brain-Computer Interfaces", Journal of Neural Engineering 4, 2007.

[12] H. Abdi and L.J. Williams, "Principal component analysis.". Wiley Interdisciplinary Reviews: Computational Statistics, 2: 433–459, 2010.

[13] P. C. Hansen, "The truncated SVD as a method for regularization". BIT 27: 534–553, 1987.

[14] Pierre Comon, "Independent Component Analysis: a new concept", Signal Processing, 36(3):287–314, 1994.

[15] Karaboga, D.: An idea based on honey bee swarm for numerical optimization. Technical Report-TR06, Erciyes University, Engineering Faculty, Computer Engineering Department, 2005.

[16] Basturk. B., Karaboga, D.: An artificial bee colony (ABC) algorithm for numeric function optimization. Proceedings of the IEEE Swarm Intelligence Symposium 2006, Indianapolis, Indiana, USA, 12-14 May 2006.

[17] Chou, C. H., Su, M. C., Lai, E.: A new cluster validity measure and its application to image compression. Pattern Analysis Application, vol. 7, no. 2, pp. 205–220, Jul. 2004.

[18] Halkidi, M., Vazirgiannis, M.: Clustering validity assessment: Finding the optimal partitioning of a data set. Proceedings of IEEE ICDM, San Jose, CA, 2001, pp. 187–194.

[19] Das, S., Abraham, A., Konar, A.: Automatic clustering using an improved differential evolution algorithm. IEEE Transactions in Systems, Man and Cybernetics, Part-A, January, 2008.

[20] BCI Competition III: http://www.bbci.de/competition/iii/

[21] Herman P., Prasad G., McGinnity T.M., Coyle D. "Comparative analysis of spectral approaches to feature extraction for EEG-based motor imagery classification." IEEE Trans. Neural sys. Rehab eng. 16(4), August 2008, pp. 317-326

[22] D.G. Childers, ed., Modern Spectrum Analysis, New York: IEEE Press, 1978

Effect of MT and VT CMOS, On Transmission gate Logic for Low Power 4:1 MUX in 45nm Technology

Meenakshi Mishra[1], Shyam Akashe[2], Shyam Babu[3]

[1] ECED ,ITM University Gwalior, India
[2] ECED ,ITM University Gwalior, India
[3] ECED ,ITM University Gwalior, India

{mishra.meenakshi13@gmail.com; shyam.akashe@itmuniversity.ac.in; itm.shyam@gmail.com}

Abstract. This paper describes the influence of leakage reduction techniques on 4:1 Multiplexer. The techniques investigated in this paper include multi-threshold (MTCMOS) and variable-threshold (VTCMOS). Impact of temperature sensitivity on power consumption is also evaluated. The CMOS transmission gate logic (TGL) is used to design a new 4:1 MUX, based on this design, it removes the degraded output, the NMOS and PMOS are combined together for strong output level with the gain in area is a central result of proposed MUX. The designed circuit is realized in 45 nm technology, with the power dissipation of 1.35pW from a 0.7V supply voltage. The MUX can operate well up to 200 Gb/s.

Keywords: Multiplexers, Low Power, transmission gate, Leakage Current, MTCMOS, VTCMOS

1 Introduction

A data multiplexer (MUX) is a key block in high-speed data communication systems. The acronym used for Multiplexer is MUX. The MUX is the heart of arithmetic circuit. MUX are a common building block for data paths and data-switching structures, and are used effectively in a number of applications including processors [2], processor buses, network switches, and DSPs with resource sharing. Several MUX circuits have been reported in technologies such as SiGe, GaAs and InP at speeds of 10 Gb/s or higher [6]-[8].Multiplexer (MUX) has become the bottleneck of speed. The speed of MUX determines the performance of the whole optic-fiber transceiver. DG-CNTFET structure uses specific properties, with structures based on conventional CMOS circuit design techniques. [3]

J. C. Bansal et al. (eds.), *Proceedings of Seventh International Conference on Bio-Inspired Computing: Theories and Applications (BIC-TA 2012)*, Advances in Intelligent Systems and Computing 202, DOI: 10.1007/978-81-322-1041-2_12, © Springer India 2013

Star-junction topology with a resonating junction is proposed for multiplexer circuit. [1]. The high-speed MUX is designed, and CMOS technology has been verified to be feasible for high speed MUX with a date rate far beyond 10 Gb/s[10], [11]. However, the power consumption is a bothering problem along with the operating rate rising, since current-model logic (CML) has to be used mostly. The reported MUX with a data rate below 5 Gb/s [3], [4] also has low power efficiency, even though CMOS logic was applied. A well-known tree-type architecture [5] is adopted for the 4:1 MUX, and high-speed with low-power dissipation can be achieved through applying dynamic CMOS logic and eliminating dispensable impedance matching.

With successive technology scaling, device feature sizes and supply voltage have shrunk to recover manufacturing cost and power of VLSI circuits. Whereas dynamic power has been recede due to the supply voltage decrease, leakage current has extremely intensified due to threshold voltage (shortly Vt) and feature size scaling down. Hence, leakage current is acquired as a major source in total power dissipation [19], and it is a key to achieve low power design, especially for mobile applications. The various approaches have been proposed to reduce power consumption of MUX trees. Some of the papers contract it at the algorithm level [13]-[15] and some at the circuit level [16]-[17]. Instantly, a new functional CMOS device called Variable Threshold Voltage MOSFET (VTCMOS) has affirmed to be, throughout the next generation of ultra-low power devices operating at low supply voltage [12][20-21].

2 MULTIPLEXER

Multiplexers are used as one method of reducing the number of integrated circuit packages required by a particular circuit design. This in turn reduces the cost of the system.

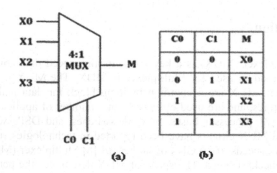

Fig. 1. Graphical Symbol 4:1 MUX.

Output=X0.$\overline{C_0}$.$\overline{C_1}$+X1.$\overline{C_0}$.C1+X2.C0.$\overline{C_1}$+X3.C0.C1   .....(1)

Output of 4:1 MUX can be calculated from equation 1.Assume that we have four lines, X0, X1,X2 and X3,which are to be multiplexed on a single line, Output The four input lines are also known as the Data Inputs. Since there are four inputs, we will need two additional inputs to multiplexer, known as the Select Inputs, to select which of the X inputs is to appear at the output, called as select lines C0 and C1. The graphical symbol (a) and truth table (b) of 4:1 MUX is shown in fig.1. Output of 4:1 can be calculated from equation 1. A multiplexer performs the function of selecting the input on any one of 'n' input lines and feeding this input to one output line.

2.1 Operation of Transmission Gate

This section describes the purpose and basic operation of a transmission gate. A transmission gate is defined as an electronic element that will selectively block or pass a signal level from the input to the output. The solid-state-switch is comprised of parallel connection of a PMOS transistor and NMOS transistor. The control gates are biased in a complementary manner so that both transistors are either ON or OFF.

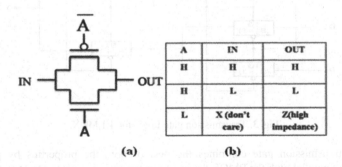

A	IN	OUT
H	H	H
H	L	L
L	X (don't care)	Z(high impedance)

(a)   (b)

Fig. 2. Transmission gate graphical symbol (a) ,truth table (b)

When the voltage on node A is a Logic 1, the complementary Logic 0 is applied to node active-low A, allowing both transistors to conduct and pass the signal at IN to OUT. When the voltage on node active-low A is a Logic 0,the complementary Logic 1 is applied to node A., turning both transistors off and forcing a high-impedance condition on both the IN and OUT nodes. The schematic diagram (Fig.2) Includes the arbitrary labels for IN and OUT, as the circuit will operate in an identical manner if those labels were reversed. The Transmission Gate graphical symbol and truth table is shown in figure 2.

2.2 Transmission CMOS logic 4:1 MUX

The transmission gate based 4:1 MUX is designed in Fig. 3.This design is the transmission gate type of MUX structure implemented with very minimum transistors compare to conventional CMOS based design. The design is implemented with minimum number of transistor. The back to back connected PMOS and NMOS arrangement acts as a switch is so called Transmission Gate. NMOS devices pass a strong 0, but a weak 1, while PMOS pass a strong 1, but a weak 0.

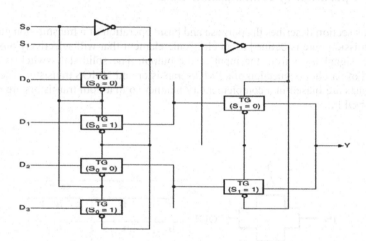

Fig. 3. Transmission gate Logic for 4:1 MUX.

The transmission gate combines the best of both the properties by placing NMOS in parallel with the PMOS device. Four transmission gates are connected as in Fig. 2 to form a MUX structure. Each transmission gate acts as an AND switch to replace the AND logic gate which is used in a conventional gate design of MUX. Hence the device count is reduced. The Transmission gate based 4:1 MUX is shown in figure 3.An advantage of the new MUX design is the remarkable gain in terms of transistors count. To the best of our knowledge, no 4:1 MUX has been realized with so few devices. Hence the gain in area is a central result for the proposed MUX.

2.3 *Temperature effect on Power Consumption*

Temperature dependence of leakage power is important, since digital very large scale integration circuits generally operate at elevated temperatures due to the power dissipation of the circuit. The power consumption increases with the rise in temperature. The effect of temperature on power consumption in TGL based 4:1 MUX is shown in Fig.4.

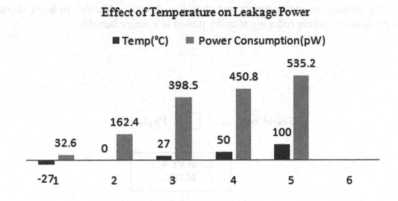

Fig. 4 .Effect of Temperature on TGL based 4:1 MUX.

3 LEAKAGE REDUCTION TECHNIQUES

In this section, two major circuit design techniques namely, MTCMOS, VTCMOS for leakage reduction in digital circuits are described:

3.1 *MTCMOS (Multi-threshold CMOS)*

Multi-Threshold CMOS (MTCMOS) is a popular power gating approach that uses high Vth devices for power switches [23]. Fig. 5 shows the basic MTCMOS structure, where a low Vth computational block uses high Vth switches for power gating. Low Vth transistor in the logic gate provides a high performance opera-tion. However, by introducing a series device to the power supplies, MTCMOS circuits incur a performance penalty compared to CMOS circuits. The basic

MTCMOS structure is shown in fig.5, where a low V_{th} computation block is gated with high V_{th} power switches.

A specialized case of dual V_{th} technology that is more effective at reducing leakage currents in the standby mode is MTCMOS (Multi-Threshold CMOS). Though, by introducing an extra series device to the power supplies, MTCMOS circuits will provoke a performance penalty compared to CMOS circuits, which declines if the devices are not sized large enough. When the high V_{th} transistors are turned on, the low V_{th} logic gates are connected to virtual ground and power, and switching is performed through fast devices [22]. When the circuit enters the sleep mode, the high V_{th} gating transistors are turned off, resulting in a very low sub threshold leakage current from V_{CC} to ground. MTCMOS is only effective at reducing standby leakage currents and therefore is most effective in burst mode type application, where reducing standby power is a major benefit.

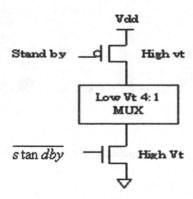

Fig. 5. MTCMOS.

3.2 VTCMOS (Variable Threshold CMOS)

Variable threshold CMOS is a body biasing based design technique Fig. 6 illustrates the VTCMOS scheme. To achieve different threshold voltages, it uses a self-substrate bias circuit to control the body bias. Regarding the speed performance of VTCMOS circuits the gate delay is the vital factor. In CMOS digital circuits, the gate delay time (t_{pd}) is given by equation 2.

$$Tpd \propto \frac{CVcc}{(Vcc-Vt)^{\alpha}} \quad \ldots.. (2)$$

α- power law model

The operating principle of VTCMOS is that its threshold voltage (V_{th}) is controlled by the applied substrate bias (- | V_{bs} |), leading to lower stand-by off current or higher active on-current. The V_{th} shift is given by: $\Delta V_{th} = \gamma | V_{bs} |$ where γ is the body effect factor [4]. Variable Threshold CMOS or (VTCMOS) is the another technique that has been developed to reduce standby leakage currents, relatively than apply multiple threshold voltage options, VTCMOS relies on a triple well process where the device V_t is dynamically is adjusted by biasing the body terminal.

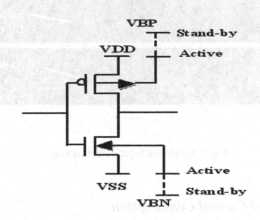

Fig. 6. VTCMOS.

4   SIMULATION RESULTS

The Simulation result is measured by CADENCE VIRTUOSO Tool.
The Simulation Conditions for 4:1 MUX circuit is summarized in TABLE I and the effect of Leakage reduction techniques (MTCMOS and VTCMOS) for 4:1 MUX is figure out in TABLE II.

4.1 Input Output Pattern for TGL 4:1 MUX

The simulation waveform of proposed 4:1 MUX is shown in Fig. 7.The resultant waveform attains a single output during power supply of 0.7v although the rise and fall time of simulation is 100 fs. The output pattern is shown in the fig.7,

S0 and S1 are the select lines for the 4:1 multiplexer and, S0bar (S0b) and S1bar (S1b) are the opposite signal of S0 and S1 respectively. Z is the output of 4:1 MUX.

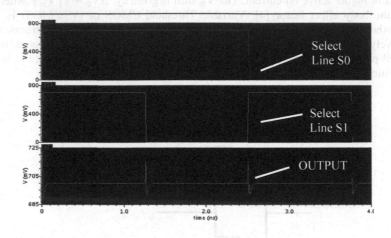

Fig. 7. Simulation Input output pattern.

4.2 Power and Current Consumption

Digital CMOS circuit may have three major sources of power dissipation namely dynamic, short and leakage power. Hence the total power consumed by every MUX style can be evaluated using the following equation 3.

$$Ptot = Pdyn + Psc + P\,leak$$
$$= CLVddVfclk + ISCVdd + I\,leakVdd \qquad \text{.......(3)}$$

Thus for low-power design the important task is to minimize $C_L\ V_{dd}\ V\ f_{Clk}$ while retaining required functionality. The first term P_{dyn} represents the switching component of power, the next component P_{sc} is the short circuit power and P_{leak} is the leakage power. Where, CL is the loading capacitance, f_{Clk} is the clock frequency which is actually the probability of logic 0 to 1 transition occurs (the activity factor). Vdd is the supply voltage; V is the output voltage swing which is equal to Vdd. The current I_{SC} is due to the direct path short circuit current. Finally, leakage current I leak, which can arise from substrate injection and sub-threshold effects, is primarily determined by fabrication technology considerations. The Leakage Current of TGL based MUX is shown in fig. 8.

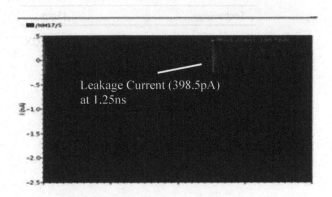

Fig. 8. Leakage current of TGL based 4:1 MUX.

4.3 Eye Diagram

Eye diagram of the output signal at a data rate of 200Gb/s. The measured eye diagram is shown in Fig.9.

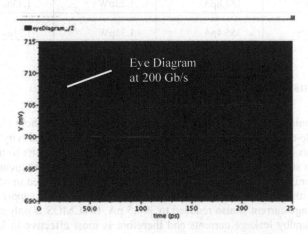

Fig. 9. Measured Eye diagrams of MUX Output at 200 Gb/s.

Table 1. Simulation Conditions of TGL based 4:1 MUX

Technology	45nm
Function	4:1 MUX
Lgate(nm)	45
Width(nm)	120
$|Vth|$ (V)	0.16
Vdd(V)	0.7
Temp.(°C)	27

Table 2. Simulation of TGL based 4:1 MUX with MTCMOS and VTCMOS

4:1 MUX	Parameters		
	Leakage Current	Leakage Power	Delay
Earlier work[5]	29.6 mA	53.3 mW	-
TGL	398.5pA	1.35pW	1.25ns
MTCMOS	182.8pA	1.23pW	1.45ns
VTCMOS	355.4pA	1.31pW	1.31ns

6 Conclusions

The techniques investigated in this paper include multi-threshold (MTCMOS) and variable-threshold (VTCMOS). Impact of temperature sensitivity on power consumption is also evaluated. The transmission gate logic results to be the efficient design styles for MUX design [18] Transistors are reduced to great extent, so that the overall area is minimized. The designed circuit is realized in 45 nm technologies, with the power consumption of 1.35 pW from a 0.7 V supply voltage at 27°C. Leakage current is also reduced to 398.5 pA. MTCMOS is only effective at reducing standby leakage currents and therefore is most effective in burst mode type application, where reducing standby power is a major benefit, while delay gets increases. For VTCMOS slightly forward substrate bias can be used to increase the circuit speed. Finally, the impact of Leakage reduction techniques (MTCMOS and VTCMOS) is analyzed in this paper; VTCMOS impart circuit designer's complete flexibility to set both V_{DD} and V_{th} during active modes to perfect balance between performance and Leakage power.

References

[1] Talal Skaik, "NOVEL STAR-JUNCTION COUPLED-RESONATOR MULTIPLEXER STRUCTURES,"Progress In Electromagnetics Research Letters, Vol. 31, 113-120, 2012.

[2] P.Metzgen "A High Performance 32-bit ALU for Programmable Logic", Proceedings of the 2004 ACM/SIGDA 12th international symposium on Field Programmable Gate Arrays. Pp 61-70. 2004.

[3] K. Jabeur, I. O'Connor, N. Yakymets, S. Le Beux, "High performance 4:1 multiplexer with am bipolar double-gate FETs,"IEEE,2011,pp-677-680

[4] T. Hiramoto and M. Takamiya., "Low power and low voltage MOSFETs with variable threshold voltage controlled by back-bias", IEIEC Trans. Electron., vol. E83- C, No.2, 2000.

[5] Xiang Sun, Jun Feng "A 10 Gb/s Low-power 4:1 Multiplexer in 0.18 µm CMOS", Proceedings of International Symposium on Signals, Systems and Electronics (ISSSE2010)

[6] M. Meghelli, "A 132-Gb/s 4:1 multiplexer in 0.13µm SiGe-bipolar technology," IEEE J. Solid-State Circuits, vol. 39, no. 12, pp. 2403-2407, Dec. 2004.

[7] S. Tanaka, and H. Hida, "120-Gb/s multiplexing and 110-Gb/s demultiplexing ICs," IEEE J. Solid-State Circuits, vol. 39, no. 12, pp. 2397-2402, Dec. 2004.

[8] Joakim Hallin, Torgil Kjellberg, and Thomas Swahn, "A 165-Gb/s 4:1 multiplexer in InP DHBT technology," IEEE J. Solid-State Circuits, vol. 41, no. 10, pp. 2209-2214, Oct. 2006.

[9] J.M.Rabaey, A.Chandrakasan, B.Nikolic, Digital Integrated Circuits A Design Perspective, PearsonEducational publishers – 2nd edition 2008.

[10] K. Kanda, D. Yamazaki, T. Yamamoto, M. Horinaka, J. Ogawa, H. Tamura, and H. Onodera, "40 Gb/s 4:1 MUX/1:4 DEMUX in 90 nm standard CMOS technology," in IEEE ISSC Tech. Dig, Feb. 2005, pp.152–153.

[11] D. Kehrer, H. D. Wohlmuth, H. Knapp, and A. L. Scholtz, "A 15 Gb/s 4:1 parallel-to-serial data multiplexer in 120 nm CMOS," in Proc. Eur. Solid-State Circuits Conf. (ESSCIRC), Firenze, Italy, Sept. 2002, pp.227–230.

[12] T. Kuroda et al, "A 0.9V, 15-Mhz, 10-mV, 4mm-2, 2-D discrete cosine transform core processor with variable threshold-voltage (VT) scheme", IEEE J. Solid-StateCircuits, vol. 31, pp. 1770-1779, 1996

[13] U. Narayanan, H. W. Leong, K.-S. Chaung, and C.L. Liu, Low power multiplexer decomposition, Int'lSymp. on Low Power Electronics and Design, pp.269-274. 1997.

[14] H.-E. Chang, J.-D. Huang, and C.-I. Chen, Input selection encoding for low power multiplexer tree, Int'l Symp. on VLSI Design, Automation, and Test,pp. 228-231, 2007.

[15] K. Kim, T. Ahn, S.-Y. Han, C.-S. Kim, and K.-H. Kim, Low power multiplexer decomposition by suppressing propagation of signal transitions, Int'lSymp. on Circuits and Systems, vol. 5, pp. 85-88,2001.

[16] Sebastian T. Ventrone, Low power multiplexer circuit, United States Patent, Patent Number:6,054,877, Apr. 2000.

[17] T. Douseki, and Y. Ohmori, BiCMOS circuit technology for a high-speed SRAM, IEEE Journal of Solid-State Circuits, vol. 23, no. 1, pp. 68-73,1988.

[18] S.Vivijayakumar, B. Karthikeyan, "Power Multiplexer Design forArithmetic Architectures using 90nm Technology", Recent Advances in Networking, VLSI and signal processing.

[19] S. G. Narendra and A. Chandrakasan, Leakage in nanometer CMOS technologies, New York: SpringVerlag, 2005.

[20] H. Mizuno, K. Ishibashi, T. Shimura, T. Hattori, S. Narita,K. Shiozawa, S. Ikeda, and K. Uchiyama, "A 18μAStandby-Current 1.8V 200MHz Microprocessor with SelfSubstrate-Biased Data Retention Mode", ISSCC Dig. Tech.Pap, pp. 280, 1999.

[21] T. Hiramoto, M. Takamiya, H. Koura, T. Inukai, H. Gymyo, H. Kawaguchi, and T. Sakurai, "Optimum device parameters and scalability of variable threshold CMOS (VTCMOS)", Ext. Abs. of 2000 SSDM, pp 372-373, 2000. To be published in Jpn. J. Appl. Phys. vol. 40, No. 4B, 2001.

[22] S. Mutoh, T. Douseki, Y. Matsuya, T. Aoki, S. Shigematsu, and J. Yamada, "1-V power supply high-speed digital circuit technology with multi-threshold voltage CMOS," IEEE J. Solid-State Circuits, vol. 30, pp. 847–854, Aug. 1995.

[23] MUTOH, S. et al. 1-V Power Supply High-speed Digital Circuit Technology with Multithreshold Voltage CMOS. IEEE J. of Solid State Circuits, New York, v.30, n.8, p. 847- 854, Aug. 1995.

Analyzing and minimization effect of Temperature Variation of 2:1 MUX using FINFET

Khushboo Mishra[1], Shyam Akashe[2], Satish Kumar

[1] ECED ,ITM University Gwalior, India

[2] ECED ,ITM University Gwalior, India

[3]ECED ,ITM University Gwalior, India

{ khushboomishra.vlsi@gmail.com; shyam.akashe@itmuniversity.ac.in,
satishkumarvatsa@gmail.com}

Abstract. —This paper proposes a Transmission gate based 2:1 MUX using FINFET (Fin Shaped Field Effect Transistor) using 45nm CMOS technology .The mobility was enhanced in devices with taller fins due to increase tensile stress. We have estimated the Optimum Power, Optimum Current, Leakage Power, Leakage Current, Operating Power and Operating Current in different voltage supply 0.3V, 0.5V and 0.7V at different temperature such as 10°C, 27°C and 50°C respectively. We have also calculated Duty cycle are 67.41%, 54.48% and 10.96%, 45.99%, rise time are 0.277ps, 0.0013ps and 0.534ps, 0.003ps, Bandwidth are 3.502GHz, 0.03THz and 3.505GHz, 0.07THz, Frequency jitter are 5.24GHz, 1.73THz and 21.51GHz, 1.199THz Period jitter are 3.424ps, 38.89ps and 21.51ps, 1.707ps in 0.7V and 0.5V supply at 27°C of FINFET as well as Transmission gate 2:1 MUX.

Keywords: MUX; CMOS; Leakage Power; Leakage Current; Frequency; FINFET; Optimum Power; Operating Current

1 Introduction

In silicon n-channel field-effect transistors (n-FETs) silicon-carbon (Si: C) source/drain (S/D) stressors may be adopted for attractive the electron mobility and drive current [1].These values are extensively higher than the doping extracted by electrical categorization, signifying doping loss in fins and/or partial activation [2]. Doped-channel FINFETs are appropriate for system-on-chip applications require various threshold voltages on the same die. For an unusual device structure for replacing the planer CMOS device structure, FINFET technology is one of the most proficient candidates [3]. In the operation of planar MOSFET scaling down

J. C. Bansal et al. (eds.), *Proceedings of Seventh International Conference on Bio-Inspired Computing: Theories and Applications (BIC-TA 2012)*, Advances in Intelligent Systems and Computing 202, DOI: 10.1007/978-81-322-1041-2_13, © Springer India 2013

of CMOS technology leads severe short channel effects. However increase in short channel effects will lead to degrade the performance. Another device structure is essential, to attain superior control over gate. They have an intrinsic force against the SCE. Therefore, many studies have focused on the FINFET SRAM cells to overcome the rapid decrease in the conventional bulk SRAM performance [4]. The effect of FINFET variability on the 2:1 MUX has been also studied [5] In sub-22-nm CMOS technology nodes due to their superior electrostatic integrity as compared to the conventional planar bulk MOSFET, .three-dimensional transistor structures such as double gate FINFET and tri gate FET are slated for adoption [6]. Even when the fin width is reduced to ~4 nm to enable gate length (Lg) scaling down to 10 nm, recent experimental results show that the FINFET performs well [7]. The process simulator within the Sentaurus technology computer-aided-design software suite [8], which uses the finite-element method, was used to perform 3-D simulations of stress within FINFETs with (100) top and (110) sidewall surfaces and [110] channel direction. Since the bulk-silicon substrate provides a pattern for epitaxial growth, so that the entire S/D regions are anxious for bulk FINFETs, the fin S/D regions are implicit to be etched away earlier to the discriminating epitaxial, [9] a gate-last (i.e., replacement metal gate) process flow, in which a dummy gate is formed earlier to the S/D epitaxial and then replaced by the metal gate [10].

2 Transmission Gate Based 2:1 MUX

This is the Transmission Gate based 2:1 MUX structure implemented with very minimum transistors (4 MOS transistors) compare to the CMOS based 2:1 MUX which has 20 CMOS devices. The back to back connected PMOS & NMOS transistors arrangement acts as a switch is so called Transmission Gate. In Transmission Gate NMOS transistor pass a strong 0, but a weak 1, while PMOS transistor pass a strong 1, but a weak 0. The CMOS based 2:1 MUX use NMOS transistor act as pull down network and PMOS transistor act as pull up network. Whereas in the transmission gate, combines the both properties by placing NMOS transistor in parallel with the PMOS transistor. Two transmission gates are connected as shows the schematic of Transmission based 2:1 MUX in Figure 1 to form a MUX structure and output waveform is shown in figure 2. Each the Transmission Gate acts as an AND switch to replace the AND logic gate which is used in a CMOS Based design of MUX. Hence the transistor count is reduced to 4 it shows that it occupies less area as compared to CMOS Based MUX. One more change when compared to CMOS Based 2:1 MUX is that there is no supply voltage applied to the circuit. It results in less operating power. It has lower gate delay and the circuit propagates faster than that of the CMOS Based 2:1 MUX. The gate delay as mentioned can be calculated as

$t_{pd} \propto C_L V_{dd} / I_{ds}$

Where, t_{pd} is the propagation delay, CL is the load Capacitance, V_{dd} is the supply voltage and I_{DS} is the drain saturation current.

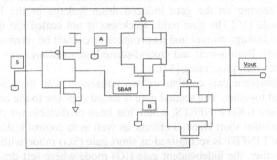

Fig. 1. Schematic of Transmission gate 2:1 MUX

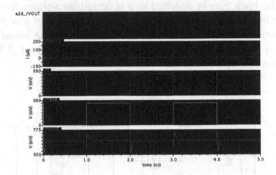

Fig. 2. Output Waveform of Transmission gate 2:1 MUX

3   Transmission gate Based 2:1 MUX using FINFET

In due to its base material the uninterrupted down in scaling of bulk CMOS creates key issues. The crucial obstacles to the scaling of bulk CMOS to 45nm gate lengths include short channel effects, optimum current, gate-dielectric leakage, and device to device variations. But FINFET based designs offers the superior control over short channel effects, low leakage and better yield [11] in 45nm helps to overcome the obstacles in scaling. Preliminary results capturing the consequence of defects manifested as cuts on the back gate were presented, demonstrating a redoubtable challenge toward the improvement of a consistent fault model [12]. However, FINFET performance is exaggerated by numerous factors, such as parasitic resistance Rp, channel stress due to the Multi Gate [13]. The author in optimizes the compensate spacer and initiate under lap on source and drain side which leads to decrease in on current [14]. In addition, for use in future

one-transistor (1T) capacitor less memory devices bulk FINFETs are also investigated. While expectant results have been obtained so far, there are some excep-
tional issues [15], such as the hot-carrier deprivation induced by the indoctrina-
tion, either relying on the gate induced drain leakage or the bipolar junction
transistor mode [16].The gate oxide thickness is not scaled too insistently to reduce the gate leakage current and its prospective contact on retention in 1T memory applications. Both p-well and ground-plane implantations have been performed
earlier to gate stack processing [17].

A parallel transistor pair consists of two transistors with their source and drain
terminals tied together. The second gate is added opposite to the conventional gate
in Double-Gate (DG) FINFETS, which has been predictable for their prospective
to superior control short channel effects, as well as to control leakage current. The
operations of FINFET is recognized as short gate (SG) mode with transistor gates
attached together, the independent gate (IG) mode where self-determining digital
signals are used to drive the two device gates, the low-power and optimum power
mode where the back gate is attached to a reverse-bias voltage to reduce leakage
power and the hybrid mode, which employs a arrangement of low power and self-
determining gate modes. The schematic of transmission gate based 2:1 MUX using FINFET is shown in figure 3 and Output waveform of 2:1 MUX using
FINFET is shown in figure 4.

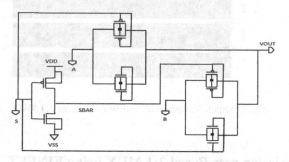

Fig. 3. Schematic of Transmission gate 2:1 MUX using FINFET

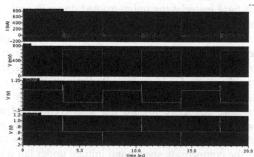

Fig. 4. Output Waveform of Transmission gate 2:1 MUX using FINFET.

4. Design parameters of 2:1 MUX

We have estimated the different design parameters of Transmission Gate as well as FINFET. Design parameters are Duty cycle, Bandwidth, Rise time, Frequency, Frequency jitter and Period jitter.

4.1. Duty Cycle

In a periodic event, duty cycle is the ratio of the duration of the event to the total period of signal [18].

Duty Cycle (D) = τ / T
Where τ is the duration that functions is active. T is the period of the function.

4.1. a Duty Cycle for Transmission Gate

The Duty Cycle of Transmission Gate is 54.48% and 45.99% in 0.7V and 0.5V at 27°C temperature respectively. From figure5 shows that at 1ns the duty cycle is 45.99% and rise in time i.e. 2ns the duty cycle become 54.48%. After 54.48% the MUX is fully saturate therefore after 3ns,4ns and 5ns the duty cycle remains constant.

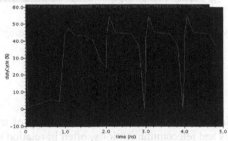

Fig.5. Duty Cycle of Transmission gate 2:1 MUX

4.1. b Duty Cycle for FINFET

The Duty Cycle of Transmission Gate is 67.41% and 10.96% in 0.7V and 0.5V at 27°C temperature respectively. The eye diagram of transmission gate based 2:1 MUX using FINFET is shown in figure 7, it shows that the wide opened eye during transmission data rates of the signal .From figure 6 shows that at 1ns the duty cycle are 50% and rise in time i.e. 10ns the duty cycle becomes 67.41%.

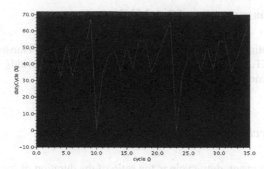

Fig.6. Duty Cycle of Transmission gate 2:1 MUX using FINFET

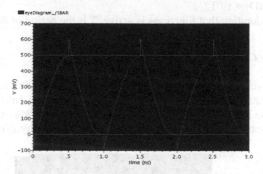

Fig.7. Eye Diagram of Transmission gate 2:1 MUX using FINFET

4.2 Jitter

Jitter is the undesired deviation from the true periodicity of an assumed periodic signal in electronics and telecommunications, often in relation to a reference clock source. Jitter may be observed in characteristics such as the frequency of successive pulses, the signal amplitude, or phase of periodic signal. Jitter can be classified in two types such as period jitter and frequency jitter.

4.2. a Period Jitter

Period Jitter is the interval between two times of maximum effect (or minimum effect) of a signal characteristic that varies regular with time.

4.2. a. 1 Period Jitter for Transmission gate

The Period Jitter of Transmission Gate is 38.89ps and 1.707ps in 0.7V and 0.5V at 27°C temperature respectively. The Output waveform is shown in figure 8.From figure 8 it shows that at 1 to 1.9ns the period jitter raises 38.89ps and at 2ns it fall and again rises 2.1 to 2.9 ns it become maximum value and at 3ns it fall. After 2ns the period jitter becomes 1.707ps and at 3ns, 4ns remains constant.

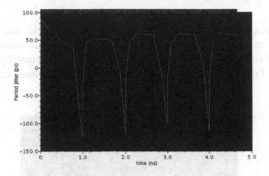

Fig.8. Period Jitter of Transmission gate 2:1 MUX

4.2. a. 2 Period Jitter for FINFET

The Period Jitter of FINFET is 3.424ps and 21.51ps in 0.7V and 0.5V at 27°C temperature respectively. The Output waveform is shown in figure 9. From figure 9 it shows that at 1ns it rises and fall, again at 3ns it becomes maximum value of period jitter is 3.424ps and remains constant with rise in time.

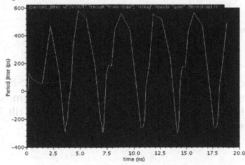

Fig.9.Peroid Jitter of Transmission gate 2:1 MUX using FINFET

4.2.b Frequency Jitter

Frequency Jitter, the more commonly quoted figure, is it inverse. Jitter frequencies below 10 Hz as wander and frequencies at or above 10Hz as Jitter.

4.2.b. 1 Frequency Jitter for Transmission Gate

The Frequency Jitter of Transmission Gate is 1.734THz and 1.199 THz in 0.7V and 0.5V at 27°C temperature respectively. The Output waveform is shown in figure 10. From figure 10 it shows that at 1ns the frequency jitter rises is 1.734THz and fall, till 1ns to 2ns the frequency jitter becomes constant and remains same.

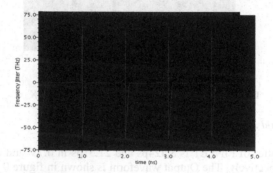

Fig.10. Frequency Jitter of Transmission gate 2:1 MUX

4.2.b. 2 Frequency Jitter for FINFET

The Frequency Jitter of FINFET is 5.24 GHz and 21.51 GHz in 0.7V and 0.5V at 27°C temperature respectively. The Output waveform is shown in figure 11. From figure 11 it shows that at 2.5 ns it rises and falls till 7.5ns. At 7.5ns it reaches its maximum value is 5.24 GHz and fall.

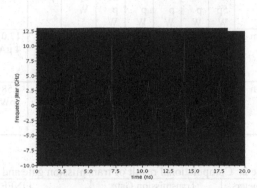

Fig.11. Frequency Jitter of Transmission gate 2:1 MUX using FINFET

5 Simulated Result Summary

To evaluate the parameters of FINFET in different power supply at various temperatures and design parameter of transmission gate and FINFET, which is shown in TABLE 1 and TABLE 2. Furthermore, accuracy of the circuit is validated by measurements in [19]. It is shown that the measured energy is in the near locality of the simulated energy dissipation. The MUX circuit often limits the operation speed of the whole system. Simulation result is calculated by CANDENCE VIRTUOSO Tool.

Table 1. Summary of 2:1 MUX using FINFET

Parameters	2:1 MUX using FINFET								
Supply Voltage	0.7V			0.5V			0.3V		
Temperature	10 °C	27 °C	50 °C	10 °C	27 °C	50° C	10° C	27°C	50°C
Operating Current	58.09μA	13.45 μA	59.03 μA	30.39 μA	27.67 μA	32.41 μA	15.43 μA	56.81 μA	15.8 1 μA
Operating Power	25.54nW	1.145 nW	26.05 nW	30.69 nW	16.02 nW	12.45nW	841.2nW	33.9 1nW	1.50 nW
Leakage Current	873.71nA	50.21 nA	599.7nA	108.1nA	103.6nA	87.76nA	49.1 nA	702.1nA	90.17nA
Leakage	21	0.1	12.	2.5	1.0	1.80	1.01	5.75	1.08

Power	.65pW	48pW	19pW	37pW	22pW	2pW	pW	5pW	pW
Optimum Current	17.54µA	3.77µA	17.34µA	9.31µA	9.04µA	9.39µA	4.05µA	17.04µA	3.91µA
Optimum Power	15.16pW	32.51pW	6.94pW	10.28pW	.303pW	15.3pW	36.30pW	7.582pW	30.27pW

Table 2. Computational result of transmission gate and FINFET

Parameters	Transmission Gate		FINFET	
Supply Voltage	0.7V	0.5V	0.7V	0.5V
Temperature	27oC	27oC	27oC	27oC
Duty Cycle	54.48%	45.99%	67.41%	10.96%
Bandwidth	30.0THz	70.7THz	3.502GHz	3.505GHz
Rise Time	0.0013ps	0.0036ps	0.277ps	0.534ps
Frequency	250.1MHz	270.3MHz	142.9MHz	287.9MHz
Frequency Jitter	1.73THz	1.19THz	5.24GHz	21.51GHz
Period Jitter	38.89ps	1.707ps	3.424ps	21.51ps

6 Conclusion

We have experimentally investigated the device performance and parameters such as operating current, operating power, leakage current, leakage power, optimum current and optimum power of transmission gate based 2:1 MUX using FINFETs with different power supply 0.3V, 0.5V and 0.7V at various temperatures such as 10°C, 27°C and 50°C respectively. Mobility was enhanced in the tall-fin devices due to increased tensile stress. We have also calculated Duty cycle are 67.41%, 54.48% and 10.96%, 45.99%, rise time are 0.277ps, 0.0013ps and 0.534ps, 0.003ps, Bandwidth are 3.502GHz, 0.03THz and 3.505GHz, 0.07THz,Frequency are 142.9MHz, 287.9MHz and 270.1MHz, 270.3MHz, Frequency jitter are 5.24GHz, 1.73THz and 21.51GHz, 1.199THz Period jitter are 3.424ps, 38.89ps and 21.51ps, 1.707ps in 0.7V and 0.5V supply at 27°C of FINFET as well as Transmission gate 2:1 MUX.

References

[1] E. R. Hsieh and S. S. Chung, "The proximity of strain induced effect to improve the electron mobility in a silicon–carbon source–drain structrue of n-channel metal–oxide semiconductor field-effect transistors," *Appl. Phys. Lett.*, vol. 96, no. 9, p. 093501, Mar. 2010.

[2] K. Akarvardar, C. D. Young, D. Veksler, K.-W. Ang, I. Ok, M. Rodgers et al., "Performance and variability in multi-V_T FinFETs using fin doping," in *Proc. VLSI TSA*, 2012, to be published.

[3] K. Kim, K. K. Das, R. V. Joshi, and C.-T. Chuang, "Leakage power analysis of 25-nm double-gate CMOS devices and circuits," *IEEE Trans. Electron Devices*, vol. 52, no. 5, pp. 980–986, May 2005.

[4] P. Zuber, M. Miranda, M. Bardon, S. Cosemans, P. Roussel, P. Dobrovolny, T. Chiarella, N. Horiguchi, A. Mercha, T. Y. Hoffmann, D. Verkest, and S. Biesemans, "Variability and technology aware SRAM Product yield maximization," in *VLSI Symp. Tech. Dig.*, Jun. 2011, pp. 222–223.

[5] E. Baravelli, L. Marchi, and N. Speciale, "VDD scalability of FinFET SRAMs: Robustness of different design options against LER-induced variations," *Solid State Electron.*, vol. 54, no. 9, pp. 909–918, Sep. 2010.

[6] K. Maitra, A. Khakifirooz, P. Kulkarni, V. S. Basker, J. Faltermeier, H. Jagannathan, H. Adhikari, C.-C. Yeh, N. R. Klymko, K. Saenger, T. Standaert, R. J. Miller, B. Doris, V. K. Paruchuri, D. McHerron, J. O'Neil, E. Leobundung, and H. Bu, "Aggressively scaled strainedsilicon-on-insulator undoped-body high- k/metal-gate nFinFETs for highperformance logic applications," *IEEE Electron Device Lett.*, vol. 32, no. 6, pp. 713–715, Jun. 2011.

[7] J. B. Chang, M. Guillorn, P. M. Solomon, C.-H. Lin, S. U. Engelmann, A. Pyzyna, J. A. Ott, and W. E. Haensch, "Scaling of SOI FinFETs down to fin width of 4 nm for the 10 nm technology node," in *VLSI Symp. Tech. Dig.*, 2011, pp. 12–13.

[8] Sentaurus Process User Guide, Version D-2011.09, Synopsys Co., Mountain View, CA, 2011.

[9] V. Moroz and M. Choi, "Strain scaling and modeling for FETs," *ECS Trans.*, vol. 33, no. 6, pp. 21–32, 2010.

[10] G. Eneman, N. Collaert, A. Veloso, A. De Keersgieter, K. De Meyer, and T. Y. Hoffmann, "On the efficiency of stress techniques in gate-last n-type bulk FinFETs," in *Proc. ESSDERC Tech. Dig.*, 2011, pp. 115–118.

[11] Jan M Rabaey, Anantha Chandrakasan, Borivoje Nikolic, "Digital Integrated Circuits: A Design perspective,"2nd Edition, Prentice-Hall, Inc. Y.Omura, S.Cristovoveanu, F. Gamiz, B-Y. Nguyen , - Silicon-onInsulator Technology and Devices 14, Issue 4 - Advanced FinFET Devices for sub-32nm Technology Nodes: Characteristics and Integration Challangesll 2009, pp.45-54.

[12] M. O. Simsir, A. N. Bhoj, and N. K. Jha, "Fault modeling for FinFET circuits," in *Proc. Int. Symp. Nanoscale Archit.*, Jun. 2010, pp. 41–46.

[13] T. Kamei, Y. X. Liu, K. Endo, S. O'uchi, J. Tsukada, H. Yamauchi, Y. Ishikawa, T. Hayashida, T. Matsukawa, K. Sakamoto, A. Ogura, and M. Masahara, "Experimental study of physical-vapor-deposited titanium nitride gate with an n+-polycrystalline silicon capping layer and its application to 20 nm fin-type double-gate metal–oxide–semiconductor field-effect transistors," *Jpn. J. Appl. Phys.*, vol. 50, no. 4, pp. 04DC14-1–04DC14-5, Apr. 2011.

[14] B. Raj, A.K. Saxena and S.Dasgupta, "Nanoscale FinFET Based SRAM Cell Design: Analysis of Performance Metric, Process Variation, Underlapped FinFET, and Temperature Effect," *IEEE Mag. Circuits and Systems*, vol. 11, no. 3, 2011.

[15] M. Aoulaiche, N. Collaert, R. Degraeve, Z. Lu, B. De Wachter, G. Groeseneken, M. Jurczak, and L. Altimime, "BJT-mode endurance on a 1T-RAM bulk FinFET device," *IEEE Electron Device Lett.*, vol. 31, no. 12, pp. 1380–1382, Dec. 2010.

[16] M. Aoulaiche, N. Collaert, A. Mercha, M. Rakowski, B. De Wachter, G. Groeseneken, L. Altimime, and M. Jurczak, "Hot hole induced damage in 1T-FBRAM on bulk FinFET," in *Proc. IEEE IRPS*, Apr. 2011,pp. 2D.3.1–2D.3.6.

[17] N. Collaert, M. Aoulaiche, B. De Wachter, M. Rakowski, A. Redolfi, S. Brus, A. De Keersgieter, N. Horiguchi, L. Altimime, and M. Jurczak, "A low-voltage biasing scheme for aggressively scaled bulk FinFET 1T-DRAM featuring 10s retention at 85 ·C," in *VLSI Symp. Tech. Dig.*, 2010, pp. 161–162.

[18] Wolaver, Dan H. 1991. Phase-Locked Loop Circuit Design, Prentice Hall, pages 211-237

[19] J. Rodrigues, O. C. Akgun, and V. Owall, "A <1 pJ sub-VT cardiac event detector in 65 nm LL-HVT CMOS," in Proc. VLSI-SoC, June 2010.

Spectral-spatial MODIS image analysis using swarm intelligence algorithms and region based segmentation for flood assessment

J. Senthilnath[a1], Vikram Shenoy H[b2], S.N. Omkar[a3], V. Mani[a4]

[a]Department of Aerospace Engineering, Indian Institute of Science, Bangalore, India.

[b]Department of Electronics and Communication Engineering, National Institute of Technology Karnataka, Surathkal, India.

{[1]snrj@aero.iisc.ernet.in; [2]vikkyshenoy@gmail.com; [3]omkar@aero.iisc.ernet.in; [4]mani@aero.iisc.ernet.in}

Abstract. This paper discusses an approach for river mapping and flood evaluation based on multi-temporal time-series analysis of satellite images utilizing pixel spectral information for image clustering and region based segmentation for extracting water covered regions. MODIS satellite images are analyzed at two stages: before flood and during flood. Multi-temporal MODIS images are processed in two steps. In the first step, clustering algorithms such as Genetic Algorithm (GA) and Particle Swarm Optimization (PSO) are used to distinguish the water regions from the non-water based on spectral information. These algorithms are chosen since they are quite efficient in solving multi-modal optimization problems. These classified images are then segmented using spatial features of the water region to extract the river. From the results obtained, we evaluate the performance of the methods and conclude that incorporating region based image segmentation along with clustering algorithms provides accurate and reliable approach for the extraction of water covered region.

Keywords: MODIS image, Flood assessment, Genetic algorithm, Particle swarm optimization, Shape index, Density index.

1 Introduction

Over the past decades, many regions around the globe have witnessed many natural hazards, flood being the most destructive one. Flood accounts for nearly $1/3^{rd}$ of the worldwide disaster damage [1]. There is also huge monetary loss involved in such natural calamity because of the unwary flood management system. Hence there is a need for efficient flood disaster management tool to

J. C. Bansal et al. (eds.), *Proceedings of Seventh International Conference on Bio-Inspired Computing: Theories and Applications (BIC-TA 2012)*, Advances in Intelligent Systems and Computing 202, DOI: 10.1007/978-81-322-1041-2_14, © Springer India 2013

endure the aftermath of flooding. Now a day, most prominent tool for evaluating the flood extent is satellite imagery because of its easy data acquisition and development of robust image processing techniques for gauging the flood map.

There has been lot of research during past decades towards the flood extent evaluation using optical imagery. One such optical image is MODIS which has attracted many researchers to work towards optical image based flood assessment because of their easy availability and cost-effectiveness. Islam *et.al* [1] presented a flood inundation mapping based on Normalized Difference Water Index (NDWI). Zhan *et.al* [2] proposed vegetative cover conversion algorithm for land cover analysis. Khan *et.al* [3] employed ISODATA algorithm for the classification of flooded and non-flooded regions using MODIS image.

However, there are several other algorithms which are more accurate compared to above conventional techniques. Genetic Algorithms (GA) are one such family of adaptive search methods and hence found its application in image segmentation in the past decades. Bosco [4] formulated image segmentation as global optimization problem used a genetic approach to solve. Particle Swarm Optimization (PSO) is one more technique which is also a population based. Many researchers have explored wide areas of applications of PSO [5, 6, 7]. PSO has been quite efficient in optimizing multi-modal problems. Nagesh *et.al* [5] presented multi-purpose reservoir system operation using PSO. Huang *et.al* [6] proposed flood disaster classification based on multi co-operative PSO. Also, Omran *et.al* [7] made use of PSO algorithm and spectral un-mixing for image classification. In their study, remote sensing data and MRI images were classified based on spectral features.

In this paper, we propose a flood extent evaluation method based on MODIS image using unsupervised techniques such as GA and PSO. In our study, the above two algorithms are used for clustering the image region into flooded and non-flooded regions based on spectral features. Time-series data for the automatic extraction of river regions (using before flood image) and for evaluating floods (using during flood image) is presented in this paper. Since GA and PSO are based on spectral features, sometimes it is unreliable for the reason that some of the non-water particles may be misclassified as water because of similar spectral information. Hence, flood inundation evaluation based on spatial features like Shape Index (SI) and Density Index (DI) is also considered. Spatial information of before flood and during flood images can be effectively analysed using the above parameters. These parameters were first proposed by Mingjun *et.al* [8] for road extraction using satellite image. Finally, the performance of these methods is evaluated by comparison with ground truth data.

2 Problem Formulation

The optimization problems often involve finding the optimal solution for a given problem. Normally clustering algorithms involve global optimization [7] and local optimization [9] to partition the given data set into *n* groups. In terms of image clustering, the data points within a group share similar spectral characteristics. Pixel values refer to each pattern in a group and image region corresponds to a cluster. The concept of fitness function is used for finding out the best solution.

A particle x is defined by its cluster centers as $x_i = \{m_{i1,}\ m_{i2}\ldots\ldots\ m_{ij}\ldots\ldots m_{iN}\}$, where N is the number of clusters, m_{ij} refers to j^{th} cluster centre of ith particle. For each particle x_i, fitness function is described as follows [7],

$$f(x_i, Z_i) = w_1 d_{max}(Z_i, x_i) + w_2(z_{max} - d_{min}(x_i)) \tag{1}$$

where Z_i is a matrix comprising the assignment of the pixels to clusters of ith particle. Z_{max} is 2^s-1 for an s-bit image, w_1 and w_2 are the inertia factors set by the user. Also,

$$d_{max}(x_i, Z_i) = max\left\{\sum\nolimits_{\forall z_p \in C_{i,j}} d(Z_{p,m_{i,j}})/|C_{i,j}|\right\} \tag{2}$$

describes the maximum Euclidean distance of particles to their associated clusters. Here, $C_{i,j}$ is nothing but j^{th} cluster of ith particle.

$$d_{min}(x_i) = \min_{\forall j_1, j_2, j_1 \neq j_2} \{d(m_{ij_1}, m_{ij_2})\} \tag{3}$$

where d is the minimum average Euclidean distance between any pair of clusters.

However, the problem associated with image clustering algorithms based on spectral features is that sometimes it leads to misinterpretation of image segments because of spectral similarities. Hence, some of the researchers have adopted region based segmentation to extract geometrical features. To extract spatial features of roads or river networks, Shape Index (SI) and Density Index (DI) are used [8, 10].

$$SI = \frac{P}{4\sqrt{A}} \tag{4}$$

where P is the perimeter of the image object and A is an area of the segmented region (or total number of pixels in the segmented image object).

$$DI = \frac{\sqrt{N}}{1 + \sqrt{var(X) + var(Y)}} \tag{5}$$

where N is the number of pixels inside the region, $Var(X)$ is the variance of X co-ordinates of all the pixels in the region and $Var(Y)$ is the variance of Y co-ordinates of all the pixels in the region. $\sqrt{var(X) + var(Y)}$ gives the value of approximate radius of the image object. Suitable thresholds of SI and DI are employed to classify the regions into water and non-water region.

2.1. Illustrative Example

Though there are several clustering algorithms which help in solving multi-modal optimization problems, in case of flood mapping, some of the non-water regions are also classified as water region due to similar spectral features. Hence, adopting SI and DI would effectively distinguish between water and non-water region.

Here, we present an illustrative example taking a sample portion of before flood MODIS image shown in fig. 1 (a) which is classified as river and non-river regions using a clustering algorithm (fig. 1 (b)). Fig. 1(c) shows the improvement over a spectrally clustered image using spatial information (SI and DI). The extracted river is shown in fig. 1(d). The following matrix is a 12x10 image portion with grayscale intensities.

Pixel-co ordinates	49	50	51	52	53	54	55	56	57	58	59	60
345	179	169	112	76	135	154	163	157	144	141	143	143
345	196	215	110	74	145	159	139	127	133	138	138	132
346	233	209	107	88	150	168	142	126	124	119	113	108
347	241	185	93	99	149	172	147	127	126	111	103	103
348	209	166	89	117	152	166	156	142	134	118	111	113
349	195	165	99	147	174	165	160	151	144	132	126	123
350	208	54	101	164	195	175	156	138	143	138	129	120
351	196	119	110	165	185	180	158	141	139	143	130	118
352	161	88	132	168	160	175	168	169	156	163	148	134
353	125	108	143	176	162	149	149	150	149	153	149	148

A clustering algorithm is applied on an image using the fitness function according to eqn (1). Regions clustered as river are represented by 1 and non-river regions are represented by 2. Consequentially, clustered matrix of above image matrix as follows:

Pixel-co ordinates	49	50	51	52	53	54	55	56	57	58	59	60
345	2	2	1	1	1	2	2	2	2	2	2	2
345	2	2	1	1	2	2	1	1	1	1	1	1
346	2	2	1	1	2	2	2	1	1	1	1	1
347	2	2	1	1	2	2	2	1	1	1	1	1
348	2	2	1	1	2	2	2	2	1	1	1	1
349	2	2	1	2	2	2	2	2	2	1	1	1
350	2	2	1	2	2	2	2	1	2	1	1	1
351	2	1	1	2	2	2	2	2	1	2	1	1
352	2	1	1	2	2	2	2	2	2	2	2	1
353	1	1	2	2	2	2	2	2	2	2	2	2

From the above image matrix, we can see that some portion is marked in red, which is misclassified as river region. To avoid the discrepancies due to similar spectral features, SI and DI as mentioned in eqn (4 and 5) are applied on clustered image. SI and DI being spatial parameters resolve the issue with spectral inconsistency. The resulting image matrix is as follows,

Pixel-co ordinates	49	50	51	52	53	54	55	56	57	58	59	60
345	2	2	1	1	1	2	2	2	2	2	2	2
345	2	2	1	1	2	2	2	2	2	2	2	2
346	2	2	1	1	2	2	2	2	2	2	2	2
347	2	2	1	1	2	2	2	2	2	2	2	2
348	2	2	1	1	2	2	2	2	2	2	2	2
349	2	2	1	2	2	2	2	2	2	2	2	2
350	2	2	1	2	2	2	2	2	2	2	2	2
351	2	1	1	2	2	2	2	2	2	2	2	2
352	2	1	1	2	2	2	2	2	2	2	2	2
353	1	1	2	2	2	2	2	2	2	2	2	2

SI and DI are chosen as spatial parameters for image segmentation because of their contrasting property in identifying a river. From the above clustered image matrix and final result image matrix, we can see that some of the non-water region pixels are also classified as water pixels because of similar spectral information.

From the above two matrices, we can see that in clustered image matrix, there are 31 misclassified water region pixels out of which, 19 pixels constitute for the perimeter. SI is calculated using the eqn (4) and it turned out to be 0.8531. Upon computing DI using eqn (5), we get DI as 1.5384. However, if we look at the region which is properly classified as water region; it has area of 19 pixels and also perimeter of 19 pixels. Hence, SI of this region is 1.0897 and DI equal to 1.0131. From the above illustration, it is evident that river regions have higher SI because of the longer perimeter and less area. Similarly, DI is lesser for river regions because of large distance coverage. Thus, suitable ratio (SI/DI) is set as threshold for classifying river and non-river regions.

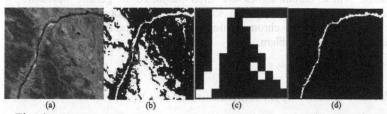

(a)   (b)   (c)   (d)

Fig. 1 (a) Original MODIS image (b) Clustered Image (c) Subset employed for the spatial analysis of flood extraction & (d) River extracted using region based segmentation.

3 Methodology

In this section, we present a flood detection and mapping in two stages: At spectral level, image clustering is done by Genetic Algorithm (GA) and Particle Swarm Optimization(PSO) and at spatial level, Shape Index (SI) and Density Index (DI) are used to classify flooded and non-flooded regions. Our proposed methodology is depicted in fig. 2

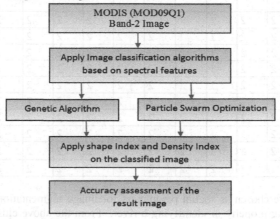

Fig. 2 Flow chart of proposed methodology

3.1 Genetic algorithm

Genetic Algorithms is a population based stochastic search and optimization techniques with inherent parallelism. For clustering, the cluster centers are encoded in the form of strings (called chromosomes). A collection of such strings is called a population. Initially, a random population of different points is created within the search space. The fitness value for each chromosome of the population is evaluated. These chromosomes are then subjected to genetic operators – reproduction, crossover and mutation to yield a better population for a fixed number of generations. The chromosomes converged to the least fitness value is the solution to the given problem.

3.2 Particle Swarm Optimization

Particle Swarm Optimization (PSO) is also a population based method, inspired by the social behaviour of bird flocks. It was first developed by Kennedy *et.al* [11]. It was then proposed by Omran *et.al.* [7] for image segmentation. Their results show that PSO outperformed K-means, Fuzzy C-means and other clustering algorithms.

Similar to the bird flock in search of its nutrition, each particle flies through the search space to find out the best solution with a velocity adjusted dynamically. There are two types of solutions are associated with each particle, one is personal best and the other is global best. Personal best is the best solution that each particle visited so far in the search space. Global best is the overall best solution found by the swarm of particles. Each particle is evaluated based on the fitness function as mentioned earlier in section 2. The main characteristics associated with each particle in the swarm are current position of the particle, current velocity of the particle and personal best position of the particle. For each iteration, particle tries to find the most optimal solution (personal best) with the dynamic adjustment of velocity. Fitness function evaluates the personal best position [7]. Global best solution takes into account all the personal best solutions [11] and it is the best solution of the entire swarm.

4   Results and discussions

In this section, we present the results obtained from image clustering done by genetic algorithm and particle swarm optimization. Accuracy assessment of the above two methods is done in terms of root mean square error (RMSE) [10] for before flood image and Receiver operating characteristics (ROC) [12] for evaluating the result of during flood image. Also, the above algorithms are compared with the two of the existing conventional unsupervised techniques.

4.1   Study area and data description

Region surrounding Krishna river near Manthralaya, Andhra Pradesh is taken as study area which is located between 16^0 38' 00"N-$77^0$09'00"W and 15^0 26' 00"S-78^0 26' 00"E. The dataset obtained from MODIS (MOD09Q1) Terra surface reflectance 8-Day L3 Global 250m^2 satellite images are used for this purpose. This dataset comprises of 2 bands. Band 1 (visible red region) lies between 620-670 mm and Band 2 (Near Infra-Red region) is centred between 841-876 mm. Band1 is more sensitive for the detection of land/cloud boundaries and NIR band is more efficient in detecting water region since water has significant low reflectance in NIR region [13].

4.2   Genetic Algorithm

All the images used for validation are clustered by GA. The pixels comprising this cluster are eventually clustered as water and those not present in the cluster are clustered as non-water. In GA, each generation has 20 chromosomes and the maximum number of generations allowed to find the best solution to the problem is 30.

The crossover operator tries to optimize the solution globally while the mutation operator searches locally. So we use the variable rates of crossover and mutation to aid swift optimization. The chromosome with best fitness value is used to cluster the image. The clustering result of the Krishna River before flood i.e. March 2009 by the GA technique is shown in Fig. 3(b). During flood classification result is depicted in fig. 4(b). As it can be seen from the results of clustering, many non-water segments have also been clustered as water because of their spectral resemblance with water. These non-water features are further removed by region based segmentation.

4.3   Particle Swarm Optimization

In case of PSO, each iteration has 20 particles and the maximum number of iterations allowed to find the best solution to the problem is 30. The best particle fitness value is used to cluster the image. The clustering result of the Krishna River before flood i.e. March 2009 by the PSO technique is shown in fig. 3(c), the clustering result of the Krishna River during flood i.e. September 2009 by the PSO technique is shown in fig. 4(c). Here also many non-water segments have also been clustered as water. These non-water features are further removed by region based segmentation.

4.4   Region based image segmentation

The failure of spectral based image clustering to extract water features is overcome by region based segmentation. For segmentation purpose, we use the geometrical features of the linear segments. As mentioned earlier, we use two indices- SI and DI with suitable thresholding to differentiate linear segments from non-linear segments.

For the image of Krishna river before the flood (March 2009), we use SI threshold of 2.0 and a DI threshold of 0.9 to extract the river course. The result of region based segmentation of the March 2009 image classified by GA is shown in fig. 3(d). Similarly, PSO classified image is segmented based on geometrical parameters whose result is shown in fig. 3(e). However, in the case of Krishna river system image during the flood i.e. September 2009, to extract flood, we use

SI threshold of 2.0 and DI threshold of 1.4. The reason behind this change in DI threshold value is that due to flood, the segment to be extracted is not perfectly linear anymore; instead it is spread on a larger area. Thus to accommodate the flooded parts, a relaxation in the DI threshold is provided. Thus, region based geometrical segmentation eventually extracts the linear features from images to a reliable extent. In case of before flood image of Krishna River 2009, the final segmented images as a result of clustering by GA are shown on a backdrop of the original image in fig. 3(f) and that of PSO is shown in fig. 3(g). In these images, river course is represented by white pixels.

Fig. 4(a) indicates the ground truth in which flooded cities are white discs and non-flooded ones have black centre. The cities flooded according to the clustering by GA and PSO are shown as white dots in the fig. 4(b) and fig. 4(c) respectively. The result images shown in fig. 4 (b) and fig. 4 (c) are evaluated in terms of true positive (TP), False positive (FP), True Negative (TN) and False negative (FN). These ROC parameters are TPA, TNR, FPR and ACC [12]. Result interpretation of cities picked by GA and PSO are presented in table 1. The ROC parameters are evaluated for accuracy assessment which is depicted in table 2.

Table 1. ROC parameters comparison for flooded cities shown in Figure 4

Terms	GA	PSO
True positive	10	11
False positive	1	1
True negative	15	15
False negative	2	1

Table 2. Evaluating features based on ROC parameters in Table 1.

Terms	GA	PSO
True positive Rate (Sensitivity)	0.83	0.92
True negative Rate (specificity)	0.94	0.94
False positive rate	0.06	0.06
Accuracy (ACC)	0.89	0.92

4.5   Comparison of unsupervised techniques

In the literature, the methods such as k-means [14] and Mean shift segmentation (MSS) [15] are widely used for image segmentation. In our study, we applied these methods to segment before flood (March 2009) image.

K-means is a parametric clustering method; here 2 clusters are generated for both classes (water and non-water) to cluster before flood image. After clustering, we have used a SI threshold value of 2.0 and a DI threshold value of 0.9 to segment the image. RMSE value for k-means to segment pre-flood image is 0.45.

One more method for image segmentation is Mean shift segmentation which is a popular non-parametric method based on kernel density estimation [15]. This method helps in identifying river network feature in the image. Initially, arbitrary point is chosen in the feature space and move towards the locally maximal density. It is an iterative procedure, where the modes of the density are the convergence points. It has been observed that RMSE value for MSS to segment before flood image is 0.26. In comparison with these two methods, RMSE value of GA is 0.18 and that of PSO is 0.15. From this result, we conclude that the approaches used in the paper are reliable techniques for linear segment extraction and thus can be successfully be used to map river courses and evaluate the flood extent in a river.

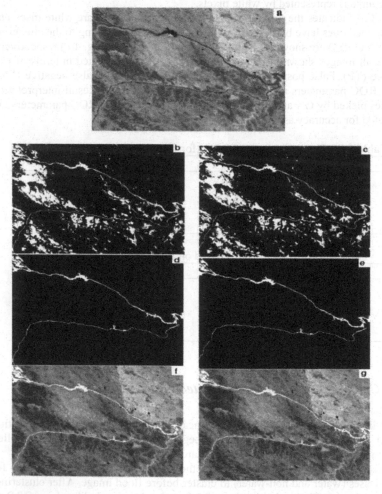

Fig. 3(a) Original MODIS image (before flood) (b) Image clustering using GA (c) Image clustering using PSO (d) Segmented image of GA clustered image (e) Segmented image of PSO clustered image (f) GA based river extraction overlaid on original image and (g) PSO based river extraction overlaid on original image

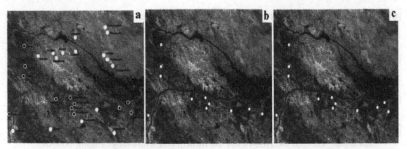

Fig. 4(a) MODIS during flood image with ground truth information – flooded (white discs) and non-flooded cities (black discs at the centre) (b) Segmented image using GA (White pts. are flooded cities) (c) Segmented image using PSO (White pts. are flooded cities)

5 Conclusions

The task of river mapping and flood extraction is accomplished successfully by the procedure of pixel based spectral information for clustering and shape information for region based segmentation as discussed above. In the clustering stage of extracting water and non-water groups, the GA and PSO are used. Results of clustering using spectral information are improved through region growing image segmentation based on geometrical features using similarity criteria resulting in effective water-covered regions.

References

[1] Islam, A.K., Bala, S.K. and Haque, A.: Flood inundation map of Bangladesh using MODIS surface reflectance data. 2nd Intl. Conf. on Water & Flood Management. (2009)

[2] Zhan, X., Sohlberg, R.A., Townshend, J.R.G., DiMiceli, C., Carroll, M.L., Eastman, J.C.: Detection of land cover changes using MODIS 250 m data. Remote Sensing of Environment. 83, 336–350 (2002)

[3] Khan, S.I., Hong, Y., Wang, J., Yilmaz, K.K., Gourley, J.J.,Adler, R.F., Brakenridge, G.R., Policelli, F., Habib, S., and Irwin, D.: Satellite remote sensing and hydrologic modelling for flood inundation mapping in lake Victoria Basin: Implications for hydrologic prediction in ungauged basins, IEEE Tran.on Geoscience and Remote Sensing. 49, 85–95 (2011)

[4] Bosco, G.L.: A genetic algorithm for image segmentation. Proc. IEEE 11thIntl Conf. on Image Analysis and Processing. 262-266 (2001).

[5]   Nagesh, K.D., Janga R.M.: Multipurpose reservoir operation using particle swarm optimization. J Water Resource Plan Manage ASCE. 133,192–201 (2007)

[6] Huang, W., Zhang, X.: Projection Pursuit Flood Disaster Classification Assessment Method Based on Multi-Swarm Cooperative Particle Swarm Optimization, Journal of Water Resource and Protection. 3, 415-420 (2011)

[7] Omran, M.G., EngelBrecht, A.P., Salman, A.A.: Particle swarm optimization for pattern recognition and image processing. Swarm Intelligence in Data Mining, 34, 125-151(2006).

[8] Mingjun, S., Daniel, C.: Road extraction using SVM and image segmentation. Photogrammetric Engineering & Remote Sensing. 70 (12), 1365–1371 (2004)

[9] Hamerly, G., Elkan, C.: Alternatives to the K-means algorithm that find better Clusterings. Proc. of the ACM Conf. on Inform and Knowledge Mgmt. 600–607 (2002)

[10] Senthilnath, J., Shivesh, B., Omkar, S.N., Diwakar, P.G., Mani, V.: An approach to Multi-temporal MODIS Image analysis using Image classification and segmentation. Advances in Space Research. 50(9), 1274 – 1287 (2012)

[11] Kennedy, J., Eberhart, R.: Particle Swarm Optimization. In Proc. of IEEE Intl. Conf.on Neural Networks. 4, 1942–1948 (1995)

[12] Fawcett, T.: ROC Graphs: Notes and Practical Considerations for Researchers. Technical Report HPL-2003-4, HP Labs. (2006)

[13] Sanyal, J., Lu, X. X.: Remote sensing and GIS-based flood vulnerability assessment of human settlements: a case study of Gangetic West Bengal, India. Hydrological Processes 19, 3699–3716 (2005)

[14] Macqueen. J.: Some methods for classification and analysis of multi-variate observations. In Proc. 5th Berkeley Symp. 281-297 (1967)

[15] Comaniciu, D., Meer, P.: Mean shift: A robust approach toward feature space analysis. IEEETrans. Pattern Anal.Mach. Intell., 24 (5), 603–619 (2002)

Parallelization of Genetic Algorithms and Sustainability on Many-core Processors

Yuji Sato

Faculty of Computer and Information Sciences, Hosei University

3-7-2 Kajino-cho, Koganei-shi, Tokyo 184-8584, Japan

yuji@k.hosei.ac.jp

Abstract. In this paper, we study and evaluate fault-tolerant technology for use in the parallel acceleration of evolutionary computation on many-core processors. Specifically, we show running evolutionary computation in parallel on a GPU results in a system that not only performs better as the number of processor cores increases, but is also robust against any physical faults (e.g., stuck-at faults) and transient faults (e.g., faults caused by noise), and makes it less likely that the application program will be interrupted while running. That is, we show that this approach is beneficial for the implementation of systems with sustainability.

Keywords: Genetic Algorithm, Many-core Processors, Fault-tolerance, Sustainability

1 Introduction

As an approach to speeding up evolutionary computation, the use of evolutionary computation methods that run on massively parallel computers has been actively researched since the 1990s [1, 2]. On the other hand, a recent trend has been towards the growing use of ordinary PCs with inexpensive multi-core processors aimed at small-scale parallelization using several cores or several tens of cores [3–5]. Research has also started on accelerating ordinary programs by using graphics processing units (GPUs) developed for the purpose of accelerating the processing of computer graphics. Against this background, the study of parallelizing evolutionary computation through the use of multi-core processors and many-core processors such as GPUs is getting under way [6–10]. A many-core processor contains multiple core processors of the same specifications. It should therefore be possible to achieve improved fault tolerance and reliability by effectively exploiting this feature. A number of fault-tolerant and enhanced reliability technologies have hitherto been proposed, principally for applications such as computers and LSIs, and there are also a good number of practical examples such as using multiplexing techniques to make computers more reliable, or using redundancy techniques to improve the yield of semiconductor memory devices. Of these conventional techniques, most of the ones that use redundancy are centered on techniques that presume a multiple module configuration, and are considered suitable for installation on many-core environments comprising multiple core processors.

On the other hand, while conventional redundant technology presumes a structure with a regular arrangement of modules with identical specifications, a many-core processor has a hierarchical structure and is configured as a system with a different architecture at each level. For example, a GPU consists of multiple streaming multi-processors (SMs), each comprising a number of core processors. This results in a hierarchical structure where the architecture inside an SM differs from the architecture between SMs. Also, for example, the GPU GTX590 consist of two GPU (GTX580) are networked together. Accordingly, there is thought to be a need for new fault-tolerant techniques for many-core processors to replace these existing techniques.

In section 2 of this paper, we present some typical conventional fault-tolerant techniques. In section 3, we investigate the relationship between fault-tolerant techniques and the parallelization of evolutionary computation on a GPU. In section 4, we experimentally evaluate situations where stuck-at faults and transient faults are assumed to occur, and finally we conclude with a summary.

2 Typical Fault-tolerant Techniques

2.1 Multiplexing

Static redundancy [11] involves using additional components to allow the effects of faults to be completely hidden (this is called "fault masking"). Typical techniques of this sort are module multiplexing and error correcting schemes, but when a fault can cause arbitrary errors to appear at the output, the fault can only be masked by a multiplexing scheme. We will consider an example of a multiplexing scheme here. To illustrate the basic concept of a multiplexing scheme, Fig. 1 shows the basic configuration of an N-Modular Redundancy (NMR) scheme. In this figure, the boxes labeled with M represent identical modules, from which the final output is produced via a majority voting element V.

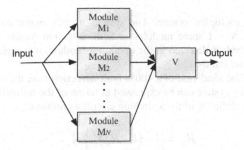

Fig. 1. Basic configuration of the multiplexing scheme

In an NMR scheme, it is possible to mask faults in up to $n = (N-1)/2$ modules. Thus, if the system reliability $R(t)$ is defined as the probability that no fault will occur before time t on condition that there are no faults in the system at time 0, then the reliability R_{nmr} of the NMR system is as follows:

$$R_{nmr} = R_v \sum_{i=0}^{(N-1)/2} \binom{N}{i} (1 - R_m)^i R_m^{(N-i)}$$ (1)

Here, R_m and R_v represent the reliability of a single module and the reliability of the voting element, respectively. For example, when it is assumed that there is no fault in the voting element, the reliability R_{3mr} of triple modular redundancy is given by the following formula:

$$R_{3mr} = 3R_m^2 - 2R_m^3$$ (2)

For each constituent module of an NMR system, if it is assumed that the early failure period has elapsed and the system has entered the period of fixed failure rate, then the failure rate R_m of a single module is given by $R_m = e^{-\lambda}$, where λ is the fixed failure rate. Substituting this value of R_m into Equation (2) yields the following formula:

$$R_{3mr} = 3e^{-2\lambda t} - 2e^{-3\lambda t}$$ (3)

2.2 Stand-by redundancy

In static redundancy schemes such as multiplexing, faults are masked by using redundancy, and the faults themselves continue to exist within the system. Therefore, the number of hidden faults gradually increases as the system continues to operate for a longer period of time. When the number of faults exceeds $(N - 1)/2$, errors will occur in the voting output and the system will fail. To deal with this problem, dynamic redundant systems [11] have been proposed. These consist of a fault detection means and a system reconfiguration means. For example, Fig. 2 shows a conceptual diagram of a stand-by redundancy system with a similar configuration to that of the above multiplex system. The system in this figure comprises a single operating module, $N - 1$ spare modules, a fault detection means, and a switching function. When a fault is detected in the operating module, it is replaced with one of the spare modules on stand-by.

Here, assuming the ideal case of a 100% fault detection rate, the reliability R_{sb} of a stand-by redundancy system can be expressed in terms of the reliability R_m of a single module and the reliability R_s of the switching circuit as follows:

$$R_{sb} = \left\{ 1 - \left(1 - R_m \right)^N \right\} R_s \tag{4}$$

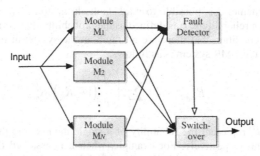

Fig. 2. Conceptual diagram of a stand-by redundancy system

3 Parallelization and Fault-tolerant Technology for EC

3.1 The problems of conventional methods

In general, multiplexing and stand-by redundancy incur costs that rise in proportion with the number of modules, so the application of these methods to real systems has chiefly involved duplex or triplex architectures. For example, the 3B20D processor developed by AT&T for electronic exchange networks [12] used duplex technology for the CPU, memory and I/O disk system. Another example is the Tandem 16 system developed by Tandem Computers for processing online transactions by organizations such as banks [13]. This was a reconfigurable multiprocessor system that achieved greater reliability by duplexing the disk devices and the buses between processors. C.vmp [14] was a multiprocessor system that could function correctly even with a mixture of permanent faults and temporary faults in the hardware. It achieved this by using a triplex configuration. These technologies are all geared towards architectures where identical modules are connected according to fixed rules. For example, if an SM is regarded as a single module, then it is suitable for implementation on a GPU.

On the other hand, in multiplexing schemes, the slowness of communication via global memory causes problems when voting circuits are implemented at the CPU end in consideration of regularity, and it is necessary to investigate how to implement voting circuits. Furthermore, it is not possible to guarantee correct results when there are faults in more than half the SMs. A problem with implementing stand-by redundancy on a GPU is the low utilization rate of the SMs that occurs as a result. Complex technology is needed to implement online processing of switching with redundant parts. When implemented with offline processing, this raises the problem of having to temporarily halt the execution of application programs. Extra hardware is needed to perform switching with redundant parts, and it is also conceivable that faults may occur in the fault detector circuits or switching circuits. Also, since a GPU has a hierarchical structure while an SM internally comprises multiple multi-core processors, the architectures inside SMs and between SMs are configured differently. We can also consider large-scale systems where multiple GPUs are networked together. Accordingly, in large-scale systems based on many-core architectures, it may become necessary to investigate new fault-tolerant technology as an alternative to conventional schemes.

3.2 Proposal of fault-tolerant technology based on parallelization of EC

The purpose of this study is to propose technology for improving the performance of active application programs and achieve greater fault tolerance by performing evolutionary computation in parallel on a GPU. As a first step, we investigate an example where the evolutionary computation of earlier studies is implemented on a GPU. Figure 3 illustrates the basic architecture of Nvidia's GTX460 GPU, and the method used to implement the parallel evolutionary computation model.

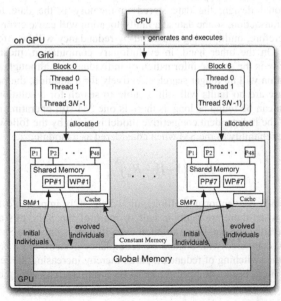

Fig. 3. GTX460 architecture and implementation of the parallel evolutionary computation model

The GTX460 consists of 7 SMs and a large (1 GB) global memory. Each SM has 48 core processors and a small (48 KB) shared memory. In each SM, it is possible to define up to 1536 threads. Communication inside the SMs can be performed at high speed. But communication between SMs is performed via the global memory and is about 100 times slower. We therefore consider a model where the same GA is run independently in each SM by changing the random initial value, and the program is terminated when an SM that has obtained a solution is found. We will refer to this as an "independent competition model".

The procedure of the parallel GA model for GPU computation is as follows.

(1) In the host machine, all individuals (population size/SM × #SM) are randomly generated and then sent to the global memory of the GPU.

(2) Each SM copies the corresponding individuals from global memory to its shared memory, and the genetic manipulation process is repeated until the termination criteria are satisfied.

(3) Finally, each SM copies the evolved individuals from its shared memory to global memory.

Since there is only a small amount of shared memory inside the SM, the number of individuals that can be stored inside the SM is limited. Also, since the number of core processors in an SM is at most a dozen or so, the search performance of a single SM is not necessarily very high. However, the total population size per GPU is same as the number in the host, therefore the number of generations required to get a feasible solution will be maintained. In an independent competition model, acceleration is achieved due to the effects of parallelization.

In conventional design, the date stored in memory is the date for the fixed calculation or transaction, so the date at a faulty location will cause errors to appear at the output. Therefore, multiplexing and stand-by redundancy will be effective for the fault masking. On the other hand, in evolutionary computation, most of the data stored in memory is the genetic information of individuals, so although a gene stored at a faulty location will no longer search effectively for a solution, the populations in SMs where there are no faults will still be able to search for a solution. Therefore, since a solution can be found as long as there is one SM still operating normally, the reliability R_p of the independent competitive model is given by the following formula, where R_m is the reliability of the SM when considered as a module:

$$R_p = \left\{ 1 - \left(1 - R_m\right)^N \right\} \tag{5}$$

Compared with conventional multiplexing where it is impossible to guarantee correct results when there are faults in half or more of the SMs, this approach has higher fault tolerance and can find a solution even if there is only one SM functioning normally. Also, compared with a stand-by redundancy system, there is no need for equipment for the switching of redundant parts, thereby increasing the reliability by a factor of $1/R_s$. There is also no need for extra hardware such as voting circuits, fault detectors or switching circuits.

4 Evaluation Experiments

4.1 Evaluation method

We performed an evaluation in which stuck-at faults and transient faults were assumed to occur. For stuck-at faults, the GTX460 was used to simulate a physical fault such as a fault in the shared memory inside the SM or in the path connecting a core processor to the shared memory. In the SM where a fault has occurred, it is considered that correct genetic operations are prevented from running, and a correct solution cannot be found. With an SM regarded as a single module, we investigated the reliability and performance (execution time needed to obtain a correct solution) of three methods — multiplexing, stand-by redundancy, and parallel evolutionary computing. The reliability comparisons were performed by comparing the relationships between time and reliability based on Equations (1), (4) and (5), with the number of modules fixed at 7. The performance comparisons were made by investigating the execution time taken to obtain a correct solution while varying the number of SMs used in the experiment from 1 to 7 (i.e., while varying the number of faulty SMs from 6 down to 0). The evaluation was performed using a Sudoku solver program based on evolutionary computation.

For transient faults, it is assumed that the faults consist of temporary non-repeating changes to data values inside each SM due to the effects of noise and the like. In the evaluation, these faults were assumed to manifest as errors whereby the IDs of parent individuals are randomly switched during the selection phase. The experiment was performed using an Intel Core i7 processor, with the error frequency varied as a parameter. Evaluations were performed using the knapsack problem which restricts the number 40 and Equation (6) below, which was chosen from the five types of function minimization problems shown by De Jong.

$$F_2 = 100\left(x_1^2 - x_2\right)^2 + \left(1 - x_1\right)^2 \qquad (6)$$

We used the tournament selection and the parameters used for genetic manipulation in these evaluation tests are shown in Tables 1.

Table 1. The parameters for genetic manipulation

Population size	Maximum generation	Tournament size	Crossover rate	Mutation rate
100	30,000	4	0.7	0.01

4.2 Experimental results and discussion

4.2.1 Evaluation of results for stuck-at faults

Comparative evaluation of reliability. Figure 4 shows the relationship between reliability and time t for a single module, multiplexing, stand-by redundancy and the evolutionary computation model. Although this figure shows the results for a single module failure rate of $\lambda = 0.001$, there is no change in the overall trends for different values of λ.

From this figure, it can be seen that the reliability R_{nmr} of multiplexing is larger than the reliability R_m of a single module up to a certain time T. The value of R_m has been observed to reverse for sufficiently large values of t, but since the reliability after a sufficiently long time had elapsed is not of major importance, it can be considered that multiplexing is effective for practical purposes. The reliability R_{sb} of stand-by redundancy is calculated by assuming a fixed value of 0.9 for the switching circuit reliability R_s. In fact, the initial value of R_s is closer to 1, but considering that there is the possibility of a fault occurring in the fault detector circuit and that the value of R_s also decays over time, we performed these calculations with a fixed value of 0.9 for the sake of convenience. The value of R_{sb} starts off close to 1, and although there is slight degradation in the reliability R_{nmr} of the multiplexing scheme during the initial stage, the reliability greater than R_{nmr} tends to be maintained as the system operation time becomes longer and the number of hidden faults gradually increases. The reliability R_p of the parallelized evolutionary computation model maintains the highest value throughout the entire period, and in terms of reliability it seems that this is an effective approach to achieving fault-tolerance in many-core architectures such as GPUs. It also has the advantage of making it unnecessary to add extra hardware such as voting circuits and switching circuits.

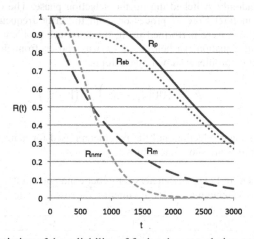

Fig. 4. Variation of the reliability of fault-tolerant techniques with time t.

Comparative evaluation of performance. Table 1 shows the relationship between the number of SMs and the performance achieved when evolutionary computation for solving Sudoku puzzles is performed in parallel on a GTX460 GPU. In Table 1, while varying the number of SMs used, we calculated the number of times a correct solution was obtained out of 100 attempts where the processes truncated at 100,000 generations (Count), the average number of generations needed to obtain a correct solution, and the computation time. However, with no set truncation point, a correct answer would have been obtained in all cases, even with just one SM.

Also, Fig. 5 shows how the execution time varies with the number of SMs in the multiplexing scheme and the parallelized evolutionary computation model. The execution times of the evolutionary computation model are the measured values shown in Table 2, and the execution times of the multiplexing scheme are the theoretical (lower bound) values for the ideal case where the time taken up by the voting logic is ignored. The execution time of the stand-by redundancy scheme was more or less the same as for the multiplexing scheme, and was constant as expected. The difference in execution times between the multiplexing and stand-by redundancy schemes corresponds to the difference in time needed to perform fault detection and switching and the time needed for the voting logic, and their relative merits depend on how they are implemented. In the multiplexing and stand-by redundancy schemes, the execution time is constant regardless of the number of SMs, whereas the parallelized evolutionary computation model has the advantage that the execution time needed to search for a solution decreases as the number of SMs increases.

Table 2. The number of generations until the correct solution was obtained, the execution time, and the rate of correct answers (SD1, Givens: 24)

	Count [%]	Average Gen.	Execution time
#SM: 1	62	57,687	16s 728
#SM: 2	80	40,820	11s 845
#SM: 4	98	19,020	5s 527
#SM: 7	100	10,014	2s 906

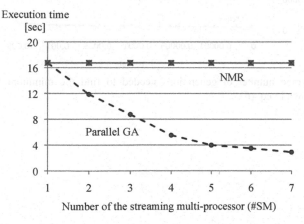

Fig. 5. Variation of execution time with number of SMs

4.2.2 Evaluation of results for transient faults

Figure 6 shows how the average number of generations needed to search for the minimum value of the function shown in Equation (6) varies with the number of threads. Figure 7 shows how the average number of generations needed to search for the solution of the knapsack problem which restricts the number 40.

Here, it can be seen that the average number of generations it takes to find a solution tends to increase gradually as the transient fault probability increases, regardless of how many threads are being used. However, in each case a solution was still obtained despite the addition of transient faults. Transient faults can also be thought to play a role in increasing diversity in the GA search process, and GA is thought to be intrinsically less susceptible to the adverse effects of transient errors.

Also, regardless of the probability of transient errors, it can be seen that the number of generations needed to find a solution decreases as the number of threads is increased (i.e., with increasing parallelism). It can thus be seen that parallelization of evolutionary computation in a many-core environment is not only robust against stuck-at faults but also against transient faults.

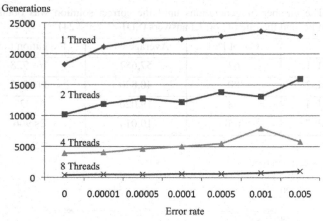

Fig. 6. Average number of generations needed to find the minimum value of a function shown in eq. (6).

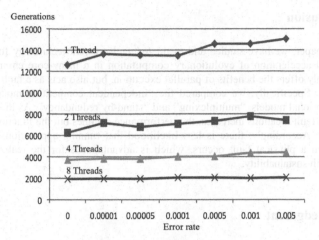

Fig. 7. Average number of generations needed to find the solution of the knapsack problem.

From the above, when parallel evolutionary computation is performed in many-core processors using a scheme based on independent competition, it seems that benefits such as higher reliability and lower susceptibility to transient errors can be achieved compared with when using conventional fault-tolerant techniques. There is also no need for extra hardware such as voting circuits or switching circuits. It is also thought to be effective at improving the performance as the degree of multiplicity or redundancy is increased. Although the performance decreases as the number of faulty physical SMs increases, a correct result can still be obtained even when there is only one functioning SM remaining, so it is though that this approach offers enhanced sustainability by increasing the performance of application programs running on the GPU and allowing application programs to continue running due to the increased fault tolerance.

In the future, it will be necessary to investigate the fault-tolerant performance in island models where modules communicate with each other as a parallelized evolutionary computation method using many-core processors. With regard to transient errors, it will be necessary to investigate in more detail whether there are any differences in trends due to the type of transient error or the problem dependencies. There is also a need for further research on the fault tolerance achieved when core processors inside the SMs are regarded as separate modules or when working with modules of two different types (core processors and SMs), and the fault tolerance of large-scale systems with multiple GPUs connected by a network.

5 Conclusion

In this paper, we have studied and evaluated fault-tolerant technology for use in the parallel acceleration of evolutionary computation in a many-core environment. This not only offers the benefits of parallel execution, but also acts as a fault-tolerant technology. Specifically, we compared the "independent competition model" with two conventional models, "multiplexing" and "stand-by redundancy". As the number of physical fault locations increases, the performance declines but the functionality is maintained. Accordingly, there is less likelihood that running applications will be halted when a physical fault occurs, which is advantageous for the realization of systems with sustainability.

Acknowledgment

This research is partly supported by the collaborative research program 2012, Information Initiative Center, Hokkaido University, and a grant from the Institute for Sustainability Research and Education of Hosei University 2012

References

[1] Mühlenbein, H.: Evolution in time and space - the parallel genetic algorithm. In Foundations of Genetic Algorithms, pp. 316–337. Morgan Kaufmann (1991)

[2] Shonkwiler, R.: Parallel genetic algorithm. In Proc. of the 5th International Conference on Genetic Algorithms, pp. 199–205 (1993)

[3] Pham, D., Asano, S., Bolliger, M., Day, M. N., Hofstee, H. P., Johns, C., Kahle, J., Kameyama, A, Keaty, J., Masubuchi, Y., Riley, M., Shippy, D., Stasiak, D., Suzuoki, M., Wang, M., Warnock, J., Weitzel, S., Wendel, D., Yamazaki, T., and Yazawa, K.: The design and implementation of a first-generation CELL processor. In 2005 IEEE International Solid- State Circuits Conference, vol. 1, pp. 184–592 (2005)

[4] Shiota, T., Kawasaki, K., Kawabe, Y., Shibamoto, W., Sato, A., Hashimoto, T., Hayakawa, F., Tago, S., Okano, H., Nakamura, Y., Miyake, H., Suga, A., and Takahashi, H.: A 51.2 gops 1.0 gb/sdma single-chip multi-processor integrating quadruple 8-way vliw processors. In 2005 IEEE International Solid-State Circuits Conference, vol. 1, pp. 194–593 (2005)

[5] Torii, S., et al.: A 600mips 120mw 70ua leakage triple-cpu mobile application processor chip. In the IEEE ISSCC Digest of Technical Papers, pp. 136–137 (2005)

[6] Byun, J.-H., Datta, K., Ravindran, A., Mukherjee, A., and Joshi, B.: Performance analysis of coarse-grained parallel genetic algorithms on the multi-core sun UltraSPARC T1. In SOUTHEASTCON'09. IEEE, pp. 301–306 (2009).

[7] Serrano, R., Tapia, J., Montiel, O., Sep´ulveda, R., and Melin, P.: High performance parallel programming of a GA using multi-core technology. In Soft Computing for Hybrid Intelligent Systems, pp. 307–314 (2008)

[8]   Tsutsui, S., and Fujimoto, N.: Solving quadratic assignment problems by genetic algorithms with GPU computation: a case study. In Proceedings of the 2009 ACM/SIGEVO Genetic and Evolutionary Computation Conference, pp. 2523–2530 (2009)

[9]   Sato, M., Sato, Y., and Namiki, M.: Proposal of a multi-core processor from the viewpoint of evolutionary computation. In Proceedings of the 2010 IEEE Congress on Evolutionary Computation, CD-ROM (2010)

[10] Sato, Y., Hasegawa, N., and Sato, M.: GPU Acceleration for Sudoku Solution with Genetic Operations. In Proceedings of the 2011 IEEE Congress on Evolutionary Computation, CD-ROM (2011)

[11] Lara, P. K.: Fault Tolerant and Fault Testable Hardware Design. Prentice-Hall International Ltd (1985)

[12] Toy, W. N., and Gallaher, L. E.: Overview and architecture of 3B20D processor. Bell Syst. Tech. J., Vol. 62, No. 1, pt. 2, pp. 181-19 (1983)

[13] Bartlet, F.: The Tandem 16; A "NonStop" operating system. In The Theory and Practice of Reliable System Design (Ed. By D. P. Siewiorek and R. S. Searz), pp. 453-460 (1982)

[14] Siewiorek, D. P., et al.: A case study of C.mmp, Cm and C.Vmp: Part 1 – Experience with fault-tolerance in multiprocessor systems. ibid., pp. 1178-1199 (1978)

Analysis and Simulation of Full Adder Design using MTCMOS Technique

Richa Saraswat[1], Shyam Akashe[2], Shyam Babu[3]

[1] ECED ,ITM University Gwalior, India

[2] ECED ,ITM University Gwalior, India

[3] ECED ,ITM University Gwalior, India

richasaraswat44@gmail.com; shyam.akashe@itmuniversity.ac.in,itm.shyam @gmail.com}

Abstract. The intention of this paper is to reduce leakage power and leakage current of 1-bit Full Adder while maintaining the competitive performance with few transistors are used (transistors count 10). A new high performance 1-bit Full Adder based on new logic approach is presented in this paper. MTCMOS technique which decreases the process variation on 1-bit Full Adder, the key of MTCMOS technique is applied on 1-bit Full Adder is to reduce the operating power, leakage power and leakage current. We investigate the use Multi-threshold CMOS (MTCMOS) technology provides low leakage and high performance operation by utilizing high speed, low Vth transistors for logic cells and low leakage, high Vth of transistor and show that it is particularly effective in sub threshold circuits and can eliminate performance variations with Low power. A 20ns access time and frequency 0.05GHz provide 45nm CMOS process technology with 0.7V power supply is employed to carry out 1-bit Full Adder.

Keywords: 1-bit Full Adder; MTCMOS; CMOS; Leakage Power; Leakage Current; Frequency.

1 Introduction

Adder is one of the most vital components of a CPU (central processing unit), Arithmetic logic unit (ALU), floating point unit and address generation like cache or memory access unit. On the other hand, increasing demand for portable equipments such as cellular phones, personal digital assistant (PDA), and notebook personal computer, arise the need of using area and Power efficient VLSI circuits. Low-power and high-speed adder cells are used in battery-operation based devices. As a result, design of a high-performance full-adder is very useful and vital [1]

One of the most well known full adders is the standard CMOS full adder that uses 28 transistors as shown in Fig.1. In this paper, we present a 1-bit full-adder circuit, which uses 10 transistor count with suitable power consumption, delay performance. The basic advantage of 10 transistors full adders are-low area compared to higher gate count full adders [2], lower power consumption, and lower operating voltage. It becomes more and more difficult and even outmoded to keep full voltage move backward and forward operation as the designs with fewer transistor count and lower power consumption are pursued [3]. In pass transistor logic, the output voltage move backward and forward may be de-graded due to the threshold loss problem Thus, attractive its presentation is significant for enhancing the overall module performance [4]-[5]. To implement probabilistic Boolean logic[6], a probabilistic gate produces a preferred value as an output that is 0 or 1 with probability p, and, hence, can produce the wrong output value with a probability (1 - p) [7]. It is supposed that probabilistic computing has potential for multimedia applications [8]. The basic disadvantage of the 10 transistors full adders are suffering from the threshold- voltage loss of the pass transistors. They all have double threshold losses in full adder output terminals [9]

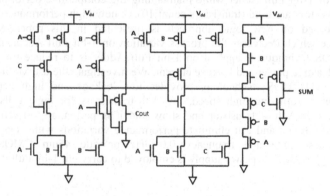

Fig.1. Schematic of conventional full adder

Multi-threshold CMOS (MTCMOS) technology provides low leakage and high performance operation by utilizing high speed, low Vth transistors for logic cells and low leakage, high Vth devices as sleep transistors. To reduce the leakage in sleep mode, sleep transistors disconnect logic cells from the power supply and/or. In this technology, also called power gating, wake up latency and power plane integrity are key concerns. The schematic of power gating technique using MTCMOS is shown in Fig. 2.The transistors having low threshold voltage are used to implement the logic. The transistors having high threshold voltage are used to isolate the low threshold voltage transistors from supply and ground during standby (sleep) mode to prevent leakage dissipation [10].

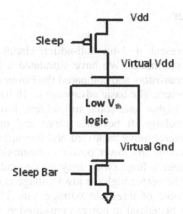

Fig.2. Power gating using MTCMOS technique

To determine the overall performance of the integrated circuits, the interconnect delay becomes the dominant factor. Since the delay of an interconnect is quadratic in its length, repeater insertion has been widely used to reduce the delay. As shown in [11] the repeaters can be optimally sized and divided to minimize the interconnect delay. The size of an optimal repeater is typically much larger than a minimum-sized repeater. To drive global interconnects, since millions of repeaters will be inserted, significant power will be consumed by these repeaters, predominantly if delay-optimal repeaters are used [12]. Several works used the extra acceptable delay for power saving in interconnects. Authors are provided investigative methods to calculate unit length power optimal repeater sizes and distances [13]. The power investigation should consider switching, leakage and short circuit correctly. This in turn increases the capacitive coupling noise on the interconnection lines, as the technology scales down wires are laid out closer to each other. This will have an effect on both delay and power consumption in interconnects. In addition to switching power on the coupling capacitances, the authors of [14] showed that the short circuit power consumption is increased indicate cantly in the being there of crosstalk noise. Therefore, one should also consider this effect in the design of power optimal repeaters. Moreover, the technology scaling has resulted in large increase in leakage current. Leakage power has grown exponentially to become a important fraction of the total chip power consumption [15]. Authors in [16] studied the applicability of MTCMOS to repeater design for leakage power saving, however they did not provide a mathematical explanation for the instantaneous optimal sizing of the sleep transistors and repeaters and the insertion length. In addition the effect of crosstalk on delay and power has not been taken into description for the optimal design.

2 1-Bit Full Adder

In this paper, we present a 1-bit full-adder circuit, with suitable power consumption, delay performance. We have simulated a 1-bit Full-adders circuit along with various 10 transistors and compared the Power dissipation, propagation delay, and other parameters. The basic advantage of 10 transistors full adders are-low area compared to higher gate count full adders, lower power consumption, and lower operating voltage. It becomes more and more difficult and even obsolete to keep full voltage move backward and forward operation as the designs with fewer transistor count and lower power consumption are pursued. In pass transistor logic the output voltage swing may be de-graded due to the threshold loss problem. That is, the output high (or low) voltage is deviated from the VDD (or ground) by a multiple of threshold voltage Vth. The reduction in voltage swing, on one hand, is beneficial to power consumption. On the other hand, this may lead to slow switching in the case of cascaded operation such as ripple carry adder. At low VDD operation, the degraded output may even cause malfunction of circuit [17]. Therefore, for designs using reduced voltage swing, special consideration must be paid to stability the power consumption and the speed. [18]. For the implementation of various 10 transistors full adder circuits we required either 4 transistors XOR circuit or 4 transistor XNOR circuit and 2-to-1 multiplexer. The schematic of 1- bit Full Adder is shown in figure 4, the output waveform is shown in figure 5.

This uses a total of 10 transistors for the implementation of following logic expressions.

Consider a 1-bit full adder. This circuit has two operands, A and B, and an input carry, Cin. It generates the sum

$$S = A \oplus B \oplus Cin \qquad (1)$$

and the output carry

$$Cout = AB + BC + AC \qquad (2)$$

Fig.3. Symbol Diagram of 1Bit full adder

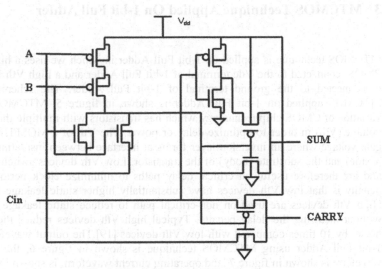

Fig.4. Schematic of 1-bit Full Adder

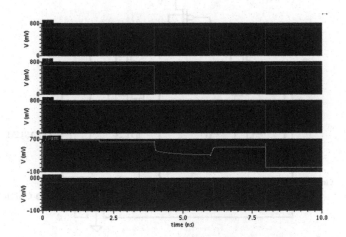

Fig.5. Output Waveform of 1-bit Full Adder

3   MTCMOS Technique Applied On 1-bit Full Adder

MTCMOS technique is applied on 1-bit Full Adder in which we uses a high Vth PMOS connected to the Vdc terminal of 1-bit Full Adder and a high Vth NMOS is connected to the ground terminal of 1-bit Full Adder. The schematic of MTCMOS applied on 1-bit Full Adder is shown in figure 5. MTCMOS is a variation of CMOS chip technology which has transistors with multiple threshold voltage (Vth) in order to optimize delay or power. The Vth of a MOSFET is the gate voltage where an inversion layer forms at interface between insulating layer (oxide) and the substrate (body) of the transistor. Low Vth devices switch faster, and are therefore useful on critical delay paths to minimize clock period. The penalty is that low Vth devices have substantially higher static leakage power. High Vth devices are used on non-critical path to reduce static leakage power without incurring the delay penalty. Typical high Vth devices reduce the static noise by 10 times compared with low Vth devices [19].The output waveform of 1-bit Full Adder using MTCMOS technique is shown in figure 6, the output waveform is shown in figure 7 ,and operating current waveform is shown in figure 8.

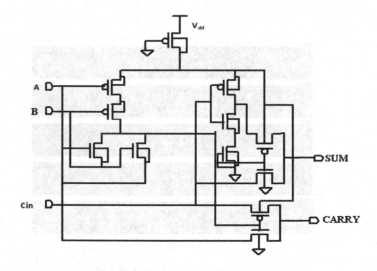

Fig.6. Schematic of 1-bit Full Adder using MTCMOS technique

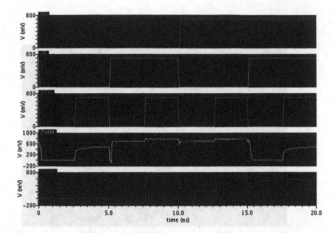

Fig.7. Output Waveform of 1-bit Full Adder with MTCMOS technique

Digital CMOS circuit may have three major sources of power dissipation namely dynamic, short and leakage power. Hence the total power consumed by every Full Adder can be evaluated using the equation 3.

$$P_{tot} = P_{dyn} + P_{sc} + P_{leak}$$

$$= CLV_{dd}Vf_{clk} + I_{SC}V_{dd} + I_{leak}V_{dd} \qquad (3)$$

Thus for low-power design the important task is to minimize CL V_{dd} V f_{Clk} while retaining required functionality. The first term P_{dyn} represents the switching component of power, the next component *Psc* is the short circuit power and P_{leak} is the leakage power. Where, CL is the loading capacitance, fClk is the clock frequency which is actually the probability of logic 0 to 1 transition occurs (the activity factor). Vdd is the supply voltage, V is the output voltage swing which is equal to Vdd; but, in some logic circuits the voltage swing on some internal nodes may be slightly less. The current I_{SC} in the second term is due to the direct path short circuit current which arises when both the NMOS and PMOS transistors are simultaneously active, conducting current directly from supply to ground. Finally, leakage current I_{leak}, which can arise from substrate injection and sub-threshold effects, is primarily determined by fabrication technology considerations. Duty cycle is also calculated in this paper for single bit full adder cell .In a periodic event, duty cycle is the ratio of the duration of the event to the total period of signal [20].

Duty Cycle (D) = τ / T
Where τ is the duration that functions is active. And T is the period of the function.

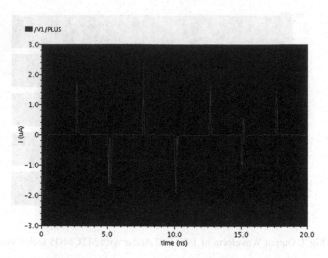

Fig.8. Operating current waveform of 1-bit Full Adder with MTCMOS

4 Simulation Result

A 1-Bit Full Adder based on MTCMOS technique have been proposed. The analysis of the simulated results confirms the feasibility of the MTCMOS technique in full adder design and shows that there is reduction of 40 to 43 percent in the value of power dissipation parameter as compared to CMOS technique at supply voltage of 0.7V. MTCMOS adders have a marginal increase in area compared to the CMOS adders; overall, we achieved the lowest power dissipation. Simulation result is measured by CANDENCE VIRTUOSO Tool .The Simulation result is summarized in TABLE 1.

Table 1. Simulated Result summary

Parameters	1-bit Full Adder	MTCMOS
Technology Used	45nm	45nm
Supply Voltage	0.7V	0.7V
Frequency Used	0.05GHz	0.05GHz
Access Time	20ns	20ns

Delay	3.94ns	2.05ns
Duty cycle	52.3 %	68.08 %
Leakage Power	1.594pW	0. 4pW
Leakage Current	652.4nA	230pA
Optimum Current	4.84 μA	73.3 nA
Optimum Power	0.53nW	2.3mW
Dynamic Current	18.02μA	1.52μA
Dynamic Power	2.086pW	1.766pW
Operating Power	51.63nW	20.03nW
Operating Current	1.804μA	2.35μA

5 Conclusion

In our investigation the 1-bit Full Adder and full adder using MTCMOS technique, we have estimated the design parameters such as operating current, operating power, leakage current, leakage power, optimum current, optimum power, dynamic current, dynamic power, delay and Efficiencies with the help of cadence virtuoso at 45nm technology. The advantage of having the same functionality with very few transistors will be beneficial in 1-bit Full Adder realization. Low Vth devices switch faster and are therefore useful on critical delay paths to minimize clock period. The penalty is that low Vth devices have substantially higher static leakage power. High Vth devices are used on non-critical path to reduce static leakage power without incurring the delay penalty. Typical high Vth devices reduce the static power by 10 times compared with low Vth devices.

References

[1] Rabaey J. M., A. Chandrakasan, B. Nikolic, Digital Integrated Circuits, A Design Perspective, 2nd 2002, Prentice Hall, Englewood Cliffs, NJ.

[2]   Dan Wang, Maofeng Yang, Wu Cheng, Xuguang Guan, Zhangming Zhu, Yintang Yang,
      "Novel Low Power Full Adder Cells in 180nm CMOS Technology", 4th IEEE
      Conference on Industrial Electronics and Applications, ICIEA 2009, pp 430-433.

[3]   Lu Junming; Shu Yan; Lin Zhenghui; Wang Ling," A Novel 10-transistor Low-power
      High-speed Full adder cell", Proceedings of 6th International Conference on Solid-State
      and Integrated-Circuit Technology, vol-2, pp. 1155-1158,2001.

[4]   Adarsh Kumar Agrawal, Shivshankar Mishra, and R. K. Nagaria, "Proposing a Novel
      Low-Power High-Speed Mixed GDI Full Adder Topology", accepted in Proceeding of
      IEEE International Conference on Power, Control and Embedded System (ICPCES), 28
      Nov.-1Dec. 2010.

[5]   N. M. Chore, and R. N. Mandavgane, "A Survey of Low Power High Speed 1 Bit Full
      Adder", Proceeding of the 12th International Conference on Networking, VLSI and
      Signal Processing, pp. 302-307,2010.

[6]   A. Bhanu. M. S. K. Lau, K. V. Ling, V. 1 Mooneylll, and A. Singh, "A more precise
      model of noise based CMOS errors," Proceedings of DELTA, 2010, pp. 99-102.

[7]   M. S K. Lau, K. V. Ling, Y C Chu, and A. Bhanu,"Modeling of probabilistic ripple-carry
      adders," Proceedings of DELTA, 2010, pp. 201-206.

[8]   J. M. Rabaey, A. Chandrakasan, and B. Nikoli' c, Digital Integrated Circuits: A Design
      Perspective, 3rd ed. Prentice Hall, 2003.

[9]   Shivshankar Mishra, V. Narendar, Dr. R. A. Mishra " On the Design   of High-
      Performance CMOS 1-Bit Full Adder Circuits," Proceedings published by International
      Journal of Computer Applications® (IJCA)2011.

[10] Hemantha S,Dhawan A and Kar H ,"Multi-threshold CMOS design for low power
      digital circuits",TENCON 2008-2008 IEEE Region 10 Conference,pp.1-5,2008.

[11] H. B. Bakoglu and J. D. Meindl, "Optimal interconnection circuits for VLSI," IEEE
      Trans. on Electron Devices, vol. ED-32, no. 5, pp. 903–909, May 1985.

[12] G. Chen and E. Friedman, "Low power repeaters driving RC interconnects with delay
      and bandwidth constraints," in Proc. Of ASIC/SOC, pp. 335–339, 2004.

[13] K. Banerjee and A. Mehrotra, "A power-optimal repeater insertion methodology for
      global interconnects in nanometer designs," IEEE Trans. on Electron Devices, vol. 49,
      pp. 2001–2007, Nov. 2002.

[14] H. Fatemi, S. Nazarian, and M. Pedram, "A current-based method for short circuit power
      calculation under noisy input waveforms," in Proc. of ASP-DAC, pp. 774-779, 2007.

[15] Semiconductor Industry Association, International Technology Roadmap for
      Semiconductors, 2003 edition.

[16] R. Rao, K. Agarwal, D. Sylvester, et al. "Approaches to run-time and standby mode
      leakage reduction in global buses," in Proc. Of ISLPED, pp. 188-193, 2004.

[17] Jin-Fa-Lin, Yin Tsung Hwang, Ming-Hwa Sheu, and Cheng-Che Ho, "A Novel High-
      Speed and Energy Efficient 10 Transistor Full Adder Design" IEEE Trans. Circuits Syst.
      I: Regular Papers, vol.54, no.5, pp.1050-1059, May 2007.

[18] H. T. Bui, Y. Wang, and Y. Jiang, "Design and analysis of low-power 10-transistor full
      adders using XOR-XNOR gates," IEEE Trans. Circuits Syst. II, Analog Digit Signal
      Process., vol. 49, no. 1, pp. 25–30, Jan. 2002.

[19] Anis, M.; Areibi, Mahmoud, Elmasry, (2002). "Dynamic and leakage power reduction in
      MTCMOS circuits". Design Automatio Conference, 2002. Proceedings 39th, pp 480–485

[20] . J. Rodrigues, O. C. Akgun, and V. Owall, "A <1 pJ sub-VT cardiac event detector in 65
      nm LL-HVT CMOS," in Proc. VLSI-SoC, June 2010.

Orientational selectivity is retained in zero-crossings obtained via stochastic resonance

Ajanta Kundu[1], Sandip Sarkar[2]

[1]Applied Nuclear Physics Division, Saha Institute of Nuclear Physics, 1/AF Bidhannagar, Kolkata-700064, India.

[2]Applied Nuclear Physics Division, Saha Institute of Nuclear Physics, 1/AF Bidhannagar, Kolkata-700064, India.

{ [1]ajanta.kundu@saha.ac.in; [2]sandip.sarkar@saha.ac.in }

Abstract: Computational theory of visual information processing suggest that the initial stages information processing consists of in part representation of zero crossing in the visual scene filtered through a suitable second order differential operator (centre-surround receptive field). These zero crossings often correspond to sharp intensity changes in the visual scene and are rich in information. We report here our investigation, through simulation study, on the role of zero crossings in orientational selectivity measurement. We show that the perceptive contrast sensitivity of zero-crossing of sub-threshold noise contaminated grating image exhibit stochastic resonance. We also show that the contrast sensitivity of test grating, in the presence of a masking grating, decreases with the increase of masking contrast.The qualitative nature of the contrast sensitivity variations are in agreement with the results of various phychophysical experiments.

Keywords: zero-crossing, contrast sensitivity, stochastic resonance, receptive field, LoG

1 Introduction

It has been demonstrated in many experiments that the addition of external noise to a weak signal can enhance its detectability by the peripheral nervous system of crayfish [1], cricket [2] and also human [3-7]. Direct evidences of performance enhancement for noise contaminated visual inputs have been demonstrated through psychophysical experiments [5-8]. It has been shown [7, 8] that the brain is capable of extracting detail in a stationary image masked with noise. The perceived image quality is strongly dependent on noise intensity. It has been observed that the detection of sub-threshold images, mediated by threshold crossing[1] (different from zero-crossing) events in the presence of noise improves non-

J. C. Bansal et al. (eds.), *Proceedings of Seventh International Conference on Bio-Inspired Computing: Theories and Applications (BIC-TA 2012)*, Advances in Intelligent Systems and Computing 202, DOI: 10.1007/978-81-322-1041-2_17, © Springer India 2013

monotonically with noise power. For all the experiments involving visual input the retinal computational network does the basic information processing which can be explained by center-surround receptive field [9, 10], Computational theory of visual information processing suggest [10-12] that the information processing, in the initial stages, consists of in part representation of zero crossings in the visual scene filtered through a suitable second order differential operator (usually $\nabla^2 G$ of different sizes where G is Gaussian). The prime motivation behind this suggestion is that the zero crossings often correspond to sharp intensity changes in the visual scene usually referred to as edges. These edges often correspond to surface discontinuities or surface reflectance or illumination boundaries of the visual environment. It has been also observed in speech experiment [13] that with only the zero-crossing information much of the information content of the speech could be retained. Studies in visual information processing [11, 14] also stresses the importance of the information content of zero-crossings in classifying visual scene.

Zero crossings (ZC) are, therefore, play an important role in the information retrieval but very little is known about its role in visual information processing that leads to noise induced performance enhancement. We investigate here the role of ZCs of sub-threshold image filtered through centre-surround receptive field in perceptive contrast sensitivity (PCS) enhancement in the presence of noise. We show that the PCS of ZC image is strongly dependent on the external noise strength and PCS attains a maximum for an optimal amount of noise via stochastic resonance (SR). We also show that the contrast sensitivity of ZCs test grating, in the presence of a masking grating, decreases in direct proportion to the masking contrast. Some important observations are that the qualitative nature of the contrast sensitivity variations is in agreement with the results of psychophysical experiment [7, 15].

2 Theoretical considerations

The very first stage of vision involves detection of intensity changes or ZC of the visual scene for the formation of the "raw primal sketch" [9, 10, 16, 17]. It is, therefore, expected that the ZC has a considerable role in visual information processing that leads to noise induced performance enhancement via a phenomenon called SR [18]. In natural image intensity changes occur over wide range of scales. At any given scale it can be detected, for an image $I(x,y)$, by finding zero values of $\nabla^2 G(x, y; t) \otimes I(x, y)$ where $G(x, y; t)$ is Gaussian function in two dimension with variance t, ∇^2 is the Laplacian and $\otimes$ represent convolution. For our analysis we revisit the proposition of Ernst Mach [19, 20] which states that the brightness sensation at any retinal point is the combined effect of the original illumination and its second differential quotient. The corresponding spatial filter function for a given scale t can then be written as

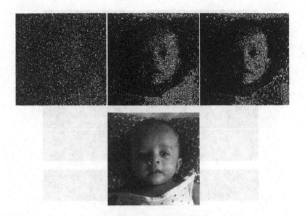

Fig. 1 This is a visual illustration of ZC, in the presence of additive noise, of a natural image S (infant portrait in the bottom row) digitized on a 0 to 1 gray scale. First the image S is pushed below threshold (defined by gray value 0), such that its mean lies Δ below the threshold resulting in the image $s - \Delta - \mu_s$ where the mean intensity of S is μ_S. After adding a random number ξ, from a zero-mean Gaussian distribution with standard deviation σ_n, to the original gray value S of every pixel, the image is convolved with filter mask (Equation (1)) resulting in $(S - \Delta - \mu_S + \xi) \otimes h(x, y; t)$. ZCs are then detected according to the following rule: if $sgn((S - \Delta - \mu_S + \xi) \otimes h(x, y; t))$ of any two consecutive pixels, either in the horizontal or in the vertical direction, is opposite, the gray value of the pixel with negative sign is made 0 (black), and the gray value of the other one is made 1 (white). For all other pixels the gray value is replaced with 0 (black). Three such cases are shown for $\sqrt{t} = 2.1$, $\sigma_0 = -0.15$, $m = 0.2$, $\sigma_n = 0.18$ and for contrast=0.1, 0.6 and 0.9 from left to right in the upper row.

$$h(x, y; t) = \delta(x, y) - m\nabla^2 G(x, y; t) \tag{1}$$

Where $\delta(x, y)$ is Kronecker delta function, m is a constant factor, and $G(x, y; t)$ is the Gaussian defined by

$$G(x, y; t) = \frac{1}{2\pi t} e^{\frac{-(x^2 + y^2)}{2t}} \tag{2}$$

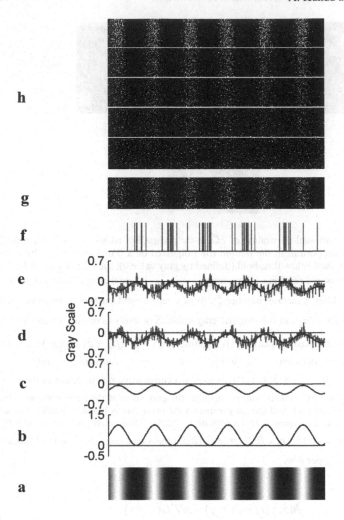

Fig. 2 (a) The sinusoidal grating image I with a spatial resolution of 512 by 512 pixels used for our experiment was generated by the spatial function $0.5A\sin(2\pi fx)+0.5$, where A is Michelson contrast[3], f is the spatial frequency in cycles/pixel and x is the horizontal coordinate in pixel. For the sake of illustration the image, with reduced spatial resolution of 400 by 64 pixels, is displayed with maximum contrast $A=1$. (b) The intensity variation along horizontal directions shown here. (c) After setting $A=0.3$, image is then pushed below the threshold (gray value 0), as depicted by the intensity profile in the figure, such that its mean lies $\Delta = 0.25$ below the threshold resulting in the image $I-(0.5+\Delta)$. (d) This is the resultant profile obtained after the addition of Gaussian noise of appropriate amount to the sub-threshold image, (e) The noise contaminated image is then convolved with the filer function $h(x,y;t)$. The resultant effect is visually represented in this figure. And in (f) the corresponding ZCs are depicted by a vertical line of equal height. (g) The resultant ZC image is shown where black dots (gray value of 0) represent ZC and the white dots (gray value of 1) represent absence of ZC. And in (h) similar ZC images with $A=0.05, 0.1, 0.2\ 0.3, 0.5$ are shown.

and t is its variance. This kind of representation is well suited for linear scale space theory of early visual operations [21-23]. This theory shows that Gaussian and its derivatives form a complete set of operators for multi-scale representation of an image. Additionally this computational paradigm is also in agreement with the biological visual computation. Studies [24, 25] have shown that superposition of Gaussian and its derivatives can be used to model the receptive field profiles of mammalian retina and visual cortex. Our focus is to study the behavior of ZC (filtered through $h(x, y; t)$) of sub-threshold images with additive noise contamination in contrast sensitivity measurement.

3 Methods

The necessary components are a filter function that mimics image enhancement in retina, a threshold (gray value 0), a threshold crossing detector, an image and an additive Gaussian noise. For demonstration a natural image of a face $S(x, y)$, as shown in Figure 1, and a sampled version[2] of the continuous filter function (Equation (1)), is used for the experiment. ZCs computed from the filtered images and the resultant output images for three cases with increasing contrast (left to right)are displayed. Though these are basically binary images containing pixels whose gray values are either 1 or 0, we could still perceive shading effect because the original intensity variation is mapped to ZC density variation through the process of ZC image formation thereby producing halftone like effect. Visually the first image, which is of lower contrast, is devoid of all shadow details. The third image, with much higher contrast, could retain comparatively better shadow details of the original face image. The middle ZC image, with intermediate contrast, on the other hand could retain best shadow detail. This simple experiment illustrates that the best representation of the original image can be achieved for an intermediate contrast level for a given amount of additive Gaussian noise contamination. We will study this phenomenon in more detail in all our experiments described below. In all the experiments, we have expressed the frequency f in cycles/pixel to avoid the explicit dependence of the visual angle. This can be converted to cycles/degree by the multiplication of an appropriate constant K_v , in pixel/degree, which is a function of the angle the image subtends at the viewer's eye.

4 Experiment 1

For a quantitative study, we choose a sinusoidal grating image generated by a function as shown in Figure 2(a) where a full contrast noise free grating image, without threshold filtering, is displayed. Its intensity variation along horizontal direction is shown in Figure 2(b).The resulting image profile, after contrast modification followed by threshold filtering by an amount Δ , will look like as shown in Figure 2(c). After the addition of a zero mean Gaussian noise a typical image profile will look like as shown in Fig. 2(d). This noise contaminated image is then convolved with the filter function given by Equation (1) resulting in the image determined by $(I - \Delta - 0.5 + \xi) \otimes h(x,y;t)$. A typical profile of this image is shown in Figure 2(e). Clearly, noise of appropriate amount helps in mediating threshold (gray value of 0) crossing events. These zero crossings (ZCs) are represented by a vertical line of unit length as depicted in Figure 2(f).The final ZC image, a binary image consisting of gray value 0 and 1, is illustrated in Figure 2(g). Here black dots (gray value of 0) represent ZC and the white dots (gray value of 1) represent absence of ZC. Starting from the grayscale pattern (Figure 2(a)) we arrive at the binary image, as in Figure 2(f), where only one bit of information marking ZC events, computed by our experimental system as described in Figure 2, is retained in every pixel. Varying only the contrast the whole process is repeated to compute the corresponding ZC image. Five such images are shown in Figure 2(h), with in-

Fig. 3 Typical image that is used for studying the contrast sensitivity variation in the presence of a mask is displayed. The background is the mask of spatial resolution 512x512 and the test grating is placed at the center of the mask. The sinusoidal grating image,with a spatial resolution of 256 by 256 pixels,is generated by the spatial function $0.5A \sin(2\pi fx) + 0.5$, where A is Michelson contrast[3], f is the spatial frequency in cycles/pixel and x is the horizontal coordinate in pixel. The background is an oriented sinusoidal grating with frequency equal to that of the test grating at the center.

creasing (bottom to top) contrast values. Visual inspection reveal that for low contrast the image looks devoid of any features and as the contrast is increased, the grating specific features begin to appear. With further increase in contrast, the images begin to look like square grating (loss of contrast detail). We can, therefore, visually identify an intermediate contrast for which the ZC images attain closest resemblance with the original sinusoidal grating. We mark this contrast as A_{opt} (optimal one) for the given noise. Numerically this is performed by inspecting the ZC images in the Fourier domain and looking for the maximum contrast for which the second harmonic does not appear. Similarly, by varying the noise strength and repeating the whole process we can identify an optimal contrast value for each of the noise strengths.

5 Experiment 2

This experiment is designed as follows. A test sinusoidal grating of resolution 256x256 pixels is placed at the center of a mask of resolution 512x512 pixels (Fig. 3). The mask is composed of another grating of the same spatial frequency but

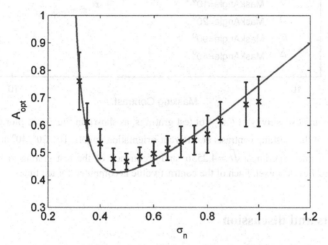

Fig. 4 Plots of optimal contrast A_{opt} versus the noise strengths σ_n for $\sqrt{t} = 2.1268$ and frequency $f = 0.10547$ cycles/pixelis shown. Each of the contrast value was computed for twenty times for twenty noise intensities. The error bars are the standard deviations of twenty A_{opt} values at each of the noise strength. The variation of optimal contrast as a function of noise strength is shown for $m = 2$, $\Delta = 0.3164$, $K = 0.8193$ and $\sigma_0 = -0.15$. The solid curve (red) is given by Equation (3) where the parameters K and σ_0 are obtained from nonlinear least square fit to the data (in blue) obtained in the experiment described in Fig 2.

oriented at angle with the test grating. By performing the experiment 1 with this image, as shown in Fig. 3, we will get an optimal contrast value for the test grating. When the experiment 1 is repeated by varying the contrast for a fixed orientation angle of the masking grating, we will get the variation of contrast sensitivity of the test grating as a function of the masking contrast. Again by repeating the experiment for different orientation angels, contrast variation for different orientation is obtained.

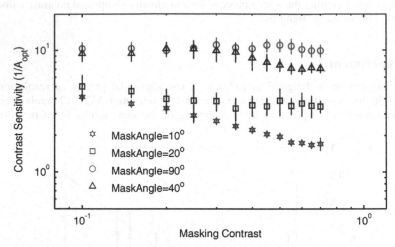

Fig. 5 Contrast sensitivity ($1 / A_{opt}$) of test gratings, as shown in Fig. 3, are plotted as a function of the masking contrasts at different orientation angels: $10°$, $20°$, $40°$ and $90°$. For the whole experiment $\sqrt{t} = 4.2536$ and frequency of the test grating is kept at $f = 0.0527$ cycles/pixel. Each of the contrast value is computed for ten times.

6 Results and discussion

The results of the study described in Fig. 2 and Fig. 3 are presented in Fig. 4 and Fig. 5. Results of Experiment 1 are summarised in Fig. 4 where the variation of A_{opt} with noise strength is plotted. The solid curve (red) given by

$$A_{opt} = K(\sigma_n + \sigma_0)\exp(\Delta^2 / 2(\sigma_n + \sigma_0)^2 \quad (3)$$

is the power spectrum of a pulse train, as shown in Fig. 2(e) and Fig. 3(g), and is taken from the threshold SR theory [7, 26]. The left hand side is the signal amplitude and K on the right hand side is proportional to signal power density. We have introduced an additional fitting parameter σ_0 in equation (3). The equation is fit to the data (blue) using K and σ_0 as the adjustable parameters. The quality of the fit to the simulated data is very encouraging. The good quality of the fit reveals that optimal contrast is proportional to the power contained in the signal (image) and therefore K can be regarded as a quantitative measure of the sensitivity to the power contained in the signal. K can also be regarded as a quantitative measure of the ability to distinguish details in images in the presence of additive Gaussian noise because A_{opt} in equation (3) is the optimal contrast value necessary for matching a ZC image with its original counterpart (as described in Fig. 2). Important findings are the following. (a) From the distribution of the data points in Fig. 4 it is evident that the optimal contrast varies non-monotonically with noise power and attains a minimum for an intermediate noise power which is a typical signature of SR. (b) The good quality fit of equation (3) (solid red curve) in Fig. 4 implies that the optimal contrast is proportional to the power contained in the image which is in line with the results obtained in [7]. (c) The fitted K value can be regarded as a quantitative measure of the ability to distinguish fine details in a subthreshold visual scene in the presence of additive Gaussian noise.

In Fig. 5contrast sensitivity of the test grating is plotted as a function of masking contrast at different orientation angles. For each of orientations angles, we get optimal contrast values, obtained via stochastic resonance, as a function of masking contrast. Inverse of measured contrast values, which is contrast sensitivity, are plotted in Fig. 5 as a function of masking contrast. It is evident from the figure that the contrast sensitivity decreases with the increase in the masking contrast. Our observation also shows that the masking effect is the maximum for the lowest orientation angle and the effect decreases with the increase of the angle. An important point to note is that the qualitative nature of these variations produced with the help of zero-crossings only, is also in agreement with the psychophysical studies [15].

7 Conclusions

Results of Experiment 1 & 2 show that the optimal contrast (contrast threshold) of images, formed from ZCs points alone, varies non-monotonically with noise power and attains a minimum, which signifies maximum contrast sensitivity, for an intermediate noise power, which is a typical signature of SR. These results are in agreement with the results of psychophysical experiments, which are evident from the goodness of fit of equation (3) to the simulated data. In other words, information content of ZCs alone can demonstrate the qualitative features of a subject's ability to distinguish fine details in noisy sub-threshold scenes via stochastic resonance. Our study also show that the measured (simulated) contrast sensitivity decreases with the increase of masking contrast and the masking effect decreases with the increase of masking angle. This behavior is qualitatively similar to the one observed in psychophysical studies [15]. An important point to note is that equation (3), which is fit to our simulated data obtained from ZCs points, is derived from the Fourier Transform of identical pulse trains similar to the neuron action potentials [26-28]. This essentially indicates that human brain may make use of similar ZC computation while processing noisy sub-threshold images. Though we do not advocate that ZCs can explain everything but all these results may provide some clues to ZC computation in our brain. These results are, therefore, may also be useful to study and analyze visual information processing.

Footnotes

[1]In case of threshold crossing all the events for which a function crosses a threshold is considered. On the contrary, for zero-crossing, only those events for which the function changes sign are considered.

[2]Use of sampled kernel can lead to implementation problems because the discrete version may not have the discrete analogs of the properties of the continuous function. For Gaussian function $G(n,t)$, truncated to give finite

impulse response, the support is generally chosen large enough such that $2 \int_{u=M}^{\infty} G(u,t) du < \epsilon$. A common choice of M is $M = C\sqrt{t} + 1$, where C is constant. For small values of ϵ ($\leq 10^{-6}$) the errors introduced by truncating the Gaussian are usually negligible. For our study $\varepsilon \leq 10^{-10}$ ($\sqrt{t}$ =2.12 and C $\geq$ 6).

[3] If I_{min} and I_{max} are the minimum and maximum intensity respectively in an image then the Michelson contrast is defined as $(I_{max} - I_{min}) / (I_{max} + I_{min})$

References

[1] Douglass, J. K., Wilkens, L., Pantazelou, E., Moss, F.: Noise enhancement of information transfer in crayfish mechanoreceptors by stochastic resonance, Nature, 365,337-340 (1993).

[2] Levin, J. E., Miller, J. P.: Broadband neural encoding in the cricket cercal sensory system enhanced by stochastic resonance, Nature, 380, 165-168 (1996).

[3] Chiou-Tan, F.Y., Magee, K.N., Robinson, L.R., Nelson, M.R., Tuel, S.S., Krouskop, T.A., Moss, F.: Enhancement of Subthreshold Sensory Nerve Action Potentials During Muscle Tension Mediated Noise. Intern. J. Bifurc. Chaos, 7, 1389. (1996).

[4] Collins, J. J., Imhoff, T. T., Grigg, P.: Noise-enhanced tactile sensation. Nature, 383, 770. (1996).

[5] Kitajo, K., Nozaki, D., Ward, L.M., Yamamoto, Y.: Behavioral Stochastic Resonance within the Human Brain. Phys.Rev. Lett. 90, 218103 (2003).

[6] Mori, T., Kai, S.: Noise-Induced Entrainment and Stochastic Resonance in Human Brain Waves, Phys. Rev. Lett. 88, 218101 (2002).

[7] Simonotto, E., Riani, M., Seife, C., Roberts, M., Twitty, J., Moss, F.: Visual Perception of Stochastic Resonance. Phys. Rev. Lett. 78, 1186-1189 (1997).

[8] Goris, R. L. T., Zaenen, P., Wagemans, J.: Some observations on contrast detection in noise. Journal of Vision 8(9):4, 1-15 (2008).

[9] Marr, D.: Vision: A Computational Investigation into the Human Representation and Processing of Visual Information. W H Freeman and Company, New York (1982).

[10] Marr, D., Hildreth, E.: Theory of Edge Detection. Proc. R. Soc. Lond. Series B, Biological Sciences. 207. 187-217 (1980).

[11] Marr, D., Ullman, S., Poggio, T.: Bandpass channels, Zero crossings, and early visual Information processing. J. Opt. Soc. Am., 69(6), 914-916 (1979).

[12] Marr, D., Ullman, S.: Directional selectivity and its use in early visual processing. Proc. Soc. Lond. B 211, 151-180 (1981).

[13] Licklider, J. C. R., Pollack, I.: Effects of differentiation, integration and infinite peak clipping upon the intelligibility of speech. J. Acoust. Soc. Amer., 20, 42-51 (1948).

[14] Curtis, S. R., Oppenheim, A. V., Lim, J. S.: Reconstruction of two-dimensional signals from threshold crossings. Acoustics, Speech, and Signal Processing, IEEE International Conference on ICASSP '85, 10, 1057 – 1060 (1985).

[15] Campbell, F. W. and Kulikowski, J. J. : Orientational selectivity of the human visual system. J. Physiol. (1966), 187, 437-445 (1966).

[16] Marr, D., Poggio, T., Ullman, S. J.: Opt. Soc. Am. 70: 868-70 (1979).

[17] Ullman, S.: Artificial Intelligence and The Brain: Computational Studies of the Visual System. Ann. Rev. Neuroscience, 9, 1-26 (1986).

[18] Sarkar, S., Ghosh, K., Bhaumik, K.: Proceedings of the 3rd Indian International Conference on Artificial Intelligence, Pune, India, December (2007), ISBN 978-0-9727412-2-4.

[19] Mach, E. (1868). On the physiological effects of spatially distributed light stimuli. Translated in F Ratliff, "Mach Bands: Quantitative Studies on Neural Networks in the Retina," Holden-Day, Sanfrancisco, 299-306, (1965).

[20] Mead, C.: Neuromorphic electronic systems. Proceedings of the IEEE, 78(10), 1639-1636 (1990).

[21] Koenderink, J. J. : The Structure of Images. Biological Cybernetics, 50, 363–370 (1984).

[22] Lindeberg, T. : Scale-Space Theory: A Basic Tool for Analyzing Structures at Different Scales. Journal of Applied Statistics 21 (2), 224–270 (1994).

[23] Yuille, A. L., and Poggio, T.A.: Scaling Theorems for Zero Crossings. IEEE Trans. Pattern Analysis & Machine Intelligence, PAMI-8(1), 15–25 (1986).

[24] Young, R. A.: The Gaussian derivative model for spatial vision: I. Retinal mechanisms. Spatial Vision2 (4), 273–293 (1987).

[25] Young, R. A., Lesperance, R. M., and Meyer, W. W.: The Gaussian Derivative model for spatial-temporal vision: I. Cortical model. Spatial Vision 14 (3-4), 261–319 (2001).

[26] Gingl, Z., Kiss, L. B., Moss, F.: Non-Dynamical Stochastic Resonance: Theory and Experiments with White and Arbitrarily Colored Noise. Europhys.Lett.29 (3), 191-196 (1995).

[27] Gingl, Z., Kiss, L., Moss, F.: Nuovo Cimento D, 17, 795 (1995).

[28] Campbell, F. W., Robson, J. G.: Application of Fourier analysis to the visibility of grating, J. Physiol. 197, 551-566 (1968).

Multi-Objective Fault Section Estimation
in Distribution Systems using Elitist NSGA

Anoop Arya[1], Yogendra Kumar[2], Manisha Dubey[3], Radharaman Gupta[4]

Deptt. of Electrical Engg. MANIT, Bhopal (MP)

{E-mail:anooparya.nitb@gmail.com}

Abstract- In this paper, a non-dominated sorting based multi objective EA (MOEA), called Elitist non dominated sorting genetic algorithm (Elitist NSGA) has been presented for solving the fault section estimation problem in automated distribution systems, which alleviates the difficulties associated with conventional techniques of fault section estimation. Due to the presence of various conflicting objective functions, the fault location task is a multi-objective, optimization problem .The considered FSE problem should be handled using Multi objective Optimization techniques since its solution requires a compromise between different criteria. In contrast to the conventional Genetic algorithm (GA) based approach; Elitist NSGA does not require weighting factors for conversion of such a multi-objective optimization problem into an equivalent single objective optimization problem and also algorithm is also equipped with elitism approach. Based on the simulation results on the test distribution system, the performance of the Elitist NSGA based scheme has been found significantly better than that of a conventional GA based method and particle swarm optimization based FSE algorithm. Multi Objective fault section estimation problem have been formulated based on operator experience, customer calls, substation & recloser data. Results are used to reduce the possible number of potential fault location which helps and equipped the operators to locate the fault accurately.

Keywords- Automatic distribution systems, Fault section estimation, Genetic Algorithms, Elitist NSGA, Particle Swarm Optimization.

1. Introduction

For estimation of fault section, the maintenance crews rely mainly on phone calls by customers and trial and error methods [1], [2]. But in this process, it takes several hours to identify the exact location of fault. Over the last three decades, due to technological progress in computers and electronics, power system has been equipped with digital relays. Traditionally, fault diagnosis is performed off line by experienced engineers. However, software tools for fault location have emerged in recent years. To improve the accuracy and speed of fault location the information is stored in a database and intelligent systems in a control centre and can be accessed for diagnosis of a fault event. Data recorded by recorders at substation, customer phone calls location and status of reclosers are used [2]. The protection system is a part of the power system responsible for fault detection and execution of automatic remedial actions. When fault appears different devices that comprise protection system are triggered. Protection system consisting of protection relays and circuit breakers (CBs) will operate in order to de-energize faulted line. Different Intelligent Electronic Devices (IEDs) and EMS located in substations for the purpose of monitoring will be automatically triggered by the fault and will record corresponding current, voltage and status signals. Those records are later used by different user groups for fault investigation. The goal of power system fault analysis is to provide enough information to utility staff to be able to understand the proper reasons for the interruption, and provide an action as quick as possible to restore the power supply. The analysis should also provide enough understanding of the status of protection system components so that a preventive set of measures can be implemented to reduce the likelihood of service interruption and damages to equipment [3].

The scope of power system fault analysis can be generally classified into two categories:
- Fault event analysis and,
- Protection system performance evaluation.

Fault event analysis focuses on determination of faulted section, fault type and fault inception angle. Protection system performance evaluation is to check whether a protection system has operated as expected, and if not, what are the causes. The first step in restoring systems after a fault is detected, is determining the fault location. The large number of candidate locations for the fault makes this a complex process. If the fault is not detected or cleared in time, it may cause cascaded events leading to major outages. Knowledge based methods have the capability to accomplish this quickly and reliably.

When a fault occurs, protection equipment initiates operation of breakers, to de- energize faulted part. This must be done before excessive currents and voltages caused by the fault inflict damage to the connected equipment. CBs have

J. C. Bansal et al. (eds.), *Proceedings of Seventh International Conference on Bio-Inspired Computing: Theories and Applications (BIC-TA 2012),* Advances in Intelligent Systems and Computing 202, DOI: 10.1007/978-81-322-1041-2_18, © Springer India 2013

the purpose to automatically connect or disconnect different parts of the power system in order to isolate the faults and/or re-route the power flow. In order to open all circuits that supply fault current, more than one CB typically reacts. Various bus arrangements are used to minimize the number of circuits that must be opened in a case of a fault [4]. Depending on a bus arrangement and status of available breakers, different breakers will automatically react in case of different faults.

Artificial Neural networks (ANN) is a solution technique with a great potential for this problem, as a demonstrated in [5]. Fuzzy logic (FL) [6] is flexible and suitable for modelling inaccuracies. Its major drawbacks is choosing the membership function, commonly define by historical data, experience or trial and error. Expert System (ES) [7], [8] are not capable of generalizing and present difficulties when validating and maintaining large knowledge bases. To solve the above mentioned limitations, GA based techniques have been proposed in the literature [9], [10]. Zhengyou He et.al. [12] proposed Binary Particle Swarm Optimization(BPSO) which takes the failure of protective relays or Circuit Breakers into account. Numerical results reveal that BPSO is superior to GA for the convergence speed and accurate estimation results. Fabio Bertequini Leao et.al. [13] proposed Unconstrained Binary Programming (UBP) model for estimating fault sections in automated distribution substations. The UBP model, established by using the parsimonious set covering theory looks for the match between the relays' protective alarms informed by the SCADA system and their expected states

In this paper, a technique is proposed which includes various rules with an optimization method called Elitist NSGA [11] developed by K.Deb. Data recorded by recorders at substation, customer phone calls location and status of reclosers are used to estimate the exact location of fault in automated distribution systems.

In this approach, initially the multi-objective optimization problem is converted into a single objective optimization without using weighting factors and subsequently, Elitist NSGA is used to solve this single objective optimization problem. As a result, the proposed methodology is generalized enough to be applicable to any power distribution network. The applicability of the proposed methodology has been demonstrated through detail simulation studies in different test systems.

2. Fault Section Estimation

When a fault occurs in a distribution system, to clear the fault, the automatic tripping-reclosing sequence of the recloser is triggered. A popular operation sequence is two fast operations followed by two delayed operations [4] .The fast operations attempt to clear temporary faults and the delayed operations allow downstream fuses to clear permanent faults. If the fault is not cleared after this sequence, which is usually the case for permanent faults lying between the fuse and the recloser, the recloser opens and locks out. An alarm is sent to the operator to inform him about this event. Substations may have recording devices that can record current and voltage values and recloser status. Analog quantities such as voltage and current waveforms may also be captured by some devices. Using customer phone calls is a widely used method for locating faults on distribution systems. Based on the call location and the configuration of the distribution system, the maintenance personnel can roughly determine the area in which the fault has occurred. In some circuits parts of these data may be absent. Hence heuristic rules are applied to make use of the date that is available to predict the fault location. In general if more data are available, then the predicted number of potential fault locations can be reduced.

3. Fault Detection

When a fault occurs in a distribution system, in an attempt to clear the fault, the automatic tripping-reclosing sequence of the recloser is triggered. A popular operation sequence is two fast operations followed by two delayed operation. The fast operations attempt to clear temporary faults and the delayed operations allow downstream fuses to clear permanent faults. If the fault is not cleared after this sequence, which is usually the case for permanent faults lying between the fuse and the recloser, the recloser opens and locks out. An alarm is sent to the operator to inform him about this event.

A. Fault Analysis

Traditionally, the computation of short circuit for unbalanced faults in a normally balanced system has been accomplished by the application of the symmetrical components method. But this method is not well suited to a distribution feeder, which is inherently unbalanced.

Thus, the Thevenin equivalent three-phase circuit is computed at the short-circuit point in order to compute the short-circuit current [2].Given the one phase and three phase circuit MVA magnitude and angle, the zero and positive sequence equivalent system impedances can be calculated as follows.

$$Z_+ = \frac{kV_{LL}^2}{(MVA_{3-phase})^*} \; \Omega \qquad \qquad \dots (1)$$

$$Z_0 = \frac{3kV_{LL}^2}{(MVA_{1-phase})^*} - 2Z_+ \Omega \qquad \qquad \ldots (2)$$

where,

kV_{LL} is the nominal line-to-line voltage (in kV) of the system

$MVA_{1-phase}$ is the one phase short circuit MVA

$MVA_{3-phase}$ is the three-phase short circuit MVA

Z_+ is the positive sequence equivalent system impedance in ohms (Ω)

Z_0 is the zero sequence equivalent system impedance in ohms (Ω)

The computed positive and zero sequence impedances can then be converted into the phase impedance matrix, Z_T. The Thevenin equivalent phase impedance matrix is the sum of the phase impedance matrices of each device in between the circuit voltage source and the fault location.

Voltage regulators are assumed to be in neutral position so they can be neglected in the short-circuit analysis. Additional complications arise if transformers are encountered. In this case, the Thevenin equivalent circuit needs to be determined at the secondary node of the transformer bank.

To calculate the short-circuit currents at all components; the fault analysis routine of commercial available modelling and simulation packages can be used [14].

5. Distribution Network Simulation Using PSCAD

A. PSCAD Simulation

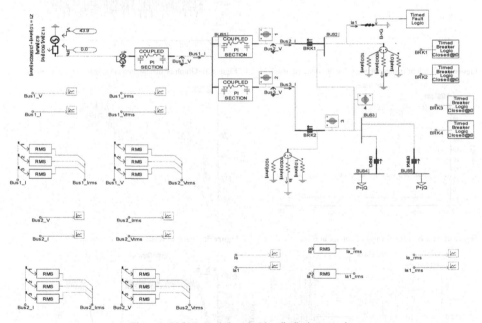

Figure 1: PSCAD simulation of a 5 bus distribution network

A PSCAD model of the distribution network develop based on circuit available. The model consist a large number of element. It is not necessary to model all the elements in PSCAD for transient simulation purposes. Components such as station poles and fuses are ignored. The substation bus is modelled as a voltage source behind equivalent impedance. Transformers are modelled with their leakage reactance. Saturation has been disabled; however an option is available to enable it with default parameters. This is not relevant at this point as most of the faulted scenarios are creating under voltage conditions. Feeders are modelled with equivalent R_L branches between load buses. Loads are modelled as fixed impedance loads. Only three phase circuits are modelled. Single phase circuits are modelled as lumped single phase loads at the branch-off point. The model has been set up such that any one of the faults can be selected at a time during a transient simulation. This is currently done by moving a custom fault module to the desired fault location when performing the simulation.The case is automated to perform various fault types at a given location. The fault duration can be varied. For each fault, voltage and current waveforms and RMS values are recorded at the fault location, substation as well as at all circuit re- closers. The record length consists of pre -fault duration, fault duration and post-fault duration. There are 7 fault types which are A-G, B-G, C-G, AB, BC, CA, and ABC. From which in this simulation only one type of fault B-G is simulated. Fig. 1 shows a custom simulation of a network in PSCAD.

B. .PSCAD Simulation Result

A simulated distribution power system network is shown in fig. 1. After compile the network using PSCAD simulation the value of instantaneous and RMS value of Current and voltage at different points in analog form are obtained, which are showing in figure.

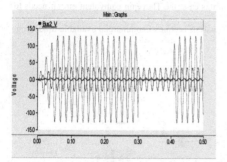

Figure 2: Graph of Instantaneous value of voltage at Bus2.

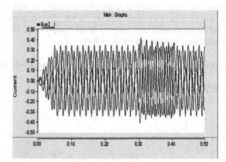

Figure 3: Graph of Instantaneous value of current at Bus2.

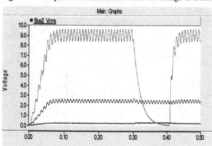

Figure 4: Graph of RMS value of voltage at Bus2

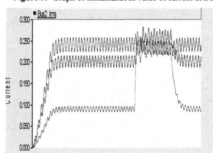

Figure 5: Graph of RMS value of current at Bus2

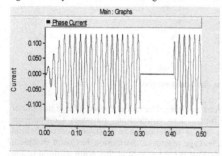

Figure 6: Graph of Instantaneous value of Phase Current.

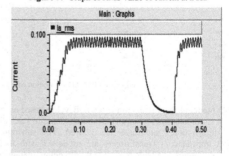

Figure 7: Graph of RMS value of Phase Current.

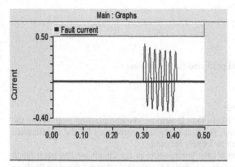

Figure 8: Graph of Instantaneous value of fault current.

Fig. 2 shows the graph of instantaneous voltage at bus 2. As a fault occur at bus 2 at time= 0.3 sec, the instantaneous value of the voltage is dropped to lower value and when this fault removed at time = 0.4 sec system will restore as previous condition. Fig. 3 shows the graph of instantaneous current at bus 2. At time = 0.3 sec during the fault current goes to a higher value as show in figure 2 and after removing the fault current flow through circuit as previous value. Fig 4 and fig 5 shows the RMS value of voltage and current at bus 2 during the fault in time 0.3 sec to 0.4 sec and pre-fault and post-fault condition. Fig 6 shows the phase current in which fault is occurred as in simulation diagram fault is occurred in phase B. So this figure show current through phase B. Fig 7 show the RMS value of this phase current. Fig 8 shows the fault current during the fault which is flow through the fault section.

6. Formulation of Fault Section Estimation Problem with Elitist NSGA

Based on operator experience a set of objective functions are formulated to predict the location of the fault. This set of objective functions will use the data collected, such as current measurements, recloser status and customer phone calls to create a list of potential fault locations.

6.1 Objective functions:

(1) Minimization of difference of measured and calculated fault current

$$f_1 = \sum_{n=1}^{N} \sum_{l=1}^{NF} (I_M - I_C) \qquad \qquad \dots (3)$$

The first objective function compares the calculated fault currents with the measured fault current value. Note that the pre-fault load current must be added to fault currents calculated by the fault analysis routine. Locations where the calculated fault current is within a specified range of the measured current are listed as potential fault locations. It is assumed that the difference between the calculated and measured values were within 10%.

(2) Minimization of distance between customer fault location and predicted fault location

$$f_2 = \sum_{n=1}^{N} \sum_{m=1}^{NC} \sum_{l=1}^{NF} L_{CC-F} \qquad \qquad \dots (4)$$

The third objective function minimizes the distance between the location of customer phone call and existing fault to further reduce the list of potential fault locations.

(3) Minimization of difference between sending end pre-fault voltage and Voltage during fault or fault location Voltage

$$f_3 = \sum_{n=1}^{N} \sum_{l=1}^{NF} (V_{sp} - V_{fl}) \qquad \qquad \dots (5)$$

where,

f_1, f_2, f_3 : the defined fitness function,

N	: number of buses
NF	: number of faults
NC	: number of customer calls
I_M	: measured fault current
I_C	: predicted fault current
$L_{CC\text{-}F}$	: distance between customer call location and predicted fault locations.
V_{sp}	: sending end pre fault voltage

V_{fl} : Voltage during fault or fault location Voltage

The minima of this function correspond to the most likely fault locations. The optimization method and the fitness function are described in detail in the following section. By decreasing the search area for the maintenance crew the time required to find the exact location of the fault can be largely reduced.

Also if we start with only the critical components rather than all the components in the circuit we can get a much smaller number of potential fault locations. The critical components are defined as the ones that are the most likely to fail, based on the historical observations by the distribution utility.

The component numbers in the distribution network are coded as binary strings. For example component number 50 is [1 1 0 0 1 1].

The GA accounts for noise and errors in measurements. As stated earlier, between the measured value and the calculated values of fault currents, a difference of 10% and 20% in other case have been assumed,

Parameters selected for Elitist NSGA-II based algorithm are as follows:

population size	: 30
crossover probability	: 0.6
mutation probability	: 0.001
generations	: 20
number of crossover points	: 1

The mechanisms related to the initialization, selection of parents, crossover and mutation are standard GA steps covered in [15].

7. Brief Description of Elitist NSGA

Elitist Non-dominated sorting genetic algorithm is essentially a modified form of conventional GA [11]. However, while conventional genetic algorithm converts the multi-objective multi-constraints optimization problem into single objective optimization problem with help of weighting factors, NSGA-II retains the multi-objective nature of the problem at hand. On the other hand, like conventional GA, NSGA-II also uses selection, crossover and mutation operator to create mating pool and offspring population. The detail philosophy and technique of Elitist NSGA is already described in great detail in [11] and hence is not repeated here. However, the step-by-step procedure of Elitist NSGA for one generation is described here for ready reference and completeness of the paper. The basic algorithm of Elitist NSGA is as follows.

Step 1: Initially a random parent population P_o of size N is created (i.e. N is the number of strings or solutions in P_o). The length of each string is L_S (i.e. L_S is the number of bits in each string).

Step 2: Create offspring population Q_o of size N by applying usual GA operators (i.e. selection, crossover, mutation) on P_o.

Step 3: Assign $P_t = P_o$ and $Q_t = Q_o$, where P_t and Q_t denote the parent and offspring population at any general 't^{th}' generation respectively.

Step 4: Create a combined population $R_t = P_t \cup Q_t$. Thus, the size of R_t is 2N.

Step 5: Perform non-dominated sorting on R_t. Non-dominated sorting divides the population in different fronts. The solutions in R_t, which do not constrained-dominate each other but constrained-dominate all the other solutions of R_t are kept in the first front or best front (called set F_1). Among the solutions not in $F = F_1$, the solutions which do not constrained-dominate each other but constrained-dominate all the other solutions, are kept in the second front (called set F_2). Similarly, among the solutions not belonging to $F = F_1 \cup F_2$, the solutions which do not constrained-dominate each other but constrained-dominate all the other solutions, are kept in the third front (called set F_3). This process is repeated until there is no solution in R_t without having its own front. Subsequently, these generated fronts are assigned their corresponding ranks. Thus, F_1 is assigned rank 1; F_2 is assigned rank 2 and so on.

Step 6: To create P_{t+1}, i.e. the parent population in the next or '$(t+1)^{th}$' generation, the following procedure is adopted. Initially, the solutions belonging to the set F_1 are considered. If size of F_1 is smaller than N, then all the solutions in F_1 are included in P_{t+1}. The remaining solutions in P_{t+1} are filled up from the rest of the non-dominated fronts in order of their ranks. Thus, if after including all the solutions in F_1, the size of P_{t+1} (let it be denoted by 'n') is less than N, the solutions belonging to F_2 are included in P_{t+1}. If the size of P_{t+1} is still less than N, the solutions belonging to F_3 are included in P_{t+1}. This process is repeated till the total number of solutions (i.e. n) in P_{t+1} is greater than N. To make the size of P_{t+1} exactly equal to N, (n-N) solutions from the last included non-dominated front are discarded from P_{t+1}. To choose the solutions to be discarded, initially the solutions of the last included non-dominated front are sorted according to their crowding distances and subsequently, the solutions having least (n-N) crowding distances are discarded from P_{t+1}.

Step 7: Create the offspring population Q_{t+1} by application of crowded tournament selection, crossover and mutation operator on P_{t+1}.

Step 8: Test for convergence. If the algorithm has converged then stop and report the results. Else, t← (t+1), P_t←P_{t+1}, Q_t←Q_{t+1} and go back to step 4.

7.1 Algorithmic Steps for implementation of Elitist NSGA based Fault Section Estimation

7.1.1 Steps for fitness function calculation:

Step .1: Compare calculated fault currents with measured /recorded fault currents (from the substation, DFRs, etc.)

Step .2: Recloser status (open/close)

Step .3: Recloser V&I rms values (if available)

Step .4: Customer (trouble) call input file (if available)

Step .5: Time synchronized phase angles and waveforms as well as transient RMS current and voltage values (if available)

7.1.2 Steps for implementation of Elitist NSGA to find the accurate location of fault and its distance from substation:

Step 1: The information available to the algorithm are, i) system data, ii) pre fault current and sending end voltages and iii) post fault configuration.

Step 2: Generate initial population, Po randomly and code the component numbers as binary strings, for example component number 50 is [1 1 0 0 1 1].

Step 3: Check the radiality of the solutions in P_o and modify them, if necessary.

Step 4: Evaluate the strings in P_o and assign P_t = P_o.

Step 5: Generate the offspring population Q_o as described in [11]

Step 6: Evaluate the strings in Q_o as described in [11] and assign Q_t = Q_o.

Step 7: Follow steps 4-7 of Elitist NSGA-II as described in [6] to obtain P_{t+1} and Q_{t+1}.

Step 8: In all the solutions of Q_{t+1}, the faulted zone is isolated and the root switch is always made 'closed'

Step 9: Check the radiality of the solutions in Q_{t+1} and modify them, if necessary

Step 10: Check for convergence as conventional NSGA, if the algorithm has converged, find the final solution. Otherwise go to step 11.

Step 11: Evaluate the strings in Q_{t+1} as described in [11].

Step 12: Update P_t←P_{t+1}, Q_t←Q_{t+1} and go back to step 7.

8. Implementation of Elitist NSGA Based Fault Section Estimation

To solve the Fault Section Estimation problem, Elitist NSGA is applied on test distribution systems and results have been compared with Binary PSO and conventional GA based approaches. The proposed approach is tested on 4 test radial distribution systems as described in Table.1.The data of IEEE 13-bus test system has been used from [16].

Table.1. A brief summary of the test systems

S. No.	Description	No. of Buses	No. of Switches	System nominal voltage(kV)	Total System Load	
					KW	KVAR
1	System-1	13	10	11	2652	866
2	System-2	10	14	13.8	5600	4080
3	System-3	33	37	12.66	3715	2300
4	System-4	173	75	33	169476	16421

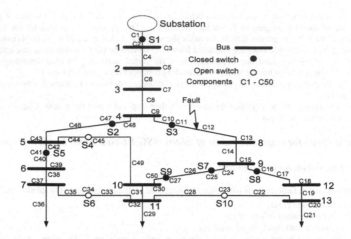

Fig 9: Component number assignment in test distribution system-1.

As shown in fig.9., every section or part of the distribution system under test have been assigned a component number and distance of the components is measured in meters from substation and transmission lines have been further divided into subcomponents at a distance of 2-3 meters. This is done to locate the accurate distance of fault location from substation.

7. Results

It is assumed that the fault has occurred between bus 4 and bus 8 and this location falls in zone 4. One of the reclosers, corresponding to the area in which the fault lies, is locked open. The distance of component C12, where the fault has occurred, is 480-604 meters.

As given in Table.2 and 3, in the first case, all 50 circuit components have been selected to start with. In the second case, only 20 critical components have been selected, in both the cases, the actual fault location was captured in this list of potential fault locations. Due to limitation of space, the results of this algorithm tested on System-I are only illustrated.

The step by step results of this algorithm, assuming that the differences between the calculated and measured currents are within 10% are presented in Table 2.

Additionally, 20% difference between the calculated and measured fault current values has been used. As seen from the results shown in Table.3, even with a large error the software tool is able to capture the exact potential fault location. It is clearly seen in the results that component C12 is faulty and the exact location of the fault is at a distance of **517** meters. It is also observed from Table.2 and Table.3 that the run time of the algorithm is lesser in the proposed method compared to Binary PSO [12] and Refined GA [10]. To develop the algorithm for optimization technique Elitist NSGA and Integer PSO, MATLAB 7.11 has been used.

Table. 2. Potential Fault Location Using Elitist NSGA (10%Difference in currents)

Technique	Number of System	Number of Components	Potential fault	Distance from Fault location to	Run Time of Algorithm(sec)
Elitist NSGA[Proposed]	50	50	C12	510	11.26
	50	20	C12	517	11.04
Binary PSO[11]	50	50	C12	510	19.68
	50	20	C12	518	19.26
Refined GA[14]	50	50	C12	512	23.81
	50	20	C12	521	22.46

Table.3. Potential Fault Location Using Elitist NSGA (20% Difference in currents)

Technique	Number of System Components	Number of Components Selected	Potential fault location	Distance from Fault location to substation(Meters)	Run Time of Algorithm(sec)
Elitist NSGA[Proposed]	50	50	C12	509	11.29
	50	20	C12	516	11.14
Binary PSO[11]	50	50	C12	508	19.78
	50	20	C12	521	19.39
Refined GA[14]	50	50	C12	515	24.13
	50	20	C12	527	22.67

8. Conclusion

In this paper, Elitist NSGA based fault section estimation algorithm has been presented .Test results demonstrate the accurate and robust fault location technique. The software tool developed should enable utility personnel to locate faults faster and thus reduce restoration time. It will aid in improved fault diagnosis for operation and planning. The proposed Elitist NSGA for FSE problem is tuned to increase the computational efficiency, ensuring robustness, accuracy and reduction of the algorithm's processing time. It is seen that for all the cases, the Elitist NSGA based approach is able to pinpoint the fault location quite faster and accurately irrespective of the fault type as compared to conventional techniques. Results show the potentiality and efficiency of the methodology for estimating fault section in Radial Distribution Systems. Fast and accurate fault section estimation will assist the maintenance crew for enhancing the restoration process.

In the future, use of other stochastic optimization methods will be used to find the optimal solution. Use of other intelligent methods such as fuzzy logic, neural networks, hybrid algorithms and latest swarm intelligence techniques will be investigated.

References

[1] Y.Sekine, Y.Akimoto, M.Kunugi, Fukui, and S.Frikui, "Fault diagnosis of power system," IEEE Transactions on Power Systems, vol.3 no. 2 pp. 673-683, May 1992.
[2] Laurentiu Nastac, Anupam.A.thatte, "A Heuristic Approach for Predicting Fault Locations in Distribution Power Systems," Proceedings of 38th North American Power Symposium, NAPS. Sept. 2006 pp.9 - 13.
[3] M. Kezunovic, "Practical application of automated fault analysis," IEEE Tutorial, IEEE Power Engineering Society, 2000, vol. 2, pp: 819 – 824.
[4] J. L. Blackburn, Thomas J. Domin, "Protective Relaying: Principles and Applications, 2nd edition. New York: Marcel Dekker, 1998.
[5] A.P. Alves da Silva, A.H.F. Insfran, P.M. da Silveira, and G. Lambert-Torres, "Neural networks for fault location in substations,", IEEE Transactions on Power Delivery, pp. 234-239, Jan 1996
[6] Z. Y. He, Q. F. Zeng, J. W. Yang, and W. Gao, "Fault Section Estimation of Electric Power Systems Based on Adaptive Fuzzy Petri Nets", International Conference on Electrical Engineering 2009
[7] C.Fukui ,J.Kawakami, "An expert system for fault section estimation using information from protective relays and circuit breakers," IEEE Transactions on Power Delivery, pp 83-90, Oct. 1986.
[8] Y.M.Park, G.W. Kim, and J.M.Sohn, "A logic based expert system (LBES) for fault diagnosis of power system," IEEE Transactions on Power Systems, vol.2, no. 2 pp. 363-369, Feb. 1997
[9] S. Toune , "Comparative study of modern heuristic algorithms to service restoration in distribution system," IEEE Trans. power Delivery, vol. 17, no. 1, pp. 173-181, Jan. 2002.
[10] F.S. Wen, and Z.X. Han, "A refined genetic algorithm for fault section estimation in power systems using the time sequence information of circuit breakers," Electric Machines and Power Systems, 1996.
[11] K.Deb, Amrit Pratap, Sameer Agrawal, "A Fast and Elitist Multiobjective Genetic Algorithm: NSGA-II", IEEE trans. on Evolutionary Computation.vol.6, no.2.April 2002.
[12] Zhengyou He, Hsiao-Dong Chiang,Chaowen Li and QingfengZeng, "Fault-Section Estimation in Power Systems Based on Improved Optimization Model and Binary Particle Swarm Optimization", IEEE Transactions on Power Delivery, 2009.
[13] Fábio Bertequini Leão, Rodrigo. A. F. Pereira and José R. S. Mantovani, "Fault Section Estimation in Automated Distribution System", IEEE Transactions on Power Delivery, pp: 1-8, 2009
[14] L.Nastac, P. Wang, R. Lascu, M. Dilck, M. Prica, and S. Kuloor, "Assessment study of modelling and simulation tools for distribution power systems," 8th IASTED International Conference on Power and Energy Systems, PES2005, October 24-26, 2005, Marina del Ray, CA, USA.
[15] D.E.Goldberg, "Genetic Algorithms in Search and machine learning", Adison Wesley.2001
[16] IEEE guide for determining fault location on AC transmission and distribution lines. IEEE standard C37.114; June 2005.

Modeling Mechanical Properties of low carbon hot rolled steels

N. S. Reddy[1], B.B. Panigrahi[2], J. Krishnaiah[3]

[1]School of Materials Science and Engineering, Gyeongsang National University, Jinju, Korea

[2]Department of Materials Science and Engineering, Indian Institute of Technology, Hyderabad, India

[2]Bharat Heavy Electrical Limited, Trichy, India

nsreddy@gnu.ac.kr; j.krishnaiah@gmail.com; bpani@gmail.com

Abstract:

Steel is the most important material and it has several applications, and positions second to cement in its consumption in the world. The mechanical properties of steels are very important and vary significantly due to heat treatment, mechanical treatment, processing and alloying elements. The relationships between these parameters are complex, and nonlinear in nature. An artificial neural networks (ANN) model has been used for the prediction of mechanical properties of low alloy steels. The input parameters of the model consist of alloy composition (Al, Al soluble, C, Cr, Cu, Mn, Mo, Nb, Ni, P, S, Si, Ti, V and Nitrogen in ppm) and process parameters (coil target temperature, finish rolling temperature) and the outputs are ultimate tensile strength, yield strength, and percentage elongation. The model can be used to calculate properties of low alloy steels as a function of alloy composition and process parameters at new instances. The influence of inputs on properties of steels is simulated using the model. The results are in agreement with existing experimental knowledge. The developed model can be used as a guide for further alloy development.

Keywords: Artificial Neural Networks, Low carbon steels, Mechanical properties, Process parameters.

1 Introduction

Microstructure of steels determines the properties of steels and the microstructural features depend on alloying elements, process parameters, and heat treatment variables. As the relationships between these are nonlinear and complex in nature, it is difficult to develop them in the form of conventional mathematical equations [1-3]. Linear regression techniques are not suitable for accurate modelling of steels data with noise which is typically the case. Regression analysis to model non-linear data necessitates the use of an equation to attempt to transform the data into a linear form. This represents an approximation that inevitably introduces a significant degree of error. Similarly, it is not easy to use statistical methods to relate multiple inputs to multiple outputs. The method using Artificial Neural Networks (ANN), on the other hand, has been identified as a suitable way for overcoming these difficulties [4-7]. ANN is mathematical models and algorithms that imitate certain aspects of the information-

J. C. Bansal et al. (eds.), *Proceedings of Seventh International Conference on Bio-Inspired Computing: Theories and Applications (BIC-TA 2012),* Advances in Intelligent Systems and Computing 202, DOI: 10.1007/978-81-322-1041-2_19, © Springer India 2013

processing and knowledge-gathering methods of the human nervous system. Although several network architectures and training algorithms are available, the feed-forward neural network with the back-propagation (BP) learning algorithm is more commonly used. Therefore, within the last decade, the application of neural networks in the materials science research has steadily increased. A number of reviews carried out recently have identified the application of neural networks to a diverse range of materials science applications[4]. The objectives of the present work are to investigate its suitability for modeling complex hot rolled steel system, to predict properties for unseen data, and to examine the effect of individual input variables on the output parameters while keeping other variables constant.

Several ANN architectures and training algorithms are available; the feed-forward neural network (FFNN) with the back-propagation (BP) learning algorithm is more commonly used. The conceptual basis of back-propagation was first presented in 1974 by Paul Werbos then independently reinvented by David Parker in 1982, and presented to a wide readership in 1986 by Rumelhart and McClelland [8-12]. Back propagation is a tremendous step forward compared to its predecessor, the perceptron. The power of back-propagation lies in its ability to train hidden layers and their bye escape the restricted capabilities of single layer networks. When two or more layers of weights are adjusted, the network has hidden layers of processing units. Each hidden layer acts as a layer of "feature detectors"-units that responds to specific features in the input pattern. These feature detectors organize as learning takes place, and are developed in such a way that they accomplish the specific learning task presented to the network[10]. Thus, a fundamental step toward solving pattern recognition problems has been taken with back-propagation. At present, the most common type of ANN used in materials science is FFNN with back propagation learning algorithm.

2   Experimental data details:

Finish rolling temperature and coil target temperature apart from chemical composition play an important role in the mechanical properties hot rolled steel strip. The hot rolled steel strip data of three days has been collected from an industry and modelled to study the effect of the said temperatures on mechanical properties. The data consists of chemical composition (Al, Al soluble, C, Cr, Cu, Mn, Mo, Nb, Ni, P, S, Si, Ti, V and Nitrogen in ppm, i.e. 15 inputs), finish rolling temperature (FRT), coil target temperature (CTT), and respective mechanical properties, namely, YS, UTS and EL. The range of the hot rolled steel strip data used for the present study is shown in the Table 1. Total 435 data sets with 17 input parameters were available and 335 sets were used for training. The best results were achieved at a learning rate of 0.7 and momentum rate of 0.6 with 2 hidden layers consisting of 34 hidden neurons in each layer. The predicted results of optimum trained NN model are within 4% of experimental values in most of the cases.

Table 1 The range of the Hot rolled steel strip data

Comp.	Al	ALS	C	Cr	Cu	Mn	Mo	N	Nb	Ni
Min.	0.021	0.02	0.02	0.014	0.005	0.17	0.001	26	0	0.008
Max.	0.065	0.063	0.06	0.039	0.012	0.38	0.003	58	0.004	0.0179

Comp.	P	S	Si	Ti	V	FRT	CTT	Mechanical Properties		
								LYS	UTS	EL
Min.	0.007	0.003	0.004	0	0.001	850.23	570	242	317	34
Max.	0.024	0.02	0.028	0.003	0.002	901.14	650	338	397	46

3   Model development and Graphical user interface design

A graphical model plays a crucial role in easy understanding and analysis of any complex systems. The present model development involves in two phases, training phase and representation of results phase. Object oriented programming language Java has been used to develop the training and representation of results phases. Training phase consists of collection of the training data, normalizing the data, selection of inputs and outputs, selection of neural networks parameters. In the training phase each of the systems are trained with various configurations and an optimum configuration is chosen which, satisfies the permissible error and minimal usage of the system resources. The final configuration is saved to files. Thus generated data is passed on to the representation phase for the further processing. The training phase takes a considerable amount of time for each process and often requires a trial and error mode of selection of hidden layers and the number of hidden neurons and the choice of learning rate, momentum rate and permissible error level is dependent on the complexity as well as the precision requirements of the system. The training phase is coded in such a manner, that the application module can be used in various technical applications and hence requires no knowledge of metallurgy to understand the software source.

Considering the time constraints and the added uncertainty in the choice of the configuration, the content generation phase is separated from the content representation. In addition, the representation phase includes the analysis of the system, which forces extensive knowledge in metallurgy; henceforth the layered structure of the isolation of the content generation and the representation is an objectified approach satisfies the principles of software development.

The required models are trained during the training phase and the network description files, the weight file along with the other required files are placed in a proper location accessible by the model. The representation phase is where model is tested for its precision, sensitivity of various inputs and outputs are determined and some inputs are plotted against the outputs and to determine outputs for the custom inputs. The model is represented as an application where each of the plots or the features is presented in separate tabs. Whenever the basic model is changed the new serialized model, which has the trained weights, is read as the current model and this will update all the plots in the tabs in turn. For the Sensitivity analysis of the input/output parameters of the lower end is send to the model to retrieve the outputs and again the inputs/output parameters of the upper end is passed onto the model to retrieve the upper limits of the outputs. From these the slope is determined and plotted in the form of blocks.

Various systems are currently under study, while these can be categorized, there are primarily two types of categories, and one is based on the type of metal being studied and other by the type of the process being used to study the system. Hence the representation is done in an application frame, which constitutes a menu bar, tool bar, status bar in addition to the actual plots for switching between various models and to have inline status help. The use of technical terms is discouraged to make the model understandable even for non-metallurgists who are not experts in process concepts or the domain knowledge. On the base of the designed FFNN model, various graphical user interfaces were created for a better and easy understanding of the system.

4 Results and Discussions

4.1 FFNN model Predictions with Test data

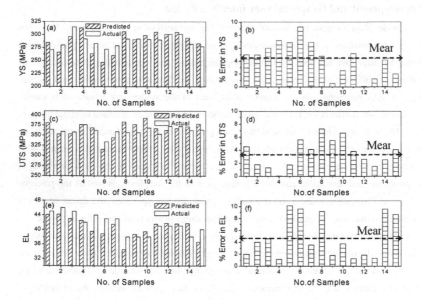

Fig. 1. FFNN model prediction for test data of Hot rolled steel strip

Total data sets available for FFNN Model training are 435. First 335 data sets have been used for training and remaining 100 data sets, which are randomly chosen from the total data sets, is used for testing. As it is difficult to represent the complete predictions, the predictions of randomly selected fifteen test data sets are shown in Fig.1. The comparison between actual and predicted properties for the 15 testing data sets of hot rolled steel strips is shown in

Fig. 1. It was observed that most of the outputs predicted by the model are within 5% of the error band (indicated in (b), (d) & (f)). For the testing data never seen by the model, the FFNN Model gives a reasonably accurate prediction. The mean percentage error in the case of YS, EL and hardness are 4.44, 3.54 and 4.84 respectively. In most of the cases the deviations are less than 5%.

4.2   Hypothetical Alloys for FRT and CTT

Sensitivity analysis studies the effects of parameter variations on the behavior of complex systems. The concept of sensitivity analysis can be extended to all essential parameters of continuous, discrete, or continuous-discrete systems. The sensitivity analysis of the trained FFNN models evolved and its application is explained in the earlier [1-3, 13-15]. The effect of chemical composition and two process parameters namely coil target temperature, finish rolling temperature individually and together on mechanical properties has been presented in the case of low carbon hot rolled steel strip. In the steels studied so far the chemical composition and heat treatment variable on properties are considered [1-3, 16-18]. However, the mechanical properties also depend on the process parameters like coil target temperature, finish rolling temperature and the forces applied during rolling of the hot rolled steel strip. In the present section, the effect of coil target temperature and finish rolling temperature variations on mechanical properties of hot roll steel strip are studied. Fig. 2 shows the variations of coil target temperature and finish rolling temperature simultaneously and their effect on mechanical properties. The figure indicates that the influence of finish roll temperature on the mechanical properties is more complex than that with coil target temperature. This is expected as the finish roll temperature (ranging between 850-910°C) influences a number of parameters such as volume fraction of proeutectoid ferrite, grain size of ferrite and the texture developed during rolling, precipitation of various carbides, etc. While the coil target temperature being in the range of 570-650°C, is much below the eutectoid temperature and hence only brings in smaller variation in microstructure and hence smaller variation in the properties. The effect of combined variation of finish rolling temperature and coil target temperature on mechanical properties is presented in Fig. 3. Table 2 shows the two temperatures for one data set and the respective properties and the model predicted properties with different input variations. The model predictions of test data are well within 5%.

Increase in coil target temperature increases strength around 605°C and then the grain size increases with further increase in temperature and the corresponding strength decreases and the respective rise in ductility. The grain growth restriction will takes place at higher coil target temperature and there-by-there is no change in the strength and ductility. The model predicted this phenomenon very well. As finish rolling temperature increases initially strength increases owing to plastic deformation and very little amount of precipitation. At the same time elongation falls drastically. This happens for increase in precipitate later with increase in temperature grain growth will takes place, which increases ductility and decreases strength. And at 890°C, strength decreases at lower value for larger grain size. Then ductility falls due to the starting of precipitation again. Thereafter precipitation and grain growth takes place simultaneously increase at a same time which results increase in strength and ductility. Table 2 shows the model predictions are well in agreement with the actual values.

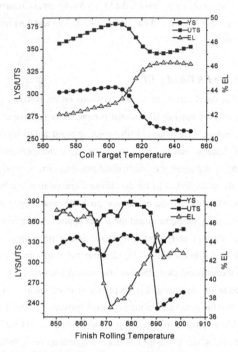

Fig 2   Effect of the variation of Finish Rolling Temperature and coil target temperature on Mechanical properties

Table 2 Comparison of actual and predicted properties of hot rolled steel strip with respective hypothetical alloys

System	LYS	UTS	EL
Actual	302	356	42
HA based on Finish Rolling Temperature (867.95°C)	313	365	41.8
HA based on Coil Target Temperature (570°C)	303	358	42.5

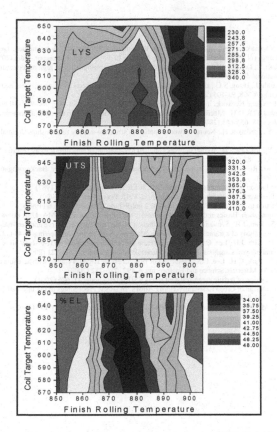

Fig 3   Combined effect of the variation of coil target temperature (a) and finish rolling temperature (b) on Mechanical properties

5 CONCLUSIONS

Neural networks model for prediction and analysis of the hot rolled steel strip data has been developed. The results demonstrated that the model can be used to examine the effects of individual inputs (Coil target temperature and finish rolling temperature) on the output parameters (mechanical properties), which is incredibly difficult to do experimentally. The present model will be helpful in reducing the experiments required for new alloys with desired properties. The user-friendly screens of the present model can make even a layman use it conveniently without the knowledge of any programming.

References

[1]   Reddy, N.S.: Study of some complex metallurgical systems by computational intelligence techniques. Department of Materials Science and Engineering, vol. Ph.D Thesis, pp. 195. Indian Institute of Technology, Kharagpur, India (2004)
[2]   Reddy, N.S., Dzhebyan, I., Lee, J.S., Koo, Y.M.: Modelling of Cr2N age-precipitation in high nitrogen stainless steels by neural networks. ISIJ International 50, 279-285 (2010)
[3]   Reddy, N.S., Krishnaiah, J., Hong, S.G., Lee, J.S.: Modeling medium carbon steels by using artificial neural networks. Materials Science and Engineering A 508, 93-105 (2009)
[4]   Bhadeshia, H.K.D.H.: Neural Networks in Materials Science. ISIJ International, Vol. 39, , 966-979 (1999)
[5]   Sourmail, T., Bhadeshia, H.K.D.H., MacKay, D.J.C.: Neural network model of creep strength of austenitic stainless steels. Materials Science and Technology 18, 655-663 (2002)
[6]   Aoyama, T., Suzuki, Y., Ichikawa, H.: Neural networks applied to structure-activity relationships. Journal of Medicinal Chemistry 33, 905 (1990)
[7]   Aoyama, T., Suzuki, Y., Ichikawa, H.: Neural networks applied to quantitative structure-activity relationship analysis. Journal of Medicinal Chemistry 33, 2583 (1990)
[8]   Rumelhart, D.E., Durbin, R., Golden, R., Chauvin, Y.: Backpropagation: The basic theory. Backpropagation: Theory, Architectures, and Applications 1 (1995)
[9]   Lippmann, R.P.: Introduction to computing with neural nets. IEEE ASSP magazine 4, 4 (1987)
[10]   Dayhoff, J.E.: Neural Network Architectures: An Introduction. VNR Press, New York (1990)
[11]   Zurada, J.M.: Introduction to Artificial Neural Systems. PWS publishing Company, Boston (1992)
[12]   Werbos, P.J.: Backpropagation: Basics and new developments. . MIT Press, Cambridge, MA, USA (1998)
[13]   Reddy, N.S., Lee, C.S., Kim, J.H., Semiatin, S.L.: Determination of the beta-approach curve and beta-transus temperature for titanium alloys using sensitivity analysis of a trained neural network. Materials Science and Engineering: A 434, 218-226 (2006)
[14]   Reddy, N.S., Lee, Y.H., Kim, J.H., Lee, C.S.: High temperature deformation behavior of Ti-6Al-4V alloy with an equiaxed microstructure: A neural networks analysis. Metals and Materials International 14, 213-221 (2008)
[15]   Reddy, N.S., Lee, Y.H., Park, C.H., Lee, C.S.: Prediction of flow stress in Ti-6Al-4V alloy with an equiaxed α + β microstructure by artificial neural networks. Materials Science and Engineering A 492, 276-282 (2008)
[16]   Bhadeshia, H.K.D.H.: Design of ferritic creep-resistant steels. ISIJ International 41, 626 (2001)
[17]   Bhadeshia, H.K.D.H.: Design of Creep Resistant Welding Alloys. In: ASM Proceedings of the International Conference: Trends in Welding Research, pp. 795-804. (Year)
[18]   Bhadeshia, H.K.D.H.: Modelling of steel welds. Materials Science and Technology 8, 123-133 (1992)

Liquid-drop-like Multi-orbit Topology vs. Ring Topology in PSO for Lennard-Jones Problem

Kusum Deep[1], Madhuri[2]

[1,2] Department of Mathematics, Indian institute of Technology Roorkee, India

{kusumfma@iitr.ernet.in; msd02dma@iitr.ernet.in}

Abstract. The Lennard-Jones (L-J) Potential Problem is a challenging global optimization problem, due to the presence of a large number of local optima that increases exponentially with problem size. The problem is 'NP-hard', i.e., it is not possible to design an algorithm which can solve it on a time scale growing linearly with the problem size. For this challenging complexity, a lot of research has been done, to design algorithms to solve it. In this paper, an attempt is made to solve it by incorporating a recently designed multi-orbit (MO) dynamic neighborhood topology in Particle Swarm Optimization (PSO) which is one of the most popular natural computing paradigms. The MO topology is inspired from the cohesive interconnection network of molecules in a drop of liquid. In this topology, the swarm has heterogeneous connectivity with some subsets of the swarm strongly connected while with the others relatively isolated. This heterogeneity of connections balances the exploration-exploitation trade-off in the swarm. Further, it uses dynamic neighborhoods, in order to avoid entrapment in local optima. Simulations are performed with this new PSO on 14 instances of the L-J Problem, and the results are compared with those obtained by commonly used ring topology in conjunction with two adaptive inertia weight variants of PSO, namely Globally adaptive inertia weight and Locally adaptive inertia weight PSO. The results indicate that the L-J problem can be solved more efficiently, by the use of MO topology than the ring topology, with PSO.

Keywords: Particle swarm optimization, neighborhood topologies, liquid-drop-like topology, multi-orbit topology.

1 Introduction

Determining the most stable configuration of atomic clusters is one of the most studied problems of cluster dynamics. The minimization of potential energy of clusters plays an important role in it, as the minimum energy usually corresponds

J. C. Bansal et al. (eds.), *Proceedings of Seventh International Conference on Bio-Inspired Computing: Theories and Applications (BIC-TA 2012)*, Advances in Intelligent Systems and Computing 202, DOI: 10.1007/978-81-322-1041-2_20, © Springer India 2013

to maximum stability for atomic clusters (Doye 1996). Among the most commonly used potential energy functions, the simplest one is the Lennard-Jones potential. With it, the mathematical model of cluster structure determination problem comes out to be a highly complex non-linear global optimization problem. Its complexity increases with the increase in number of atoms as the number of local minima increases with an increase in the size of the problem (Hoare 1979). Various approaches have been successfully applied to solve this problem. The earliest efficient and successful approaches to Lennard-Jones (L-J) cluster optimization were demonstrated in (Northby 1987) and later in (Xue 1994). Thereafter, many heuristic approaches such as Particle Swarm Optimization (PSO) (Hodgson, 2002), differential evolution (Moloi and Ali 2005), genetic algorithms (Hartke 2006) etc have also been used due to their well established reputation for efficiently handling complex optimization problems. Also, hybridization of GA with Monte Carlo approach (Dugan and Erkoc 2009) is being used for geometry optimization of atomic clusters. Evolutionary algorithms (Marques and Pereira 2010, Guocheng 2011, Deep et al. 2011) have also been used for finding the global minima of atomic clusters.

This paper attempts to solve the L-J Problem using one of the most popular global optimization techniques, namely, PSO by incorporating a newly designed multi-orbit dynamic neighborhood topology in it. This topology is inspired from the network structure of cohesive interconnections of molecules in a drop of liquid. The MO topology with two inertia weight variants of PSO, namely, Globally Adaptive Inertia weight (GAIW) and Locally Adaptive Inertia weight (LAIW) PSO (Deep et al. 2011), is used to solve 14 instances of L-J Problem. The results have been compared with those obtained by ring topology with GAIW and LAIW PSOs. As a result there are 4 algorithms in all, namely, GAIW PSO with MO topology (GMO), GAIW PSO with ring topology (GR), LAIW PSO with MO topology (LMO), and LAIW PSO with ring topology (LR), which are used for simulations.

Rest of the paper is organized as follows: Section 2 gives mathematical description of L-J Problem. Section 3 describes PSO variants used. Experimental results and discussions are presented in section 4 and the paper is concluded in section 5.

2 Lennard-Jones Problem

The L-J potential is an important part of most empirical energy models. A system containing more than one atom, whose Vander Waals interaction can be described by L-J potential, where the Vander Waals potential characterizes the contribution of the non-bonded pair wise interactions among atoms, is called an L-J cluster. The L-J potential function for a single pair of neutral atoms is a unimodal function. The minimum of this function can be found, but in a more complex system, many atoms interact with one another. Hence the potential energy for each pair of atoms in a cluster has to be summed up. This gives rise to a complicated landscape with numerous local minima. The potential energy surface for an atomic cluster consisting of 8 atoms is shown in Fig. 1.

In its simplest form the potential energy of an L-J cluster is defined by the sum of pair-wise interactions (i.e., L-J potential) among its atoms. For an n atoms cluster, this potential energy, V, is given by the following equation:

$$V = \sum_{i=1}^{N-1} \sum_{j=i+1}^{N} \left(r_{ij}^{-12} - 2r_{ij}^{-6} \right) \qquad (1)$$

where r_{ij} is the Euclidean distance between atom i and atom j. The L-J Problem for an N atoms cluster, consists of determining the 3-D positions t_1, t_2, ... t_N of atoms that will minimize its potential energy function V given by equation (1). Since each atom's position has three real numbers x, y and z, representing its Cartesian coordinates, associated with it, so the number of variables, to be determined, is three times the number of atoms. The problem turns out to be an unconstrained nonlinear, continuous global optimization problem, whose objective function is non-convex and highly nonlinear in nature. The number of local minima increases exponentially with the size of cluster. The problem has been stated as a real life benchmark for testing the performance of evolutionary computation algorithms by Das and Suganthan (2010).

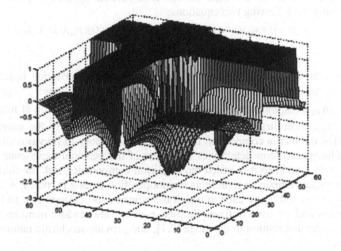

Fig. 1. The landscape of the L-J potential energy surface for a cluster of 8 atoms

3 Particle Swarm Optimization (PSO)

Particle Swarm Optimization (PSO) is a swarm intelligence paradigm (Kennedy and Eberhart 1995), which is found to be increasingly popular due to its ease of implementation, simple mathematical operations, quick convergence and a small number of parameters to fine-tune. It mimics the social intelligence of birds in flocks and fish in schools. The search for global optimum is performed by a swarm of individuals, called particles.

3.1 Basic PSO

In PSO a swarm of n particles, each of which is represented by a d-dimensional vector $X_{i1}(t) = (x_{i1}(t), x_{i2}(t), ...x_{id}(t))$ in search space and also, each has a d-dimensional velocity vector $V_{i1}(t) = (v_{i1}(t), v_{i2}(t), ...v_{id}(t))$ $(i=1,2,...n)$ is randomly generated initially. Then, it searches for the optimum in the search space, and while searching, each particle keeps track of the best position it has visited so far, called the particle's personal best position ($pbest$), and also of the best position visited so far by any particle in its neighborhood, called the local best position ($lbest$). At each iteration, the swarm moves in a way that it is accelerated towards the $pbest$ and the $lbest$. In the inertia weight version each particle updates its position and velocity using the following two equations:

$$v_{id}(t+1) = v_{id}(t) + c_1 r_1(t)(p_{id}(t) - x_{id}(t)) + c_2 r_2(t)(p_{lid}(t) - x_{id}(t)) \tag{2}$$

$$x_{id}(t+1) = x_{id}(t) + v_{id}(t+1) \tag{3}$$

This process is then iterated until a predefined stopping criterion is satisfied. Here $P_{best}(t) = (p_{i1}(t), p_{i2}(t), ...p_{id}(t))$ is i^{th} particle's $pbest$ position, $P_{lbest}(t) = (p_{li1}(t), p_{li2}(t), ...p_{lid}(t))$ is the ith $(i=1,2,...n)$ particle's $lbest$ position at iteration t $(t=1,2,..., t_{max})$. The first term of equation (2) provides inertia to the swarm and hence called the inertia component. It serves as the memory of previous flight direction. The second term determines the effect of the personal experience of the particle on its search, so it is called the cognition component. And, the third term called the social component represents the cooperation among particles. Acceleration constants c_1 and c_2 are used to weigh the contribution of cognition and cooperation terms and are usually fixed. Further, r_1 and r_2 are random numbers drawn from a uniform distribution in the range [0,1], and provide stochastic nature to the algorithm.

The inertia weight w, in the PSO variants GAIW and LAIW (Deep et al. 2011), used here, is defined by

$$w(t+1) = \begin{cases} 0.9 & \text{if } t = 0 \\ f(t-1) - f(t) & \text{if } t > 0 \end{cases}$$

$$\begin{aligned} and \\ w = w_{start} - (w_{start} - w_{end}) * t / t_{max}; & \quad \text{if } w = 0 \text{ for } M \\ & \quad \text{successive iterations} \end{aligned} \tag{4}$$

Where $f(t)$ is global (personal) best fitness of the swarm (particle) at iteration t in the GAIW (LAIW) PSO, w_{start} is the initial value of the inertia weight, and w_{end} is its value at iteration t_{max}, and t is the current iteration of the algorithm while t_{max} is the maximum number of iterations, fixed a priori by the user.

In order to keep the particles within the search space their velocities are clamped as follows:

$$\begin{aligned} v_{id} = -V_{d\max} & \quad \text{if } v_{id} < -V_{d\max} \\ v_{id} = V_{d\max} & \quad \text{if } v_{id} > V_{d\max} \end{aligned}$$

Where $V_{d\max}$ is the maximum allowed velocity in dimension d.

3.2 The liquid-drop-like Topology

The sharing of social information is crucial to the search in PSO, and it is determined by the swarm topology, which assigns to each particle, a subset of the swarm (called its neighborhood) which consists of the particles that will have direct communication with it. Swarm topology greatly affects the search performance of PSO (Kennedy 1999). Though a large number of topologies have been proposed and investigated in literature, the most traditional topologies, namely, gbest (star) and the ring topology are the most widely used ones. The two most important characteristics of swarm topologies are *the degree of connectivity of particles* (Watts 1998), and *the dynamism of neighborhoods*.

In gbest topology each particle has the entire swarm as its neighborhood i.e., it is fully connected, and due to this connectivity the flow of information is very fast which causes the swarm to converge quickly. But it may sometimes lead to premature convergence. On the other hand, in the ring topology, each particle is connected to two particles which are immediately next to it - one on either side - when all the particles in the swarm are arranged in a ring (Fig. 2(a)). Due to this low connectivity, the information processing is slow in ring topology. It helps in maintaining the diversity of swarm for longer and gives it the ability to flow around local optima (Kennedy and Eberhart 2001). But it suffers from the disadvantage of very slow convergence.

The dynamism of neighborhoods plays a key role in avoiding entrapment in local optima and makes the search more effective resulting in reduced computational cost.

It may be inferred that by developing a dynamic neighborhood topology that provides a heterogeneous network of particles with some parts of the swarm tightly connected while the others relatively isolated, may combine the advantages of both the gbest and the ring topologies. In nature, a network with these desired characteristics, is formed by cohesive interconnections among molecules in a liquid-drop. It is found that in a liquid-drop the molecules on the surface feel cohesive force only inwards the drop, while the molecules inside the drop are attracted by others from all directions, in this way giving rise to a heterogeneous cohesive network (Fig. 2 (b)). Inspired from this structure, a multi-orbit (MO) topology with dynamic neighborhoods has been proposed (Deep and Madhuri 2012). This topology is conceptualized as a set of virtual concentric spherical orbits with 2^r particles in the rth orbit ($r=0,1,2,...n_r-1$), having a total of $(2^{n_r} -1)(= n)$ particles in the entire swarm, where n_r is the total number of orbits (Fig. 2(c)). The neighborhoods of the particles are defined as follows:

(1) For the inner most orbit $r=0$, there is only one particle ($2^0 = 1$) and its neighborhood consists of particles 0,1,2.

(2) For the orbits $r=1,2,...$ n_r-2, the neighborhood of each particle i consists of 6 particles, namely, the particle itself, two particles adjacent to it- one on either side in its own orbit, one particle in the consecutive inner orbit and two particles in the consecutive outer orbit.

(3) For the outer most orbit $r=n_r-1$, the neighborhood of each particle i consist of 4 particles namely the particle itself, two particles adjacent to it- one on either side in its own orbit, one particle in the consecutive inner orbit.

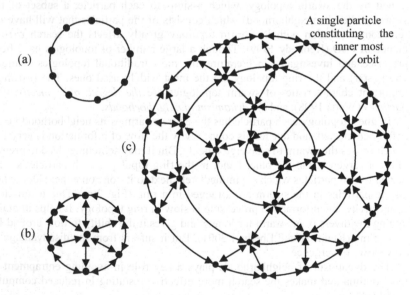

Fig. 2. (a) Ring topology, (b) Cohesive interconnection network in a drop of liquid, (c) Multi-orbit structure with 5 orbits

Clearly, the particles in the outermost orbit have no connections outwards and so have low degree of connectivity, whereas the particles in the inner orbits have connections inwards, as well as outwards and so are tightly connected, hence giving rise to heterogeneous interconnections among particles in the swarm. Initially the neighborhoods are assigned to all the particles as described above and the dynamism is then introduced viz. if at any iteration, the neighborhood best fitness for the ith particle does not improve, its neighborhood is replaced by the neighborhood of the particle with index $(i-1)\%n$.

The effect of a topology on the performance of PSO is problem dependent (Kennedy 1999) i.e., some topology may perform better than the other for the problem at hand. So, the performance of PSO with MO and ring topologies is compared for L-J Problem.

4 Simulations

For comparing the performance of MO topology and ring topology, in PSO, simulations have been performed with 14 instances of L-J Problem for clusters containing 3 to 20 atoms, using GAIW and LAIW PSOs. So, there are 4 algorithms in all. The known global minima and search space for this problem is available in the Cambridge Cluster Database (http://www.wales.ch.cam.ac.uk/CCD.html).

4.1 PSO Parameters used

$c_1 = c_2 = 2$, M=25 (Deep et al. 2011) are used here. The algorithm is terminated if any one of the following three criteria is satisfied: (i) Maximum number of iterations (t_{max}=20000), (ii) The absolute difference between the known minimum fitness (f_{min}) and the global best fitness obtained by the algorithm, becomes smaller than the maximum admissible error (0.000001), and (iii) There is no significant improvement in the global best fitness value for a prefixed number (600 here) of iterations. As the swarm size for MO is to be taken so as to comply with $n = 2^{n_r} - 1$; 31 particles have been used for problems upto 7 atoms (21 variables) and 63 particles for all others. For each problem instance, the same swarm size is taken for all the four algorithms.

4.2 Computational Results and Discussions

This section presents the results of 100 runs for each problem-algorithm pair and compares them on the basis of the following performance evaluation criteria:

(1) Proportion of runs that reached the tolerance limit of the known global optima i.e., Success Rate (SR).
(2) Best function value obtained after maximum number of iterations. It is particularly important for those problem instances for which tolerance limit of the known global minima could not be attained, even in the maximum number of iterations in any of the 100 runs.
(3) Average number of Function evaluations (AFE) for successful runs.
(4) Average error (AE) i.e., the absolute difference between obtained and known optimum, and Standard Deviation (SD) of function value, of successful runs.

[1]**Table 1.** Minimum potential energy obtained

N	GMO	GR	LMO	LR
3	-3	-3	-3	-3
4	-6	-6	-6	-5.9951
5	-9.10385	-9.10385	-9.10385	-9.01114
6	-12.7121	-12.7121	-12.7121	-12.003
7	-16.5054	-16.5054	-16.5054	-13.8759
8	-19.8215	-19.8215	-19.8215	-16.0366
9	-24.1134	-24.1134	-24.1134	-14.1853
10	-27.5452	-27.6645	-27.5547	-15.6188
11	-32.6843	-31.7794	-30.7267	-20.5428
12	-35.8374	-35.091	-33.6998	-21.7622
13	-38.4703	-38.7069	-38.5118	-23.1675
14	-43.0016	-44.389	-41.5151	-23.3953
15	-45.2224	-44.4407	-43.5066	-22.8961
20	-60.6132	-61.7846	-56.0512	-30.8103

Table 2. Success rate (%)

N	GMO	GR	LMO	LR
3	100	100	100	100
4	93	96	84	7
5	90	82	84	0
6	12	7	10	0
7	5	9	9	0
8	20	15	8	0
9	8	4	3	0

---

[1]   GMO- GAIW PSO with MO topology,   GR- GAIW PSO with ring topology
LMO- LAIW PSO with MO topology, LR- LAIW PSO with ring topology

The values for all these performance measures have been recorded and analyzed statistically. For this purpose, paired t-test has been applied to the minimum potential energy obtained (Table 1), and box-plots have been drawn for all other performance measures (Fig. 3,4,5,6). When two tailed t-test is applied, for GAIW, the p-value (=0.8439) at 5% level of significance shows that there is no significant difference between the minimum energy values obtained by ring and MO topologies, whereas that for LAIW gives a p-value of 0.0011, which is highly significant at 1% level of significance and consequently shows (in view of Table 1) that MO topology gives significantly better minima than the ring topology with LAIW. Table 2 shows the success rate of all the 4 algorithms for 3 to 9 atom clusters, beyond which none of the simulations could produce a solution within the admissible range of known minimum.

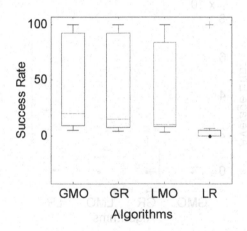

Fig. 3. Box-plot showing comparison of all 4 PSOs with respect to Success Rate

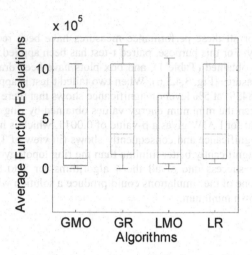

Fig. 4. Box-plot showing comparison of all 4 PSOs with respect to Average Function Evaluations

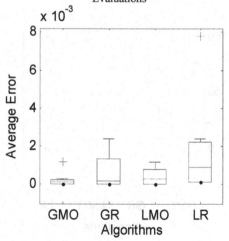

Fig. 5. Box-plot showing comparison of all 4 PSOs with respect to average Error

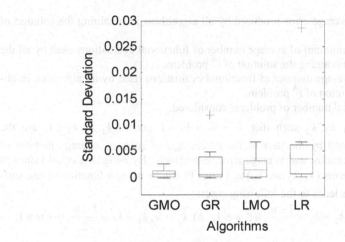

Fig. 6. Box-plot showing comparison of all 4 PSOs with respect to Standard deviation.

Box-plots have been drawn for each algorithm, taking all the problem instances at a time. In Fig. 3, 4, 5 and 6, the area of boxes for MO topology is significantly smaller than that for corresponding ring topology, for all performance measures except SR. It clearly shows that MO topology performs better than ring topology in terms of accuracy (measured by AE and SD) and computational cost (measured by AFE). LAIW PSO with ring topology could not solve the L-J problem for more than 4 atoms, whereas the use of MO topology made it to solve problems upto 9 atoms. It is a big gain.

To study the combined effect of SR, AFE and AE on all the four algorithms simultaneously, Performance Index (PI) analysis as given by Bharti (1994) and used by (Bansal and Deep, 2012), has been applied. This index gives prescribed weighted importance to SR, AFE and AE. For each of the algorithms, the value of PI is computed as follows:

$$PI = \frac{1}{N_p} \sum_{i=1}^{N_p} (k_1 \alpha_1^i + k_2 \alpha_2^i + k_3 \alpha_3^i),$$

$$where \quad \alpha_1^i = \frac{Sr^i}{Tr^i}, \quad \alpha_2^i = \begin{cases} \dfrac{Me^i}{Ae^i}, & if \ Sr^i > 0 \\ 0, & if \ Sr^i = 0 \end{cases} \quad and \quad \alpha_3^i = \begin{cases} \dfrac{Mf^i}{Af^i}, & if \ Sr^i > 0 \\ 0, & if \ Sr^i = 0 \end{cases}$$

$i = 1,2,...N_p$

Sr^i is the number of successful runs of ith problem.

Tr^i is the total number of runs of ith problem.

Me^i is the minimum of the average error produced by an algorithm in obtaining the solution of ith problem.

Ae^i is the average error produced by all algorithms in obtaining the solution of ith problem.

Mf is the minimum of average number of functional evaluations used by all the algorithms in obtaining the solution of ith problem.

Af is the average number of functional evaluations used by an algorithm in obtaining the solution of ith problem.

N_p is the total number of problems considered.

Further, k_1, k_2, k_3 such that $k_1 + k_2 + k_3 = 1$ and $0 \le k_1, k_2, k_3 \le 1$, are the weights assigned by the user to the percentage of success, average number of function evaluations, and average error respectively. By assigning equal values to two of these terms (k_1, k_2 and k_3) at a time, PI can be made a function of one variable only. This leads to the following cases:

a) $k_1 = w$, $k_2 = k_3 = \dfrac{1-w}{2}$, $0 \le w \le 1$; b) $k_2 = w$, $k_1 = k_3 = \dfrac{1-w}{2}$, $0 \le w \le 1$;

c) $k_3 = w$, $k_1 = k_2 = \dfrac{1-w}{2}$, $0 \le w \le 1$.

PI in all these cases have been plotted for all the four algorithms (Fig. 7, 8, 9).

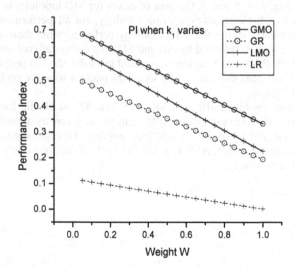

Fig. 7. Performance index plots when variable weight k_1 is assigned to SR

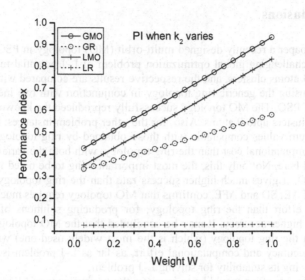

Fig. 8. Performance index plots when variable weight k_2 is assigned to AFE

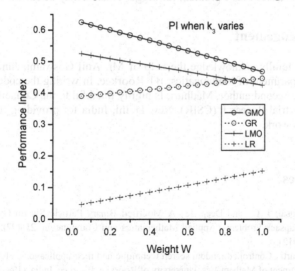

Fig. 9. Performance index plots when variable weight k_3 is assigned to AE.

It may be observed from these plots that for both GAIW and LAIW PSOs, MO topology gives higher PI than ring topology. So, the better performance of MO over ring topology is confirmed by PI analysis also.

5 Conclusions

In this paper a recently designed multi-orbit (MO) topology in PSO is used to solve the challenging global optimization problem of L-J potential minimization for 3 to 20 atoms clusters, and the respective results are compared with those obtained by using the generic ring topology in conjunction with two inertia weight variants of PSO. The MO topology successfully reproduced the known global minima for clusters upto 9 atoms. Also, for the other problem instances, it produced the minimum values comparable with those obtained by ring topology, in much smaller computational cost than the ring topology, with both the variants of PSO, considered here. Not only this, the most important thing to be noted is that, with LAIW PSO, it gives much higher success rate than the ring topology. Also, the analysis of AE, SD and AFE, confirms that MO topology requires much less computational effort than the ring topology, for producing solutions of similar or somewhat higher accuracy. It may be concluded that the MO topology performs better than the ring topology (which is the most widely used one) with PSO, in terms of accuracy and computational effort, as far as L-J problem is concerned, hence showing its suitability for solving L-J problem.

In future, our effort would be to improve the performance of MO topology, so that other instances of L-J problem for bigger clusters may be solved with it.

Acknowledgement

Authors gratefully acknowledge the help of Mr. Anil Goswami, Junior Research Fellow, Department of Mathmeatics, IIT Roorkee, in writing the codes for the topology. The second author, Madhuri, is highly thankful to the Council of Scientific and Industrial Research (CSIR), New Delhi, India for providing financial support for this work.

References

[1] Bansal, J. C., and Deep, K.: A Modified Binary Particle Swarm Optimization for Knapsack Problems. Applied Mathematics and Computation, 218(22), 11042–11061 (2012).

[2] Bharti , Controlled random search technique and their applications, Ph.D. Thesis, Department of Mathematics, University of Roorkee, Roorkee, India, (1994)

[3] Das S., and Suganthan, P. N.: Criteria for CEC 2011 Competition on Testing Evolutionary Algorithms on Real World Optimization Problems, Technical report (2010)

[4] Deep K., and Madhuri: Liquid drop like Multi-orbit Dynamic Neighborhood Topology in Particle Swarm Optimization, Applied Mathematics and Computation (2012) (Communicated).

[5] Deep, K., Madhuri, and Bansal, J. C.: A Non-deterministic Adaptive Inertia Weight in PSO. In proceedings of the 13th annual conference on Genetic and evolutionary computation (GECCO'11), ACM: vol. 131, pp. 1155-1161 (2011)

[6] Deep, K., Shashi, and Katiyar, V. K.: Global Optimization of Lennard-Jones Potential Using Newly Developed Real Coded Genetic Algorithms. In proceedings of IEEE International Conference on Communication Systems and Network Technologies, 614-618 (2011)

[7] Doye, P. K.: The Structure: Thermodynamics and Dynamics of Atomic Clusters. Department of Chemistry, University of Cambridge (1996)

[8] Dugan, N., and Erkoc, S.: Genetic algorithm Monte Carlo hybrid geometry optimization method for atomic clusters. Computational Materials Science 45, 127-132(2009)

[9] Guocheng, Li: An effective Simulated Annealing-based Mathematical Optimization Algorithm for Minimizing the Lennard-Jones Potential. Adv. Materials and Computer Science. Key Engineering Materials Vols. 474-476, pp 2213-2216 (2011).

[10] Hartke, B.: Efficient global geometry optimization of atomic and molecular clusters. Global Optimization 85, 141-168 (2006)

[11] Hoare, M.R.: Structure and dynamics of simple microclusters. Adv. Chem. Phys. 40, 49-135 (1979)

[12] Hodgson, R. J. W.: Particle Swarm Optimization Applied to the Atomic Cluster Optimization Problem. In Proceedings of the Genetic and Evolutionary Computation Conference 2002, 68-73 (2002)

[13] Kennedy, J., and Eberhart, R. C.: Particle Swarm Optimization. In proceedings of 1995 IEEE International Conference on Neural Networks, 4, 1942–1948 (1995)

[14] Kennedy, J., and Eberhart, R. C.: Swarm Intelligence, Morgan Kaufmann, Academic Press (2001)

[15] Kennedy, J.: Small worlds and mega-minds: Effects of Neighborhood Topology on Particle Swarm Performance. In proceedings of Congress on Evolutionary Computation,3, 1931-1938 (1999)

[16] Marques, J. M. C., and Pereira, F. B.: An evolutionary algorithm for global minimum search of binary atomic clusters. Chemical Physics Letters 485, 211-216 (2010)

[17] Moloi N. P., and Ali, M. M.: An Iterative Global Optimization Algorithm for Potential Energy Minimization. J. Comput. Optim. Appl. 30(2), 119-132 (2005)

[18] Northby, J. A.: Structure and binding of Lennard-Jones clusters: $13 \leq n \leq 147$. J. Chem. Phys. 87, 6166–6178 (1987)

[19] Watts, D. J., and Strogatz, S. H.: Collective Dynamics of 'Small-World' Networks. Nat. 393, 440-442 (1998)

[20] Xue, G. L.: Improvements on the Northby Algorithm for molecular conformation: Better solutions. J. Global Optimization 4, 425–440 (1994)

Modeling and Simulation of High Speed 8T SRAM cell

Raj Johri[1], Ravindra Singh Kushwah[2], Raghvendra Singh[3] Shyam Akashe[4]

[1] M Tech VLSI ITM University Gwalior

[2] M Tech VLSI ITM University Gwalior

[3] M Tech VLSI ITM University Gwalior

[4] Associate Professor, ECED, ITM University Gwalior

{johrirj@yahoo.com, rsk.ravindrasingh@gmail.com, raghav.raghvendra7@gmail.com, shyam.akashe@yahoo.com}

Abstract. SRAM cells are known for their high speed operation and low power consumption, and have got considerable attention in research works. Different cell topologies have been developed and proposed to improve various important parameters of the cell to make them favorable for practical operations. These parameters include power consumption, leakage current, stability and speed of response. The paper here describes new 8T SRAM cell which is quite different from recently proposed 8T cells. The modification has been done to increase the speed of cell by reducing the write access time. The other parameters especially power consumption of cell has also been kept in consideration.

Keywords: 8T SRAM, High Speed, SRAM cell, Power consumption.

1 Introduction

SRAM cells are the memory cells used to achieve high speed data access (read or write) and are the fastest among other categories of memory. SRAM cells have got considerable attention over improvement of power consumption of the cell, stability, packaging density and other parameters. Since SRAM cell imparts significant effect on the performance of digital electronic devices like microprocessors and microcontrollers [2] [3], it is necessary to improve the performance of the cell to make these circuits reliable to be employed practically.

In conventional 6T structure the cell employs only one access transistor for each bit line for accessing the latch. The transistors supply current to the gates of the respective inverter to switch the status of the cell (during write operation) and to read the status of the cell (during read operation).

The innovative approach in this paper is to design the SRAM cell with the aim of improvement of speed [4] [5]. This paper will introduce, in brief, about conventional 6T SRAM cell followed by detailed analysis of proposed 8T SRAM cell, waveform produced and comments on the speed of the cell. The cell proposed in this paper has been simulated at 45nm Technology using Cadence Virtuoso tool.

2   The 6T SRAM Cell

The conventional 6T SRAM cell has been illustrated in fig.1. The cell consists of two access transistors and the cross-coupled inverters as the basic memory element. The gates of the two access transistors are connected to the word line and the source to bit and bit-bar line respectively. Whenever the memory element is to be accessed for read or write operation the access transistors must be switched on. The write and read operation in 6T SRAM is discussed in brief as under.

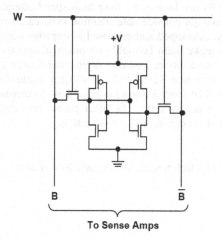

Fig. 1. The conventional 6T SRAM cell.

2.1   Write Operation

The word line is pulled high to activate the access transistors. This enables the memory element to be accessed for read or write. Now required bit '1' or '0' is written by the bit line (BL) and its complement is written in the bit bar line (BBar) and the word line is pulled low. This confirms to the write operation.

2.2 Read Operation

The bit and bit bar lines are pre-charged, and the word line is pulled high to activate the access transistors. At this moment one of the lines remain high and other will go low as the state of the memory element may be. Now the bit and bit bar line is fed to difference amplifier which amplifies the output of the cell to a significant value. This confirms to the read operation.

3 The 8T SRAM Cell

The proposed 8T cell structure is quite different from the conventional 8T cells proposed [1]. It consists of 2 extra nMOS transistors in parallel with each access transistor as shown in fig.2. This modification has been done in order to reduce the write access time [6]. All the access transistors commonly share the same word line as was the case with 6T SRAM cell. However only difference here is that there are two access transistors placed in parallel on each side of the memory element.

It may seem that the new 8T SRAM structure may increase the size of SRAM cell, but at the same time one should notice the most appreciable advantage of the structure that we are achieving writing speed as fast as 80ps/bit. This means if we make an array of 8 bit with independent sources for each bit, we can write 1 byte at the same speed that means 80ps/byte.

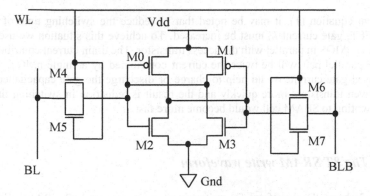

Fig. 2. Proposed 8T SRAM Schematic.

3.1 Role of Parallel Access Transistors

We know that nMOS transistors are faster as compared to the pMOS transistors and that is the reason why most of us prefer to employ nMOS transistors as access transistors. In conventional 6T cell there is only one nMOS transistor used to access the latch. However in case of proposed cell, we have used two nMOS transistors in parallel. We may express the drain source current for nMOS transistor as shown below:

$$I_D = K \left(\frac{W}{L}\right) (V_D - V_T)^2$$

Where, K = constant

$\frac{W}{L}$ = Width to length ratio

V_D = Drain Voltage

V_T = Threshold Voltage of nMOS

The switching speed of the MOSFET depends on the rate of charging and discharging of the gate capacitance [7]. The gate current is described as the function of charge of gate capacitance and the charge time, also termed as switching time, as follows:

$$I_G = \frac{Q_g}{t_s}$$

Here I_G, represents the gate current, Q_g denotes gate capacitance charge and t_s is the switching time of the MOSFET.

The above equation can be written as,

$$t_s = \frac{Q_g}{I_G} \tag{1}$$

From equation (1), it may be noted that to reduce the switching time of the MOSFET, gate current I_G must be increased. To achieve this situation we use an identical nMOS in parallel with the access transistor. The drain current contributed by the parallel pair will be twice the current contributed by a single nMOS. The increased gate current would help to charge or discharge the gate capacitance of the driven transistors more quickly and the result is reduction in switching time. Thus writing to SRAM cell would become more fast.

3.2 The 8T SRAM write waveform

Fig.3 shows the waveform for write operation in proposed 8T SRAM cell. In the waveform net023 and net019 refer to bit and bit bar lines respectively. The voltage level on every source has been kept at 700mV. All the MOSFETs have been configured to have W/L ratio of 2 for optimum power consumption. Clearly we notice when word line is high the states of Q and Qb change according to the bit and bit bar lines and when word line goes low the SRAM retains the data.

The pulse width of word line is only 250ps, which means the cell is accessed only for the time of 250ps the bit required to store in the cell is successfully written within this period and when the word line is pulled low, the SRAM retains the data.

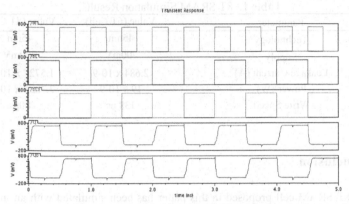

Fig. 3. Proposed 8T SRAM write waveform with high frequency Word Line.

Conversely to test whether the cell can hold the data when bit and bit bar lines change with this frequency, we have simulated the cell by interchanging the timings of word line and bit lines. Fig. 4 shows this situation. Here the word line is set high for 1 ns while bit and bit bar lines have pulse width of 250ps. The waveform clearly shows that cell states Q and Qb change according to the bit and bit bar lines till the word line is pulled high, while they retain their states when word line is low.

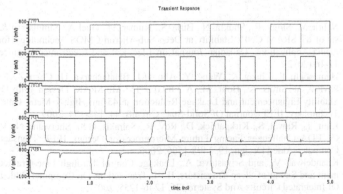

Fig. 4. Proposed 8T SRAM write waveform with high frequency Bit Lines.

4 Result

The proposed SRAM cell has been simulated using Cadence Virtuoso Tool, all the waveforms have been generated on Cadence SPECTRE simulator. Clearly we see that the proposed 8T cell shows good performance efficiency in terms of write speed of the cell as compared to the conventional 6T cell. The results obtained from the simulator have been summarized in the table 1.

Table 1. 8T SRAM Simulation Results.

Parameter	Value (6T Cell)	Value (8T Cell)
Technology	45nm	45nm
Supply	700mV	700mV
Leakage Current (A)	2.684 x 10-9	1.5727 x 10-12
Power (W)	19 x 10-9	7.7025 x 10-12
Write Speed	138 ps	80ps

5 Conclusion

The 8T SRAM cell proposed in this paper has been simulated with an innovative approach with the aim of increasing the write speed of the cell. The waveforms generated show the results clearly. The idea may be utilized to increase the speed of the cell incredibly just by introducing to more transistors in practical SRAMs. The overhead of size of the cell and packaging density can be compromised against such a high speed and low voltage cell to work with.

References

[1] Nehra, V., Singh, R., Shukla, N., Birla, S., Kumar, M., Goel A., Simulation & Analysis of 8T SRAM Cell's Stability at Deep Sub-Micron CMOS Technology for Multimedia Applications, Canadian Journal on Electrical and Electronics Engineering, 3, 11-16 (2012).

[2] Hamzaoglu, F., Zhang, K., Wang, Y., Ahn, H.J., Bhattacharya, U., Chen, Z., Ng, Y.G., Pavlov, A., Smits, K., Bohr, M., A 3.8 GHz 153 Mb SRAM Design With Dynamic, Stability Enhancement and Leakage Reduction in 45 nm High-k Metal Gate CMOS Technology, IEEE Journal Of Solid-State Circuits, 44, 148-154, 2009.

[3] Kam, T., Rawat, S., Kirkpatrick, D., Roy, R., Spirakis, G.S., Sherwani, N., Peterson, C., EDA Challenges Facing Future Microprocessor Design, IEEE Transactions on Computer-Aided Design, 19, 1498-1506, 2000

[4] Khandelwal, V., and Srivastava, A., Leakage Control Through Fine-Grained Placement and Sizing of Sleep Transistors, IEEE Transactions On Computer-Aided Design Of Integrated Circuits And Systems, 26, 1246-1255, 2007.

[5]   Zhu, H. and Kursun, K., Data Stability Enhancement Techniques for Nanoscale Memory Circuits: 7T Memory Design Tradeoffs and Options in 80nm UMC CMOS Technology, SoC Design International Conference (ISOCC), 158-161, 2010.

[6]   Charan Thondapu, Praveen Elakkumanan, and Ramalingam Sridhar, RG-SRAM: A Low Gate Leakage Memory Design, Proceedings of the IEEE Computer Society Annual Symposium on VLSI, New Frontiers in VLSI Design, 295-296, 2005.

[7]   McArthur, R., Making Use of Gate Charge Information in MOSFET and IGBT Data Sheets, Advanced Power Technology, 1-8, 2001.

[8]   Seevinck, E., Van Beers, P.J., Ontrop, H., Current mode techniques for high speed VLSI circuits with application to current sense amplifiers for CMOS SRAMs, IEEE Journal of Solid Sstate Circuits, 26.4, 525-536, 1991.

[9]   Dutta, S., Nag, S., Roy, K., ASAP: A Transistor Sizing tool For Speed, Area and Power Optimization of Static CMOS Circuits, IEEE International Symposium on Circuits and Systems, 61-64, 1994.

[10] Kumar, A., Shalini, S., Khan, I.A., Optimized SRAM Cell Design for High Speed and Low Power Application, IEEE Conference on Information and Communication Technologies (WICT), 1357-1362, 2011.

[11] Ahmadimehr, A.R., Evrahimi, B., Afzali-Kusha, A., A High Speed Sub-Threshold SRAM Cell Design, IEEE Conference on Quality of Electronic Design (ASQED), 9-13, 2009.

[12] Tzartzanis, N., and Athas, W.C., Energy Recovery for The Design of High Speed Low Power Static RAMs, IEEE International Symposium on Low Power Electronics and Design, 55-60, 1996

Optimum Design of Turbo-Alternator Using Modified NSGA-II Algorithm

K.V.R.B. Prasad[1], P.M. Singru[2]

[1] MITS, Madanapalle

[2] BITS, Pilani – K.K. Birla Goa Campus

{prasad_brahma@rediffmail.com[1]; pravinsingru@gmail.com[2]}

Abstract. This paper presents a method to select the optimum design of turbo-alternator (TA) using modified elitist non-dominated sorting genetic algorithm (NSGA-II). In this paper, a real-life TA used in an industry is considered. The probability distribution of simulated binary crossover (SBX-A) operator, used in NSGA-II algorithm, is modified with different probability distributions. The NSGA-II algorithm with lognormal probability distribution (SBX-LN) performed well for the TA design. It found more number of optimal solutions with better diversity for the real-life TA design.

Keywords: Convergence, design optimization, diversity, genetic algorithm, turbo-alternator.

1 Introduction

The turbo-alternator (TA) runs at high speeds and is directly connected to a prime mover like steam turbine, diesel engine etc. An efficient TA design improves its performance with a reduction in its cost. This is useful to meet the extra demand of electrical load. The TA design includes stator-design, rotor-design, air-gap and cooling. A classical TA design methodology is used as a first step [1, 2]. The computer-aided design (CAD) program is developed to obtain the TA design. In this design power rating, operating voltage, current delivered by the machine, operating power factor, operating frequency, operating speed, number of phases, and type of stator winding connection are taken as input. The CAD program gives efficiency and cost (normalized) of TA from the input data. In this paper, a real-life TA working in industry is considered to optimize its design.

Due to high rotating speeds of TA, two types of rotor designs, radial slot-slot type and parallel-slot type are used. Washing of air gives clean and cool air for ventilation. Mica is used on buried parts of the coil against high temperature and insulation. Subdividing the conductors into a number of conductors in each individual slot, or providing parallel conductors in two halves of a complete coil reduces the eddy current losses [3]. In large TA having superconducting field windings, the damping and natural frequencies of rotor oscillations are affected by the different parameters. The parameters having significant effect on the damping are inertia constant, rotor screen time constant, synchronous reactance, transformer and transmission line reactance, coupling between stator and rotor screen, and environmental screen [4]. A computationally efficient and easily tunable multi-objective optimization process is obtained for the axial flux permanent magnet synchronous generator (AFPMSG). Firstly, a design analytical mode which is based on a coupled electromagnetic, thermal and mechanical model is obtained. Later, the multi-objective optimization method is used for optimization [5].

The classical optimization techniques are used to optimize the multi-objective optimization problems (MOOPs). Most of the classical multi-objective optimization algorithms convert the MOOP into single-objective optimization problem using different procedures like weighted sum optimization etc. [6]. These algorithms may have to be used many times, each time finding a different Pareto-optimal solution (POS). They also involve guessing initial solutions besides having high computational time and hence not used nowadays. The genetic algorithms (GAs) are optimization algorithms based on the mechanics of natural genetics and natural selection. The GA gives near global population of optimal solutions. It works with a population of solutions and gives multiple optimal solutions in one simulation run. It has two distinct operations – selection and search. It is flexible enough to be used in a wide variety of problem domains. The operators use stochastic principles [7].

The elitist non-dominated sorting GA (NSGA-II) used an elite preservation strategy along with an explicit diversity preservation mechanism [8]. This allows a global non-dominated check among the offspring and parent solutions. The diversity among non-dominated solutions is introduced by using the crowding comparison procedure which is used in the tournament selection and during the population reduction phase. Since, solutions compete with their crowding distance; no extra niching parameter is required here. Although the crowding distance is calculated in the objective function space, it can also be implemented in the parameter space, if so desired [9]. However, in all simulations performed in this study, the objective function space niching is used. The elitism mechanism does not allow an already found POS to be deleted [7, 8, 10]. In GA, reproduction operator makes duplicates of good solutions while crossover and mutation operators create new solutions by recombination [7, 11]. The performance of NSGA-II is improved by modifying the probability distribution of simulated binary crossover (SBX-A) operator [11, 12]. The performance of NSGA-II algorithm is improved by using lognormal probability distribution to crossover operator (SBX-LN). The NSGA-II (SBX-LN) algorithm found better optimal solutions for various functions having continuous and discontinuous solutions, unconstrained and constrained functions, unimodal and multimodal functions, functions with different number of variables, and more epistasis functions. The NSGA-II (SBX-LN) algorithm also found more number of optimal solutions for TA design with better diversity when compared with the optimal solutions obtained by NSGA-II (SBX-A) algorithm [13]. The CAD program of TA design is developed and MODE-I algorithm is used to find the optimal solutions of it. These results are compared with the optimal solutions obtained by NSGA-II (SBX-LN) algorithm. From this, it is observed that the NSGA-II (SBX-LN) algorithm found more number of optimal solutions with better diversity. In MODE-I algorithm, the constraints are handled by using penalty function method. The constraint violated solutions are penalized by assigning a very high value (10000) [14].

In this paper, a CAD program of TA design is developed. The performance of NSGA-II (SBX-A) algorithm is used to obtain the optimum design of TA. The NSGA-II algorithm is modified by using different probability distributions for SBX-A operator. The modified NSGA-II algorithm is used to obtain the optimum design of TA. These results are compared to find the performance of NSGA-II algorithm. The POS of TA design and best optimization algorithm, for the real-life machine, are obtained.

2 Problem formulation

2.1 Design of turbo-alternator

In the TA design, its name plate ratings (NPR) such as power rating (P_0), operating voltage (V_L), current delivered by the machine (I_L), operating power factor (pf), operating frequency (f), operating speed (N_r), number of phases (N_{sp}), and type of stator winding connection (W_{cs}) are given as inputs to obtain the conventional design using CAD program. The efficiency and cost, along with other design parameters, are obtained from the above data [1, 2]. The total loss in TA (P_{TLoss}) is divided into five major types such as stator copper loss (P_{L1}), stray load loss (P_{L2}), friction and windage loss (P_{L3}), stator iron loss (P_{L4}), and excitation loss (P_{L5}). The total weight of copper (W_1) includes weight of copper in stator and weight of copper in rotor. The total weight of iron (W_2) includes weight of iron in stator and weight of iron in rotor. The cost of insulation and frame are neglected while computing the cost of TA design. The relations for efficiency and cost of TA design, obtained by using the CAD program, are as follows [14].

The efficiency of TA design, in terms of P_0 and P_{TLoss}, is

$$\eta = \left[\frac{(P_0)}{(P_0 + P_{TLoss})} \right] \tag{1}$$

The cost of TA design, in terms of cost of copper (C_1), cost of iron (C_2), W_1 and W_2, is

$$C_T = \left[(C_1 * W_1) + (C_2 * W_2) \right] \tag{2}$$

The conventional design of TA is obtained from the CAD program, by specifying all the input parameters at their rated values.

2.2 Optimization problem formulation

The CAD program, developed to obtain the TA design, is reformulated as a MOOP [13]. In this MOOP, five design parameters such as P_0, V_L, I_L, pf and f are considered as variables which will affect the design. The other design parameters such as N_{sp}, N_r and W_{cs} are kept constant in the design. Six constraints are formulated to obtain the required TA design. They are formed to maintain the stator slot pitch (λ) to be within maximum stator slot pitch (SP_{sm}), the temperature difference between stator copper and iron (TD_{sci}) to be within the maximum temperature difference between stator copper and iron (TD_{scim}), the rotor critical speed (N_{rc}) to be within rotor maximum critical speed (N_{mrc}), the rotor exciting current (I_{rme}) to be within its lower bound of rotor exciting current (I_{rmel}) and upper bound of rotor exciting current (I_{rmeu}), and rotor shaft deflection (DFL_{rs}) to be within its maximum rotor shaft deflection (DFL_{rsm}).

The optimization problem formulation of TA design is as follows.

$$Maximize, \eta = \left[\frac{(P_0)}{(P_0 + P_{TLoss})} \right] \tag{3}$$

$$Minimize, C_T = \left[(C_1 * W_1) + (C_2 * W_2) \right] \tag{4}$$

subjected to

$$\left[1 - \left(\frac{\lambda}{\lambda_m} \right) \right] \geq 0 \tag{5}$$

$$\left[1 - \left(\frac{TD_{sci}}{TD_{msci}} \right) \right] \geq 0 \tag{6}$$

$$\left[1 - \left(\frac{N_{rc}}{N_{mrc}} \right) \right] \geq 0 \tag{7}$$

$$\left[\left(\frac{I_{rme}}{I_{rmel}} \right) - 1 \right] \geq 0 \tag{8}$$

$$\left[1 - \left(\frac{I_{rme}}{I_{rmeu}} \right) \right] \geq 0 \tag{9}$$

$$\left[1 - \left(\frac{DFL_{rs}}{DFL_{rsm}} \right) \right] \geq 0 \tag{10}$$

By using the duality principle, both the objective functions are brought to same kind i.e., minimizing the objective functions. The five design variables are allowed to vary within their allowable range from their rated values. The value of P_0 is allowed to vary from zero to its rated value. The value of V_L is allowed to vary ±5% from its rated value as per the Indian Electricity Rules (IER) [15].

The value of I_L is allowed to vary from zero to its rated value. The value of pf is allowed to vary from its rated value to unity power factor. This is because, in all practical applications, the operating power factor should not be less than its rated value. The value of f is allowed to vary from 97% to 100% of the rated value, though as per the IER the operating frequency is allowed to vary from 97% to 103 % of the rated value [15]. The variation of f beyond its rated value will lead to instability and hence, it is not considered in this work. The other ratings of TA such as N_r, N_{sp} and W_{cs} are maintained at their rated values. The shaft deflects due to the weight which it carries so that as it revolves, it is bent to and fro once in a revolution. If the speed at which the shaft revolves is such that the frequency of this bending is same as the natural frequency of vibration of the shaft laterally between its bearings, then the equilibrium becomes unstable, the vibrations are excessive and the shaft is liable to break unless very stiff. The speed at which this takes place is called the critical speed and should not be within 20% of the actual running speed [1]. Hence, N_{rc} should be less than or equal to N_{mrc} which is 80% of the rated speed.

During the normal operation, the I_{rms} will be increased by 25% from no-load to full-load, to maintain normal voltage [1]. In this work, the I_{rme} is allowed to vary by 20% i.e., from 80% to 100% of the rated fixed excitation current.

3 Modified NSGA-II algorithm

3.1 NSGA-II algorithm

The NSGA-II (SBX-A) algorithm uses normal probability distribution for its crossover operator. The SBX-A operator, used in this algorithm, works with parent solutions and creates two offspring. This operator simulates the working principle of the single-point crossover on binary strings. This operator respects the interval schemata processing, in the sense that common interval schemata between parents are preserved in children [7, 11].

3.2 Modifying the NSGA-II algorithm

In GA, the reproduction operator makes multiple copies of good solutions, to eliminate the bad solutions from the population. The crossover and mutation operators create new solutions, by recombination [7, 11]. The crossover operator is the main search operator in GA. The role of mutation is to restore lost or unexpected genetic material into population to prevent the premature convergence of GA to suboptimal solutions. The performance of NSGA-II algorithm is improved by modifying the probability distribution of SBX-A [11, 12]. In this work, different probability distributions such as Cauchy probability distribution (SBX-C), Fisher-Tippett probability distribution (SBX-F), logistic probability distribution (SBX-L), Rayleigh probability distribution (SBX-R), uniform probability distribution (SBX-U), and SBX-LN are used for the simulated binary crossover (SBX) operator. The NSGA-II algorithm with different crossover probability distributions such as SBX-A, SBX-C, SBX-F, SBX-L, SBX-R, SBX-U, and SBX-LN is used to obtain the optimum design of TA. In this paper, a real-life TA design is considered and the results are presented in the next section.

4 Test results

The NPR of real-life TA design is as follows. A 7999.35 kW, 3300 V, three-phase, 1646 A, 50 Hz, star-connected, 3000 rpm TA to be operated at 85 percent power factor, the stator has 36 slots with 36 coils. The rated field excitation current is 252 A and the field excitation voltage is 125 V. Two conflicting design parameters such as η and C_T are considered as two objective functions, while optimizing the TA design. The value of C_T is normalized by 1000000 units to maintain its optimum value within 1. The value of N is 2, because the stator has equal number of slots and coils, which represents the two layer winding.

The conventional design parameters of TA are obtained by using the CAD program. These are useful in analyzing the design parameters of TA design. In the next stage, the NSGA-II algorithm with different crossover probability distributions such as SBX-A, SBX-C, SBX-F, SBX-L, SBX-R, SBX-U, and SBX-LN are used to obtain the optimum design of TA. The five design variables considered from NPR are allowed to vary within certain limits, which are specified in section 2.2. The six constraints are used to obtain the required design, which are specified in equations (3) to (10). The extreme values of design parameters, considered for constraints, are SP_{sm} = 0.07 m, TD_{scim} = 19 °C, N_{mrc} = 2400 rpm, I_{rmel} = 201.6 A, I_{rmeu} = 2552.0 A, and DFL_{rsm} = 5% of air-gap thickness (δ).

The suitable design parameters, within the allowable range, are input to the CAD program. The two objective functions and six constraints computed by the CAD program, for a given set of parameters, are input to NSGA-II algorithm to obtain the optimum design of TA. In the case of NSGA-II (SBX-LN) algorithm, while finding the optimum design of TA, crossover probability (p_c) = 0.8, mutation probability (p_m) = 0.2, crossover index (η_c) = 0.05, and mutation index (η_m) = 0.5 [11]. In the case of NSGA-II algorithm with other crossover probability distributions, while finding the optimum design of TA, p_c = 0.9, p_m = 0.2, η_c = 5, and η_m = 15. Five best runs are made, for each combination, with different random seeds [7]. Initially, the population size and number of generations are chosen as 44 and 200 respectively. The best optimum solutions of TA design are obtained when the population size is 1000 and number of generations is 10000, for a random seed of 0.9749. In obtaining the optimum design of TA, the NSGA-II (SBX-LN) algorithm gave best optimal solutions with good diversity. The number of optimal solutions obtained by this algorithm is also more. Hence, these solutions are considered as POS, for this case study. The performance of NSGA-II algorithm, with other crossover probability distribution, is tested by computing the variance generational distance (GD) of convergence and diversity metrics. These results, for a population of 1000 and a generation of 10,000, are shown in Table 1. From Table 1, it is observed that the NSGA-II (SBX-A) algorithm found optimal solutions with good convergence while the NSGA-II (SBX-C) algorithm found optimal solutions with better diversity. The optimal solutions of TA design, obtained by the NSGA-II algorithm with SBX-A and SBX-C probability distributions, are shown in Fig. 1 and 2 respectively. The POS is also shown in these figures. From Table 1, it is also observed that the variance GD of convergence and diversity metrics for SBX-A and SBX-U are equal. Hence, the results obtained by NSGA-II (SBX-A) algorithm are considered for the analysis. The design solutions of TA at four different points, shown in Fig. 1 and 2, are shown in Table 2. The POS results, at these four points, are also shown in this table. The conventional design parameters of TA, obtained by CAD program at rated operating condition, are also shown in this table. In the conventional design of TA obtained by CAD program, the values of W_1, effective weight of iron (WT_{ei}), and total weight of mica used in stator and rotor slots (WT_{mica}) are found at rated operating condition. The mica is used as an insulating material in stator and rotor slots. The total weight of machine (WT_{total}) is found from these three weights. From this, the percentage of each material is obtained. These results are shown in Fig. 3, in a pie chart. From Fig. 3, it is observed that the total weight of iron in TA design is more. The total weight of mica insulation in TA design is very small. Hence, the weight of insulation is ignored while obtaining the total cost of TA design.

Table 1. Convergence and diversity metrics

Crossover operator	Mean GD of convergence metric	Variance GD of convergence metric	Mean GD of diversity metric	Variance GD of diversity metric
SBX-A	5.50E-05	**3.03E-05**	1.94	2.78E-03
SBX-C	1.29E-04	3.77E-05	1.94	**1.38E-03**
SBX-F	4.57E-05	5.54E-05	1.94	4.94E-03
SBX-L	6.66E-05	6.65E-05	1.93	5.90E-03
SBX-R	1.06E-04	5.35E-05	1.92	7.05E-03
SBX-U	5.50E-05	3.03E-05	1.94	2.78E-03

The discussion on the results obtained by NSGA-II algorithm with different crossover probability distributions is presented in next section.

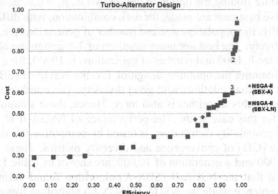

Fig. 1. Results obtained by SBX-A and SBX-LN probability distributions.

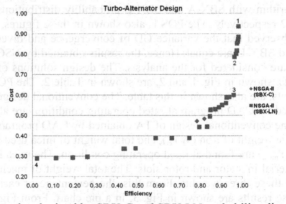

Fig. 2. Results obtained by SBX-C and SBX-LN probability distributions.

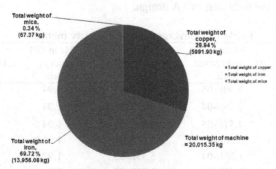

Fig. 3. Pie chart of different weights obtained by CAD program.

Table 2. Parameters of turbo-alternator design

S.No.	Variable/ Constraint/ Objective function	NSGA-II (SBX-A)				NSGA-II (SBX-C)				NSGA-II (SBX-LN)				Actual values (CAD Prog.)	Remarks
		1	2	3	4	1	2	3	4	1	2	3	4		
1	Output Power (MW)	7.99935	2.63	1.76	1.1E-03	7.99935	2.63	1.76	1.1E-03	7.99935	2.63	1.76	1.1E-03	7.99935	Variable (0.0 to 7.99935)
2	Line Voltage (kV)	3.14	3.14	3.14	3.14	3.14	3.14	3.14	3.14	3.14	3.14	3.14	3.14	3.3	Variable (3.135 to 3.465)
3	Line Current (kA)	0.67	0.58	0.38	0.16	0.67	0.58	0.38	0.16	0.67	0.58	0.38	0.16	1.646	Variable (0.0 to 1.646)
4	Power Factor	1.0	1.0	1.0	1.0	1.0	1.0	1.0	1.0	1.0	1.0	1.0	1.0	0.85	Variable (0.85 to 1.0)
5	Frequency (Hz)	49.88	50.0	50.0	49.85	49.88	50.0	50.0	49.89	49.88	50.0	50.0	49.85	50.0	Variable (48.5 to 50.0)
6	Stator slot pitch (m)	6.9E-02	6.9E-02	6.8E-02	6.6E-02	6.9E-02	6.9E-02	6.8E-02	6.6E-02	6.9E-02	6.9E-02	6.8E-02	6.6E-02	6.9E-02	Constraint (≤ 0.07)
7	Temp. diff. b/w cu & iron (°C)	16.55	15.52	18.79	14.22	16.55	15.52	18.79	14.22	16.55	15.52	18.79	14.22	10.99	Constraint (≤ 19)
8	Rotor Cri. Speed (rpm)	1730	1710	1605	2393	1730	1710	1605	2393	1730	1710	1605	2393	1575	Constraint (≤ 2400)
9	Field Exci. Current (A)	229.32	230.82	207.89	201.72	229.32	230.82	207.89	201.72	229.32	230.82	207.89	201.72	248.12	Constraint (201.6 ≤ I_f ≤ 252.0)
10	Rotor shaft deflection (m)	2.9E-04	2.9E-04	3.3E-04	1.2E-04	2.9E-04	2.9E-04	3.4E-04	1.2E-04	2.9E-04	2.9E-04	3.3E-04	1.2E-04	3.5E-04	Constraint (≤ 0.058)
11	Air-gap thickness (m)	3.22E-02	2.89E-02	2.82E-02	1.47E-02	3.22E-02	2.89E-02	2.89E-02	1.47E-02	3.22E-02	2.89E-02	2.82E-02	1.47E-02	3.29E-02	Parameter (δ)
12	Efficiency	0.98	0.965	0.961	0.02	0.98	0.965	0.961	0.02	0.98	0.965	0.961	0.02	0.91	Objective function
13	Cost (normalized by 1000000 units)	0.94	0.79	0.60	0.29	0.94	0.79	0.60	0.29	0.94	0.79	0.60	0.29	2.35	Objective function

5 Discussion of results

From the results presented in section 4, it is observed that the NSGA-II algorithm found more number of design solutions with better diversity. The major observations of the real-life TA design are as follows.

1. From the design solutions obtained by the NSGA-II algorithm, the variation in η from point 2 to point 3 is small, but the variation in C_T is large. This is due to the considerable variation in P_0 and I_L, which gives a considerable variation in TD_{sci} and I_{rme}. This gives a considerable change in the C_T of TA design.

2. The conventional design parameters of TA, obtained by the CAD program, are having low efficiency with more cost whereas the design solutions obtained by NSGA-II algorithm are having better efficiency with low cost, besides having the optimum design parameters of TA.

3. More number of design solutions with good convergence and better diversity are obtained by the NSGA-II algorithm. The performance of NSGA-II algorithm is improved by modifying the crossover operator probability distribution.

4. The NSGA-II (SBX-LN) algorithm found best design solutions of TA with good diversity, when the population size is 1000 and number of generations is 10000. Hence, these solutions are considered as POS for the TA design.

5. The design solutions of TA obtained by NSGA-II (SBX-A) algorithm have good convergence with the POS while the design solutions of TA obtained by NSGA-II (SBX-C) have better diversity with the POS, when the population size is1000 and number of generations is 10,000.

6. The stator has 36 slots with 36 coils. To satisfy this condition, each stator slot has two coil sides with two-layers in stator slot (depth). For multilayer (more than one layer) winding, the size of alternator is reduced, which in turn reduces the cost considerably, while reducing the efficiency marginally.

The validation of results for the real-life TA design, considered in this paper, is presented in the next section.

6 Validation of results

The best optimal design parameters of TA obtained by NSGA-II (SBX-LN) algorithm are compared with the conventional design parameters obtained by CAD program and the actual design parameters. These results are shown in Table 3.

Table 3. Parameters of turbo-alternator design

S. No.	Parameter	Actual value (manual)	CAD program	NSGA-II (SBX-LN)
1	P_0 (MW)	7.99935	7.99935	7.99935
2	V_L (kV)	3.3	3.3	3.3
3	I_L (kA)	1.56	1.646	0.67
4	pf	0.9	0.85	1.0
5	f (Hz)	50	50	49.88
6	λ (m)	---	6.9E-02	6.9E-02
7	TD_{sci} ($^{\circ}$C)	---	10.99	16.55
8	N_{rc} (rpm)	---	1575	1730
9	I_{rme} (A)	145	248.12	229.32

10	DFL$_{rs}$ (m)	---	3.5E-04	2.9E-04
11	δ (m)	2.0193E-02 (TE)	3.29E-02	3.22E-02
		2.0174E-02 (GE)		
12	η	0.90	0.91	0.98
13	C$_T$ *	20.0	2.35	0.94
14	d$_s$ (m)	0.307	0.34	0.17
15	D$_r$ (m)	0.68	0.73	0.73
16	L$_{sb}$ (m)	3.14	3.32	1.58
17	L$_r$ (m)	1.53	1.20	0.56

* The value of C$_T$ is normalized by 1000000 units.

From table 3, it is observed that most of the TA design parameters obtained by conventional design, developed using CAD program, are matching with actual design. The NSGA-II (SBX-LN) algorithm found best design solutions of TA with improved efficiency and low cost, while satisfying all the constraints. The major conclusions drawn from this research work are presented in the next section.

7 Conclusions

The following conclusions are drawn from this work.
1. The conventional design of TA developed using CAD program is helpful to find the total efficiency and total cost, along with other design parameters. This is helpful to find the optimum design of TA, using the optimization algorithms.
2. The rotor exciting current, considered as a constraint in optimization problem formulation, influences the design solutions to a large extent. When it is maintained at its rated value, very few design solutions are obtained. When it is allowed to vary within certain allowable range from its rated value, many near global design solutions are obtained. Hence, this constraint is identified as a hard constraint. The other four design parameters, considered as constraints, are identified as soft constraints.
3. The NSGA-II (SBX-LN) algorithm is identified as the best optimization algorithm to obtain the optimum design of TA. The performance of NSGA-II algorithm is improved by the SBX-LN crossover probability distribution.
4. The NSGA-II (SBX-LN) algorithm found more number of design solutions with good convergence and better diversity. These design solutions are taken as Pareto-optimal solutions, for the real-life TA design.
5. The number of conductor layers in stator slot (depth) influences the size of TA which in turn affects the cost and efficiency of TA design. Hence, a proper value of the number of conductor layers in stator slot (depth) is to be selected with suitable values of cost and efficiency, for a specific power rating.

The important contribution of this work is to find the optimum design of TA using NSGA-II algorithm with different crossover probability distributions, comparing these results, identifying the Pareto-optimal solutions and best probability distribution which improves the performance of NSGA-II algorithm.

References

[1]   Gray, A.: Electrical Machine Design – The Design and Specification of Direct and Alternating Current Machinery. Mcgrawhill Book Company, Inc. (1913)

[2]   Sawheny, A.K.: Electrical Machine Design. Dhanpat Rai & Sons, Newdelhi (1991)

[3]   Lamme, B.G.: High speed turbo-alternator-designs and limitations. Proceedings of the IEEE. 72, 494-526 (1984)

[4]   Ula, A.H.M.S., Stephenson, J.M., Lawrenson, P.J.: The effect of design parameters on the dynamic behavior of the superconducting alternators. IEEE Trans. Energy Conversion. 3, 179-186 (1988)

[5]   Azzouri, J., Karim, N.A., Barkat, G., Dakyo, B.: Axial flux permanent magnet synchronous generator design optimization: Robustness test of the genetic algorithm approach. ESPE 2005, Dresden. P1-P10 (2005)

[6]   Deb, K.: Optimization for Engineering Design. Prentice-hall of India Private limites, Newdelhi (1998)

[7]   Deb, K.: Multi-Objective Optimization using Evolutionary Algorithms. John wiley & sons limited, Chichester (2002)

[8]   Deb, K., Agarwal, S., Pratap, A., Meyariven, T.A.: A fast and elitist mutli-objective genetic algorithm: NSGA-II. IEEE Trans. Evolutioanary Computation. 6, 182-197 (2002)

[9]   Deb, K.: Multi-objective genetic algorithms: Problems difficulties and construction of test functions. Evolutionary Computation. 7, 205-230 (1999)

[10] Deb, K., Pratap, A., Moitra, S: Mechanical component design for multiple objectives using elitist non-dominated sorting GA. Kangal Technical Report no. 200002, Indian Institute of Technology Kanur, Kanpur, India (2000)

[11] Raghuwanshi, M.M., Singru, P.M., Kale, U., Kakde, O.G.: Simulated binary crossover with lognormal distribution. Complexity International. 12, 1-10 (2008)

[12] Price, K.V., Storn, R.M., Lampinen, J.A.: Differential Evolution: A Practical Approach to Global Optimization. Springer, Verlag Berlin Heidelberg (2005)

[13] Prasad, K.V.R.B., Singru, P.M.: Performance of lognormal probability distribution in crossover operator of NSGA-II algorithm. SEAL 2010, Kanpur. 514-522 (2010)

[14] Prasad, K.V.R.B., Singru, P.M.: Identifying the optimum design of turbo-alternator using different multi-objective optimization algorityhms. ITC 2010, Kochi. 150-158 (2010)

[15] Central Electricity Board: Indian Electricity Rules, 1956 (as amended up to 25th Nov., 2000). Under section 37 of the Indian Electricity Act. 1910 (9 of 1910). Ministry of power, Government of India.
      http://www.powermin.nic.in/acts_notification/pdf/ier1956.pdf (2000).

Implementation of 2:4 DECODER for low leakage Using MTCMOS Technique in 45 Nanometre Regime

Shyam Akashe[1],Rajeev sharma[2] , Nitesh Tiwari[3], Jayram Shrivas[4]

[1] Associate Professor

ITM University Gwalior, India

[2] M-Tech VLSI Design

ITM University Gwalior, India

[3] M-Tech VLSI Design

ITM University Gwalior, India

[4] M-Tech VLSI Design

ITM University Gwalior, India

{shyam.akashe@itmuniversity.ac.in;rajeev2007sharma@yahoo.co.in;nitesh1109@gmail.com
; jrm.shrivas@gmail.com; }

Abstract. A newly high performance and low leakage 2:4 decoder is proposed in this paper. We are compare the MTCMOS techniques (Multi Threshold Complementary Metal Oxide Semiconductor) with voltage scaling technique in 45 nm technology .after simulation we can see the result with the MTCMOS technique is better than the voltage scaling technique we can also reduced the leakage current and leakage power by MTCMOS technique in 2:4 decoder effectively. The main objective of this paper is to provide new low power solution for very large scale integration (VLSI) designers. MTCMOS works with low and high threshold voltage can improve the performance. In the MTCMOS technique small power dissipation as compared to traditional CMOS. But in the case of voltage scaling the circuits are simulated at a wide supply voltage down to their minimal operating point .the circuit is simulated with the help of cadence software.

Keywords: 2:4 decoder, High Speed, Low Power, MTCMOS, CMOS.

J. C. Bansal et al. (eds.), *Proceedings of Seventh International Conference on Bio-Inspired Computing: Theories and Applications (BIC-TA 2012),* Advances in Intelligent Systems and Computing 202, DOI: 10.1007/978-81-322-1041-2_23, © Springer India 2013

1.   Introduction

It is the time of technology our maximum work is supported CMOS technology. CMOS is also sometimes referred to as complementary-symmetry metal–oxide–semiconductor (or COS-MOS)[1]. The words "complementary-symmetry" refer to the fact that the typical digital design with the help of n-channel and p-channel in the form of complementary and symmetrical pairs of p-type and n-type metal oxide semiconductor field effect transistor(MOSFETs) for logic functions. The material is polysilicon is dominant in the CMOS technology. Other metal gates have made a comeback with the advent of high-k dielectric materials in the CMOS process [2]. With the help of CMOS technology we can generate the integrated circuit. It was first time made at Fairchild Semiconductor in [3], [4]. The growth of the semiconductor industry driven by the advancements of the integrated circuit (IC) technology and the market dynamics was predicted by Gordon Moore in 1965 [3], [5]-[6].

In order to illustrate the compare of MTCMOS technology with technology scaling and impact of supply voltage variation on the behaviour of Half 2:4 decoder circuit design in CMOS technology. 2:4 decoder circuits is implemented with the inverter, AND gate in 45nm CMOS technology at multi threshold and various supply voltages or scaled voltage are optimized for minimum leakage current and leakage power. The basic principle of MTCMOS is to use low v_{th} transistors to design the logic gates where the switching speed is essential. While the high v_{th} transistors also called sleep transistors are used to effectively isolate the logic gates in stand by state and limit the leakage dissipation [14].

The remainder of this paper is organized as follows. Section2 and section3 gives a brief description of designing 2:4 decoder using CMOS gates also the transistor level circuit which made for simulation. Section4 and section5 describes MTCMOS technique and supply voltage variation on decoder for leakage current and leakage power optimization.section6 describes the definition of leakage current and leakage power. Section7 discuss simulation results of 2:4 decoder for leakage current and leakage power. Conclusions are drawn in section8.

2. Implementation of 2:4 decoder using CMOS Technology

Decoder circuits are used to decode encoded information. A decoder is logic circuits with n inputs and 2^n outputs. Only one input is asserted at a time and each output corresponds to one valuation of the inputs. The decoder also has an enable input, when $E_n=1$ which determines the output. Enable inputs must be on for the decoder to function, otherwise its outputs assume a single "disabled" output code word. Here we discussed the 2:4 decoder circuits with two inputs and four outputs .The enable is also work as an input. In digital circuit theory, combinational logic (sometimes also referred to as combinatorial logic) is a type of digital logic which is implemented by Boolean circuit and also expressed by truth table. The implementation of the circuit is done by with the help of truth table and Boolean logic. The decoder is also classified in Analog and Digital [7] Analog decoders are envisioned to reduce the total power consumption of the receiver in two ways: First, computation and processing in the analog domain is much less power demanding than digital implementations. Second, they can take the analog soft data and directly produce the decoded digital data. Some techniques which reduces the power consumption in the decoder circuit the variable length code (VLC) is effective at minimizing the average code length, and is thus widely used in many multimedia compression systems [8].

3. Implementation of 2:4 decoder using Logic gates

A decoder is logic circuits with n inputs and 2^n outputs. Here we discussed the 2:4 decoder circuits with two inputs and four output .The enable is also work as a input. The input is given in the form of binary number (0 or 1). These decoders use NAND gates instead of the AND gates we have seen in the decoder 1 example. In these cases the output to an inverted 2-4 line decoder would be like this. Fig 1 shows the symbol of decoder and fig 2 schematic of decoder gate level diagram.

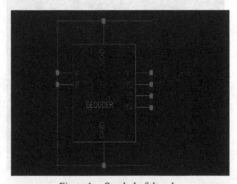

Figure 1.   Symbol of decoder

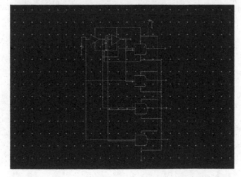

Figure 2.   :   gate level diagram of 2:4 decoder

TABLE I.   TRUTH TABLE OF DECODER

A	B	E	Y1	Y2	Y3	Y4
X	X	0	0	0	0	0
0	0	1	1	0	0	0
0	1	1	0	1	0	0
1	0	1	0	0	1	0
1	1	1	0	0	0	1

The simplified Boolean functions for the outputs can be obtained directly from the truth table. The simplified sum of products expressions are

$$Y_1 = A'B' \qquad Y_3 = A B'$$

$$Y_2 = A'B \qquad Y_4 = A B$$

Where A and B are the inputs.

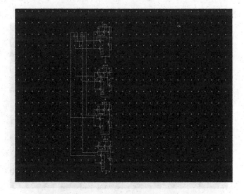

Figure 3.   Schmatic of 2:4 Decoder

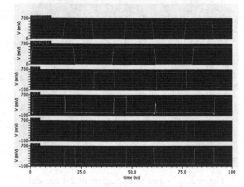

Figure 4.   Waveform of 2:4 decoder

A. Logic gates description for Implementation of 2:4 decoder:

NOT gate:
CMOS logic gate is designed by using one NMOS and one PMOS transistor [9]. PMOS transistor works as pull-up network and NMOS transistor works as a pull-down network. In this combination PMOS transistor is connected to power supply and NMOS transistor is connected to ground. The output of the NOT gate is inverted.i.e, the output is HIGH (1) if the inputs are not alike otherwise the output is LOW (0).

AND gate:
In the AND gate when both two inputs are high then the output of the AND gate is also high. But if only one input is low then the output of AND gate is also low. The truth table is shown below.

TABLE II.   TRUTH TABLE OF AND GATE

Input		Output
A	B	
0	0	0
0	1	0
1	0	0
1	1	1

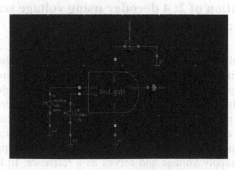

Figure 5.   Symbol of AND gate

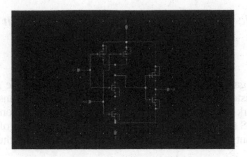

Figure 6.   Schamatic of AND gate

Fig 5 shows the symbol of AND gate fig 6 shows the schematic of AND gate. in the fig6 the two NMOS connected in serial in the form of pull down network and two PMOS is connected in parallel in the form of pull up network .the output of this combination is connected to an inverter

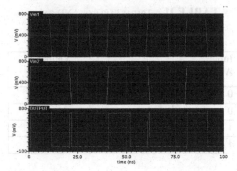

Figure 7.   Waveform of AND gate

4. Implementation of 2: 4 decoder using voltage scaling Technique :

The voltage scaling techniques with multiple supply voltages which required for low leakage current and leakage power .leakage power is directly proportional to the square of supply voltage so that the supply voltage is major contribution of leakage power. The technique is based on regulation of the supply voltage of an equivalent critical path a small circuit with delay V $_{dd}$ properties proportional to these of the actual critical path. The output of the equivalent critical path is compared with the output of a second identical equivalent critical path which is connected to the full supply voltage and serves as a reference. In the first order approximation the ratio of the delay of a critical path to the period of a ring oscillator is a constant that depends only on the number of gates the dimensions of the transistors and the load capacitances

Finally [10] we required the minimum voltage for reduction of leakage power. In [11]-[13] the leakage power was reduced with multiple supply voltages at function level. The leakage power was minimize with two or more supply voltages at the gate of the transistor. The physical layout also simple in with multiple supply voltages.

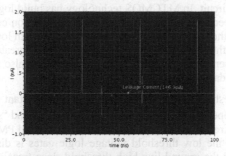

Figure 8.   Waveform of leakage current 2:4 decoder at 0.7

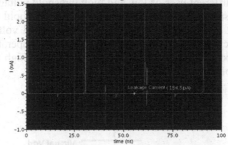

Figure 9.   Waveform of leakage current 2:4 decoder at 0.8v

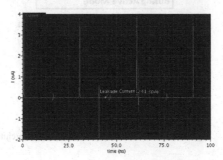

Figure 10.   Waveform of leakage current 2:4 decoder at 1v

5. Implementation of 2:4 decoder by using MTCMOS techniques:

Fig. has shown the general MTCMOS architecture. Power dissipation can be reduced by the voltage scaling techniques but the voltage scaling techniques affects the speed of the circuit. In MTCMOS technology has build logic gates which operated at a high speed with relatively small power dissipation compare voltage scaling technique .in the MTCMOS techniques we can use both low and high threshold transistor [15].In the voltage scaling when the voltage is greater than the particular operating point then the power dissipation is more which also responsible for overheating and degradation of performance.

The multi threshold CMOS technology has two important characteristic. One it has support two operational mode called "active mode" and "sleep mode". Second is two different threshold voltages are used for NMOS and PMOS in a single chip. In this technique the low threshold voltage logic gates is disconnected from the power supply and the ground line high threshold sleep transistor is also known as power gating. The different uses of low and high threshold voltages. The high threshold voltage transistor are used to isolate and the low threshold voltage to prevent leakage dissipation. In the active mode the high threshold voltage transistor are turned on. And logic consisting low threshold voltages can operate with high speed and low switching power dissipation. In the sleep mode the high threshold voltages transistor is turned off.

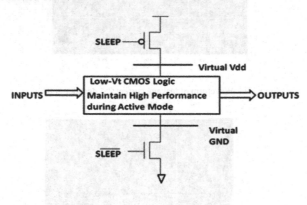

Figure 11. General architecture of MTCMOS architecture

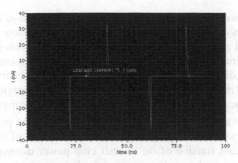

Figure 12. Waveform of Leakage Current using MTCMOS techniques

6. Leakage current and leakage power:

Leakage current is a main factor for any circuit. The leakage current is directly related to the electric field of the device. By reducing the supply voltages, we can decrease the leakage current. Gate length scaling increases device leakage exponentially across technology generations. In other words we can say that Leakage current/power is a waste charge of any device which is regularly discharging from the device even the device in off state. Leakage current is the main source of standby power for a SRAM cell. There are three major components of leakage current are the sub-threshold leakage, gate direct tunneling leakage, and the reverse biased band-to-band tunneling junction leakage. Leakage current calculated at the node which transistor is turn off.

Sub Threshold Leakage- Sub threshold leakage is the drain-source current of a transistor when the gate-source voltage is less than the threshold voltage happens when transistor is operating in weak inversion region. For the reduction of sub threshold leakage of an SRAM cell by increasing the threshold voltage of all or some of the all or some of the transistor in the cell. $Vgs < Vth$, is the condition of sub threshold leakage. The drawback of this technique is an increase in read/write delay of the cell

Junction Tunnelling Leakage- The junction tunnelling current is an exponential function of junction doping and reverse bias voltage across the junction. Since junction tunnelling current is a minimal contributor to total leakage current .

Gate Tunnelling Leakage- Gate tunnelling current is mainly due to three components ;(1) gate to source and gate to drain overlap current (2) gate to channel current, one part goes to source and the rest goes to drain (3) gate to substrate current.

Leakage Power: the leakage power is the multiplication of leakage current and voltage supply. The main contribution factor of leakage power of voltage supply because it is proportional to the square of supply voltage. Power dissipation in CMOS circuits can be categorized into two main components - dynamic and static power dissipation. Dynamic power dissipation occurs due to switching transient current (referred to as short-circuit current) and charging and discharging of load capacitances (referred to as capacitive switching current). Static dissipation is due to leakage currents drawn continuously from the power supply. Leakage power dissipation is roughly proportional to the area of the circuit. It is expected to significant fraction of the overall chip power dissipation in nanometer CMOS design process .In many processors cache occupy about 50% of the chip area. The leakage power of cache is one of the major sources of power consumption in high performance microprocessor. Thus, for low-power devices, e.g. sensor nodes, standby leakage power reduction is crucial for device-operation within the scavenging power limit.

7. Simulation Results

Implementation of 2:4 decoder by using MTCMOS technique have been done in 45 nm technology. The compare of Leakage current and leakage power of 2:4 decoder using voltage scaling and MTCMOS technique. After simulation we can see the table the leakage current and leakage power reduced effectively in MTCMOS compare than voltage scaling. the performance result is also better in MTCMOS technology. The simulated result In the MTCMOS technology the leakage current is reduced by 48.73% and leakage power is reduced by 49.85%.The overall performance improved by the MTCMOS technology.

TABLE III. SIMULATION RESULT OF 2:4 DECODER

2:4 Decoder	Voltage Scaling			MTCMOS
	0.7v	0.8V	1V	0.7V
Leakage current(pA)	146.9	184.5	241	75.31
Leakage Power (nW)	12.98	15.28	18.28	6.509

8. Conclusion

This paper we proposed a MTCMOS technique that greatly reduces the leakage current and leakage power Of the 2:4 decoder. Finally it is concluded that MTCMOS technique is better as compared to voltage scaling techniques. MTCMOS is an effective circuit level technique that enhances the performance and provides low leakage by using both low and high threshold voltage transistors. The simulated results confirms the advantages of MTCMOS techniques over the simple CMOS and voltage scaling technique also the marginal increases in area compared to the voltage reduction or voltage scaling and CMOS logic because of extra transistors. But finally we achieved the high speed and low power dissipation. From the simulation result it is cleared that after applying this technique we have reduced 48.73% in leakage current and 49.85% in leakage power.

Acknowledgment

This work was supported by ITM University Gwalior, with the support of Cadence tool System Bangalore.

References:

[1] V. Kursun and E. G. Friedman, Multi-Voltage CMOS Circuit Design, John Wiley & Sons Ltd., 2006, ISBN # 0-470-01023-1.

[2] G. E. Moore, "The Role of Fairchild in Silicon Technology in the Early Days of Silicon Valley," Proceedings of the IEEE, Vol. 86, Issue 1, pp. 53-62, January 1998.

[3] G. E. Moore, "No Exponential is Forever: But 'Forever' Can be Delayed!," Proceedings of the IEEE International Solid-State Conference, Vol. 1, pp. 20-23, February 2003.

[4] S. Borkar, "Design Challenges of Technology Scaling," IEEE Micro, Vol. 19, Issue 4, pp. 23- 29, (July–August) 1999.

[5] G. E. Moore, "Progress in Digital Integrated Electronics," Proceedings of the IEEE International Electron Device Meeting, pp. 11-13, December 1975.

[6] V. Kursun, Supply and Threshold Voltage Scaling Techniques in CMOS Circuits, Ph.D Thesis, University of Rochester, 2004.

[7] R. Kumar, Temperature Adaptive and Variation Tolerant CMOS Circuits, Ph.D Thesis, University of Wisconsin-Madison, 2008.

[8] Nirmal U., Sharma G., Mishra Y., "Low Power Full Adder using MTCMOS Technique" in proceeding of International conference on advances in Information, Communication Technology and VLSI Design, Coimbatore, India, August 2010.

[9] Kang S, and Leblebici Y., " CMOS Digital Integrated Circuit", TMGH 2003

[10] Mutoh S et al "1-V, Power Supply High Speed Digital Circuit Technology with Multithreshold-Voltage CMOS" IEEE J. Solid State Circuits, Vol.30, pp 847-854,August 1995.

[11] Yu et al., "Limits of gate oxide scaling in nano-transistors," in Proc. Symp. VLSI Technol., 2000, pp. 90–91

[12] System Drivers, "International Technology Roadmap for Semiconductors," http://www.itrs.net, pp. 1–25, 2005.

[13] M. Sheets, B. Otis, F. Burghardt, J. Ammer, T. Karalar, P. Monat, and J. Rabaey, "A (6x3)cm2 self-contained energy-scavenging wireless sensor network node," in Wireless Personal Multimedia Communications, WPMC, Abano Terme, Italy, 2004.

[14] S. Mutah, T. Douseki, Y. Matsuya, T. Aoki, S. Shigematsu, and J. Yamada. 1-V Power Supply High-Speed Digital Circuit Technology with Multi-Threshold Voltage CMOS. IEEE Journal of Solid-State Circuits,30(8):847–853, August 1995.

[15] O. Thomas, "Impact of CMOS Technology Scaling on SRAM Standby Leakage Reduction techniques", ICICDT, May2006.

An Improved Bayesian Classification Data mining Method for Early Warning Landslide Susceptibility Model Using GIS

Venkatesan.M[1], Arunkumar .Thangavelu[2], and Prabhavathy.P[3]

[1]School of Computing Science & Engineering, CAMIR[*], VIT University, Vellore, India

[2] School of Computing Science & Engineering, CAMIR[*], VIT University, Vellore, India

[3]School of Information Technology & Engineering, VIT University, Vellore, India

venkivit@yahoo.co.in, arunkumar.thangavelu@gmail.com, pprabhavathy@vit.ac.in

[*]Centre for Ambient Intelligence and Advanced Networking Research

Abstract –Landslide causes huge damage to human life, infrastructure and the agricultural lands. Landslide susceptibility is required for disaster management and planning development activities in mountain regions. The extent of damages could be reduced or minimized if a long-term early warning system predicting the landslide prone areas would have been in place. We required an early warning system to predict the occurrence of Landslide in advance to prevent these damages. Landslide is triggered by many factors such as rainfall, landuse, soil type, slope and etc. The proposed idea is to build an Early Warning Landslide Susceptibility Model (EWLSM) to predict the possibilities of landslides in Niligri's district of the Tamil Nadu. The early warning of the landslide susceptibility model is built through data mining technique classification approach with the help of important factors, which triggers a landslide. In this study, we also compared and shown that the performance of Bayesian classifier is more accurate than SVM Classifier in landslide analysis.

Keywords – Bayesian, classification, GIS, Landslide, Susceptibility, Spatial

1 Intrduction

Natural disasters like hurricanes, earthquakes, erosion, tsunamis and landslides cause countless deaths and fearsome damage to infrastructure and the environment. Between geological risks, landslides are the important disaster which causes most damages, producing thousands of deaths every year and material losses of billions of dollars in all around the world. Landslide susceptibility can be identified using different methods based on the GIS technology. In the last few years, many research papers were published in order to solve deficiencies and difficulties in assessment of the landslide susceptibility. Huge landslide happened recently at Nilgiri hills in tamilnadu enforce us to propose this idea. Landslide disaster could have been reduced, if more had been known about forecasting and mitigation. Due to more sliding and flooding in many regions around the world in recent years, there is a need to improve the ability to deal with the hazards and risks. So far, few attempts have been made to predict these landslides or prevent the damages caused by them. In the previous studies, the Artificial Neural Network(ANN) and Decision tree model applied to these problems shows that it is difficult to understand and tricky to predict. Therefore, the Early Warning Landslide Susceptibility Model (EWLSM) is proposed to predict landslide susceptibility, which gives simple and accurate results.

However, there are many influencing factors for landslide and the effects of the same factors in various areas are different. The mathematical relationship between the factors which impact landslide and the landslide stability prediction is hard to obtain. Therefore, it is a comparatively accurate method to get a statistical analysis model with the historical data. To get a stable and reliable prediction result of landslides, we have to analyze the geological and environmental conditions and consider certain factors such as the slope, aspect, elevation, Stream power index, and distance from drainage, geology, land use, topography, and rainfall. These factors are used to construct Early Warning Landslide Susceptibility Model (EWLSM) using Naïve Bayesian Classification technique for the prediction of landslides in the area of Nilgiri.

J. C. Bansal et al. (eds.), *Proceedings of Seventh International Conference on Bio-Inspired Computing: Theories and Applications (BIC-TA 2012)*, Advances in Intelligent Systems and Computing 202, DOI: 10.1007/978-81-322-1041-2_24, © Springer India 2013

2 Literature Survey

Several different approaches for the assessment of landslide susceptibility have been employed. Almost all of these approaches are data driven. Geographical Information Systems (GIS) Technology has raised great expectations as potential means of coping with natural disasters, including landslides. In spite of recent achievements, the use of GIS [1], [5] in the domain of prevention and mitigation of natural catastrophes remains a pioneering activity. The use of remote sensing and GIS technologies have tremendously helped the preparation of susceptibility maps with greater efficiency and accuracy in nowadays. These techniques are used to collect, extract and analyze a variety of spatial and non-spatial data about the important factors such as geology, landcover and geomorphology. Robust landslide hazard predictions achieved through an integration of GIS, fuzzy k-means, and Bayesian modeling techniques [2][11]. The continuous fuzzy k-means classification provides a significant amount of information about the character and variability of data, and seems to be a useful indicator of landscape processes relevant for predicting landslide hazard. Application of two bivariate statistical methods [9],[15] namely information value(LF) and landslide nominal susceptibility factor(LNSF) are most suited for the regional –scale landslide susceptibility zonation. LNSF method is better than infoval method, since it uses the statistical basis of dividing the area into very high, high, moderate, low and very low susceptibility zones.

Both frequency analysis and logistic regression [12],[21] methods were applied for the landslide hazard mapping. Frequency ratio model[4],[10] has better predication accuracy compare than the logistic regression model, because the frequency ratio model doesn't require data conversion, but logistic regression requires conversion of the data to ASCII or other format to be used in the statistical package. A neural network [23] method was used for landslide susceptible mapping using integration of various topographic, geological, structural, land use, and other datasets in GIS. Each factor's weight was determined by the back-propagation training[7] method and then the landslide susceptibility indices were calculated using the trained back-propagation weights. SLIDE [13] suggested a physical model which links water content in the soil column and rainfall series. It has been integrated into an early warning system and applied to landslides susceptibility analysis. The analytical hierarchic process (AHP) [14] is used to evaluate the susceptibility, which assigns. scores to each factor of micro-topography of landslide-prone areas identified in aerial photographs, and assesses the susceptibility of landslide from the total score.

The application of computational intelligence tools [16] on the real-world data sets using both supervised and unsupervised methods gives reasonable results in landslide susceptibility. Remote sensing [17],[18] and Geographic Information System (GIS) are well suited for landslide studies. Adaptive neuro-fuzzy inference system (ANFIS) is another successful method for shallow landslide- susceptibility mapping in a tropical hilly area. The study included three main stages such as preparation of the landslide inventory, susceptibility analyses using the ANFIS and validation of the resultant susceptibility maps. The applied ANFIS [19] model learns the if–,then rules between landslide-related conditioning factors and landslide location, for generalization and prediction. Landslide occurrence and behavior are governed by numerous spatial factors that can be useful for the purpose of regional susceptibility assessment. The results of the univariate statistics proved to be useful in assessing the importance of each individual factor. The univariate statistic's results have indicated the same important factors as the multivariate analyses, but they cannot be simply used for susceptibility model development, since neglecting the existing interactions between factors results in wrong predictions. Multivariate statistics [20] were used instead of univariate and proved to be the most effective multivariate statistical tool in developing a landslide susceptibility model. The Mamdani FIS [22] is an expert-based system, it had previously been applied to a limited number of landslide susceptibility assessment studies. However, it is a suitable system for landslide susceptibility assessment because it has a considerable capacity to model complex and nonlinear systems. A methodology for landslide susceptibility assessment to delineate landslide prone area is presented using factor analysis and fuzzy membership functions [24] and Geographic Information Systems (GIS).

3 Early Warning Landslide Susceptibility Model (EWLSM)

Several qualitative and quantitative methods have been implemented to analyze landslide susceptibility as per figure 1. The distribution analysis is only useful for direct mapping of landslide locations from field surveys or aerial photographic interpretation and which do not provide information on predictive behavior of future landslide activity. In this type of analysis, GIS is used to digitize landslides prepared from field survey maps, aerial photographs and remote sensing images. In qualitative analysis, subjective decision rules are applied to define weights and ratings based on the experience of experts [9]. The logical analytical method is used for the field survey data on slope deformation, which helps to decide the numerical weights. Remote sensing and GIS techniques may be utilized for thematic map preparation and overlay analysis. To remove subjectivity in qualitative analysis, various statistical methods have been employed for Landslide susceptibility analysis. These methods can broadly be classified into three types: Bivariate, Multivariate and Probabilistic prediction models[15]. The bivariate models consider each individual thematic map in terms of landslide distribution and can be easily implemented in GIS. Multivariate methods consider various thematic at once into a complex and time-consuming data analysis processes. The probabilistic prediction models [1] provide quantitative estimates of future landslide activity based on prediction from past events. Recently, data mining [2],[3],[6],[8] approaches have been used in landslide susceptibility analysis.

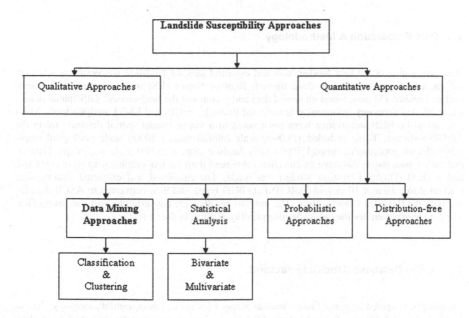

Figure 1. Landslide Susceptibility Approaches

3.1 Proposed Architecture

The proposed idea is to predict the possibilities of landslides in mountain areas by using Early Warning Landslide Susceptibility Model (EWLSM). A new data mining approach Bayesian Classification is applied to construct Early Warning Landslide Susceptibility Model (EWLSM) to predict the occurrence of the landslide using the detected landslide locations and the constructed spatial database. The landslide-susceptibility analysis results are verified using the landslide location test sites for each studied area. The results obtain are verified for accuracy and further fine tuned using hybrid computing techniques.

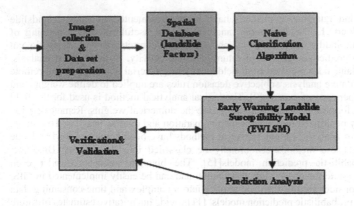

Figure 2.The Proposed Early Warning Landslide Susceptibility Model (EWLSM) Architecture

3. 2 Data Preparation & Methodology

Data preparation is the first fundamental and essential step for landslide susceptibility analysis. GIS based image collection is done through Remote Sensor (RS) monitoring of land, which utilizes Landsat TM data, based on spatial data and pictures of the land survey. GIS spatial data is analyzed, and necessary information is retrieved through ArcGIS and ENVI analysis tools. Maps relevant to landslide occurrence were constructed in a vector format spatial database using the ArcGIS software. These included 1:5,0000 scale rainfall map, 1:50000 scale geological maps, 1:50000 scale topographic maps,1:50000 scale landuse map, 1:25,000 scale soil maps. Contour and survey base points that have an elevation value read from the topographic map were extracted, and a DEM (Digital Elevation Model) was made. The calculated and extracted factors were converted to a 10 m x 10 m grid (ARC/INFO GRID type), and then converted to ASCII data for use with the Bayesian Classification. The analysis results were converted to grid data using GIS. The data requirement for the landslide susceptibility is given in the table1.

3.3 Spatial Database (landslide Factors)

Landslide is triggered by many factors such as Slope, Elevation, Aspect, rainfall, Geology, Stream power index and distance from drainage. Slope is an important parameter in landslide studies, since its stability forms the basis for the frequency and intensity of landslide. Lee[10] reported that landslide was maximum when the slope angle is between 35° and 40° and it decreases when slope is > 40°.Occurrence of landslide increased dramatically in areas where slope is between 6° and 45°. In this study, slope angle derived and used as one thematic layer for landslide analysis.

Classification		GIS Data Type	Scale of Resolution	
Spatial Database	Factor	Spatial Database	Spatial Database	Factor
Landslide	Landslide	ARC/INFO Polygon coverage	1:50000	
Rainfall Map	Rainfall data	ARC/INFO Polygon coverage	1:50000	
Geologic al Map	Litholoty	ARC/INFO Polygon coverage	1:50000	
	Acid intrusive			
	Acid intermediate volcanics			
	Limestone /marble			10m X 10m
	Sandsone/ mudstone			
Topograp hic Map	Slope	ARC/INFO Polygon coverage	1:50000	
	Aspects			
	Curvature			
Landuse Map	Land use type	ARC/INFO Grid	1:50000	
Soil Map	Texture	ARC/INFO Polygon coverage	1:25000	
	Drainage			
	Material			
	Effective Thickness			
	Diameter			
	Age			
	Density			

Table 1. Various spatial Layers for landslide study

The removal of vegetation increased the occurrence of landslides as slope increases. Steep slopes, at the same time, could not fail in response to rainfall because it requires a sufficiently thick mantle of residual soil. Drainage is a major factor that contributes to landslide in hill areas. As the distance from the drainage line increases, landslide frequency generally decreases because the seepage of water near the drainage network is more as compare to far away from them, so shear strength is reduced near the drainage network. Another highly significant factor in slope stability and in turn landslide is the aspect map. It represents the direction of slope and refers to the direction in which a mountain / hill slope faces a possible amount of sunshine and shadow. Slope aspect influence terrain exposure to storm fronts. The aspect of the slope is one of the major contributing factors for a landslide. The elevation variable can provide information on potential energy. The parameter "Geology" is related to the resistance to landslides. The most landslide sensitive formations are considered to be those with a mollasic character, i.e., sandstone, siltstone, marl alteration, due to the different degree of weathering. Limestones and quartzites are considered to be resistant to landsliding formations, although this is highly dependent on various aspects such as erosion degree and structural fabric, which may alter the initial character of a geological formation.

Elevation	Slope	Aspect	SPI	Distance from drainage	Geology	RainFall	Landslide
560–600	0–5	Flat	0–0.367	0–150	Alluvium	1000-1200	Low
600–675	5–10	N	0.367–0.734	150–300	Sedimentary	1200-1400	Medum
675–750	10–15	NE	0.734–1.183	300–450	Debris	1400-1600	Medum
750–825	15–20	E	1.183–1.652	450–600	Volcanics	1600-2000	Medum
825–900	20–25	SE	1.652–2.080	600–750	Alluvium	2000-2400	High
900–975	25–30	S	2.080–2.468	750–900	Sedimentary	2400-2800	High
975–1050	30–35	SW	2.468–2.835	900–1050	Debris	1000-1200	Low
1050–1125	35–40	W	2.835–3.223	1050–1200	Volcanics	1200-1400	Medum
1125–1200	40–45	NW	3.223–3.692	1200–1350	Alluvium	2000-2400	High
1200–1449	45–52	N	3.692–5.202	1350–1604	Sedimentary	2400-2800	High
825–900	5–10	SE	1.183–1.652	750–900	Sedimentary	2000-2400	High
1050–1125	45–52	N	0–0.367	1200–1350	Alluvium	1000-1200	Low
600–675	0–5	E	0.367–0.734	1050–1200	Debris	1200-1400	Medum
560–600	45–52	Flat	0.367–0.734	1200–1350	Alluvium	1400-1600	Low

Table 2.Spatial database

The heavy rainfall induces more landslides in mountain areas. So the rainfall is another important factor for the prediction of landslides. The last parameter considered in the present study is stream power index SPI. It is a measure of erosive power of water flow based on the assumption that discharge(q) is proportional to specific catchment area (As) SPI = As tan β Where As is the specific catchment area while β is the slope gradient in degree.

3.4 Naive Bayesian Classification

Classification Data mining technique have number of algorithms like Decision Tree, Back propagation Neural Network, Support Vector machine(SVM),Rule based Classification and Bayesian Classification. In the present scenario, landslide prediction study was done by using Neural Network, Decision tree, but it is difficult to understand and tricky to predict .Bayesian Classifiers to be comparable in performance with other classification techniques such as high accuracy, Speed, minimum error rate. The Bayesian Learning just reduces the probability of an inconsistent hypothesis. This gives the Bayesian Learning a bigger flexibility. This classifier can accept any number of either continuous or categorical variables. In fact, the Naive Bayesian classifier technique is particularly suited when the number of variables (dimensionality) is high. Bayesian analysis is an approach to statistical analysis that is based on the Bayes's law, which states that the posterior probability of a parameter p is proportional to the prior probability of parameter p multiplied by the likelihood of p derived from the data collected.

$$P(h \mid D) = \frac{P(D \mid h)P(h)}{P(D)}$$

P(h)   = Prior Probability of hypothesis h
P(D)   = Prior Probability of training data D
P(h|D) = Posterior Probability of h given D
P(D|h) = Posterior probability of D given h

Suppose that we have m layers of spatial map data containing "causal" factors, which are known to correlate with the occurrences of future landslides in the study area. The important landslide factors from the study area are extracted from the relevant map sources and represented in table 1. Assume that we have three classes of landslide susceptibility C1=High, C2=Low, C3= medium.

Let D be a training set of tuples and their associated class labels. each tuple is represented by an n-dimensional attribute   vector, $X = (x_1, x_2, ..., x_n)$, depicting n- measurements made on the tuple from n attributes, respectively $A_1, A_2, ..., A_n$. Suppose that there are m classes $C_1, C_2, ..., C_m$. Given a tuple, X, the classifier will predict that X belongs to the class having the highest posterior probability, conditioned on X. That is, the naïve Bayesian classifier predicts that tuple X belongs to the class C_i if and only if

$$P(C_i / X) > P(C_j / X) \text{ for } 1 \leq j \leq m, j \neq i$$

Thus we maximize $P(C_i / X)$. The class C_i for which $P(C_i / X)$ is maximized is called the maximum posteriori hypothesis. Modified Bayesian classifier, By Bayes' theorem is

$$P(C_i / X) = \frac{P(X / C_i)(P(C_i)}{P(X)}$$

If $P(X)$ is constant for all classes, only $P(X / C_i)(P(C_i)$ need to be maximized. If the class prior probabilities are not known, then it is commonly assumed that the classes are equally likely, that is, $P(C_1) = P(C_2) = .. = P(C_m)$, and we would therefore maximize $P(X / C_i)$. Otherwise, we maximize $P(X / C_i)(P(C_i)$. Data sets with many attributes, it would be extremely computationally expensive to compute $P(X / C_i)$. In order to reduce computation in evaluating $P(X / C_i)$, the naive assumption of class conditional independence is made. This presumes that the values of the attributes are conditionally independent of one another, given the class label of the tuple. Thus,

$$P(X / C_i) = \prod_{k=1}^{n} P(x_k | C_i) = P(x_1 | C_i) \times P(x_2 | C_i) \times \ldots \times P(x_n | C_i)$$

here x_k refers to the value of attribute A_k for tuple X. For each attribute, we look at whether the attribute is categorical or continuous-valued. If A_k is categorical, then $P(X / C_i)$ is the number of tuples of class C_i in D having the value x_k for A_k, divided by $|C_{i,D}|$, the number of tuples of class C_i in D. If A_k is continuous-valued, A continuous-valued attribute is typically assumed to have a Gaussian distribution with a mean μ and standard deviation σ, defined by

$$g(x, \mu, \sigma) = \frac{1}{\sqrt{2\pi}\sigma} e^{\frac{-(x-\mu)^2}{2\sigma^2}}$$

$$P(x_k | C_i) = g(x_k, \mu c_i, \sigma c_i)$$

In order to predict the class label of X, $P(X / C_i)(P(C_i)$ is evaluated for each class C_i. The classifier predicts that the class label of tuple X is the class C_i if and only if

$$P(C_i / X) > P(C_j / X) \text{ for } 1 \leq j \leq m, j \neq i$$

In other words, the predicted class label is the class C_i for which $P(X / C_i)(P(C_i)$ is the maximum. Naïve Bayesian classification approach is applied and constructed early warning landslide susceptibility model to predict the occurrence of landslide.

3.5 Prediction Analysis, Verification and Validation

The landslide-susceptibility analysis results are verified using the landslide location test sites for the study area. The verification method is then performed by comparing the landslide test data. Two basic assumptions are then needed to verify the landslide susceptibility calculation methods. One is that landslides are related to spatial information such as topography, land cover, soil, and geology. The other is that future landslides will be swift by a specific impact factor such as rainfall.

4 Implementation & Result

GIS spatial data is analyzed and necessary information is retrieved through ArcGIS and ENVI analysis tools. . Maps relevant to landslide occurrence were constructed in a vector format spatial database using the ArcGIS software. There were seven landslide-related factors considered in this analysis such as Slope, Elevation, Aspect, rainfall, Geology, Stream power index and distance from drainage .Contour and survey base points that have an elevation value read from the topographic map were extracted, and a DEM (Digital Elevation Model) was made. The calculated and extracted factors were converted to a 10 m x 10 m grid (ARC/INFO GRID type), and then converted to ASCII data for use with the Bayesian Classification. The modeling of the Bayesian has been developed in STATISTICA data miner tool. The probabilities of occurrence of landslides were calculated based on the various input parameters specified in Table 2 .

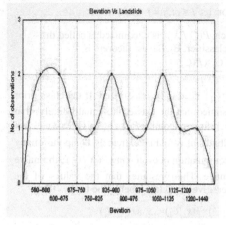

(a). Impact of Elevation factor in Landslide

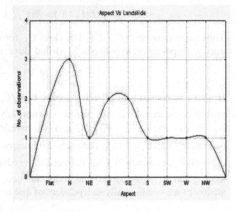

(b). Impact of Aspect factor in Landslide

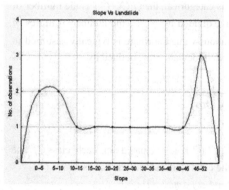

(c). Impact of Contribution of Slope factor in Landslide

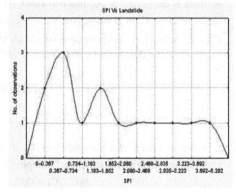

(d). Impact of SPI factor in Landslide

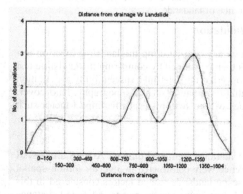

(e).Impact of Distance from drainage factor in Landslide

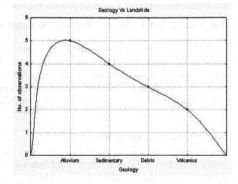

(f).Impact of Geology in Landslide

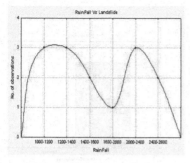

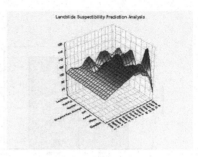

(g).Impact of Rainfall factor in Landslide
Figure 3. Impacts of various factors in
Landslide

Figure 4. Early Warning Landslide
Susceptibility Model (EWLSM) Prediction
for different cases

The Bayesian model (figure 3) result shows that slope is an important parameter for landslide studies. Landslide is maximum when the slope angle is > 45° and it decreases when slope is between 15° and 40°. In the case of aspect and stream power index, the occurrence of landslide is high when it is flat and 0.367-0.764 respectively. Heavy rainfall and more distances from drinage induce the possibilities of landslide. The most landslide sensitive formations are considered to be those with a alluviam and sedimentary. The level of landslide is predicted based on the spatial database as shown in Table 2. The landslide occurrence is classified into 3 classes such as high, medium and low. In our study, the Bayesian approach predicts the possibility of landslide accurately. Figure 4 shows the early warning landslide susceptibility model prediction for different cases of input. The landslide susceptibility analysis result was validated using known landslide locations.

5 Performance analysis

Bayesian Classifiers to be comparable in performance with other classification techniques in terms of high accuracy, speed, minimum error rate. The Bayesian Learning just reduces the probability of an inconsistent hypothesis. This gives the Bayesian Learning a bigger flexibility. This classifier can accept any number of either continuous or categorical variables. In fact, the Naive Bayesian classifier technique is particularly suitable for high dimension data.

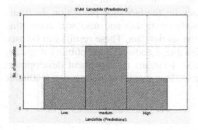

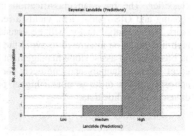

Figure 5. SVM classifier landslide prediction

Figure 6. Bayesian classifier landslide prediction

The Bayesian approach predicts (figure 5 and 6) the occurrence of landslide very accurately either high or low. But the SVM classifiers predict the occurrence of landslides in to three cases like high, medium and low.

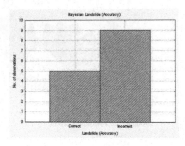

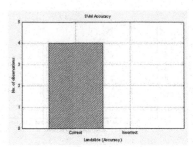

Figure 7. Bayesian classifier prediction accuracy

Figure 8. SVM classifier prediction accuracy

SVM was a new general-purpose machine learning method based on statistical learning theory, and it was built under the theory framework and general approach of machine learning with limited samples. Its basic thought was to transform the input space to a high-dimension one by using the non-linear transformation defined by inner product function and to find a non-linear relationship between the input variables and the output ones in the high-dimension. SVM had a better generalization than neural network which used empirical risk minimization principle. SVM is poor in speed to training and testing the data set, so that it's not suitable for very large datasets. In figure 7 and 8, the Bayesian classifies the result accurately in terms of correct and incorrect, but the SVM classifies only correct and its fails to consider all the possibility. Bayesian Classifier predicts the landslide susceptibility accurately compare than SVM classifier.

6 Conclusion

In this study, an improved Bayesian classification approach is used to build Early Warning Landslide Susceptibility Model (EWLSM) to estimate areas susceptible to landslides using a spatial database and GIS. We also compared the performance of Bayesian approach in landslide susceptibility with SVM Classifier. Bayesian approach is more suitable and accurate than SVM classifier. The bayesian classification modeling approach, combined with the use of remote sensing and GIS spatial data, yields a reasonable accuracy for the landslide prediction. In future work, more spatial factor will be considered for landslide prediction and also soft computing techniques can be integrated to avoid the uncertainty in the spatial data. These results can be used as basic data to assist slope management and landuse planning. Landslide susceptibility maps are great help to planners and engineers for choosing suitable locations to implement development action plans. In order to obtain higher prediction accuracy it is recommended to use a suitable dataset of landslide data.

References

1.A.Carrara, F.Guzzetti, M.Cardinali and P. Reichenbach, "Use of GIS Technology in the Prediction and Monitoring of Landslide Hazard" Natural Hazards ,20: 117–135, 1999

2.Qiang Yang, "10 CHALLENGING PROBLEMS IN DATA MINING RESEARCH", International Journal of Information Technology & Decision Making, Vol. 5, pp 597–604,2006,

3. S. Wan , T. C. Lei , T. Y. Chou, " A novel data mining technique of analysis and classification for landslide problem" , Nat Hazards (2010) 52: 211–230, Springer.

4. Is-ık Yilmaz, "Landslide susceptibility mapping using frequency ratio, logistic regression, artificial neural networks and their comparison: A case study from Kat landslides (Tokat—Turkey)", Computers & Geosciences 35 (2009) 1125–1138

5. Wang Jiana,b, Peng Xiang-guo, "GIS-based landslide hazard zonation model and its application", Procedia Earth and Planetary Science 1 (2009) 1198–1204

6. Colin H Aldridge, "Discerning Landslide Hazard Using a Rough Set Based Geographic Knowledge Discovery Methodology", The 11th Annual Colloquium of the Spatial Information Research.

7.Biswajeet Pradhan, Saro Lee, "Regional landslide susceptibility analysis using Back-propagation neural network model at Cameron Highland, Malaysia", Landslides (2010) 7:13–30 Springer-Verlag

8.Phil Flentje1, David Stirling2 & Robin Chowdhury "Landslide Susceptibility And Hazard Derived From A Landslide inventory Using Data Mining" – An Australian Case Study

9.Ashis K. Saha · Ravi P. Gupta · Irene Sarkar · Manoj K. Arora · Elmar Csaplovi "An approach for GIS-based statistical landslide susceptibility zonation—with a case study in the Himalayas" Landslides (2005) 2:61–69

10. S Lee ,Jasmi Abdul Talib, "Probabilistic landslide susceptibility and factor effect analysis" Environ Geol (2005) 47: 982–990

11. Pece V. Gorsevski, Paul E. Gessler, Piotr Jankowski ,"Integrating a fuzzy k-means classification and a Bayesian approach for spatial prediction of landslide hazard", Journal of Geographical Systems (2003) 5:223–251

12. Saro Lee . Biswajeet Pradhan , "Landslide hazard mapping at Selangor, Malaysia using frequency ratio and logistic regression models", Landslides (2007) 4:33–41

13.Zonghu Liao, Yang Hong, Jun Wang, Hiroshi Fukuoka , Kyoji Sassa, Dwikorita Karnawati and Faisal Fathani , "Prototyping an experimental early warning system for rainfall-induced landslides in Indonesia using satellite remote sensing and geospatial datasets" Landslides (2010) 7:317–324

14. H. Yoshimatsu · S. Abe, "A review of landslide hazards in Japan and assessment of their susceptibility using an analytical hierarchic process (AHP) method", Landslides (2006) 3: 149–158

15. Paolo Magliulo & Antonio Di Lisio & Filippo Russo, "Comparison of GIS-based methodologies for the landslide susceptibility assessment" Geoinformatica (2009) 13:253–265

16.M.D. Ferentinou, M.G. Sakellariou "Computational intelligence tools for the prediction of slope performance", Computers and Geotechnics 34 (2007) 362–384

17.R Rajakumar, S. Sanjeevi, S. Jayaseelan, G. Isakkipandian, M. Edwin, P. Balaji And Ct Ehanthalingam ,"Landslide Susceptibility Mapping In A Hilly Terrain Using Remote Sensing And GIS", Journal of the Indian Society of Remote Sensing, Vol. 35, No. 1, 2007

18. S. Prabu . S.S. Ramakrishnan Combined use of Socio Economic Analysis, "Remote Sensing and GIS Data for Landslide Hazard Mapping using ANN " , J. Indian Soc. Remote Sens. (September 2009) 37:409–421

19. Hyun-Joo Oh, BiswajeetPradhan, "Application of a neuro-fuzzy model to landslide-susceptibility mapping for shallow landslides in a tropical hilly area", Computers & Geosciences 37 (2011) 1264–1276

20. Marko Komac, "A landslide susceptibility model using the Analytical Hierarchy Process method and multivariate statistics in perialpine Slovenia", Geomorphology 74 (2006) 17–28

21.Saro Lee ,Touch Sambath, "Landslide susceptibility mapping in the Damrei Romel area, Cambodia using frequency ratio and logistic regression models" ,Environ Geology (2006) 50: 847–855

22. Akgun, A., et al., "An easy-to-use MATLAB program (MamLand) for the assessment of landslide susceptibility using a Mamdani fuzzy algorithm", Computers & Geosciences (2011)

23. Biswajeet Pradhan , Saro Lee, "Landslide susceptibility assessment and factor effect analysis: back propagation artificial neural networks and their comparison with frequency ratio and bivariate logistic regression modeling", Environmental Modelling & Software 25 (2010) 747–759

24. A. Gemitzi,G. Falalakis,P. Eskioglou,C. Petalas, "Evaluating Landslide Susceptibility Using Environmental Factors, Fuzzy Membership Functions And GIS" ,Global Nest Journal, Vol 12, No 4,2010

Detection of Plasmodium Falciparum in Peripheral Blood Smear Images

Feminna Sheeba [1], Robinson Thamburaj [2], Joy John Mammen [3], Atulya K. Nagar [4]

[1] Department of Computer Science, Madras Christian College, Chennai, India

fsheeba@gmail.com

[2] Department of Mathematics, Madras Christian College, Chennai, India

robin.mcc@gmail.com

[3] Department of Transfusion Medicine & Immunohaematology, Christian Medical College, Vellore, India

joymammen@cmcvellore.ac.in

[4] Centre for Applicable Mathematics and Systems Science, Department of Computer Science

Liverpool Hope University, Liverpool, UK

nagara@hope.ac.uk

Abstract. *Malaria* is a mosquito-borne infectious disease caused by the parasite Plasmodium, which requires accurate and early diagnosis for effective containment. In order to diagnose malaria in a patient, timely detection of malaria parasites in blood smear images is vital. The traditional methods are time–consuming, tedious and the quality of detection is highly subjective to the individual who performs the analysis. These results can clearly be improved upon by using image processing techniques. The malaria parasite appears in four stages, namely the ring, trophozoite, schizont, and gametocyte. The ring and the gametocyte stage are the ones seen in a peripheral blood smear and hence detecting these two stages, would help in the accurate diagnosis of malaria. The proposed work aims at automating the analysis of the blood smear images using appropriate segmentation techniques, thereby detecting infected red blood cells as well as the gametocytes found in the blood.

Keywords: Plasmodium, Chromatin Dots, Gametocytes, Segmentation, Morphology.

J. C. Bansal et al. (eds.), *Proceedings of Seventh International Conference on Bio-Inspired Computing: Theories and Applications (BIC-TA 2012),* Advances in Intelligent Systems and Computing 202, DOI: 10.1007/978-81-322-1041-2_25, © Springer India 2013

1 Introduction

Malaria is an infectious disease caused by a parasite called Plasmodium, which is transmitted via the bites of infected mosquitoes. There are four species of Plasmodium that infect man: P. falciparum, P. vivax, P. ovale, and P. malariae, out of which P, vivax and P. falciparum are the common ones. The parasite appears in four stages in blood - ring, trophozoite, schizont, and gametocyte. The four stages of the parasite as seen in blood smear images are shown in Fig. 1.

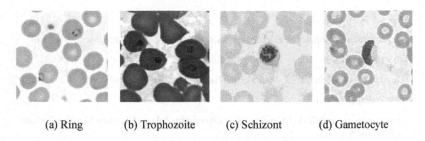

    (a) Ring   (b) Trophozoite   (c) Schizont   (d) Gametocyte

Fig. 1 Stages of Malaria Parasites in Blood Smear Images

Trophozoite and schizont are seldom seen in peripheral blood smear images, which means that it would suffice to look for the Ring and Gametocyte stage to diagnose a case of malaria.

Malaria can be life threatening if not treated properly. It can be diagnosed by prompt and effective detection of malaria parasites in blood smear. Traditional methods like Quantitative Buffy Coat Parasite Detection System are currently used for identifying the malaria parasites in the peripheral blood. Such methods are cumbersome, time consuming, monotonous and results are prone to be affected by the operator's fatigue. Therefore effective image analysis and segmentation methods are analysed for diagnoses of malaria. Segmentation based on gray values is a very simple method [1]. In order to segment overlapping objects watershed transform can be used [2]. Another method of segmentation is to use tissue like P systems [3].

In segmenting malaria parasites, a simple method is by using multiple thresholding to identify infected erythrocytes and the components of parasites inside the erythrocytes [4]. Malaria parasites are detected using certain masks and clump splitting [5]. Screening of malaria infection in microscope images of thin blood film smears is also done using image analysis and pattern classification [6]. Classification techniques like Bayesian and distance weighted K nearest neighbour classifiers are used to detect malaria parasites (*Plasmodium spp*) in blood smear images [7]. Artificial Neural Network can be used for detection, where various image analysis techniques and a feed forward back propagation neural network [8] or two back propagation Artificial Neural Network models [9] are used to detect red blood cell disorders and intracellular malarial parasites. The parasitaemia measure is carried out by partitioning the uninfected and infected cells using an unsupervised and a training-based technique [10]. This paper focuses on segmentation of chromatin dots in the ring structure and thereby detecting the infected RBCs. It also focuses on detecting gametocytes including those that are touching or overlapping the RBCs.

2 Image Preprocessing

2.1 Image Acquisition

The images for this study are obtained using a digital camera DFC280 attached to a compound microscope. It uses a choice of 1.3 megapixel standard resolution or 2.9 megapixel high resolution. The image is digitized with a 10-bit AD conversion that features a 700:1 dynamic range. The TWAIN interface is used to transfer images to the Leica IM50 Image Manager. The images acquired are 24 bit colored tiff images with a resolution size of 1280x1024 pixels. Seventy five such images were used for the study. The software for the segmentation was written using MATLAB R2009b.

2.2 Image Adjusting

The RGB images are first converted to HSV color models and then converted to gray scale images. The images thus acquired may have poor illumination. Therefore, the contrast of the images is improved by equalizing the histogram of the image. Filtering was used to remove noise from the images.

3 Detection of Malaria Parasites in Blood Smear Images

3.1 Segmentation

Segmentation refers to the process of partitioning a digital image into multiple segments [12] [13]. The goal of segmentation is to simplify and/or change the representation of an image into something that is more meaningful and easier to analyze. More precisely, image segmentation is the process of assigning a label to every pixel in an image such that pixels with the same label share certain visual characteristics. The result of image segmentation is a set of segments that collectively cover the entire image, or a set of contours extracted from the image. In order to count the gametocytes in an image, they need to be first segmented. An automated segmentation method is used to find out the infected RBCs by segmenting the chromatin dots in rings and also detect the gametocytes.

3.2 Detection of Infected RBCs

A malaria patient could have the ring stage of the parasite in the RBCs of his/her blood. P. falciparum rings have delicate cytoplasm and one or two small chromatin dots as seen in Fig. 1(a). Chromatin dots are more easily distinguished from the RBC by the cytoplasm which is a delicate structure that shows great variability. In order to segment the chromatin dots, WBCs are first identified in the input color images using methods discussed in [2] and are eliminated. The chromatin dots are then segmented using the color segmentation method discussed in [11], by choosing appropriate ranges of hue and saturation. This results in binary images consisting of segmented chromatin dots and artefacts. In order to differentiate chromatin dots from artefacts, it is checked to see if the dots are found inside or on the RBCs. The centroid of all objects (dots) in the image was found by applying granulometry and extracting the centroid property. The presence of the centroid in any of the RBCs is checked. On finding a centroid in the RBC, the corresponding RBC is marked in the source image. In the resulting image, certain RBCs are found overlapping each other which may result in faulty findings while locating the Chromatin Dots inside them. The overlapping RBCs are separated by applying watershed segmentation techniques discussed in [2]. Thus an RBC is marked as infected if it contains a chromatin dot within, and all other dots lying outside the RBCs are assumed to be artefacts.

3.2 Detection of Gametocytes

Segmentation of Normal Gametocytes: Normal Gametocytes are those whose outlines are clearly defined and do not overlap other objects in the vicinity. Using appropriate morphological operations, segmentation of various objects in the images is carried out. The images are segmented such that all the objects (WBCs and RBCs), apart from platelets and gametocytes, are eliminated. The platelets are then eliminated based on the metric of the blobs, leaving behind only the gametocytes.

Segmentation of Gametocytes touching other Cells: However with the method of segmenting normal gametocytes, the ones that were touching the RBC, were not segmented as expected. In order to segment such overlapping gametocytes, the distance transform of the binary form of the input image is obtained. The holes are then filled and the gametocytes obtained are smoothed using morphological operations.

Differentiating Gametocytes from Sickle Shaped RBCs: Due to some diseases, certain RBCs may turn out to be sickle shaped. Therefore, during segmentation, such RBCs may be mistaken for gametocytes. In order to eliminate such RBCs, the presence of a nucleus in the segmented object is detected. All the objects without nuclei are eliminated to obtain only the gametocytes.

4 Results and Findings

Forty blood smear images containing ring stage parasites and thirty four images containing malaria gametocytes were used to test the algorithms for segmentation of the parasites.

4.1 Results – Infected RBCs

A source image with malaria parasites in the form of a ring with a chromatin dot are found in four RBCs as shown in Fig 2.

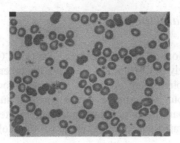

Fig. 2 Blood Smear Image with Chromatin Dots

Color segmentation was done and the segmented image with chromatin dots is shown in Fig. 3. The centroid of all objects in the image were found and plotted in the binary image as shown in Fig. 4.

Fig. 3 Segmented Chromatin Dots   **Fig. 4** Centroid of Objects in the Image

The RBCs (that are not touching the border) were segmented and labeled, which is shown in Fig. 5(a). The infected RBCs (having the chromatin dots) are alone marked as shown in Fig. 5(b).

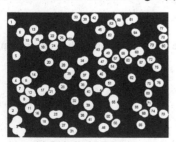
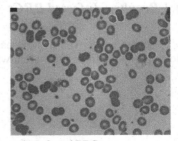

(a) Labelled RBCs   (b) Infected RBCs

Fig. 5 Detection of Infected RBCs

4.2 Results – Gametocyte Detection

One of the input images used for gametocyte detection is shown in Fig. 6(a). Otsu's method of thresholding was applied to the inverse of the input image to obtain a binary image. The inverse image and binary image are shown in Fig. 6(b) and Fig. 6(c) respectively.

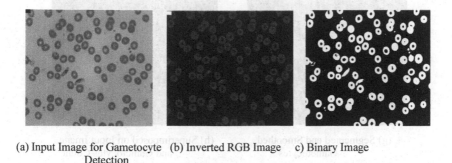

(a) Input Image for Gametocyte   (b) Inverted RGB Image   c) Binary Image
                 Detection

Fig. 6 Conversion to Binary Image

The holes in the RBCs were filled and smaller objects were eliminated (I1). Round shaped blood cells were then extracted (I2). Absolute difference between Images I1 and I2 gives an Image, eliminating all cells and showing parasites alone. I1, I2 and the difference image are shown in Fig. 7(a), Fig. 7(b) and Fig. 7(c) respectively.

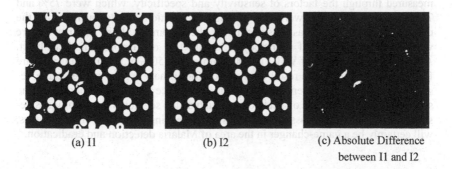

(a) I1   (b) I2   (c) Absolute Difference
                                                   between I1 and I2

Fig. 7 Detection of Gametocytes with Artefacts

Sometimes platelets that were too small to be eliminated under the morphological opening were also displayed. In that case a metric analysis was done to eliminate them and the parasites alone were displayed as shown in Fig. 8(a). The image was smoothed using morphological operations and superimposed on the original Image. Finally the count of the segmented gametocytes was obtained, which is shown in Fig. 8(b).

(g) Segmented and Smoothed Gametocytes

(h) Superimposed on Input Image and Gametocytes Count

Fig 8 Segmentation of Gametocytes

4 Conclusion

The study shows the vast potential of harnessing image analysis techniques for the detection and study of malaria parasites. The success rate of this project was measured through the factors of sensitivity and specificity, which were 75% and 78% respectively. Future work will include the detection of cytoplasm around the chromatin dot of the parasite, surrounding the chromatin dots. Another feature would be the detection of Gametocytes, completely overlapping another object. According to the report of WHO in 2009, more than 50% of the malaria cases are from India. Therefore, the resources required for primary screening are scarce, resulting in an enormous demand for quality assurance during the conduct of the screening tests. The usage of Technology, such as the segmentation techniques described in this paper, to assist the process of screening the blood smear images, will certainly be a game-changer in the area of Malaria detection and eradication.

Acknowledgement

The authors would like to thank the Centre for Applicable Mathematics and Systems Science (CAMSS), Department of Computer Science, Liverpool Hope University, UK for the support and funding towards this project work and the Department of Pathology, CMC, Vellore, India for providing them with sample images for the study. The authors also thank Miss Maqlin P. and Dickson Jebaraj, Madras Christian College, Chennai, for their contribution towards the development of the system that performs automatic segmentation of malaria parasites in blood smear images.

References

[1] Sheeba, F., Thomas, Hannah, M.T.T., Mammen. J.J.: Segmentation and Reversible Watermarking of Peripheral Blood Smear Images. Proc. of the IEEE Fifth International Conference on Bio-Inspired Computing: Theories and Applications (BIC-TA), 1373-6.IEEE BIC-TA (2010).

[2] Sheeba, F., Thamburaj. R,, J.J. Mammen , Hannah.M.T.T., Nagar,A.K.: White Blood Cell Segmentation and Watermarking. Proc. of the IASTED International Symposia Imaging and Signal Processing in Healthcare and Technology , Washington DC, USA. ISPHT (2011)

[3] Sheeba, F., Thamburaj, R., Nagar, A.K., Mammen, J.J.: Segmentation of Peripheral Blood Smear Images using Tissue-like P Systems. The 6th International Conference on Bio-Inspired Computing pp. 257-261. BICTA (2011).

[4] Anggraini, D., Nugroho, A.S., Pratama, C., Rozi, I.E., Iskandar, A.A. , Hartono, R.N.: Automated Status Identification of Microscopic Images Obtained from Malaria Thin Blood Smears. International Conference on Electrical Engineering and Informatics. 1-6. ICEEI (2011).

[5] Selena W.S. et al, Malaria Count.: An image analysis-based program for the accurate determination of parasitemia. Journal of Microbiological Methods. Elsevier (2006). doi:10.1016/j.mimet.2006.05.017

[6] Tek, F.B., Dempsterb, A.G., Kale, I.: Computer Vision for Microscopy Diagnosis of Malaria. Malaria Journal. 8:153. 2009. doi:10.1186/1475-2875-8-153

[7] Tek, F.B., Dempsterb, A.G., Kale, I.: Malaria Parasite Detection in Peripheral Blood Images. In: British Machine Vision Conference 347-356. BMVC (2006).

[8] Premaratne, S.P., Karunaweera, N.D., Fernando, S., Perera, W.S.R., Rajapaksha, R.P.A. : A Neural Network Architecture for Automated Recognition of Intracellular Malaria Parasites in Stained Blood Films, http://journalogy.net/Publication/10403644

[9] Hirimutugoda,Y.M, Wijayarathna, G. : Image Analysis System for Detection of Red Cell Disorders Using Artificial Neural Networks, Journal of Bio-Medical Informatics. 1(1): 35-42. (2010)

[10] Halim, S., Bretschneider, T.R. , Yikun Li., Preiser, P.R. , Kuss, C.: Estimating Malaria Parasitaemia from Blood Smear Images.

9th International Conference on Control, Automation, Robotics and Vision. 1-6.
ICARCV (2006).

[11] Makkapati, V.V., Rao R.M..: Segmentation of Malaria Parasites in Peripheral Blood
Smear Images. In: IEEE Conf. on Acoustics, Speech and Signal Processing 1361-
1364. ICASSP (2009).

[12] Gonzalez, R.C., Woods, R.E.: Digital Image Processing 3ed. by Prentice Hall. (eds).
(2008).

[13] Gonzalez, R.C., Woods, R.E., Eddins, S.L.: Digital Image Processing using Matlab.
ed, by Pearson Education.(2009).

Supervised Opinion Mining of Social Network data using a Bag-of-Words approach on the Cloud

S. Fouzia Sayeedunnissa[1], Adnan Rashid Hussain[2], Mohd Abdul Hameed[3]

[1] Department of Information Technology, M.J. College of Engineering & Technology,
Sultan-ul-Uloom Education Society, Hyderabad, India
researcher.fouzia@gmail.com
[2] Research & Development, Host Analytics Software Pvt. Ltd., Hyderabad, India
adnanrashid.ar@gmail.com
[3] Department of Computer Science & Engineering, University College of Engineering (A),
Osmania University, Hyderabad, India
researcher.hameed@gmail.com

Abstract. Social networking and micro-blogging sites are stores of opinion-bearing content created by human users. Sentiment analysis extracts and measures the sentiment or "attitude" of documents as well as the topics within documents. The attitude may be the person's judgment (e.g., positive vs. negative) or emotional tone (e.g., objective vs. subjective). In this paper we define a supervised learning solution for sentiment analysis on twitter messages which gives profound accuracy using Naïve Bayes Classification. It classifies the tweets either positive or negative. Our dataset consists of tweets containing move reviews retrieved from twitter using certain keywords on a cloud platform. The experiment and analysis has 3 major steps. In the first step, the algorithm used is Naïve Bayes which performed Boolean classification on bag of words, resulting in 71% of accuracy. In the second step, Naïve Bayes algorithm is applied on bag of words without stop words which gave 72% accuracy. In the third step, using the concept of Information gain, high value features are selected using Chi-Square, which gave maximum accuracy. The evaluating metrics used in this work are Accuracy, Precision, Recall and F-Measure for both the sentiment classes. The results show that Naïve Bayes algorithm with the application of feature selection using the minimum Chi-Square value of 3 gave an Accuracy of 89%, Positive Precision 83.07%, Positive Recall 97.2%, Positive F-Measure 89.9%, Negative Precision 96.6%, Negative Recall 81.2% and Negative F-Measure 88.2%. The significant increase in Positive Recall specifies that the classifier classifies positive words with more probability when compared to the negative words. This paper also describes the preprocessing steps needed in order to achieve high accuracy.

Keywords: Sentiment Analysis, Social Network, Twitter, Cloud Computing

J. C. Bansal et al. (eds.), *Proceedings of Seventh International Conference on Bio-Inspired Computing: Theories and Applications (BIC-TA 2012)*, Advances in Intelligent Systems and Computing 202, DOI: 10.1007/978-81-322-1041-2_26, © Springer India 2013

1   Introduction

Sentiment analysis has recently received a lot of attention in the Natural Language Processing (NLP) community. Polarity classification, whose goal is to determine whether the sentiment expressed in a document is "thumbs up i.e. positive" or "thumbs down i.e. negative", is arguably one of the most popular tasks in document-level sentiment analysis. Various sentiment measurement platforms employ different techniques and statistical methodologies to evaluate sentiment across the web [3][6][7][8][9]. Some rely 100% on automated sentiment, some employ humans to analyze sentiment, while others use a hybrid system. The various applications include Public Opinion, Customer Satisfaction, Targeted marketing strategy and many others.

Because the internet provides so many opportunities for people to publish their views on various products as such, companies and organizations are under increasing pressure to track and analyze opinions and attitudes about their products and services [1]. These opinions from customers, prospects, reviewers, analysts and employees are expressed in various web forums, from consumer review sites to Twitter as well as in emails, call logs, and web-based surveys and forms. By monitoring and analyzing opinions, companies can gather intelligence on a customer's defection risk and other business metrics such as the Churn rate. There is a very high impact if an influential user shares his opinion on the level of satisfaction or intensity of complaint, as their opinion becomes a contributing driver to other people's purchase decision. Capturing and analyzing these opinions is a necessity for proactive product planning, marketing and customer service. It is also critical in maintaining brand integrity. Thus sentiment analysis proves to be of great help in wide areas [2][3][4].

The challenge for leveraging sentiment is tracking disparate sources and then accurately capturing the meaning in the opinion in time to effectively analyze and act. Opinions are expressed in many different ways; accurately analyzing and measuring this diverse content produces quantitative values that improve the usefulness of the data on which companies rely to run their business [2][10]. In addition, entity-level sentiment helps refine the data to give insight into specific topics within documents, allowing deeper and more accurate analysis.

2   Literature Survey

2.1   NAÏVE BAYES

The Naïve Bayes algorithm is based on conditional probabilities. It uses Bayes Theorem, a formula that calculates a probability by counting the frequency of values and combinations of values in the historical data. Bayes' Theorem finds the probability of an event occurring given the probability of another event that has already occurred. If B represents the dependent event and A represents the prior event, Bayes' theorem can be stated as follows:

$$Prob(B \ given \ A) = \frac{Prob(A \ and \ B)}{Prob(A)}$$

The Naïve Bayes algorithm makes the simple assumption that the features consti-tuting the items are conditionally independent of the given class. It affords fast, highly scalable model building and scoring. The Naïve Bayes classifier uses the prior proba-bility of each class which is the frequency of each class in the training set, and the contribution from each feature. The classifier then returns the class with the highest probability and thus can be termed as positive class or negative class given the docu-ment. It scales linearly with the number of predictors and rows. The build process for Naïve Bayes is parallelized. Naïve Bayes can be used for both binary and multiclass classification problems.

2.2   Bag-of-words

Each document is a bag of words, meaning an unordered collection of words, disre-garding grammar and even word order. Models have been successfully used in the text community for analyzing documents and are known as "bag-of-words" models, since each document is represented by a distribution over fixed vocabulary. The bag-of-words model is used in some methods of document classification. When a Naïve Bayes classifier is applied to text, for example, the conditional independence assump-tion leads to the bag-of-words model.

2.3   Information Gain

Information gain for classification is a measure of how common a feature is in a par-ticular class compared to how common it is in all other classes. A word that occurs primarily in positive class and rarely in negative class is high information. Thus, the amount of information associated with an attribute value is related to the probability of occurrence. If we have a set with N different values in it, we can calculate the en-tropy as follows:

$$Entropy(S) = -\sum_{i=1}^{N} P_i . \log_2(P_i)$$

Where, P_i is the probability of getting the ith value when randomly selecting one from the set.

3   Corpora

Our objective was to perform sentiment analysis on messages from twitter of movie reviews. Twitter is a micro-blogging site, in which messages shared called tweets have a limit of 140 characters. We selected the top movies based their box-office performance using the Internet Movie Review Database (IMDB) as reference. We then identified the official twitter accounts or hash-tags representing the tweets for these movies. We then short-listed those accounts and hash-tags which were pub-lished as trending by twitter, thus ensuring that we would be able to extract sufficient content for our analysis. Using these keywords, we queried twitter using the Twitter Search API and collected the tweets. We analyzed the collected tweets and prepared a selection/transformation process to prepare the dataset for our experiment.

Our first step was to extract tweets which were in English and discarded the ones which were in foreign languages. Although our experiment was targeted towards tweets in English, the approach can also be used for other languages. We then identified the emotional tone of the tweets (i.e. objective vs. subjective) and discarded the objective tweets. We found that a large portion of the collected tweets were objective in nature such as "Planning to see #inceptionfilm this weekend" and did not convey any sentiment. The remaining tweets were then hand labeled for supervised training, based on the understanding of the English language and using a Thesaurus as a reference. We hand labeled tweets as either "positive" or "negative" using a custom web form of our cloud application. Since we needed to train the model appropriately for both positive and negative tweets, we chose equal amount of tweets for both classes so as to prevent over-training for any one class. Using the above parameters our dataset constituted of 5000 tweets, of which 75% were used for training and 25% were used for testing. Using this corpus we applied the 10-fold cross-validation technique for training the model to achieve the best results. The advantage of this technique over repeated random subsampling is that all observations are used for both training and testing, and each observation is used for validation exactly once.

4   Architecture

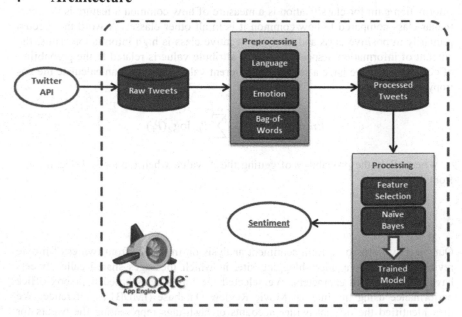

Fig. 1. Architecture Diagram

The application was developed in Python using Google's Platform-as-a-Service (PaaS) offering called Google App Engine (GAE). We used various services from GAE such as Task Queues, Backends, etc. which aided in the implementation of the sentiment analysis process workflow. As discussed in the corpora section, we prepared a list of keywords for movies which we wanted to collect information for and queried the Twitter Search API. The twitter search API provides results in various data formats; XML, JSON, RSS and Atom, and we chose JSON as GAE Datastore (which is a NoSQL schema-less object storage) stores objects in JSON format. To connect to the twitter API using python the PyTwitter package was used, which is a command-like tool to connect to the different available twitter REST and search APIs.

As twitter permits up to 350 API calls per hour as per its rate limiting policy, we used GAE task queues to schedule search queries to retrieve information through the API. We also monitored the rate limit status to ensure information retrieval within fair usage limits and dynamically throttle requests if necessary. The raw tweets retrieved with this process were stored in a GAE Datastore model. We then used the GAE Backends API which allows executing long running processes, to select and pre-process the raw tweets as described in the corpora section. We also developed custom web forms, to allow user input for tweets such as hand labeling the sentiment class. The resultant tweets were stored as a bag-of-words along with its hand-tagged sentiment class label. Using this data as baseline, our experiment was conducted by modifying the parameters of the Feature Selection process. Once the appropriate features were selected for training we used a Naïve Bayes algorithm to train a sentiment classification model. This trained model was then used to verify the results on the test dataset and calculate and compare the performance of the model using various metrics such as Accuracy, Precision, Recall and F-Measure.

5   Metrics

For classification tasks, the terms true positives, true negatives, false positives and false negatives compare the results of the classifier under test with trusted external judgments. The terms positive and negative refer to the classifier's prediction, and the terms true and false refer to whether that prediction corresponds to the external judgment (or expectation).

5.1   Accuracy

Accuracy is used as a statistical measure of how well a binary classification test correctly identifies or excludes a condition. The accuracy of a measurement system is the degree of closeness of measurements of a quantity to that quantity's actual (true) value. That is, the accuracy is the proportion of true results (both true positives and true negatives) in the population.

$$Accuracy = \frac{t_p + t_n}{t_p + t_n + f_p + f_n}$$

Where, t_p is the number of true positives, t_n is the number of true negatives, f_p is the number of false positives and f_n is the number of false negatives.

5.2   Precision

The precision of a measurement system is the degree to which repeated measurements under unchanged conditions show the same results. It is the fraction of retrieved instances that are relevant. Thus precision is defined as the proportion of the true positives against all the positive results (both true positives and false positives) in the population.

$$Precision = \frac{t_p}{t_p + f_p}$$

Where, t_p is the number of true positives and f_p is the number of false positives.

5.3   Recall

Recall is the fraction of relevant instances that are retrieved. Both precision and recall are therefore based on an understanding and measure of relevance.

$$Recall = \frac{t_p}{t_p + f_n}$$

Where, t_p is the number of true positives and f_n is the number of false negatives.

5.4   F-Measure

F-Measure considers both precision and recall of the test. The F-Measure score can be interpreted as the weighted average of the precision and recall, where an F-Measure score reaches its best value at 1 and worst score at 0.

$$F - Measure = 2 \cdot \frac{precision \cdot recall}{precision + recall}$$

6   Experiment

In this experiment we use several feature selection techniques and apply the same to the bag-of-words model to achieve higher accuracy, precision and recall. Our intent is to eliminate low information features and thus give the model clarity by removing the noisy data. Since we are dealing with a high volume of data (words), the dimensional space greatly increases causing the available data to be sparse and less useful. Thus we needed to avoid over-fitting and the curse of dimensionality, by decreasing the size of the model which ultimately improves the performance.

6.1 Bag-of-words

We first trained a Naïve Bayes classifier using a simple Bag-of-Words model, which yielded the following test results as shown:

Table 1. Bag-of-Words Feature Extraction

Metric	Score
Accuracy	0.718
Positive Precision	0.643
Positive Recall	0.976
Negative Precision	0.950
Negative Recall	0.460
Positive F-Measure	0.775
Negative F-Measure	0.619

Since the classifier gave a low accuracy with just a score of 0.718, we then proceeded to reduce the model size by filtering the stop words.

6.2 Bag-of-words eliminating Stop Words

It is widely accepted that stop words can be considered useless, as they are so common that including them would greatly increase the size of the model, without any improvement to the precision or recall and thus can be ignored. To improve the training features, we ignored the following stop words:

Table 2. List of Stop Words

i	her	the	of	through	further	such
me	hers	who	was	do	then	no
my	herself	and	at	during	once	nor
myself	it	whom	were	before	here	not
we	its	but	by	after	there	only
our	itself	this	be	above	when	own
ours	they	if	for	below	where	same
ourselves	them	that	been	to	why	so
you	does	or	with	from	how	than
your	their	these	being	up	all	too
yours	did	because	about	down	any	very
yourself	theirs	those	have	in	both	can
yourselves	doing	as	against	out	each	will
he	themselves	am	has	on	few	just
him	a	until	between	off	more	dont
his	what	is	had	over	most	should
himself	an	while	into	under	other	now
she	which	are	having	again	some	

After filtering the above stop words, the classifier was trained again with the remaining features, which yielded the following test result:

Table 3. Bag-of-words without Stop Words

Metric	Score
Accuracy	0.726
Positive Precision	0.645
Positive Recall	0.976
Negative Precision	0.950
Negative Recall	0.464
Positive F-Measure	0.776
Negative F-Measure	0.623

Using this approach we achieved a slightly higher score on almost all the metrics, yet found that they still remain insignificant. Thus we needed to further improve the feature selection process to achieve better performance results for the classifier.

6.3 High Information Bag-of-words

As the classification model has hundreds or thousands of features, there is very good probability that many (if not most) of the features are low information. These are features that are common across both sentiment classes and therefore contribute little information to the classification process. Individually they are harmless, but in aggregate low information features can decrease the classifier performance.

To find the highest information features, we need to calculate information gain for each word. One of the best metrics for information gain is Chi-Square. To use it, first we need to calculate a few frequencies for each word: its overall frequency and its frequency within each class. This is done with a Frequency Distribution for overall frequency of words, and a Conditional Frequency Distribution where the conditions are the class labels. Once we have those numbers, we can score words with the chi-square function, then sort the words by score and select the words with a minimum score value (ex. value >= 3). We then put these words into a set, and use a set membership test in our feature selection function to select only those words that appear in the set. Now each message is classified based on the presence of these high information words. Using this technique, we again trained the Naïve Bayes classifier and following are the test results:

Table 4. High Information Bag-of-words

Metric	Score
Accuracy	0.892
Positive Precision	0.837
Positive Recall	0.976
Negative Precision	0.966
Negative Recall	0.812
Positive F-Measure	0.901
Negative F-Measure	0.882

As shown, there is a considerable increase in all the metrics indicating that the classifier performs the best when trained using a model which excludes stop words and only contains high information features.

7   Graphs

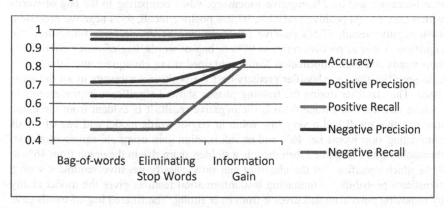

Fig. 2. Accuracy vs. Precision vs. Recall

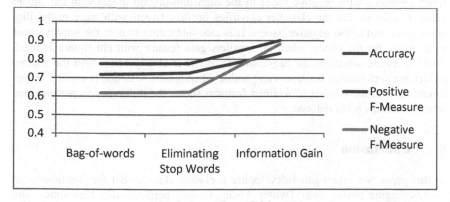

Fig. 3. Accuracy vs. F-Measure

The evaluation metrics used here are Accuracy, Positive Precision, Positive Recall, Positive F-Measure, Negative Precision, Negative Recall, and Negative F-Measure. These metrics are plotted for bag-of-words, Bag-of-words without stop words and high information bag-of-words. The performance of the classifier is compared over the positive and negative classes.

As expected, the classifier performed best when dealing with high information features with an accuracy of 89.2% i.e. analyzing the sentiments of tweets whether positive or negative have almost reached the human accuracy. The model scored best for the information gain features measured using Chi Square. The bag-of-words shows an accuracy of 71.8% which increases to accuracy of 72.6% when stop words were filtered which is less in comparison with information gain feature measured using chi-square. These values are evident in the graph as we see accuracy remained almost same in both bag-of-words and bag-of-words eliminating stop words but there is a sharp rise in the graph from 72% to 89% as information gain with chi-square is applied The high information features gave a maximum of 83.07% positive precision, 97.2% positive recall, 96.6% negative precision, 81.2% negative recall, 89.9% positive f-measure and 88.2% negative f-measure; when comparing to the bag-of-words which gave 64.3% positive precision, 97.6% positive recall, 95% negative precision, 46% negative recall, 77.5% positive f-measure and 61.9% negative f-measure. The significant value of positive recall as 97% in bag-of-words, bag-of-words eliminating stop words and high information feature obtained using chi-square and information gain specifies that the classifier predicts best for positive sentiments in all the three cases. The classifier during the training phases showed significant improvements in various metrics, one among them is the negative recall. It is evident from the graph that negative recall had a very low value in bag-of-words model and bag-of-words eliminating stop words i.e. 46% and as information gain using chi-square minimum threshold value 3 is applied then there is a sudden sharp rise in the graph from 46% to 81% which specifies that the classifier can now classify negative sentiment with a significant probability. Eliminating low information features gives the model clarity by removing noisy data and saves it from over fitting. The filtered bag-of-words gave minimum positive precision of 64.98%. The significant increase in positive recall when compared with negative recall in the high information model with chi square value 3 specifies that the classifier classifies positive tweets with more probability when compared to the negative tweets. It is also interesting to note the improvement in accuracy for the tweets when information gain feature with chi square is used. Without feature selection, the bag-of-words model is clearly better than the filtered model, but performance becomes comparable when feature selection is applied. Experimenting with the number of retained features shows the difference in performance when the feature set is reduced.

8   Conclusion

In this paper, we investigated developing a classification model for sentiment in micro-blogging posts from Twitter. Using various preprocessing techniques, and applying various feature selection techniques to the Naïve Bayes classifier, we were able to achieve reasonably good performance for the training set used. Ultimately we also noticed that all the classifiers trained were performing slightly better for classifying the positive class compared to the negative class. We show that Naïve Bayes algorithm with application of Information Gain measured using Chi square with minimum value of 3 to select high information features, gives accuracy above 89%. This study can further be extending by applying other techniques such as feature selection and aggregation using n-gram collocations and Parts-Of-Speech tagging to improve the feature model by eliminating word disambiguation.

References

1. Hussain, A.R., Hameed, M.A., Hegde, N.P.: Mining Twitter using Cloud Computing. In: World Congress on Information and Communication Technologies, pp. 187-190 (2011)
2. Pang, B., Lee, L.: Opinion Mining and Sentiment Analysis. In: Foundations and Trends in Information Retrieval (2008)
3. Das, S.R., Chen, M.Y.: Yahoo! for Amazon: Sentiment Extraction from Small Talk on the Web. In: Management Science, pp. 1375-1388 (2007)
4. Forman, G.: An extensive empirical study of feature selection metrics for text classification. In: The Journal of Machine Learning Research, pp. 1289-1305 (2003)
5. Rennie, J.D.M, Shih, L., Teevan, J., Karger, D.R.: Tackling the Poor Assumptions of Naïve Bayes Text Classifiers. In: Proceedings of the Twentieth International Conference on Machine Learning, pp. 616-63 (2003)
6. Agarwal, A., Xie, B., Vovsha, I., Rambow, O., Passonneau, R.: Sentiment Analysis of Twitter Data. In: Proceedings of the Workshop on Languages in Social Media, pp. 30-38 (2011)
7. Jiang, L., Yu, M., Zhou, M., Liu, X., Zhao, T.: Target-dependent Twitter sentiment classification. In: Proceedings of the 49th Annual Meeting of the Association for Computational Linguistics: Human Language Technologies, pp. 151-160 (2011)
8. Pandey, V., Iyer, C.: Sentiment analysis of microblogs. Technical Report, Stanford University (2009)
9. Lewin, J. S., Pribula, A.: Extracting Emotion from Twitter. Technical Report, Stanford University (2009)
10. Pang, B., Lee, L., Vaithyanatan, S.: Thumbs up?: sentiment classification using machine learning techniques. In: Proceedings of the ACL-02 conference on Empirical methods in natural language processing, pp. 79-86 (2002)

Automatic detection of Tubules in Breast Histopathological Images

Maqlin P[1], Robinson Thamburaj[1], Joy John Mammen[2], and Atulya K. Nagar[3]

[1]Department of Mathematics, Madras Christian College, Chennai, India

{maqlinparamanandam@yahoo.com, robin.mcc@gmail.com}

[2]Department of Transfusion Medicine & Immunohaematology, Christian Medical College,

Vellore, India

{joymammen@cmcvellore.ac.in}

[3]Centre for Applicable Mathematics and Systems Science, Department of Computer Science

Liverpool Hope University, Liverpool, L16 9JD, UK

{nagara@hope.ac.uk}

Abstract. Histopathological examination of tissues enables pathologists to quantify the morphological features and spatial relationships of the tissue components. This process aids them in detecting and grading diseases, such as cancer. Quite often this system leads to observer variability and therefore affects patient prognosis. Hence quantitative image-analysis techniques can be used in processing the histopathology images and to perform automatic detection and grading. This paper proposes a segmentation algorithm to segment all the objects in a breast histopathology image and identify the tubules in them. The objects including the tubules and fatty regions are identified using K-means clustering. Lumen belonging to tubules is differentiated from the fatty regions by detecting the single layered nuclei surrounding them. This is done through grid analysis and level set segmentation. Identification of tubules is important because the percentage of tubular formation is one of the parameters used in breast cancer detection and grading.

Keywords: Breast histopathology, tubules, cancer grading and segmentation

J. C. Bansal et al. (eds.), *Proceedings of Seventh International Conference on Bio-Inspired Computing: Theories and Applications (BIC-TA 2012)*, Advances in Intelligent Systems and Computing 202, DOI: 10.1007/978-81-322-1041-2_27, © Springer India 2013

1 Introduction

In clinical medicine, histopathology is the golden standard used by pathologists for studying the various manifestations of diseases. During histopathological examination, a pathologist quantifies the appearance of the tissue, and based on the visual quantification, detects and grades diseases. In spite of the experience and expertise of the pathologist, this manual system is subject to a considerable amount of inter- and intra- observer variability which could affect diagnosis and treatment planning. An investigation of the inconsistencies prevailing in the histological diagnosis of breast lesions was made by a panel of breast tumor pathologists [1]. This study showed that the consistency in diagnosis was good when only two categories of tumor: benign and malignant were considered, but it was greatly affected by observer variability when cases of border line lesions were considered. These issues caused by the limitations of human visual interpretation, can greatly affect the prediction of patient prognosis and treatment planning. This problem can be alleviated by developing quantitative image analysis methods to automatically process the digital images of histopathology tissue sections and quantitative computational tools to quantify the appearance of various tissue components [2].

The proposed work aims at image segmentation methods to automatically segment the various components in a breast histopathology image and identify the tubules in them. Automatic detection of tubules is important as the percentage of tubular formation in the cancer tissue is one of the key factors in the grading of breast cancer tissues[3]. The proposed segmentation methodology uses a combination of color based segmentation and grid analysis to identify the nuclei regions in the image. This step is followed by segmentation using level sets which uses the grid co-ordinates as the initial level contour. This is used to derive the level set function in order to detect the boundary of the nuclei. The tubules are identified from other non-tubular structures by detecting the presence of a single layer of nuclei around them and also by measuring the evenness of the nuclei positioning, around the lumen.

A background on grading breast cancer histopathology sections, the structure of tubule formations and its importance in grading breast cancer is given in section 2. Section 3 provides a Literature survey on the related works. Section 4 and 5 present the proposed segmentation methodology and experimental results respectively. Discussions on this work and the Conclusion are placed in section 6.

2 Background

Cancer can be termed as a broad group of diseases caused by uncontrollable growth of abnormal cells in a tissue, which may invade other parts of the body. Cancers are named depending on the organ from where they start. Breast cancer is one of the common types of cancer among the female population worldwide. It is estimated that 226,870 women will be diagnosed with and 39,510 women will die of cancer of the breast in 2012[4].In a normal breast tissue the group of cells are organized together to form a particular function. Cancer causes a break down in the normal functioning of the cells and causes changes in the appearance and organization of the cells in the breast tissues. The histopathology examination enables the pathologists to classify the cancer as low grade, intermediate or high grade cancer. Cancer grading determines how much the tumor resembles the tissue of their origin. The Bloom Richardson grading system is presently, the most widely used system in the grading of breast cancer tissues. It uses three parameters for grading: tubular formation (percentage of tubules found in the tumor), variation in cell size and shapes (nuclear pleomorphism) and the mitotic count (number of cells that undergo division)[4]. Points are awarded according to whether the three of the histological factors are present in slight, moderate or marked degree.

Prior to the microscopic examination, the tissues are stained with hematoxylin and eosin (H&E) stains. The purpose of the stains is to reveal the various components in the tissue. Hemotoxylin stains cell nuclei blue and Eosin stains cytoplasm and connective tissues pink. In an H&E, stained breast cancer tubules are characterized by a white region called lumen, surrounded by a single layer of nuclei around each of them. Fig. 2.1 shows a normal breast histopathology image containing tubules in it.

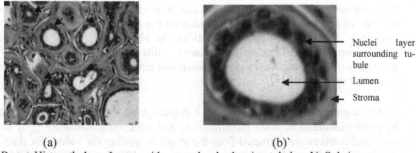

Nuclei layer surrounding tubule

Lumen

Stroma

(a)   (b)

Fig.1.a) Breast Histopathology Image with arrow heads showing tubules. **b)** Sub image showing a tubule in the tissue

3   Related Works

The first step in the segmentation of histopathological images is to detect the various histological structures in the tissue. The size, shape, count, extent and other morphological features of these structures influence the presence or severity of diseases. For example, the presence of a large number of lymphocytes in breast cancer histopathology is strongly suggestive of poor disease outcome and survival

[5]. Identifying these histological structures is a pre-requisite for detecting and classifying diseases. Segmentation algorithms of histopathological images can either have a local, structural segmentation approach or a global scene segmentation approach [6]. This approach targets specific objects in the image and identifies them. Numerous works have been conducted on segmentation of various structures in breast histopathology images using methodologies such as thresholding, K-means, fuzzy c-means clustering and adaptive thresholding [7]. The drawback of thresholding is that it can work only on uniform images and does not produce consistent results, in case there is considerable variability within image sets. Watershed algorithms tend to pose the same problem [8]. Active contours are widely used in image segmentation; however, contours enclosing multiple overlapping objects pose a limitation. In addition, inclusion of other irrelevant objects from the background further complicates the possibility of obtaining a viable segmentation. Nuclear segmentation in breast and prostate cancer histopathology was achieved by integrating a Bayesian classifier driven by image color and texture and a shape-based template-matching algorithm [9]. Several nuclei lie adjacent to each other and hence template matching is used to extricate the individual nuclei. A unified segmentation algorithm for sub-cellular compartmentalization is done in which, quantization of biomarkers at sub-cellular resolution requires segmentation of sub-cellular compartments such as nuclei, membranes, and cytoplasm [10].

After the segmentation process the various features of the histological structures are extracted from the regions of interest. These features play a vital role in detecting and grading diseases. Images can be analyzed by examining the objects in them. Object level analysis mainly relies on the results of the segmentation methodology. Segmentation process identifies the objects of interest in an image. Comparisons of Classification performance, while distinguishing between different grades of prostate cancer, using manual and automated gland and nuclear segmentation were studied [9].

A cell-graph is generated for breast histopathology images using segmentation by color quantization, node identification, and edge establishment [11]. A set of structural features are extracted from the graph by getting the values for their different metrics. These feature sets are used for classification through a SVM classifier. Tissue images can be represented using a color graph and based on the metrics of the graph; structural features of various tissue components are extracted and used for classification [12].

4 Proposed Segmentation Methodology for identifying the Tubules

4.1 Segmentation using K-means clustering

In the input histopathology images the various tissue components are revealed by their color variability caused by the H&E stain (cytoplasm and connective – pink, Lumen and fat-white and cell nuclei-purple). As a first step, each pixel in the image is classified into one of the three component groups: stroma, cytoplasm and lumen. This is done using the K-means clustering technique. The resulting cluster images as seen in Fig. 2, shows the blue pixels representing the cell nuclei in the tissue and white pixels representing the region of lumen or fat.

4.2 Detecting tubules from the segmented results

In this phase, the objects of interest are the tubules which are characterized by the presence of a white region (lumen) surrounded by a single layer of cell nuclei. But the resulting white cluster image (Fig.3 (a)) represents white tissue components like lumen, fat and background pixels from which the tubules alone should be detected. In order to detect the tubules, white regions surrounded by a well defined single layer of nuclei, have to be identified. This is done by the following steps:

Step 1: Identifying the nuclei nearer to each white region

In order to find out if a white region is surrounded by a layer of nuclei (tubule), the first step is to find out all the nuclei that are nearer to any white region (lumen and fat) in the image. This is done by applying a distance constraint as proposed by [13]. In this method all the nuclei which are within a threshold value from the center of mass of a white region are identified nuclei nearer to that white region. The resultant image of this step is shown in Fig. 3.b

Step 2: Contour detection of the nuclei near white regions using Grid analysis and Level set method

The K-means clustering algorithm results in clusters of clouded nuclei which need to be separated for further analysis. A level set segmentation method is applied to identify the boundaries of the nuclei that are closely surrounding the lumen. This helps in separating the clouded cells.

The level set segmentation method is a numerical method that represents an evolving contour using a level set function, where its initial level corresponds to the actual contour. Then, according to the motion equation of the contour, one can easily derive a similar flow for the implicit surface, so that when applied to the ze-ro-level will reflect the propagation of the contour. It converges and finds the contour of the region to be segmented.

The zero level set is represented by the boundary, $B = \{(x, y)| \varphi\ (t, x, y) = 0\}$ of a level set function φ, evolving in time t where $i = (x, y)$. φ takes positive values inside the area delimited by the Boundary B. The evolution of level set function is given by the level set formulation explained in [14]

$$\frac{\partial \varphi}{\partial t} + V|\nabla \varphi| = 0 \qquad (1)$$

Where the function V defines the speed of the evolution, t is the time of evolution of a boundary B in a 2D grid of pixels I.

The level set segmentation method needs initial contours $\varphi 0 = \varphi\ (0, x, y)$ to be specified around the object of interests. In this work the initial level contour is determined by a grid based method. A grid of fixed height and width is determined based on the size of the nuclei in the image. The grid is then placed on the image which shows all the nuclei that are nearer to the white region (Fig.4). If a particular grid entry contains the pixels of the nuclei, it is retained and the other grid entries are eliminated. Thus grid entries containing the nuclei pixels alone are obtained. The boundary pixels of these grid entries are given as initial level contours as shown in Fig 5.a.). The algorithm is run until level set algorithm attains convergence showing the contours of the nuclei Fig 5.d)..Illustration of this step is shown in Fig. 4 & 5.

Step 3: Identifying nearest nucleus to the lumen in each direction

The nuclei belonging to the single layer surrounding the white lumen area can be identified by finding the nearest nucleus to the lumen in each direction. A direction constraint proposed by [13] is used to eliminate nuclei lying right behind the nuclei surrounding the lumen .As a result of the steps mentioned above, all the white regions with a single layer of nuclei around them, i.e. potential lumen, are identified. Illustration of this step is shown in Fig. 6.

Step 4: Identifying tubules from other white regions

It is understood from the original image that the closeness and evenness in the spacing of the nuclei surrounding the lumen makes the tubules distinguishable from other white regions in the image. The evenness in the distribution of the nuclei surrounding the lumen can be analyzed by finding the distance between the adjacent nuclei in the string. The standard deviation of these values is less than an empirical determined threshold value for a tubule.

5   Experimental Results

The proposed algorithm was tested for a set of twenty nine histopathology images. As a first step, K–means clustering was done to the original image to classify the objects in the image based on the color. For a healthy breast tissue shown in Fig. 2(a), three clusters were obtained (purple, white and pink). As the analysis of nuclei and lumen is included in the study, the purple (nuclei) and white (lumen and fat) clusters are used. Images with purple and white objects are shown in Fig. 2(b) and 2(c) respectively.

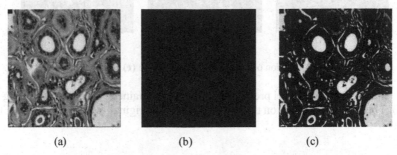

(a)   (b)   (c)

Fig. 2. (a) H&E stained Normal breast histopathology image, **(b)** Cluster image with pixels of nuclei, **(c)** Cluster image with white regions in image (lumen, fat)

The purple and white clusters are merged and converted to a binary image as shown in Fig. 3 (a). The lumen and the objects surrounding them are retained and the rest of the objects were eliminated, which is shown in Fig. 3 (b).

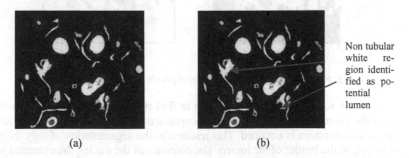

Non tubular white region identified as potential lumen

(a)   (b)

Fig. 3. (a) Purple and White Clusters representing the nuclei, lumen or fat respectively, **(b)** Nuclei nearer to the white regions (applying the distance constraint)

White objects which were sufficiently large, along with their surrounding nuclei were further analyzed by placing a grid in the image. The objects of interest and the image with the grid are shown in Fig. 4 (a), Fig. 4 (b) and Fig. 4 (c) respectively.

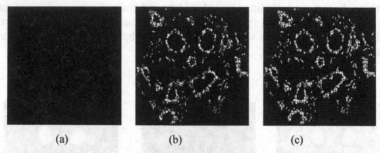

(a)   (b)   (c)

Fig. 4 (a) Extraction of Nuclei **(b)** Binary image **(c)** Grid on the image

The pixels of the nuclei present in the grids are retained. This is shown in Fig. 5 (a). This is superimposed on the gray image of the original, which is shown in Fig. 5 (b).

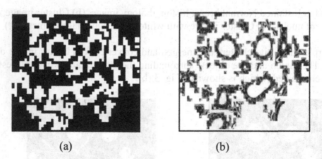

(a)   (b)

Fig. 5. (a) Grid entry containing nuclei pixels **(b)** Superimposed image

Level set segmentation was then used to find the contours of the nuclei within the grids. Using the direction constraint explained in [13] the closest nucleus to the lumen in a direction is obtained. This resulted in the segmentation of only the row of nuclei, in the border of the lumen. The contours of the nuclei, the extracted nuclei and the nuclei in the border of the lumen are shown in Fig. 6(a), (b) and (c).

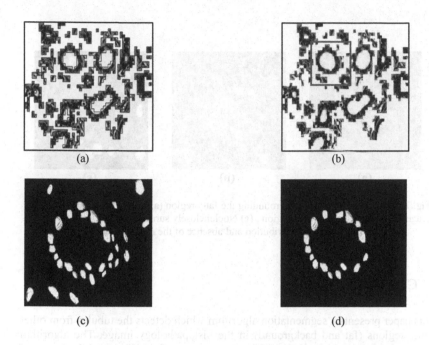

(a)   (b)

(c)   (d)

Fig. 6. Segmentation of nuclei surrounding the lumen (**a**) Initial Level Contour for Level Set segmentation (**b**) Detection of Nuclei boundaries (**c**) Extracted nuclei of a subimage showing a lumen area (**d**) The single layer of nuclei surrounding the lumen region (direction constraint applied)

Often fatty regions in histopathology images are mistakenly counted as tubules because they appear similar to the central white lumen area of the tubules. However fatty regions can be differentiated from the tubules by the fact that they do not have a layer of nuclei arrangement around them. The experiment was conducted for a fatty region which looks like a lumen of a tubule. It was observed that unlike a tubule, the extracted nuclei closely surrounding the fatty region were not uniformly distributed in a single layer (shown in Fig 7.). In the case of tubules the standard deviation of the distance between the adjacent nuclei in the single nuclei layer around the lumen, was below a predetermined threshold value. Whereas for a fatty region the standard deviation was above the threshold value indicating that the nuclei were not uniformly distributed in the layer.

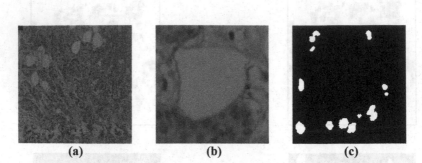

Fig. 7. Segmentation of nuclei surrounding the fatty region (**a**) Breast Histopathology Image, (**b**) Sub image of a fatty region , (**c**) Nuclei closely surrounding the fatty region (shows uneven distribution and absence of the nuclei layer)

6 Conclusion

This paper presents a segmentation algorithm which detects the tubules from other white regions (fat and background) in the histopathology image. The algorithm uses the background knowledge of the arrangement of nuclei around the lumen of a tubule, which makes it distinguishable from other white regions in the image. This finding was tested on 29 images of breast histopathology images. The algorithm gave accurate results for 90% of the white objects, identifying them to be tubular lumen. The algorithm presented inaccurate results by identifying a non-tubular white object to be a tubular lumen, only when it has evenly distributed nuclei around it. Therefore, considering the parameter, evenness in nuclei arrangement around the tubule alone, has its limitations. In the next phase of this study, it is planned that the focus will be on using all the domain specific metrics of the nuclei around the lumen to make the tubule detection accurate.

Acknowledgement

This work was supported by the Centre for Applicable Mathematics and Systems Science (CAMSS), Department of Computer Science, Liverpool Hope University, UK. The authors would like to thank the Department of Pathology, CMC, Vellore, India for providing the sample images for the study.

References

1. Beck, J.S.:Observer variability in reporting of breast lesions. J ClinPathol. 38, 1358-1365 (1985)

2. Gurcan, M.N, Boucheron, L.E., Can, A., Madabhushi, A., Rajpoot, N.M., Yener, B.:Histopathological image analysis: A review. IEEE Reviews in Biomedical Engineering. 2, 147 –171 (2009)

3. Bloom, H. J., Richardson, W. W.:British journal of cancer11 (3). 359–377 (1957).

4. Siegel, R., Naishadham, D., Jemal, A.: Cancer Statistics,CA Cancer J. Clin (2012). 62, 10–29. doi: 10.3322/caac.21149

5. Alexe, G., Dalgin,G.S., Scanfeld, D., Tamayo, P., Mesirov, J.P., DeLisi, C., Harri, L., Barnard, N., Martel, M., Levine, A.J., Ganesan, S., Bhanot, G.,High expression of lymphocyte-associated genes in node-negative HER2+ breast cancers correlates with lower recurrence rate. Cancer Res. 2007, 67(2). 10669–76(2007)

6. Gurcan M.N., Boucheron, L., Can, A., Madabhushi, A., Rajpoot, N.M. and andYener, B, Histopathology Image Analysis: A review, IEEE Rev Biomed Engn. 2:147-171 (2009)

7. Petushi S, Garcia FU, Haber MM, Katsinis C, Tozeren A. 'Large-scale computations on histology images reveal grade-differentiating parameters for breast cancer'. BMC Med Imaging, 6:14.(2006) doi:10.1186/1471-2342-6-14

8. Karvelis, P.S., Fotiadis, D.I., Georgiou, I., Syrrou, M.: A watershed based segmentation method for multispectral chromosome images classification. In:IEEEEng Med BiolSoc, 1. 3009–12 (2006)

9. Naik, S., Doyle, S., Agner, S., Madabhushi, A., Tomaszeweski, J., Feldman, M.: Automated gland and nuclei segmentation for grading of prostate and breast cancer histopathology. InBiomedical Imaging: From Nano to Macro, pp. 284 – 287. 5th IEEE ISBI, Paris, France (2008)

10. Tosun, A.B., Kandemir, M., Sokmensuer, C., and Gunduz-Demir, C.: Object-oriented texture analysis for the unsupervised segmentation of biopsy images for cancer detection. Pattern Recognition, 42(6):1104-1112 (2009)

11. Bilgin C, Demir C, Nagi C, Yener B. Cell-Graph Mining for Breast Tissue Modeling and Classification. In: Engineering in Medicine and Biology Society, pp416 - 438. 29thAnnual International Conference of the IEEE, Lyon, France.(2007)

12. Altunbay, D., Cigir, C., Sokmensuer, C., and Gunduz-Demir, C.: Color graphs for automated cancer diagnosis and grading. In: IEEE Transactions on Biomedical Engineering, 57(3):665-674 (2010)

13. Basavanhally, A., Yu, E., Xu, J., Ganesan, S., Feldman, M.D., Tomaszewski, J.E.andMadabhushi, A.: Incorporating domain knowledge for tubule detection in breast histopathology using O'Callaghan neighborhoods. In: SPIE Medical Imaging (2011).doi: 10.1117/12.878092

14. Li,C., Xu, C., Gui, C. and Fox, M.D.: Level set evolution without re-initialization:a new variational formulation. In: IEEE Computer Society Conference on Computer Vision and Pattern Recognition, vol. 1, pp. 430–436. IEEE CVPR (2005)

Artificial bee colony algorithm integrated with differential evolution operators for product design and manufacturing optimization

R. S. S. Prasanth[1], K. Hans Raj[1]

[1] Dayalbagh Educational Institute, Dayalbagh, Agra, INDIA

{ rss.prasanth@gmail.com ; khansraj@rediffmail.com}

Abstract: Artificial bee colony (ABC) algorithm is a nature-inspired algorithm that mimics the intelligent foraging behavior of honey bees and it is steadily gaining popularity. It is observed that convergence of ABC algorithm in local minimum is slow. This paper presents an effort to improve the convergence rate of ABC algorithm by integrating differential evolution (DE) operators into it. The proposed ABC-DE algorithm is first tested on three product design optimization problems and the results are compared with co-evolutionary differential evolution (CDE), hybrid particle swarm optimization-differential evolution (PSO-DE) and ABC algorithms. Further, the algorithm is applied on three manufacturing optimization problems, and the results are compared with genetic algorithm (GA), real coded genetic algorithm (RCGA), and RCGA with Laplace Crossover and Power Mutation (LXPM) algorithm and ABC algorithm. Results indicate that ABC-DE algorithm is better than the state of the art algorithms for the aforesaid problems on selected performance metrics.

Keywords: Artificial bee colony; differential evolution, design optimization; manufacturing.

1 Introduction

The implicit financial restraint involved in all product design and manufacturing processes essentially demand intelligent and robust optimization algorithms which can address mutually conflicting objectives, i.e., cost and quality. But the optimization of exceedingly complex, non linear product design and manufacturing processes, which are associated with many variables (discrete,

J. C. Bansal et al. (eds.), *Proceedings of Seventh International Conference on Bio-Inspired Computing: Theories and Applications (BIC-TA 2012)*, Advances in Intelligent Systems and Computing 202, DOI: 10.1007/978-81-322-1041-2_28, © Springer India 2013

continuous, mixed, etc), constraints and conditions, is a challenging task. Mathematical optimization techniques often fail to give optimal results and struck in local optima and converge very slowly in large search spaces [1, 2]. Here, nature–inspired algorithms or soft computing techniques [3, 4] can be very useful as they can handle complexity and uncertainty and can help manufacturing sector. A swarm intelligence algorithm is an important class of nature-inspired algorithms [5]. Swarm intelligence is as defined by Bonabeau, as " ...any attempt to design algorithms or distributed problem solving devices inspired by the collective behavior of social insect colonies and other animal societies ..." [6]. One among such recent swarm intelligence inspired evolutionary algorithms is artificial bee colony (ABC) algorithm [7] which is increasingly attracting many engineers and researchers as its performance is better than GA, PSO and DE [8, 9] on certain test functions. In a comprehensive survey, Dervis Karaboga et al. [10] reviewed the wider applicability of ABC algorithm and the different efforts made to improve its performance. However, like all other evolutionary algorithms, ABC also has some drawbacks. For example since it does not uses an operator like crossover as employed in GA or DE the distribution of good information between solutions is not at a required level. This causes the convergence performance of ABC for local minimum to be slow [10]. As it is also observed that, no one algorithm can find best solution to all optimization problems, the advantageous features of two different algorithms may be used to form a hybrid or augmented technique so as to improve the quality of the solutions [11]. More recently, ABC is simply cascaded with DE in one study [12] and in another study [13] only neighborhood operator of ABC is replaced with the mutation operator of DE. But such variants undermine the importance of crossover operator of DE in their study, which controls the diversity and maintains uniform distribution of the population [14]. Therefore, this paper proposes, ABC algorithm integrated with DE operators, named ABC-DE algorithm. The paper is further organized as follows. The section two briefly describes ABC and DE algorithms. In section three, proposed ABC-DE algorithm is presented. In section four details of experiments conducted on ABC-DE are presented. In section five results of experiments on product design and manufacturing optimization problems are presented. Conclusions are drawn accordingly that are followed by references.

2 Artificial bee colony (ABC) and differential evolution (DE) algorithms

2.1 ABC algorithm

Karaboga, D. [7] has presented the ABC algorithm that simulates the behavior of real honey bees, for solving multidimensional and multi modal optimization problems. In this optimization technique, bees are classified into three groups as employed bees, onlookers and scouts.

Pseudo code of ABC algorithm [15]

1. Initialize the Colony Size, Number of Food Sources/Solutions (SN), Number of dimensions to each solution (D), Modification Rate(MR), SPP(Scout Production Period-limit)
2. Initialize the population of solutions $x_{i,j}$ where i= 1 .. SN and j= 1 .. D
3. Evaluate the population
4. cycle = 1
5. **REPEAT**
6. Produce a new solution v_i for each employed bee by using (1) and evaluate it

$$V_{ij} = X_{ij} + \emptyset_{ij} (X_{ij} - X_{kj}), \qquad \text{if } R_j < MR, \text{ otherwise } X_{ij} \quad \ldots\ldots (1)$$

[$\emptyset_{ij}$ - is a random number in the range $[-1, 1]$. k $\in$ {1, 2 ,... SN} (SN: Number of solutions in a colony) is randomly chosen index. Although k is determined randomly, it has to be different from i . R_j is a randomly chosen real number in the range $[0, 1]$ and j $\in$ {1, 2 ,... D} (D: Number of dimensions in a problem). MR, modification rate, is a control parameter.]

7. Apply greedy selection process for the employed bees between the v_i and x_i
8. Calculate the probability values Pi using (2) for the solutions x_i

$$P_i = Fitness_i / \sum_{N=1}^{SN} (Fitness_N) \quad \ldots\ldots (2)$$

9. For each onlooker bee, produce a new solution v_i by using (1) in the neighborhood of the solution selected depending on P_i and evaluate it.
10. Apply greedy selection process for the onlooker bees between the v_i and x_i
11. If Scout Production Period (SPP) is completed, determine the abandoned solutions by using "limit" parameter for the scout, if it exists, replace it with a new randomly produced solution using (3)

$$X^j_i = X^j_{min} + \text{rand} (X^j_{max} - X^j_{min}) \quad \ldots\ldots (3)$$

12. Memorize the best solution achieved so far
13. cycle = cycle +1
14. **UNTIL** (Max Cycle Number or Max CPU time)

In the initialization phase, the basic control parameters such as number of food sources, maximum cycle number, modification rate and the limit of scout production period are initialized and the artificial scouts are allowed to randomly generate the population of food sources and assign one employee bee to a randomly generated food source. In the employee bee phase, the employee bee searches for the better neighborhood food sources applies the greedy selection strategy to find the best known between two food sources and also shares this information with the onlooker bees waiting at the hive, by dancing on the dancing area. The onlooker bee gathers the information from the dancing pattern, tempo and the duration of the employee bee and adopts the employee bee food source probabilistically. Once the onlooker bee chooses its food source, the onlooker bee agents also searches for the better neighborhood solution and applies the greedy selection process to find the best known between two food sources. The food sources, whose solutions were not improved, after certain predefined number of iterations, will be assigned to scouts for random search. All the three phases are repeated until a termination condition (maximum cycle number) is met.

2.2 DE algorithm

DE algorithm [16] is a popular global optimization algorithm that is similar to GA, as it employs mutation, crossover, and selection operators. But DE is different from GA, since it mainly relays on mutation operator, while GA is based on crossover operator. In DE, mutation operator provides good search mechanism that produces difference of randomly selected pairs of solutions in the population, while a crossover operator maintains the uniform distribution of population. The selection process ensures that the fitness value of the population improves or remains unchanged but does not get worse. The main steps of differential evolution [15] algorithm are:

- Initialize population
- Evaluation
- Repeat
  - o Mutate
  - o Recombination
  - o Evaluation
  - o Selection
- Until requirements are met

Mutation: A mutant solution vector $\hat{x}_i$ is produced by $\hat{x}_i = x_{r1} + F(x_{r2} - x_{r3})$ ...... (4). Where F is the scaling factor in the range of [0, 1], and solution vectors x_{r1}, x_{r2}, x_{r3} are randomly chosen and must satisfy the condition $x_{r1}, x_{r2}, x_{r3} \mid r_1 \neq r_2 \neq r_3 \neq I$, where I is the index of the current solution.

Crossover: The parent vector is mixed with mutant vector to produce a trail vector using $y^j_i = \hat{x}^j_i$ if $R_j \leq CR$, otherwise x^j_i. ...... (5). CR is a crossover constant which controls the recombination and varies between [0, 1]. And R_j is a random number between [0, 1].

Selection: All solutions in the population have the same chance of being selected as parents without dependence of their fitness value. The child produced after the mutation and crossover operations is evaluated. Then, the performance of the child vector and its parent is compared and the better one wins the competition. If the parent is still better, it is retained in the population [15].

3 ABC algorithm integrated with DE operators (ABC-DE)

Balancing between exploration and exploitation is always a big concern for any stochastic search technique and as it is vital for its success. ABC algorithm exhibits good balance between exploration and exploitation but it also has certain limitations like all other evolutionary algorithms. As ABC algorithm does not uses an operator like crossover that is employed in GA or DE, the distribution of good information between solutions is not at the required level. This causes the convergence performance of ABC for local minimum to be slow [10]. The DE algorithm has a crossover operator which makes it converge faster in initial generations than ABC but which may result in premature convergence in case of multi-modal problems since the DE algorithm does not have an operator maintaining sufficient diversity [17]. A comprehensive survey of state of the art of DE [18] indicates some of its drawbacks, for example, for some problem landscapes DE get stuck in local optima and has limited ability to move its population to larger distances of solution spaces. Therefore, DE also needs further improvement. Although, the DE and ABC algorithms both appear to have similar operators that weigh the difference of solutions, but the DE algorithm uses a constant scaling factor while the ABC algorithm uses a random number in the range [−1, 1], therefore they are different to each other. So a neighborhood operator of ABC algorithm in employee bee phase, and mutation and crossover operators of DE algorithm in onlooker bee phase of ABC algorithm, together may result in better convergence speed of the existing ABC algorithm. Therefore, DE's mutation and crossover operators are integrated into onlooker phase of ABC algorithm, since onlooker phase represent global search. The proposed ABC-DE algorithm incorporates the following modifications to the original ABC algorithm keeping its all other steps unchanged.

1. In step 9, for each onlooker bee produce new solutions v_i using differential mutation operator and crossover operators, i.e., using equations (4) and (5), in the neighborhood of the solutions selected depending on p_i and evaluate it.

2. The scaling factor F, employed in mutation operator of DE which is a constant and greater than 0 (F>0) has been changed to vary in the range of [0, 1] and thus making the algorithm self adaptive.

Constraint handling: Constraint handling is the key concern in optimizing the constrained product design and manufacturing optimization problems, as the choice of the constraint handling technique tends to have a definite impact on quality of search technique. Among the number of constraint handling techniques reviewed comprehensively [19], because of its simplicity and effectiveness Deb's feasibility rules [20] are adopted for constraints handling, in this study. In this method tournament selection is employed, that is two solutions are compared with each other, according to the following order of preference of feasible solution. If both solutions are feasible, the one with better objective function value wins. If one solution is feasible and the other unfeasible, the feasible one wins. If both solutions are infeasible, the one with lower constraint violation wins.

4 Experiments

Both ABC algorithm and ABC-DE are implemented in MATLAB 7.0, and tested on a Laptop machine equipped with Intel Centrino Duo processor with 512 MB RAM, and 150 GB HDD, and that runs on Windows XP platform. ABC-DE is tested on four product design optimization problems and three manufacturing optimization problems. Experimental settings of ABC and ABC-DE are presented in table no. 1.

Table 1 Experimental parameters

Parameters	ABC	ABC-DE
Colony size	40	40
Maximum Cycle Number (MCN)	500	500
Modification rate (MR)	0.9	0.9
Limit	0.25*MCN	0.25* MCN
Cross over rate	–	0.5

Product design optimization problems
In this study, three standard constrained product design optimization problems that are considered from [21]. The first design problem is to design a welded beam for minimum cost, considering four design variables that are subject to two linear and five nonlinear constraints, such as shear stress, bending stress in the beam, buckling load on the bar, and end deflection on the beam. The second design problem is to design a cylindrical pressure vessel capped at both ends by hemispherical heads, i.e., compressed air storage tank with a working pressure of 3,000 psi and a minimum volume of 750 ft3. This is design problem has four design variables and three linear and one nonlinear constraints that are required to be handled to achieve minimum manufacturing cost. The third design problem is to minimize the weight of a tension/compression spring, subject to one linear and three nonlinear constraints of minimum deflection, shear stress, surge frequency, and boundary constraints of design variables such as the wire diameter, the mean coil diameter, and the number of active coils.

Manufacturing optimization problems

Three manufacturing optimization problems are considered for this study. The first model [22] is for multi pass turning operation of mild steel work piece using a carbide tool. The objective is to minimize the production cost in dollars/piece. In this model n is umber of passes, d is depth of cut in mm, V is cutting speed in m/min, and f is feed rate in mm/rev. For comparison values $n=2$ and $d=2.5$ are adopted.

$$\text{Min. Cost} = n * (3141.59 * V^{-1}f^{-1}d^{-1} + 2.879 * 10^{-8} V^4 f^{0.75} d^{-0.025} + 10) \quad .... (6)$$

Subject to:
$$50 \leq V \leq 400 \text{ m/min} \quad .... (7)$$
$$0.30 \leq f \leq 0.75 \text{ mm/rev} \quad .... (8)$$
$$1.20 \leq d \leq 2.75 \text{ mm} \quad .... (9)$$

Cutting force constraint $\quad F_c = (28.10* V^{0.07} - 0.525 * V^{0.5}) * d$

$$* f\{1.59 + 0.946 * (1+x) / ((1-x)^2 + x)^{0.5}\}$$

$$\text{Where } x = \{V/142 * exp(2.21f)\}^2 \quad ..... (10)$$

Cutting power constraint $\quad P_c = 0.746 * F_c V / 4500 \quad ..... (11)$

Tool life constraint $\quad TL = 60 * \{(10^{10})/ V^5 f^{1.75} d^{0.75}) \quad ..... (12)$

Temperature constraint $\quad T = 132 * V^{0.4} f^{0.2} d^{0.105} \quad ...... (13)$

The second model [23] is for single pass turning operation and the objective is to minimize the production cost in dollars/piece. In this model V is in m/min, and cutting speed and f is feed rate in mm/rev.

$$\text{Min. cost} = 1.25 V^{-1}f^{-1} + 1.8 * 10^{-8} V^3 f^{0.16} + 0.2$$

Subject to:
$$0 \leq V \leq 400 \text{ m/min} \quad .... (14)$$
$$0 \leq f \leq 0.01 \text{ in /rev} \quad .....(15)$$

The surface finish constraint $\quad SF = 1.36 \times 10^8 V^{-1.52} f^{1.004} \leq 100 \text{ μin} \quad .... (16)$

The horse power constraint $\quad HP = 3.58 \, 10^8 V^{0.91} f^{0.78} \leq 2hp \quad ....(17)$

The third model [24] is to minimize tool wear in turning operation using a combination of high carbon steel–tungsten carbide tool. In this model there are three variables namely d is depth of cut in mm, V is cutting speed in m/min, and f is feed rate in mm/rev.

Min. Tool wear $(t_w) = 0.0512 *V^{0.588} f^{0.034} +1.8 * 10^{-8} V^3 f^{0.16} d^{0.667}$

Subject to: $30 \leq V \leq 190$ m/min   .... (18)

                                                         $0.01 \leq f \leq 2.5$mm/rev   .... (19)

                                                         $0.5 \leq d \leq 4.0$ mm   .... (20)

Cutting force constraint $f_c = 844 * V^{-0.1013} f^{0.725} d^{0.75} \leq 1100$ .... (21)

Temperature constraint t in ^{0}c $t = 75.0238 V^{0.527} f^{0.836} d^{0.156} \leq 500$ .... (22)

5 Results

For product design optimization problems 30 independent experimental runs are conducted to ascertain the performance of the algorithm on standard metrics are such as best, mean, standard deviation and maximum number of evaluations that represent the quality, robustness and rate of convergence respectively. And the performance of ABC-DE is compared with CDE, PSO-DE and ABC in table no 2.

Table 2 Performance comparison on design optimization problems

Problem	Statistics	CDE [25]	PSO-DE[26]	ABC[27]	ABC-DE
Welded Beam	Best	1.733461	1.7248531	1.724852	**1.724852**
	Mean	1.768158	1.7248579	1.741913	1.8131
	Stdev	2.2E-02	4.1E-06	3.1E-02	0.11997
	Eval.	240000	33000	30000	**20000**
Pressure Vessel	Best	6059.7340	6059.7143	6059.714736	**2685.753407**
	Mean	6085.2303	6059.714335	6245.308144	**2685.753407**
	Stdev	4.3E+01	3.1E+2	2.05E+02	0
	Eval.	240000	42,100	30000	**20000**
Tension/ Compression Spring	Best	0.0126702	0.012665233	0.012665	**0.0126652**
	Mean	0.012703	0.012665244	0.012709	0.0126729
	Stdev	2.7E-05	1.2E-08	0.012813	3.0657E-005
	Eval.	240000	24950	30000	**20000**

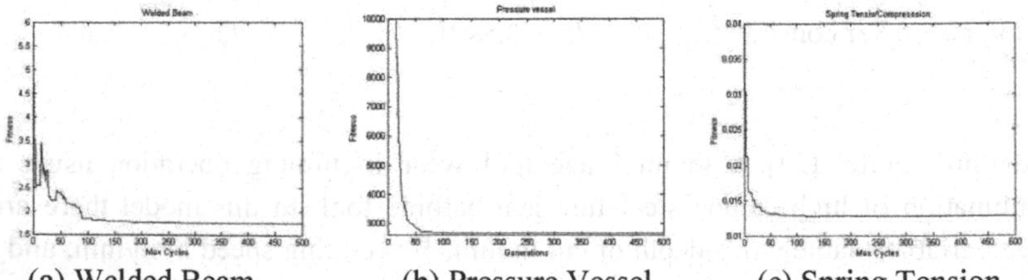

(a) Welded Beam   (b) Pressure Vessel   (c) Spring Tension

Figure 1: The convergence behavior of ABC-DE on design optimization problems

The convergence behavior of ABC-DE algorithm for the three product design optimization problems is presented for the five hundred cycles in figure 1. The best values for pressure vessel and tension/compression spring design problems, ABC-DE achieved improved results over CDE, PSO-DE and ABC, which is a metric of quality of the solution. For all the product design optimization problems ABC-DE has outperformed in terms of no. of fitness evaluations, a convergence metric. The stability of ABC-DE is also highly comparable to that of CDE, PSO-DE and ABC, for all the three product design optimization problems.

For manufacturing optimization problems 100 independent experimental runs have been carried out on both ABC and ABC-DE algorithm so as to compare the results with results reported in literature i.e., with genetic algorithm (GA) [24, 28], real coded genetic algorithm (RCGA) [29] RCGA with Laplace cross over and power mutation (LXPM) [30] algorithm. ABC-DE performance is compared with the state of the art computing techniques in table no 3.

Table 3 Performance comparison on manufacturing optimization problems

Problem	Algorithm	Objective function		Standard deviation	Eval.
		Best	Worst		
Multi pass turning	GA	79.569	NA	NA	50000
	RCGA	79.554	NA	NA	14306
	RC/LXPM	79.542	79.6488	0.03074	399
	ABC*	79.5448	79.5876	0.00956423	294
	ABC-DE	**79.5422**	**79.54223**	**1.57107E-013**	**230**
Single pass turning	GA	6.2758	NA	NA	65500
	RCGA	6.255718	NA	NA	11412
	LXPM	6.254948	7.6912	0.41213	743
	ABC*	6.24079	**6.3605**	0.027187	254
	ABC-DE	**6.23637**	6.39385	**0.0246696**	**50**
Tool wear min.	GA	0.244	0.328	NA	NA
	ABC*	0.20401	0.212083	0.00151163	306
	ABC-DE	**0.203718**	**0.203771**	**1.59443E-007**	**50**

*Present study

For the three manufacturing optimization problems, i.e., minimizing the production cost in dollars/piece for single pass and multi pass turning and minimizing the tool wear in turning operations, ABC-DE algorithm has outperformed GA, RCGA, RCGA with LXPM and ABC on selected performance metrics such as best and worst, the quality indicators of the solution, standard deviation a metric for stability of the algorithm and no. of fitness evaluations may be a metric for convergence speed.

Conclusions

This paper proposes, artificial bee colony (ABC) algorithm integrated with differential evolution (DE) operators, to improve the convergence rate of ABC algorithm. The ABC-DE algorithm is first applied on three product design optimization problems and its performance has been found to be better than co-evolutionary differential evolution (CDE), hybrid particle swarm optimization-differential evolution (PSO-DE) and ABC algorithms on standard performance metrics. Further the ABC-DE algorithm is applied on three manufacturing optimization problems and its performance has been found to be better than genetic algorithm (GA), real coded genetic algorithm (RCGA), and RCGA with Laplace Crossover and Power Mutation (LXPM) algorithm and ABC algorithm on selected performance metrics. This comparative study exhibits the potential of the ABC-DE algorithm to handle complex intelligent manufacturing optimization problems. The future work may be focused on further study of ABC-DE algorithm on other different intelligent manufacturing optimization problems.

Acknowledgements

Authors gratefully acknowledge the inspiration and guidance of Most Revered Prof. P. S. Satsangi, the Chairman, Advisory Committee on Education, Dayalbagh, Agra, India.

References

[1] Rao RV, Pawar PJ: Grinding process parameter optimization using non-traditional optimization algorithms. Proc Inst Mech Eng Part B-J Eng Manuf 224(B6):887–898 (2010)

[2] Oduguwa V., Tiwari A., Roy R.: "Evolutionary computing in manufacturing industry: an overview of recent applications", Applied Soft Computing, (2005), 5:281–299. DOI:10.1016/j.asoc.2004.08.003

[3] Deb S, Dixit US: Intelligent machining: computational methods and optimization. In: Davim JP (ed) Machining: fundamentals and recent advances. Springer, London (2008)

[4] James M. Whitacre: "Recent trends indicate rapid growth of nature-inspired optimization in academia and industry", Computing, (2011), 93:121–133. DOI 10.1007/s00607-011-0154-z

[5] Kennedy J, Eberhart R: Particle swarm optimization. In: Proceedings of the IEEE International Conference on Neural Networks (ICNN'95), Perth, Australia, (1995)

[6] Bonabeau E, Dorigo M, Théraulaz G.: Swarm intelligence: from natural to artificial systems. Oxford University Press; (1999)

[7] Karaboga, D.: An idea based on honey bee swarm for numerical optimization. Technical report TR06, Erciyes University, Engineering Faculty, Computer Engineering Department. (2005)

[8] Karaboga, D., Basturk, B: A powerful and efficient algorithm for numerical function optimization: Artificial bee colony (abc) algorithm. Journal of Global Optimization, 39(3), 459–471. (2007)

[9] Karaboga, D., Basturk, B: On the performance of artificial bee colony (abc) algorithm. Applied Soft Computing, 8(1), 687– 697 (2008)

[10] Dervis Karaboga, Beyza Gorkemli Celal Ozturk, Nurhan Karaboga: A comprehensive survey: artificial bee colony (ABC) algorithm and applications, Artif Intell Rev, (2012), DOI 10.1007/s10462-012-9328-0

[11] Grosan C., Abraham A.: "Hybrid Evolutionary Algorithms: Methodologies, Architectures, and Reviews", Studies in Computational Intelligence (SCI), (2007), 75:1-17.

[12] Ajith Abraham, Ravi Kumar Jatoth, and A. Rajasekhar: Hybrid Differential Artificial Bee Colony Algorithm, Journal of Computational and Theoretical Nanoscience, Vol. 9, 1–9, (2012)

[13] Bin Wu and Cun hua Qian: Differential Artificial Bee Colony Algorithm for Global Numerical Optimization, Journal of Computers, VOL. 6, No. 5, May (2011)

[14] Corne, D., Dorigo, M., & Glover, F: New ideas in optimization. New York: McGraw-Hill. (1999)

[15] Karaboga, D., Akay, B: A comparative study of artificial bee colony algorithm. Applied Mathematics and Computation, 214, 108–132 (2009)

[16] Storn, R., Price, K: Differential evolution—a simple and efficient heuristic for global optimization over continuous spaces, Journal of Global Optimization, 11 pp341–359. (1997)

[17] Bahriye Akay, Dervis Karaboga: Artificial bee colony algorithm for large-scale problems, and engineering design optimization, J Intell Manuf (2010) DOI 10.1007/s10845-010-0393-4

[18] Swagatam Das, Ponnuthurai Nagaratnam Suganthan: Differential Evolution: A survey of the state of the art, IEEE Transactions on Evolutionary Computation, Vol. 15, No. 1, pp 4-31, Feb. (2011)

[19] Efrén Mezura-Montesa, Carlos A. Coello Coello: Constraint - handling in nature-inspired numerical optimization: past present and future, Swarm and Evolutionary Computation, 1: 173-194, (2011)

[20] Deb, K: An efficient constraint handling method for genetic algorithms. Computer Methods in Applied Mechanics and Engineering, 186, 311–338. -17(2000)

[21] Rao, S. S.: Engineering optimization. New York: Wiley (1996)

[22] Hati SK and Rao SS: Determination of machining conditions probabilistic and deterministic approaches. Transactions of ASME, Journal of Engineering for Industry, Paper No.75-Prod-K. (1975)

[23] Ermer DS: Optimization of the constrained maching economics problem by geometric programming. Transactions of ASME, 93, pp. 1067-1072 (1971)

[24] C Felix Prasad, S Jayabal & U Natrajan: Optimization of tool wear in turning using genetic algorithm, Indian Journal of Engineering & materials Sciences, Vol. 14, pp 403-407 (2007)

[25] F.Z. Huang, L. Wang, Q. He: An effective co-evolutionary differential evolution for constrained optimization, Applied Mathematics and Computation 186 (1) pp 340–356. (2007)

[26] Hui Liu, Zixing Cai, Yong Wang: Hybridizing particle swarm optimization with differential evolution for constrained numerical and engineering optimization, Applied Soft Computing 10 pp 629–640 (2010)

[27] Bahriye Akay, Dervis Karaboga: Artificial bee colony algorithm for large-scale problems, and engineering design optimization, J Intell Manuf (2010) DOI 10.1007/s10845-010-0393-4

[28] Duffuaa SO, Shuaib AN, Alam A: Evaluation of optimization methods for machining economic models. Computers and Operation Research, 20, pp. 227-237. (1993)

[29] Kim SS, Kim H-Il, Mani V, Kim HJ: Real-Coded Genetic algorithm for machining condition optimization. The International Journal of Advanced Manufacturing Technology, 38, pp. 884-895 (2008)

[30] Deep K, Singh KP, Kansal M S: Optimization of machining parameters using a novel real coded genetic algorithm. Int. J. of Appl. Math and Mech. 7 (3): 53-69, (2011)

A Framework for Modeling and Verifying Biological Systems using Membrane Computing

Ravie Chandren MUNIYANDI, Abdullah MOHD. ZIN

Research Center for Software Technology and Management (SOFTAM), Faculty of Technology and Information Science, University Kebangsaan Malaysia, 43600 UKM Bangi, Selangor, Malaysia.

{ravie@ftsm.ukm.my; amz@ftsm.ukm.my}

Abstract. Membrane computing can abstract biological structures and behaviors, and formally represent them without disregarding their biological characteristics. However, there is the lack of a proper framework to model and verify biological systems with membrane computing that could act as a guideline for researchers in computational biology or systems biology in using and exploring the advantages of membrane computing. This paper presents a framework for modeling and verifying biological systems using membrane computing. The framework processes are made up of biological requirement and property specification, membrane computing model, membrane computing simulation strategy, and model checking approach. The evaluation of the framework with biological systems shows that the proposed framework can be used as the first step to further improve the modeling and verification approaches in membrane computing.

Keywords: membrane computing, framework, simulation strategy, model checking, biological systems.

1 Introduction

Membrane computing [1] is a theoretical paradigm that abstracts the structure and functionality of the living cell to become models and simulators of cellular phenomena. Growing interest in the use of membrane computing has been observed recently [2].

The research attempts so far in modeling biological systems with membrane computing can be categorized into three, based on the context of biological systems. First is the modeling of biological systems involved in molecular reactions that react to changes in the environment of the system or reactions caused by external stimulations of a certain pressure or s pecific stimulus. The model of

J. C. Bansal et al. (eds.), *Proceedings of Seventh International Conference on Bio-Inspired Computing: Theories and Applications (BIC-TA 2012),* Advances in Intelligent Systems and Computing 202, DOI: 10.1007/978-81-322-1041-2_29, © Springer India 2013

photosynthesis in chloroplast dealing with light reaction and photo-inhibition modeled by Nishida [3] using membrane computing belongs to this category. Second is the modeling of biological systems that prioritize the executions of interconnected processes in sequences, wherein a process should be completed before the next is executed. The research attempts in this category include descriptions of mechanosensitive channel behavior in *Escherichia coli* [4] and definitions of sodium-potassium exchange pump in animal cells [5]. Third is the modeling of biological systems that involve emergent behaviors, wherein the interactions of processes within or between compartments generate a global behavior. Under this category, the biological systems that were modeled in membrane computing include the comprehensive approaches on Quorum Sensing aspects in *Vibrio fischeri* [6], dynamics of HIV infection [7] and synthetic Autoinducer-2 signaling system in genetically engineered *E. coli* [8].

To date, studies regarding the use of membrane computing in modeling biological systems have addressed the following biological elements: process interactions and compartment interconnections; discrete evolution of objects and processes; non-deterministic and parallel execution of rules; emergent behaviors based on the various process interactions within a compartment or communications among compartments; representation of the structure of biological systems; topological organizations based on structural arrangements of compartments; and, characterization of the stochastic behaviors of biological systems that involve the interactions of processes among the small number of objects or the microscopic elements of biological systems.

The categorization of biological elements shows that membrane computing can abstract biological structures and behaviors, and formally represent them without disregarding their biological characteristics. This categorization also demonstrates that membrane computing could act as a better computational tool for investigations in systems biology.

Although membrane computing is theoretically proven for its computational completeness and computational efficiency, several of its aspects are yet to be practically tested in real problems. One reason is the lack of a proper framework to model and verify biological systems with membrane computing that could act as a guideline for researchers in computational biology or systems biology in using and exploring the advantages of membrane computing. Studies on modeling biological systems using membrane computing proposed some guidelines to model and simulate this method. However, these studies are limited in terms of specific biological systems or processes.

Therefore, a general framework for modeling and verifying biological systems using membrane computing is required to guide and encourage biologists to use this mechanism. In this paper, a framework for modeling and simulating biological systems using membrane computing formalism and approach is proposed. The framework is intended to outline the courses of action or to present a preferred approach, and to serve as a guideline for modeling and verifying biological systems. This framework also indicates the methods that can or should be used and how they would interrelate.

The framework is categorized into four parts. The first task in the framework is to extract from a biological system the elements and properties essential for modeling the system in membrane computing. Second is the building of the membrane computing model of the biological system, wherein the membrane computing formalism is used accordingly to represent the elements and attributes extracted from the system. The third task is to verify the correctness of the membrane computing model by simulating it with the membrane computing simulation strategy. The results of the simulations are compared with the results of in vivo or in vitro experiments. Finally, the membrane computing model is formally validated to ascertain that it sustained the fundamental behavior or property of the biological system. The processes in the framework are further elaborated in the following sections.

2 Biological Requirement and Property Specification

The biological system contains a large amount of information. The information can be categorized into two parts. The first is the information about the elements involved in the structure and processes of the biological systems. The common elements present in most biological systems include membrane compartments, chemical species, and reactions. Table 1 gives the descriptions of these elements.

Species concentration is measured discretely in membrane computing. Conversion into molecules should be performed if the species are measured in moles. In this case, the Avogadro's number is used to convert one mole of chemical species into 6.022×10^{23} molecules [9]. The number of compartments and the structure that determine the links to other compartments should also be identified. The species is also specified according to compartment. The parameters and the initial concentrations of the species used in the system are also identified.

The reaction in each compartment is involved in either the transformation of a set of species into a new set of species, or the transport of a set of species to another compartment. A reaction is converted into a discrete system if it is not discretely modeled. If the reactions are modeled in ordinary differential equations (ODE), the continuous system is converted into a discrete system using rewriting rules.

The method is simplified using differential equations (1) and (2), which generally represent the ODE system. Let the concentrations of species A and B in a compartment be measured in the ODE with kinetic constants k_1, k_2, k_3 and k_4. Thus,

$$\frac{dA}{dt} = k_1 A - k_2 AB \tag{1}$$

$$\frac{dB}{dt} = k_2 AB - k_3 B + k_4 \tag{2}$$

The conversion steps are illustrated in Fig. 1.

Table 1. Descriptions of the basic components in a biological system

Component	Description
Compartment	A container enclosed by a membrane such as mitochondria, chloroplasts, cytoplasm, endoplasmic reticulum, nucleus, and Golgi apparatus. These organelles have their own chemical reactions and species to carry out different metabolic activities.
Species	A collection of chemically identical molecular structures, such as genes, proteins, ions, and molecules, that perform reactions to characterize certain behavior based on their concentrations.
Modifier	An activator or inhibitor of reaction such as an enzyme that enhances or inhibits the reaction without changing their concentration.
Reactant	A species that acts as a substance consumed in the course of a chemical reaction.
Product	A species formed during a chemical reaction.
Reaction	A process that transforms one set of species acting as reactants to another set of species acting as products.

If membrane computing can be used to model biological systems modeled in ODE, then biological systems can also be modeled directly into membrane computing from in vivo or in vitro experiments. The reason is that the information extracted from the ODE model that transformed the system into a membrane computing model was also obtained from in vivo or in vitro experiments. Therefore, if in vivo or in vitro biological systems are modeled in membrane computing, then the similar information gathered from the ODE model should also be gathered from in vivo or in vitro biological systems.

Lastly, the behavior or specific properties of the biological system are determined through the results provided in published in vivo or in vitro experiments. The properties can be specified based on the following observations from in vivo or in vitro experiments:

- The expected quantity of certain objects in a specific compartment at time step t.
- The probability that the objects achieve the lowest and highest concentration at time step t.

- The expected reaction that determines the concentration of a specific object at time step t.
- The probability that k object elements are present at time step t.
- The probability that a particular object is bound to another object at time step t.
- The expected percentage of activated objects at time step t
- The expected number of reactions between two objects by time step t.

However, the behavior or property of the specific biological system is unique to that system only. This behavior or property should be determined from in vivo or in vitro experiments and published literature.

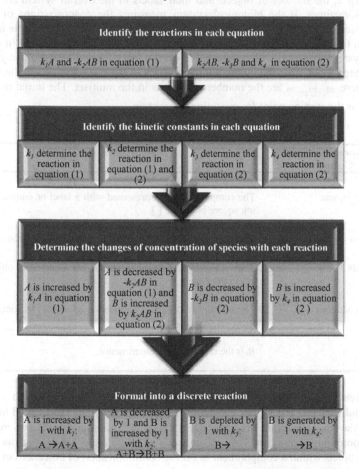

Fig. 1. Conversion of ODE reactions into discrete reactions.

3 Membrane Computing Model

In this part, two tasks should be performed. First, the extracted information in Section 2 is transformed into membrane computing formalism. Second, the biological system is modeled with membrane computing. Table 2 describes the main components in membrane computing and its representation.

Based on the information of the aforementioned components of membrane computing, the model is built according to the steps in Fig. 2. In step 1, system contents such as objects, initial multisets, reactions, and structure are signified in general. The following steps describe each of the components in further detail.

In step 2, the number of objects and their labels in the overall system are defined. Sometimes, if the biological system measures the concentrations of the same object available in different compartments, the compartment's label is indicated on the object's label to differentiate it from the other compartments. In step 3, the system structure is illustrated based on the links formed between compartments. Step 4 describes the initial multiset in each of the compartments in the system, where $m_1, m_2...m_i$ are the number of objects in the multiset. The initial multiset acts as the starting point for executing the system.

Table 2. Description of the components in membrane computing

Membrane Computing Component	Representation
Compartment	The compartment is represented with a label or number using square brackets, [ ]$_i$
System structure	The structure of the system (μ) is represented by considering the links between compartments.
Objects	The species including the modifiers are assumed as objects (V) in the system.
Initial Multiset	The multiset (ω_i) is the combination of objects in compartment i at step 0.
Reaction	R_i is the reaction in compartment i.

The reactions in step 5 are defined based on the compartments, in which u and v are multisets and k is kinetic constant. The discrete reactions represented in rewriting the rules generated from the ODE in Section 2 are transformed into membrane computing formalism. The reactions can be based either on the transformation of objects within a compartment or on the communication of objects from one compartment to another.

If the reactions are not based on these two processes, they should be converted into either one of them. If the reaction is a transformation of objects within compartment i, such as $X + Y \overset{k}{\rightarrow} Y + Y$, where k is the reaction rate, then the reaction is represented in membrane computing as $[X, Y]_i \overset{k}{\rightarrow} [Y, Y]_i$. If the reaction is a communication of objects from compartment i to compartment j, such as $X \overset{k}{\rightarrow} \bar{X}$, where X and $\bar{X}$ are similar objects in compartments i and j, respectively, and k is the reaction rate, then the communication of objects from one compartment to another can be represented in two ways. First is the movement of object X from an external compartment i into an internal compartment j, which can be represented as $X[\,]_j \overset{k}{\rightarrow} [X]_j$. Second is the movement of object X from an internal compartment j into an external compartment i, which can be represented as $[X]_j \overset{k}{\rightarrow} X[\,]_j$.

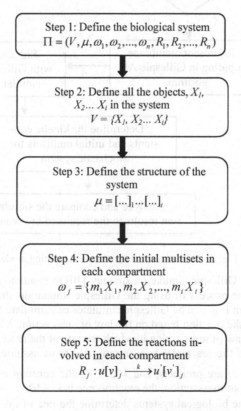

Fig. 2. Steps in membrane computing modeling.

4 Membrane Computing Simulation Strategy

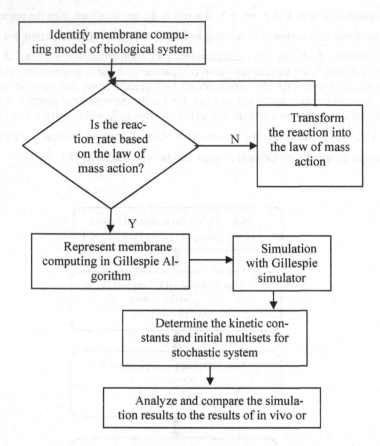

Fig. 3.   Simulation of membrane computing model

This part uses Gillespie simulation strategy [10] to simulate a membrane computing model. The process of using the Gillespie simulation strategy is described in the flowchart in Fig. 3. The Gillespie simulator can simulate the biological system that defines the reaction based on the law of mass action. Most chemical reactions follow the law of mass action in which the rate of the reaction is proportional to the product of the reactants' concentrations. Let us assume that A and B are reactants, C and D are products, and k is a kinetic constant as $A + B \xrightarrow{k} C + D$. Based on the law of mass action, the reaction rate is $k*A*B$.

However, some biological systems determine the rate of the reaction based on specific formulas and parameters. Therefore, an assumption is made to transform the reaction into the law of mass action. The assumption is based on how the reaction and parameters are represented in the formula. The membrane computing model is represented in the systems biology markup language (SBML) to extract information on the biological system into the Gillespie simulator. The SBML converter tool embedded in the Gillespie simulator converts the SBML into the Gillespie simulator representation in the form of text files comprising information regarding the rules, objects, kinetic constants, and compartments. This information is used to perform the simulations with the Gillespie simulator. Currently, no specific method can convert the kinetic constant of ODE into the kinetic constant for stochastic system to sustain the behavior of the biological system. Moreover, the dissimilar behavior of one biology system to another further complicates this process.

Therefore, the value of the initial multisets and kinetic constants extracted in Section 2 from the continuous and deterministic approach of ODE model should be adjusted to emulate the discrete and stochastic behavior of the membrane computing model through black box testing [11]. In this approach, the value of the initial multisets and kinetic constants for membrane computing simulation strategies are determined through black box testing, in which the inputs are selected based on the expected output of the system.

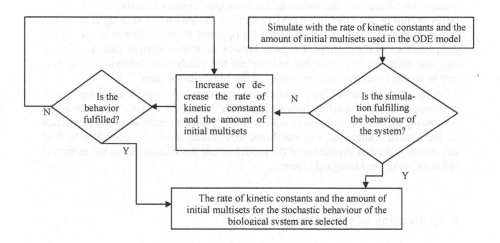

Fig. 4. Selection processes of the rate of kinetic constants and the amount of initial multisets for the stochastic behavior of the biological system.

The steps in Fig. 4 are used to determine the amount of the initial multiset and the rate of kinetic constant required for stochastic behavior of the biological system. The process starts by using the data of the initial multisets and kinetic constants gathered in the ODE model. When the system behavior is not fulfilled, the amount of the initial multisets and the rate of kinetic constants are either increased or decreased accordingly to meet the stochastic behavior of the system. This process is repeated until the intended amount of initial multisets and the rate of kinetic constants are selected. Lastly, the simulation results are compared and analyzed with the in vivo or in vitro behavior of the biological system.

5 Approach for Model Checking a Membrane Computing Model

Model checking formally validates a model to determine whether the properties of the system are conserved in the model. Probabilistic and Symbolic Model Checker (PRISM) [12] validates various properties of a membrane computing model that behave in a stochastic and discrete manner.

The PRISM formalism, in which the membrane computing model is interpreted to model check, is assumed to emulate the concept of membrane computing based on the investigation of Romero-Campero et al. [13].

The properties of the biological system are identified in the first part of this framework where the information of the biological system is extracted. The properties should be specified in PRISM formalism. Based on the investigation in this research, the reward-based properties with operator R are sufficient in specifying the properties of the biological system. However, other operators such as P and S can also measure the probability and explain the steady-state behavior of the system to model check specific properties of the biological system.

In the analysis part of the model checking process, the model checking results are compared with the related properties of the biological system. The membrane computing model is concluded as correct if the results meet the required properties of the system. Otherwise, the membrane computing model is rechecked to find any errors in the representation of the model or any insufficiencies in the extracted information of the biological system.

6 Evaluation of the Framework

The framework has been evaluated by Muniyandi and Mohd. Zin with two biological systems: Ligand-receptor network of protein TGF-β [14] and Hormone-induced Calcium Oscillations in Liver Cell [15].

The results shows that the framework provides better options to symbolize the structure and processes involved in a biological system compared with the ODE approach. The reason is that the framework provides the ability to distinguish the processes and the movement of objects between compartments in this hierarchical system. Membrane computing can abstract the structure and processes of multi-compartment biological systems, such that they are represented in a formal manner without disregarding their biological characteristics. The results generated by simulation and model checking approaches in the framework have demonstrated this ability.

The simulation results show that the membrane computing model provides results that are approximately similar to those generated by the ODE model. However, with the membrane computing model, the non-deterministic and parallel executions of the processes in the different compartments of the biological system are more obvious compared with the deterministic approach of the ODE model.

However, the membrane computing model has difficulty in differentiating the small changes in the reactions. The appropriate parameters or kinetic constants used to prioritize the stochastic selection of the rules to determine the execution of specific process at certain time steps have to be correctly selected based on the behavior of the system. The attempt in this investigation to use the same value of parameters as in the ODE model for the implementation of the membrane computing model could not generate the intended results in this case. Therefore, the selection of parameters should be based on the stochastic characteristics of the system.

Nonetheless, the basic properties of the biological systems have been preserved in the membrane computing model, as shown by the model checking results. These results demonstrate how the different processes in each compartment play a role in contributing to the behavior of the overall system. They show that the membrane computing model not only can capture the structure and processes involved in multi-compartment biological systems but also can preserve the behavior and properties of the system.

5 Conclusion

The proposed framework is capable of providing an alternative modeling and simulation approach to the conventional mathematical models, such as the ODE. The procedures in the framework allow biologists or other researchers to systematically model and simulate biological systems for in silico experiments and analyses.

Although the stages in the framework provide sufficient guidelines, this framework still needs enhancement for application in specific processes. One such enhancement is the process of selection of kinetic constants for the stochastic system. Future studies should focus on finding a better selection approach. More research using different case studies of biological systems would also further strengthen the framework.

Nonetheless, the proposed framework could be used as the first step to further improve the modeling and simulation approaches in membrane computing.

Acknowledgement

This work supported by the Young Researcher Grant of the National University of Malaysia (Grant code: GGPM-2011-051).

References

[1] Paun, G.: Computing with Membranes. Journal of Computer and System Sciences 61(1), 108-143 (1998)

[2] Ciobanu, G., Perez-Jimenez, M. J., Paun, G.: Application of Membrane Computing. Springer Verlag, Berlin (2006)

[3] Nishida, T.Y.: Simulations of photosynthesis by a K-subset transforming system with membranes. Fundam. Inform. 49, 249-259 (2002)

[4] Ardelean, I.I., Besozzi, D., Garzon, M.H., Mauri, G., Roy, S.: P-System Models for Mechanosensitive Channels. Springer Verlag, Berlin (2005).

[5] Besozzi, D., Ciobanu, G.: A P-system description of the sodium-potassium pump. Lecture Notes Comput. Sci. 3365, 210-223 (2004)

[6] Romero-Campero, F.J, Perez-Jimenez, M.J.: A Model of the Quorum Sensing System in Vibrio fischeri Using P Systems. Artificial Life 14(1), 99 – 109 (2008)

[7] Corne, D.W. & Frisco, P. 2008. Dynamics of HIV infection studied with cellular automata and conformon-P systems. Biosystems 91(3), 531-544 (2008)

[8] Esmaeili, A., Yazdanbod, I., Jacob, C.: A Model of the Quorum Sensing System in Genetically Engineered E.coli Using Membrane Computing. International Genetically Engineered Machine (iGEM) competition (2009)

[9] Pauling, L.: General chemistry. Courier Dover Publications, London (1998)

[10] Gillespie, D.T.: Stochastic Simulation of Chemical Kinetics. Annual Review of Physical Chemistry 58, 35-55 (2007)

[11] Beizer, B.: Black-Box Testing: Techniques for Functional Testing of Software and Systems. Wiley, London (1995)

[12] Kwiatkowska, M., Norman, G., Parker, D: Stochastic Model Checking. Lecture Notes in Computer Science 4486, 220-270 (2007)

[13] Romero-Campero, F.J, Gheorghe, M., Bianco, L., Pescini, D., Pérez-Jiménez, M.J. , Ceterchi, R. Towards Probabilistic Model Checking on P Systems Using PRISM. Lecture Notes Comput. Sci. 4361, 477-495 (2006)

[14] Muniyandi, R.C., Abdullah, M.Z.: Evaluating Ligand-Receptor Networks of TGF-β with Membrane Computing. Pakistan Journal of Biological Sciences 14(24), 1100-1108 (2011)

[15] Muniyandi, R.C., Abdullah, M.Z.: Modeling Hormone-induced Calcium Oscillations in Liver Cell with Membrane Computing. Romanian Journal of Information Science and Technology 15(1), 63-76 (2012)

Optimal Promotional Effort Control Policy for Segment Specific New Product Growth

Prerna Manik[1], Kuldeep Chaudhary[2], Yogender Singh[3], P. C. Jha[4]

[1,2,3,4] Department of Operational Research, University of Delhi, Delhi

{prernamanik@gmail.com;chaudharyiitr33@gmail.com ;aeiou.yogi@gmail.com; jhapc@yahoo.com}

Abstract. We have incorporated the concept of market segmentation in the mixed influence diffusion model to study the impact of promotional effort on segment specific new product growth, with a view to arrive at the optimal promotional effort rate in a segmented market. Evolution of sales rate for each segment is developed under the assumption that practitioner may choose both differentiated market promotional effort and mass market promotional effort to influence the sales in unsaturated portion of the market. Accordingly, we have formulated the optimal control problem incorporating impact of differentiated market promotional effort as well as mass market promotional effort on sales rate for each segment, where mass market promotional effort influences each segment with a fixed spectrum. We have obtained the optimal promotional effort policy for the proposed model. To illustrate the applicability of the approach, a numerical example has been discussed and solved using differential evolution algorithm.

Keywords: Differentiated market promotional effort, Mass market promotional effort, Spectrum effect, Promotional effort policy, Differential evolution algorithm.

1 Introduction

Diffusion theory is extensively used in marketing to capture the dynamics involved in life cycle of a product. Many researchers have lucratively applied innovation diffusion models to study the adoption behavior of the product over its life cycle and make imperative decisions related to product modification, price differentiation, resource optimization etc. Diffusion models aim to depict the growth of an adoption process of a product and predict its continued development in the future times over the product life cycle. The most commonly and widely used model in this field is Bass advertising model [2,3] as it captures the interaction between advertising and sales in an instinctively satisfying manner.

Modern marketing focuses on customer satisfaction as the "center of the Universe". Firms design customer-driven marketing strategies depending on needs and wants of the customers with an aim to achieve profit maximization. But in today's competitive environment, where each company is trying to do better than its competitors, it is becoming increasingly difficult for firms to create customer satisfaction profitably due to continuously changing tastes of the customers.

J. C. Bansal et al. (eds.), *Proceedings of Seventh International Conference on Bio-Inspired Computing: Theories and Applications (BIC-TA 2012)*, Advances in Intelligent Systems and Computing 202, DOI: 10.1007/978-81-322-1041-2_30, © Springer India 2013

Present globalized markets comprise multi-cultural customers with diversified tastes and needs. As a result firms need to constantly innovate and promote their products for sustainable growth. However, it is seldom possible to satisfy every customer demand by treating them alike. This necessitates the need for segregating markets into various segments comprising of customers with similar demand characteristics. Market segmentation [14] help firms to understand and satisfy their customers better. Markets can be segregated based on geographical (nations, region, state, countries, cities and neighborhoods); demographical (age, gender, income, family size, occupation, education); psycahographical (social class, life style, personality, value) and behavioral (user states, usage rate, purchase occasion, attitude towards product) characteristics of the customers. After identifying various segments, firms develop independent promotional strategies for each segment. At the same time, firms may also develop promotion strategy for the mass market that reaches different segments with a fixed spectrum.

While market segmentation has long been a popular topic for academic research in marketing, only a few papers on dynamic advertising model deal with market segmentation (Little and Lodish [15]; Seidmann et al. [20]; Buratto et al. [5,6]). Little and Lodish [15] analyzed a discrete time stochastic model of multiple media selection in a segmented market. Seidmann et al. [20] proposed a general sales-advertising model in which the state of the system represents a population distribution over a parameter space and they show that such models are well poised and that there exists an optimal control. Buratto et al. [5,6] discussed the optimal advertising policy for a new product introduction in a segmented market with Narlove-Arrow's [16] linear goodwill dynamics. Jha et al. [13] used the concept of market segmentation in diffusion model for advertising a new product and studied the optimal advertising effectiveness rate in a segmented market. They discussed the evolution of sales dynamics in the segmented market under two cases. Firstly, they assumed that the firm advertises in each segment independently and further they took the case of a single advertising channel, which reaches several segments with fixed spectrum.

In this paper, we assume that firm has defined its target market in a segmented consumer population and that it wants to plan the differentiated market and mass market promotion strategies with the aim to maximize its profit. We amalgamate the two cases proposed by Jha et al. [13] and study the differentiated market and mass market promotion effectiveness rates. Optimal control theory is applied to solve the problem. The problem has been further extended to incorporate budgetary constraint. The budgetary problem is then discretized and solved using differential evolution algorithm.

One of the fascinating features of optimal control is the extraordinarily wide range of its possible applications [1,6,7,8,11,12,13]. The application of optimal control theory in sales-advertising control analysis is possible due to its dynamic behaviour. Continuous optimal control models provide a powerful tool for understanding the behavior of sales-advertising system where dynamic aspect plays an important role. Several papers have been written on the application of optimal control theory in sales-advertising problem [5,9,10,14,20,22].

Rest of the paper is organized as follows. In section 2, we introduce the diffusion model and discuss its optimal control formulation and develop segmented sales rate under the assumption that the firm promotes its product independently in each segment as well as there is a mass market promotional campaign that reaches several segments with a fixed spectrum. The problem is discussed and solved using Pontryagin's Maximum Principle with a particular case. Differential evolution to solve the discretized problem is presented in section 3. Numerical illustration has been discussed in section 4. Section 5 concludes the paper.

2 Model Formulation

We begin our analysis by stating the following assumptions that M denotes the total number of market segments and the function $\overline{N}_i$, $i = 1, 2, ..., M$, denotes the number of potential customers in i^{th}, $i=1,2,...,M$ segment. Further, we assume that segmentation is a discrete variable and the value $\sum_{i=1}^{M} \overline{N}_i$ represents the total number of potential customers of the product.

Let $x_i(t)$, $i=1,2,...,M$ be the promotional effort rate for each segment and $x(t)$ be the mass market promotional effort rate at time t, that influence the sales in unsaturated portion of the market $(\overline{N}_i - N_i(t)), i = 1, 2, ..., M$. Therefore, evolution of sales intensity under the joint effect of differentiated market and mass market promotion (that reaches each segment with a fixed spectrum α_i, $i=1,2,...,M$) follows the differential equation

$$\frac{dN_i(t)}{dt} = b_i(t)\left(x_i(t) + \alpha_i x(t)\right)\left(\overline{N}_i - N_i(t)\right) \quad i = 1, 2, ..., M \tag{1}$$

where, $b_i(t)$ is the adoption rate per additional adoption and $\alpha_i > 0$, $\sum_i \alpha_i > 0 \ \forall \ i = 1....M$, is the segment specific spectrum, and the elements α_i, provide different promotion expenditure rate to the market segments.

Two distinct approaches have been used to represent $b_i(t)$. One has been to represent $b_i(t)$ as a function of time; the other has been to represent $b_i(t)$ as a function of the number of previous adopters. Because the latter approach is by far the most common, it is the one employed here. Therefore, we assume the adoption rate per additional adoption $b_i(t) = \left(p_i + q_i \frac{N_i(t)}{\overline{N}_i}\right)$ [2], and then sales intensity is described by the following differential equation

$$\frac{dN_i(t)}{dt} = \left(p_i + q_i \frac{N_i(t)}{\overline{N}_i}\right)\left(x_i(t) + \alpha_i x(t)\right)\left(\overline{N}_i - N_i(t)\right) \quad i = 1, 2, ..., M \tag{2}$$

Under the initial condition $\qquad N_i(0) = N_{i0}, \ i = 1, 2, ..., M \tag{3}$

The promotion expenditure (differentiated as well as mass market) is the degree to which the firm achieves the goal of sales or profit. Therefore, we are led to formulate the problem of determining the optimal differentiated market and mass market promotional effort rates $x_i(t)$, $x(t)$ policy for the new product as the following optimal control problem

$$J = \int_0^T e^{-\rho t}\left(\sum_{i=1}^M\left[(P_i-C_i(N_i(t)))N_i'(t)-u_i(x_i(t))\right]-v(x(t))\right)dt \tag{4}$$

subject to system equation (2) and (3).

where, $u(x_i(t))$ and $v(x(t))$ are the differentiated market promotional effort and mass market promotional effort costs respectively and ρ is the discounted profit. Further, P_i is a constant sales price which depend upon segments and $C_i(N_i(t))$ is total production cost, which is continuous and differentiable with the assumption that $C'(.)>0$ and $P_i-C'(N_i(t))>0$ for all segments.

For the sake of simplicity we assume that $C_i(N_i(t)) = C_i$ (a constant); $i = 1, 2, ..., M$

Here, we have an optimal control problem with $M+1$ control variables $(x_i(t), x(t))$ and M state variables $(N_i(t))$. From the Maximum Principle, we define the Hamiltonian as

$$H=\left(\sum_{i=1}^M\left[(P_i-C_i)N_i'(t)-u_i(x_i(t))\right]-v(x(t))\right)+\sum_{i=1}^M\lambda_i(t)\left(\left(p_i+q_i\frac{N_i(t)}{\overline{N}_i(t)}\right)(x_i(t)+\alpha_i x(t))\left(\overline{N}_i-N_i(t)\right)\right) \tag{5}$$

A simple interpretation of the Hamiltonian is that it represents the overall profit of the various policy decisions with both the immediate and the future effects taken into account. Assuming the existence of an optimal control solution, the maximum principle provides the necessary optimality conditions, there exists a piecewise continuously differentiable function $\lambda_i(t)$ for all $t\in[0,T]$, where $\lambda_i(t)$ is known as adjoint variable and the value of $\lambda_i(t)$ at time t describes the future effect on profits upon making a small change in $N_i(t)$.

For the optimality conditions

$$\frac{\partial H^*}{\partial x(t)} = 0,$$

$$\frac{\partial H^*}{\partial x_i(t)} = 0, \tag{6}$$

$$\frac{d\lambda_i(t)}{dt} = \rho\lambda_i(t) - \frac{\partial H^*}{\partial N_i(t)}, \ \lambda_i(T) = 0$$

Hence, the Hamiltonian H of each of the segments is strictly concave in $x_i(t)$ and $x(t)$ and according to the Mangasarian Sufficiency Theorem [20,22]; there exists unique values of promotional effort control $x_i^*(t)$ and $x^*(t)$ for each segment respectively. From equation (5) and the optimality conditions (6), we get

$$x_i^*(t) = \phi_i\left(\left\{(P_i-C_i)\frac{\partial N_i}{\partial x_i}\right\}+\lambda_i(t)\left(p_i+q_i\frac{N_i(t)}{N_i(t)}\right)\left(\overline{N}_i(t)-N_i(t)\right)\right) \tag{7}$$

$$x^*(t) = \varphi\left(\sum_{i=1}^M\left[\left((P_i-C_i)\frac{\partial N_i}{\partial x}\right)+\lambda_i(t)\alpha_i\left(p_i+q_i\frac{N_i(t)}{N_i(t)}\right)\left(\overline{N}_i(t)-N_i(t)\right)\right]\right) \tag{8}$$

where, $\varphi i(.)$ and $\varphi(.)$ are the inverse functions of u_i and v respectively. This optimal control promotional policy shows that when market is almost saturated, then differentiated market promotional expenditure rate and mass market promotional expenditure rate respectively should be zero (i.e. there is no need of promotion in the market).

For optimal control policy, the optimal sales trajectory using optimal values of differentiated market promotional effort ($x_i^*(t)$) and mass market promotional effort ($x^*(t)$) rate are given by

$$
N_i^*(t) = \frac{\overline{N}_i\left(\left(\dfrac{p_i + q_i \dfrac{N_i(0)}{\overline{N}_i}}{\overline{N}_i - N_i(0)}\right)\exp\left((p_i+q_i)\int_0^t\left((x_i(t)+\alpha_i x(t))\right)dt\right) - p_i\right)}{\dfrac{q_i}{\overline{N}_i} + \left(\dfrac{p_i + q_i \dfrac{N_i(0)}{\overline{N}_i}}{\overline{N}_i - N_i(0)}\right)\exp\left((p_i+q_i)\int_0^t\left((x_i(t)+\alpha_i x(t))\right)dt\right)} \qquad \forall i \qquad (9)
$$

If we define $X^*(t) = \int_0^t (x_i(t) + \alpha_i x(t))\, d\tau$ is the cumulative promotional effort, the above equation can be rewritten as

$$
N_i^*(t) = \frac{\overline{N}_i\left(\left(\dfrac{p_i + q_i \dfrac{N_i(0)}{\overline{N}_i}}{\overline{N}_i - N_i(0)}\right)\exp\left((p_i + q_i)X^*(t)\right) - p_i\right)}{\dfrac{q_i}{\overline{N}_i} + \left(\dfrac{p_i + q_i \dfrac{N_i(0)}{\overline{N}_i}}{\overline{N}_i - N_i(0)}\right)\exp\left((p_i + q_i)X^*(t)\right)} \qquad \forall i \qquad (10)
$$

If $N_i(0) = 0$, then we get the following result

$$
N_i^*(t) = \overline{N}_i\left(\frac{1 - \exp\left(-(p_i + q_i)X^*(t)\right)}{1 + \dfrac{q_i}{p_i}\exp\left(-(p_i + q_i)X^*(t)\right)}\right) \qquad \forall i \qquad (11)
$$

and adjoint trajectory can be shown as

$$
\frac{d\lambda_i(t)}{dt} = \rho\lambda_i(t) - \left\{(p_i - C_i + \lambda_i(t))\left(\frac{\partial \dot{N}_i}{\partial N}\right) - \dot{N}_i\left(\frac{\partial C_i}{\partial N}\right)\right\} \qquad (12)
$$

with transversality condition $\lambda_i(T) = 0$.

Integrating (12), we have the future benefit of having one more unit of sale

$$\lambda_i(t) = e^{-\rho t} \int_t^T e^{-\rho s} \left(\left\{ (p_i - C_i + \lambda_i) \left(\frac{\partial \dot{N}_i}{\partial N_i} \right) - \dot{N}_i \left(\frac{\partial C_i}{\partial N_i} \right) \right\} \right) ds \qquad (13)$$

2.1 Particular Case of General Formulation

When Differentiated Market Promotional Effort and Mass Market Promotional Effort Costs are Linear Functions

If differentiated market promotional effort and mass market promotional effort costs respectively are $u_i(x_i(t)) = k_i x_i(t)$, $v(x(t)) = ax(t)$ and $0 < x_i(t) < \overline{A}_i$, $0 < x(t) < \overline{A}$, where $\overline{A}_i$, A_i are positive constants which are maximum permissible promotional effort rates ($\overline{A}_i, \overline{A}$ are determined by the promotional budget etc.) and κ_i is the per unit cost of promotional effort per unit time towards i^{th} segment, and a is the per unit cost of promotional effort per unit time towards mass market. Now, Hamiltonian can be written as

$$H = \left(\sum_{i=1}^{M} \left[(P_i - C_i) N_i(t) - k_i x_i(t) \right] - ax(t) \right) + \sum_{i=1}^{M} \lambda_i(t) \left(\left(p_i + q_i \frac{N_i(t)}{N_i(t)} \right) (x_i(t) + \alpha_i x(t)) \left(\overline{N}_i - N_i(t) \right) \right) \qquad (14)$$

Because Hamiltonian is linear in $x_i(t)$ and $x(t)$ we obtain optimal differentiated market promotional effort and mass market promotional effort by the maximum principle and it is given by

$$x_i^*(t) = \begin{cases} 0 & if & W_i \le 0 \\ \overline{A}_i & if & W_i > 0 \end{cases} \qquad (15)$$

$$x^*(t) = \begin{cases} 0 & if & B \le 0 \\ \overline{A} & if & B > 0 \end{cases} \qquad (16)$$

where, $W_i = -k_i + (P_i - C_i + \lambda_i) \left(p_i + q_i \frac{N_i(t)}{\overline{N}_i} \right) (\overline{N}_i - N_i(t))$

and $B = -a + \sum_{i=1}^{M} \left\{ \alpha_i (P_i - C_i + \lambda_i) \left(p_i + q_i \frac{N_i(t)}{\overline{N}_i} \right) (\overline{N}_i - N_i(t)) \right\}$

W_i and B are promotional effort switching functions. This type of control is called "Bang-Bang" control in the terminology of optimal control theory. However, interior control is possible on an arc along $u(t,a)$ Such an arc is known as "Singular arc" [20]. There are four set of optimal control values of differentiated market promotional effort $(x_i(t))$ and mass market promotional effort $(x(t))$ rate: 1) $x_i^*(t) = 0$, $x^*(t) = 0$; 2) $x_i^*(t) = 0, x^*(t) = B$; $x_i^*(t) = A_i, x^*(t) = 0$; $x_i^*(t) = A_i, x^*(t) = B$. And the optimal sales trajectory and adjoint trajectory respectively using optimal values of differentiated market promotional effort ($x_i^*(t)$) and mass market promotional effort ($x^*(t)$) rate are given by

$$N_i^*(t) = \dfrac{\overline{N}_i\left[\left(\dfrac{\left(p_i+q_i\dfrac{N_i(0)}{\overline{N}_i}\right)}{\overline{N}_i-N_i(0)}\right)\exp\left({}^{(p_i+q_i)\int_0^t\left(\left(\overline{A}_i+\alpha_i\overline{A}\right)\right)dt}\right)-p_i\right]}{\dfrac{q_i}{\overline{N}_i}+\left(\dfrac{\left(p_i+q_i\dfrac{N_i(0)}{\overline{N}_i}\right)}{\overline{N}_i-N_i(0)}\right)\exp\left({}^{(p_i+q_i)\int_0^t\left(\left(\overline{A}_i+\alpha_i\overline{A}\right)\right)dt}\right)}\quad \forall i \tag{17}$$

If $N_i(0)=0$, then we get the following result

$$\dot{N}_i^*(t)=\overline{N}_i\left(\dfrac{1-\exp\left(-(p_i+q_i)\left(\overline{A}_i+\alpha_i\overline{A}\right)t\right)}{1+\dfrac{q_i}{p_i}\exp\left(-(p_i+q_i)\left(\overline{A}_i+\alpha_i\overline{A}\right)t\right)}\right)\quad \forall i \tag{18}$$

which is similar to Bass model [2] sales trajectory and the adjiont variable is given by

$$\lambda_i(t)=e^{-\rho t}\int_t^T e^{-\rho s}\left(\left\{(P_i-C_i+\lambda_i)\left(\partial\dot{N}_i\Big/\partial N_i\right)-\dot{N}_i\left(\partial C_i\Big/\partial N_i\right)\right\}\right)ds \tag{19}$$

In order to demonstrate the numerical results of the above problem, the discounted continuous optimal problem (4) is transformed into equivalent discrete problem [19] which can be solved using differential evolution. The discrete optimal control can be written as follows:

$$J=\sum_{k=1}^{T}\left[\left[\left(\sum_{i=1}^{M}(P_i(k)-C_i(k))(N_i(k+1)-N_i(k))-u_i\left(x_i(k)\right)\right)-v\left(x(k)\right)\right]\left(\dfrac{1}{(1+\rho)^{k-2}}\right)\right] \tag{20}$$

s.t. $N_i(k+1)=N_i(k)+\left(p_i+q_i\dfrac{N_i(k)}{\overline{N}_i(k)}\right)\left(x_i(k)+\alpha_ix(k)\right)\left(\overline{N}_i-N_i(k)\right)$ $i=1,2,...,M$ $\tag{21}$

Generally, firms promote their products in order to spread awareness and sway customers to buy their products. Although promotion is indispensable for sales growth of a firm's product, firms cannot continue to promote its products generously and indefinitely due to availability of limited resources and short product life cycles. Also the consumer adoption process change with time. In light of these facts, it is necessary to incorporate a budget constraint to make the problem more realistic. The budgetary problem can be stated as

Max $J=\displaystyle\int_0^t e^{-\rho t}\left(\sum_{i=1}^{M}\left[(P_i-C_i)N_i'(t)-u_ix_i(t)\right]-\alpha x(t)\right)dt \tag{22}$

s.t. $\dfrac{dN_i(t)}{dt}=\left(p_i+q_i\dfrac{N_i(t)}{\overline{N}_i(t)}\right)\left(x_i(t)+\alpha_ix(t)\right)\left(\overline{N}_i-N_i(t)\right)$ $i=1,2,...,M \tag{23}$

$\left(\displaystyle\sum_{i=1}^{M}\int_0^T u_ix_i(t)\,dt+\int_0^T \alpha x(t)\,dt\right)\leq Z_0 \tag{24}$

$N_i(0)=N_{i0}$, $i=1,2,...,M \tag{25}$

where, Z_0 is the total budget for differentiated market promotion and mass market promotion. The equivalent discrete optimal control of the budgetary problem can be written as follows

$$J=\sum_{k=1}^{T}\left[\left[\left(\sum_{i=1}^{M}(P_i(k)-C_i(k))(N_i(k+1)-N_i(k))-u_i\left(x_i(k)\right)\right)-v\left(x(k)\right)\right]\left(\frac{1}{(1+\rho)^{k-2}}\right)\right] \quad (26)$$

$$\text{s.t. } N_i\left(k+1\right)=N_i\left(k\right)+\left(p_i+q_i\frac{N_i(k)}{\overline{N}_i(k)}\right)\left(x_i(k)+\alpha_i x(k)\right)\left(\overline{N}_i-N_i(k)\right) \quad i=1,2,...,M \quad (27)$$

$$\sum_{i=1}^{M}\left(\sum_{k=0}^{T}(x_i(k)+\alpha_i x(k))\right)\leq Z_0 \quad (28)$$

This discrete problem can be solved using differential evolution (DE). Procedure for applying DE is presented in the following section.

3 Differential Evolution Algorithm

Differential evolution is an exceptionally simple evolutionary algorithm, which is rapidly growing field of artificial intelligence. This class also includes genetic algorithms, evolutionary strategies and evolutionary programming. DE was proposed by Price and Storn [17]. Since then it has earned a reputation as a very powerful and effective global optimizer. The basic steps of DE are as follows
Start
Step 1: Generate an initial population of random individuals
Step 2: Create a new population by repeating following steps until the stopping criterion is achieved
- **[Selection]** Select the random individuals for reproduction
- **[Reproduction]** Create new individuals from selected ones by mutation and crossover
- **[Evolution]** Compute the fitness values of the individuals
- **[Advanced Population]** Select the new generation from target individual and trial individuals
End steps

3.1 Initialization

Suppose we want to optimize a function of D number of real parameters. We must select a population of size NP. NP parameter vectors have the form
$$X_{i,G}=(x_{1,i,G},x_{2,i,G},...,x_{D,i,G})$$
where, D is dimension, i is an individual index and G represents the number of generations.

First, all the solution vectors in a population are randomly initialized. The initial solution vectors are generated between lower and upper bounds $l=\{l_1,l_2,...,l_D\}$ and $u=\{u_1,u_2,...,u_D\}$ using the equation
$$x_{j,i,0}=l_j+rand_{i,j}[0,1]\times(u_j-l_j)$$

where, i is an individual index, j is component index and $rand_{i,j}[0,1]$ is a uniformly distributed random number lying between 0 and 1. This randomly generated population of vectors $X_{i,0}=(x_{1,i,0}, x_{2,i,0},..., x_{D,i,0})$ is known as target vectors.

3.2 Mutation

Each of the NP parameter vectors undergo mutation, recombination and selection. Mutation expands the search space. For a given parameter vector $X_{i,G}$, three vectors $X_{r_1,G}, X_{r_2,G}, X_{r_3,G}$ are randomly selected such that the indices i, r_1, r_2, r_3 are distinct. The ith perturbed individual, $V_{i,G}$, is therefore generated based on the three chosen individuals as follows

$$V_{i,G} = X_{r_1,G} + F*(X_{r_2,G} - X_{r_3,G})$$

where, $r_1, r_2, r_3 \in \{1,2,...,NP\}$ are randomly selected, such that $r_1 \neq r_2 \neq r_3 \neq i$, $F \in (0, 1.2]$ and $V_{i,G}$ is called the mutation vector.

3.3 Crossover

The perturbed individual, $V_{i,G} = (v_{1,i,G}, v_{2,i,G},..., v_{D,i,G})$ and the current population member, $X_{i,G} = (x_{1,i,G}, x_{2,i,G},..., x_{D,i,G})$ are then subject to the crossover operation, that finally generates the population of candidates, or "*trial*" vectors, $U_{i,G} = (u_{1,i,G}, u_{2,i,G},..., u_{D,i,G})$, as follows

$$u_{j,i,G} = \begin{cases} v_{j,i,G} & if\ rand_{i,j}[0,1] \le C_r \vee j = j_{rand} \\ x_{j,i,G} & \text{otherwise} \end{cases}$$

where, $C_r \in [0,1]$ is a crossover probability, $j_{rand} \in \{1,2,...,D\}$ is a random parameter's index, chosen once for each i.

3.4 Selection

The population for the next generation is selected from the individuals in current population and its corresponding trial vector according to the following rule

$$X_{i,G+1} = \begin{cases} U_{i,G} & if\ f(U_{i,G}) \ge f(X_{i,G}) \\ X_{i,G} & \text{otherwise} \end{cases}$$

where, $f(.)$ is the objective function value. Each individual of the temporary population is compared with its counterpart in the current population. Mutation, recombination and selection continue until stopping criterion is reached.

3.5 Constraint Handling in Differential Evolution

Pareto ranking method is used to handle constraints in DE. The value of constraints is calculated at target and trial vectors. The method is based on the following three rules

1) Between two feasible vectors (target and trial), the one with the best value of the objective function is preferred

2) If out of target and trial vectors, one vector is feasible and the other is infeasible, the one which is feasible is preferred
3) Between two infeasible vectors, the one with the lowest sum of constraint violation is preferred

3.6 Stopping Criterion

DE algorithm stops when either
  1) Maximum number of generations are reached or
  2) Desired accuracy is achieved i.e.,

$$\left|f_{\max} - f_{\min}\right| \leq \varepsilon$$

4 Numerical Illustration

Specifically in multi-cultural countries like India, firms such as Honda, Hyundai, Sony, Samsung to name a few promote their products in national as well as local regional languages to influence a large customer base where national language influence each region with a fixed spectrum. Hence it is imperative to allocate atleast 30-40% of the total budget to mass market promotion in such a situation.

This discrete optimal control problem is solved by using DE. Parameters of DE are given in Table 1. Total promotional budget is assumed to be ₹1,50,000 of which ₹55,000 is allocated for mass market promotion and rest for differentiated market promotion. We further assume that the time horizon has been divided into 10 equal time periods. The number of market segments are assumed to be six (i.e. M=6). Value of parameters a and ρ are taken to be 0.2 and 0.095 respectively and the values of rest of the parameters are given in Table 2.

Table 1. Parameters of Differential Evolution

Parameter	Value	Parameter	Value
Population Size	100	Scaling Factor (F)	.7
Selection Method	Roulette Wheel	Crossover Probability (C_r)	.9

Table 2. Parameters

Segment	$\overline{N}_i$	P_i	C_i	u_i	p_i	q_i	α_i
S1	52000	10000	9850	0.0016	5.21E-05	6.26E-04	0.1513
S2	46520	12500	12360	0.0019	4.93E-05	5.26E-04	0.2138
S3	40000	11000	10845	0.0022	6.10E-05	6.31E-04	0.1268
S4	29100	13200	13055	0.0017	5.51E-05	5.50E-04	0.2204
S5	35000	12000	11841	0.0021	5.41E-05	5.50E-04	0.1465
S6	25000	10250	10108	0.0018	5.71E-05	5.68E-04	0.1412
Total	227620						1

Optimal allocation of promotional effort resources for each segment are given in Table 3 and the corresponding sales with these resources and the percentage of adoption (sales) for each segment out of total potential market are tabulated in Table 4.

Table 3. Optimal Segment-wise Promotional Effort Allocation

Seg	T_1	T_2	T_3	T_4	T5	T6	T7	T8	T9	T10	Total
S1	1738	1608	1348	1829	1686	1673	1582	1218	1699	1816	16200
S2	1348	1530	1712	1699	1608	1842	1751	1569	1608	1712	16382
S3	1478	1569	1322	1968	1608	1309	1439	1712	1465	1673	15546
S4	1218	1218	1439	1608	1426	1803	1309	1647	1322	1699	14692
S5	1738	1478	1738	1803	1491	1829	1829	1322	1803	1452	16486
S6	1608	1283	1296	1803	1712	1569	1309	1491	1751	1868	15693
MPA	6230	5040	5320	6420	5810	5040	5600	4970	5390	5180	55000

*MPA- Mass Market Promotional Allocation

Table 4. Sales (in Thousands) and Percentage of Adoption for Each Segment from Total Potential Market

Seg	T1	T2	T3	T4	T5	T6	T7	T8	T9	T10	Total	% of Cap Mkt Size
S1	3.9	3.4	2.9	4.8	3.7	3.6	3.3	2.9	3.7	4.2	36.40	70.00
S2	2.6	2.7	3.4	4.0	3.0	3.9	3.5	2.9	3.0	3.4	32.38	69.60
S3	2.2	2.5	2.2	7.0	2.6	2.2	2.2	2.9	2.2	2.8	28.80	72.00
S4	1.6	1.6	1.6	5.4	1.6	2.3	1.6	2.0	1.6	2.1	21.39	73.50
S5	2.6	1.9	2.6	3.3	2.0	2.9	2.9	1.9	2.8	1.9	24.85	71.00
S6	1.6	1.4	1.4	4.2	1.8	1.6	1.4	1.4	1.9	2.1	18.75	75.00
Total											162.57	71.42

5 Conclusion

One of the most striking developments in marketing is the amount of interest shown in market segmentation. The goal of the management is to use market segmentation in promotion of the product to have competitive advantages over its competitors. The purpose for segmenting a market is to allow promotion programs to focus on the subset of prospective customers that are "most likely" to purchase the product. In this paper, we use the concept of promotional effort in innovation diffusion modeling for promotion of a new product in segmented market with the assumption that discrete market promotional effort and mass market promotional effort influence the sales in unsaturated portion of the market. Here, we have formulated an optimal control problem for the innovation diffusion model and have made the problem more realistic by adding the budgetary constraint and solution of proposed problem has been obtained using maximum principle. Using the optimal control techniques, the main objective here was to determine optimal promotional effort policies. A special case of the proposed optimal control problem has also been discussed and a numerical example has been solved using differential evolution algorithm to illustrate the applicability of the approach. Optimal control promotional effort policies show that when market is almost saturated, promotional effort diminishes.

References

[1]   Amit, R.: Petroleum reservoir exploitation: switching from primary to secondary recovery. Operations Research, 34 (4), 534-549 (1986)

[2]   Bass, F.M.: A new product growth model for consumer durables. Management Science, 15, 215-227 (1969)

[3]   Bass, F.M., Krishnan, T.V., Jain, D.C.: Why the Bass model fits without decision variables. Marketing Science, 13(3), 203–223 (1994)

[4]   Bryson, A.E., Ho, Y.C.: Applied Optimal Control. Blaisdell Publishing Co.,Waltham, Mass (1969)

[5]   Buratto, A., Grosset, L., Viscolani, B.: Advertising a new product in segmented market. EJOR, 175, 1262-1267 (2005)

[6]   Buratto, A., Grosset, L., Viscolani, B.: Advertising for the introduction of an age-sensitive product. Optimal Control Applications and Methods, 26, 157-167 (2006)

[7]   Davis, B.E., Elzinga, D.J.: The solution of an optimal control problem in financial modeling. Operations Research, 19, 1419-1433 (1972)

[8]   Derzko, N.A., Sethi, S.P.: Optimal exploration and consumption of a natural resource: deterministic case. Optimal Control Applications and Methods, 2(1), 1-21 (1981)

[9]   Elton E., Gruber, M.: Finance as a Dynamic Process. Prentice Hall, Englewood Cliffs, New Jersey (1975)

[10]   Feichtinger, G. (Ed.): Optimal Control Theory and Economics Analysis-Vol. 3. North-Holland, Amsterdam (1988)

[11]   Feichtinger, G., Hartl, R.F., Sethi, S.P.: Dynamic optimal control models in advertising: recent developments. Management Science, 40(2), 195-226 (1994)

[12]   Heaps, T.: The forestry maximum principle. Journal of Economic and Dynamics and Control, 7, 131-151 (1984)

[13]   Jha, P.C., Chaudhary K., Kapur P.K.: Optimal advertising control policy for a new product in segmented market. OPSEARCH, 46(2), 225-237 (2009)

[14]   Kotler, P.: Marketing Management, 11th Edition. Prentice Hall, Englewood Cliffs, New Jersey (2003)

[15]   Little, J.D.C., Lodish, L.M.: A media planning calculus. Operations Research, 17, 1-35 (1969)

[16]   Nerlove, M., Arrow, K.J.: Optimal advertising policy under dynamic conditions. Economica, 29,129-142 (1962)

[17]   Price, K.V., Storn, R.M.: Differential Evolution-a simple and efficient adaptive scheme for global optimization over continuous space. Technical Report TR-95-012, ICSI, March 1995. Available via the Internet: ftp.icsi.berkeley.edu/pub/techreports/1995/tr-95-012.ps.Z, (1995)

[18]   Price, K.V.: An introduction to Differential Evolution. In: Corne, D., Marco, D. and Glover, F. (eds.), New Ideas in Optimization, McGraw-Hill, London (UK), 78–108 (1999)

[19]   Rosen, J.B.: Numerical solution of optimal control problems. In: Dantzig, G.B., Veinott, A.F. (eds.), Mathematics of decision science: Part-2, 37-45. American Mathematical society (1968)

[20]   Seidmann, T.I., Sethi, S.P., Derzko, N.: Dynamics and optimization of a distributed sales-advertising models. Journal of Optimization Theory and Applications, 52, 443-462 (1987)

[21]   Seierstad A., Sydsaeter, K.: Optimal Control Theory with Economic Applications. North-Holland, Amesterdam (1987)

[22]   Sethi, S.P.: Optimal control of the Vidale-Wolfe advertising model. Operations Research, 21, 998-1013 (1973)

[23]   Sethi, S.P., Thompson, G.L.: Optimal Control Theory: Applications to Management Science and Economics. Kluwer Academic Publishers, Dordrecht (2000)

[24]   Vidale, M. L., Wolfe, H.B.: An operations research study of sales response to advertising. Operations Research, 5(3), 370-381 (1957)

θ - Bordered Infinite Words

C. Annal Deva Priya Darshini[1], V. Rajkumar Dare[2]

Dept. of Mathematics, Madras Christian College, Tambaram, Chennai – 600 059.

{[1]annaldevapriyadarshini@gmail.com; [2]rajkumardare@yahoo.com}

Abstract. In this paper, we define θ - bordered infinite word and unbordered words and study their properties. We give a characterization of θ - bordered infinite words for an antimorphic involution θ. We show that the limit language of the set of all θ - bordered words is a ω - regular language for an antimorphic involution θ.

Keywords: θ - bordered infinite word, θ - unbordered infinite word, θ – bordered infinite language, limit language.

1 Introduction

The concept of using DNA for computation as studied by Adleman [4] has opened a wide area of research. The application of formal language theory in molecular biology has solved many problems in language theory. The DNA strand has been treated as a finite word over the four letter alphabet {A, T, C, G} and the relations between the alphabets is modeled as a morphic or antimorphic involution on the set of alphabets. A general overview of this field can be seen in [3].

The study of finite bordered words is useful in the context of DNA computations since DNA strands are finite in nature. This paper is a theoretical study of the generalisation to the case of infinite words. In this paper, we introduce involutively bordered infinite words. By definition, a finite word u is θ - bordered, for a morphic or antimorphic involution θ if there exists $v \in \Sigma^+$ that is a proper prefix of u while $\theta(v)$ is a proper suffix of u. A word u is called θ - unbordered if it is not bordered. Thus, θ - bordered words are defined with the help of prefixes and suffixes. However, the ω - word does not have a suffix and hence we cannot find its border on the right. This problem has been overcome by using the concept of limit of a language. The infinite θ - bordered word is constructed as the limit of an increasing sequence of θ - bordered words which are prefixes of the infinite word. This definition carries the properties of θ - bordered words to infinite words in a natural way.

J. C. Bansal et al. (eds.), *Proceedings of Seventh International Conference on Bio-Inspired Computing: Theories and Applications (BIC-TA 2012)*, Advances in Intelligent Systems and Computing 202, DOI: 10.1007/978-81-322-1041-2_31, © Springer India 2013

The paper is organised as follows: Section 2 gives the preliminaries and definitions. Section 3 gives the definition and the characterization of θ - bordered and unbordered infinite words. The language of θ - bordered infinite words is proved to be the limit language of the involutively bordered words. Hence we are able to show that the set of all θ - bordered infinite words is a ω - regular set for an antimorphic involution θ.

2 Preliminaries and Basic Definitions

In this section, we introduce some basic definitions used in this paper. Let Σ be a finite alphabet and Σ^* be the collection of all words over Σ including the empty word λ. Let $\Sigma^+ = \Sigma^* - \{\lambda\}$. For $w \in \Sigma^+$, $alph(w)$ is the elements of Σ found in w. The length of a word w is denoted by $|w|$.

A mapping $\theta : \Sigma^* \to \Sigma^*$ such that $\theta(xy) = \theta(x)\,\theta(y)$ is a morphism on Σ^*. If $\theta(xy) = \theta(y)\theta(x)$ then θ is an antimorphism on Σ^*. An involution θ is a mapping such that $\theta(\theta(x)) = x$ for all $x \in \Sigma^*$. We recall the definition of involutively bordered and unbordered words proposed in [1].

Definition 2.1. [1]
Let θ be a morphic or antimorphic involution on Σ^*. A word $u \in \Sigma^+$ is said to be θ - bordered if there exists a $v \in \Sigma^+$ such that $u = vx = y\theta(v)$ for some $x, y \in \Sigma^+$. A nonempty word which is not θ - bordered is called θ - unbordered.

3 θ - Bordered Infinite Words and their Properties

In this section we introduce the θ - bordered infinite word as a generalization of a θ - bordered word. Bordered words are also called as overlapping words and they were studied as words with a prefix which is also a suffix. The extension of this was involutively bordered words where the prefix is the θ - image of the suffix. We define θ - bordered infinite word as a generalization of this word to the infinite case. We make use of prefixes alone to define θ - bordered infinite words.

Definition 3.1.
Let $x \in \Sigma^\omega$ and $x[n]$ be the prefix of x of length n. Let θ - be a morphic or antimorphic involution on Σ. We say that x is θ - bordered infinite word if for each $i \in N, \exists n_i > i$ such that $x[n_i]$ is a θ - bordered word.

We give some simple examples of infinite words that are θ - bordered.

Example 3.1.

Consider the word $w = at^\omega$ on $\Sigma = \{a,t\}$ over which a morphic involution is defined as $\theta(a) = t, \theta(t) = a$. w is θ - bordered infinite word since the prefixes $w[n]$ $= at^{(n-1)}$ are θ - bordered as n tends to infinity.

Example 3.2.

The word $w = (ab)^\omega$ on $\Sigma = \{a,b\}$ over which an antimorphic involution is defined as $\theta(a) = b, \theta(b) = a$ is θ - bordered infinite word since the prefixes $w[2n]$ $= (ab)^n$ are θ - bordered as n tends to infinity.

Example 3.3.

The infinite word $w = abaabaaab\grave{a}ba^5b\ldots$ on $\Sigma = \{a,b\}$ with an antimorphic involution defined as $\theta(a) = b, \theta(b) = a$ is θ - bordered as the prefixes $w[n_i] = abaab\ldots b^n{}_i$ are θ - bordered as i tends to infinity.

The examples show that a θ - bordered infinite word is the limit of an increasing sequence of finite θ - bordered words. Hence the properties of θ - bordered words carry over to the limit word.

Example 3.4.

Consider the word $w = ac^\omega$ on $\Sigma = \{a,t,c,g\}$ over which a morphic involution is defined as $\theta(a) = t, \theta(c) = g$. w is θ - unbordered infinite word since the prefixes $w[n] = ac^{(n-1)}$ are θ - unbordered as n tends to infinity.

We observe that $u \in \Sigma^\omega$ is θ - unbordered infinite word if $\exists i \in \mathrm{N} : \forall n_i > i$ $u[n_i]$ is a θ - unbordered word.

Now, we can study the properties of θ - bordered infinite words in two directions:

(i)   we can examine whether the properties of θ - bordered words extend to θ - bordered infinite words,

(ii)   we can study how the properties of words are affected by the involution.

Moreover, we are especially interested in studying unbordered words.

We state some results for involutively bordered words and then give its extension.

Lemma 3.1. [2]

Let θ - be an antimorphic involution on Σ^*. Then for all $u \in \Sigma^+$ such that $|u| \geq 2$ we have u is θ - unbordered iff $\theta(Pref(u)) \cap Suff(u) = \phi$.

Lemma 3.2. [2]

Let θ - be a morphic involution on Σ^*. Then for all $u \in \Sigma^+$ such that $|u| \geq 2$ and $u \neq \theta(u)$ we have u is θ - unbordered iff $\theta(Pref(u)) \cap Suff(u) = \phi$.

The two results can be extended to ω - words to give a characterization of θ - unbordered infinite words.

Proposition 3.1.

Let θ - be a morphic or antimorphic involution on Σ (not the identity mapping). Then $w \in \Sigma^\omega$ is θ - unbordered infinite word if and only if there is a prefix $u \in \Sigma^*$ of w such that $\theta(Pref(u)) \cap Suff(u) \neq \phi$ and for every other prefix $v \in \Sigma^*$ of w with $|v| > |u|$ we have $\theta(Pref(v)) \cap Suff(v) = \phi$

Proof.

Let $w \in \Sigma^\omega$ be θ - unbordered infinite word. Then $\exists i \in N : \forall n_i > i$ $w[n_i]$ is a θ - unbordered word. We form the required prefixes u and v of w as follows: Take $v = w[n_{i+1}]$. The word v is θ - unbordered word. By Lemma 3.1 and 3.2 we have $\theta(Pref(v)) \cap Suff(v) = \phi$. In fact, the words $v_1 = w[n_{i+1}]$, $v_2 = w[n_{i+2}]$, ... all unbordered. To construct u we search for a bordered prefix of $w[n_i]$. If we find a prefix which is bordered we take that as u. Then $\theta(Pref(u)) \cap Suff(u) \neq \phi$. If not, u is the empty word.

Conversely, let $\theta(Pref(w[n_i])) \cap Suff(w[n_i]) \neq \phi$ for some prefix $w[n_i]$ of $w \in \Sigma^\omega$ and $\theta(Pref(w[n_k])) \cap Suff(w[n_k]) = \phi$ for all $k > n_i$. This implies $w[n_i]$ is a θ - bordered word and all words of length greater than $w[n_i]$ are θ - unbordered. For this $w[n_{i+1}]$ we have $\exists i+1 \in N : \forall n_k > i+1$, $w[n_k]$ is a θ - unbordered word. Hence the result follows.

We give a characterization of θ - bordered infinite words for an antimorphic involution θ.

Lemma 3.3. [2]

Any $x \in \Sigma^*$ is θ - bordered for an antimorphic involution θ if and only if $x = ay\theta(a)$ for some $a \in \Sigma$ $y \in \Sigma^*$.

Proposition 3.2.

Let θ be an antimorphic involution on Σ^{ω}. Then u is θ - bordered infinite word if and only if $u = ay$; $y \in \Sigma^{\omega}$ and $\theta(a)$ is repeated infinitely often in y, $a \in \Sigma$.

Proof.

Let u be a θ - bordered infinite word. By definition, for each $i \in N, \exists n_i > i$ such that $u[n_i]$ is a θ - bordered word. By lemma 3.3, $u[n_1] = ay\theta\,(a)$ for $a \in \Sigma$ $y \in \Sigma^*$. Now, $u[n_2]$ is also θ - bordered word and $n_2 > n_1$. Therefore, $u[n_2] = ty'\theta\,(t)$ for $t \in \Sigma$ $y' \in \Sigma^*$. But $u[n_1]$ is a prefix of $u[n_2]$. Therefore, $u[n_2] = u[n_1]\, r$ for some $r \in \Sigma^+$. That is,

$$ty'\theta\,(t) = ay\theta\,(a)r \qquad (1)$$

where $a \in \Sigma$ $y \in \Sigma^*$ $t \in \Sigma$ $y' \in \Sigma^*$. Now, $a \in \Sigma$, $t \in \Sigma$ implies that $a = t$. Then we must have $\theta\,(a) = \theta\,(t)$. Substituting this relation in (1) we have, $ay'\theta\,(a) = ay\theta\,(a)r$ This implies $r = r'\theta\,(a)$where $r' \in \Sigma^*$. Thus, $u[n_2] = ay\theta\,(a)\,r'\theta\,(a)$. Similarly, $u[n_3] = ay\theta\,(a)\,r'\theta\,(a)\,r''\theta\,(a)$. Thus, we have $u = ay\theta\,(a)\,r'\theta\,(a)\,r''\theta\,(a)\,r'''\theta\,(a)...$ where $\theta\,(a)$ is repeated infinitely often. Hence we write u as $u = ay$ where $a \in \Sigma$ and $\theta\,(a)$ is repeated infinitely often in y.

We can form sets of θ - bordered infinite words or ω - languages of θ - bordered words. The set of all θ - bordered infinite words form a language which we denote by B_{θ}^{ω}. Subsets of B_{θ}^{ω} are θ - bordered ω - languages. For example, $L = \{a^*b^{\omega}\}$ is a θ - bordered ω - language on $\Sigma = \{a,b\}$ for an antimorphic involution θ defined as $\theta(a) = b, \theta(b) = a$.

Recall that a language L is θ-stable if $\theta(L) \subseteq L$. We prove that B_{θ}^{ω} is stable.

Proposition 3.3.

B_{θ}^{ω} is θ-stable for a morphic or antimorhic involution θ.

Proof.

Let $x \in B_{\theta}^{\omega}$. Then for each $i \in N, \exists n_i > i$ such that $x[n_i]$ is a θ - bordered word. That is, $x[n_i] = t\alpha = \beta\theta(t)$ for $t, \alpha, \beta \in \Sigma^+$. Now, $\theta(x[n_i]) = \theta(t\alpha) = \theta(\beta\theta(t))$. For a morphic involution θ, we have $\theta(t)\theta(\alpha) = \theta(\beta)\theta(\theta(t)) = \theta(\beta)t$. This implies that $\theta(x[n_i])$ is a θ-bordered word.

If θ is a antimorphic involution, we have $\theta(t\alpha) = \theta(\alpha)\theta(t) = \theta(\theta(t))\theta(\beta) = t\theta(\beta)$. This implies that $\theta(x[n_i])$ is a θ-bordered word.

Thus, for each $i \in N, \exists n_i > i$ such that $\theta\ (x[n_i])$ is a θ - bordered word. $\theta(x) \in B_\theta^\omega$. Since x is arbitrary, we have $\theta(B_\theta^\omega) \subseteq B_\theta^\omega$. Therefore, B_θ^ω is θ_- stable.

We recall that a language $L \subseteq \Sigma^\omega$ is the limit language of $L_1 \subseteq \Sigma^*$ if for each $x \in L$, there is a sequence of words $x_1 < x_2 < x_3 < ... < x_i ...$ in $L_1 \subseteq \Sigma^*$ such that $x_1 < x_2 < x_3 < ... < x_i - \rightarrow x$ in $L \subseteq \Sigma^\omega$.

Next, we give a result which shows that B_θ^ω is the limit language of a finite language of θ - bordered words. This gives the limit language all the properties of the finite language.

Proposition 3.4.

Let B_θ be the set of all θ - bordered words in for an antimorphic involution θ. Let B_θ^ω be the set of all θ-bordered infinite words over Σ^ω. Then lim $B_\theta = B_\theta^\omega$.

Proof.

Let $x \in \lim B_\theta$. Then there exists an increasing sequence $\{x_n\}$ of elements in B_θ such that $x_1 < x_2 < x_3 < ... < x_i - \rightarrow x$. Since all the x_i's are θ - bordered. x is θ - bordered infinite word. Hence $x \in B_\theta^\omega$ which implies that for each i there exists n_i such that $x[n_i] \in B_\theta$ implies $x[n_1] < x[n_2] < x[n_3] < ...$ $x \in \lim B_\theta$. Thus, lim $B_\theta = B_\theta^\omega$.

By the above result, the language of θ - bordered infinite words is the limit language of θ - bordered words. The limit language possesses the properties of the finite language. We can generalise the properties of θ - bordered words to θ - bordered infinite words.

It has been proved in [2] that B_θ is regular for an antimorphic involution θ. Since the limit language of a regular language is regular, B_θ^ω is ω - regular. We state this as a theorem.

Theorem 3.1.

The limit language B_θ^ω of the set of all θ - bordered words B_θ is a ω - regular language for an antimorphic involution θ.

4 Conclusion

Thus the generalization of θ - bordered words to the infinite case has been defined and studied. The main result is that the limit language of the set of θ - bordered words is also ω - regular as in the finite case.

References

[1] Kari, L., Mahalingam, K.: Involutively bordered words. International Journal of Foundations of Computer Science. 18, 1089–1106 (2007).

[2] Kari, L., Mahalingam, K.: Watson Crick bordered words and their syntactic monoid. International Journal of Foundations of Computer Science. 19(5) (2008) 1163 – 1179 (2008).

[3] Jonoska, N.: Computing with Biomolecules: Trends and Challenges. XVI Tarragona Seminar on Formal Syntax and Semantics. 27 (2003).

[4] Adleman, L.M.: Computing with DNA. Scientific American (1998).

A Two Stage EOQ Model for Deteriorating Products Incorporating Quantity & Freight Discounts, under Fuzzy Environment

Kanika Gandhi[1], P. C. Jha[1], Sadia Samar Ali[2]

[1] Department of Operational Research, Faculty of Mathematical Sciences
University of Delhi, Delhi-110007, India
[2] Fortune Institite of International Business, New Delhi-110057, India

{gandhi.kanika@gmail.com; jhapc@yahoo.com; sadiasamarali@gmail.com}

Abstract. As the industrial environment becomes more competitive, supply chain management (SCM) has become essential. Especially in the case of deteriorating products, demand is an imprecise parameter and leads to uncertainty in other parameters like holding cost and total cost. The objective of the current study is to manage procurement & distribution coordination, who faces many barriers because of the imprecise behaviour of the parameters discussed above while calculating economic order quantity (EOQ), which moves from one source to an intermediate stoppage (Stage I) and further to final destination (Stage II) incorporating quantity and freight discounts at the time of transporting goods in stage I and using truckload (TL) and less than truckload (LTL) policy in stage II. Finding solutions for such class of coordination is highly complex. To reduce the complexity and to find the optimal solution, differential evolution approach is used. The model is validated with the help of a case problem.

Keywords: Supply Chain Management, Fuzzy Optimization, Quantity & Freight Discounts, Truck Load & Less than Truck Load, Differential Evolution.

1 Introduction

The classical EOQ model developed in 1915 had the specific requirements of deterministic costs and demand and lack of deterioration of the items in stock. Gradually the concept of deterioration in inventory system caught up the mind of inventory researchers [1]. In order to fulfill the needs of market and customers, to manage inventory properly has become an important operation activity and a source of profits [2, 3]. When developing inventory model which have the traits of deterioration, it is very important to understand the traits and categories of deterioration [4, 5]. Deterioration means the traits of wear away, lower utility and loss due to physical depletion phenomenon of objects or usage.

J. C. Bansal et al. (eds.), *Proceedings of Seventh International Conference on Bio-Inspired Computing: Theories and Applications (BIC-TA 2012)*, Advances in Intelligent Systems and Computing 202, DOI: 10.1007/978-81-322-1041-2_32, © Springer India 2013

Literature discusses EOQ and inventory models for deteriorating products under fuzzy and non-fuzzy environments as [6] formulated an inventory model with a variable rate of deterioration, a finite rate of production. Several researchers like [7, 8, 9] have developed the inventory models of deteriorating items in different aspects. A comprehensive survey on continuous deterioration of the on-hand inventory has been done first by [10]. [11], develops a two-echelon inventory model with mutual beneficial pricing strategy with considering fuzzy annual demand; single vendor and multiple buyers in this model. This pricing strategy can benefit the vendor more than multiple buyers in the integrated system, when price reduction is incorporated to entice the buyers to accept the minimum total cost. [12], develops an optimal technique for dealing with the fuzziness aspect of demand uncertainties. Triangular fuzzy numbers are used to model external demand and decision models in both non-coordination and coordination situations are constructed.

In this research paper an integrated inventory-transportation two stage supply chain model with one source (supplier), one intermediate stoppage & one destination (buyer), incorporating both quantity & freight discounts and transportation policies is discussed. The transportation of goods from source to destination takes place in two stages, where in the first stage; goods are moved from source to intermediate stoppage. At the stoppage, unloading of goods and their further processing takes a specified time for which the halting cost is fixed, which increases with very high rates after first period. Movement of goods from intermediate stoppage to destination is the second stage of model which is completed through modes of transportation as truck load (TL) and less than truck load (LTL). In Truck load transportation, the cost is fixed of one truck up to a given capacity. In this mode company may use less than the capacity available but cost per truck will not be deducted. However in some cases the weighted quantity may not be large enough to substantiate the cost associated with a TL mode. In such situation, a LTL mode may be used. LTL is defined as a shipment of weighted quantity which does not fill a truck. In such case transportation cost is taken on the bases of per unit weight. The problem is faced at source not being able to forecast demand because of the deteriorating nature of the products. In the process, buyer at destination is playing a major role, which will provide an holistic approach by integrating all the holding, procurement, inspection and transportation activities such that, in stage I, holding cost at source; ordered quantity to source & its purchase cost; transported weights from source to destination & its freight cost; halting cost at intermediate stoppage and in stage II, transportation cost using TL & LTL policies; inspection cost on ordered quantity & holding cost at destination. In the model buyer at destination avails quantity discounts on bulk order and freight discounts on bulk transported quantity. Quantity discounts are provided by the supplier, in which supplier has fixed the quantity level beyond which discount would be given. The paper presents a two stage fuzzy optimization model, who integrates inventory, procurement and transportation mechanism to minimize all the costs discussed above. The total cost of the model becomes fuzzy because of fuzzy holding cost and demand. Such types of models are highly complex in nature, even after converting fuzzy problem in crisp form. Lingo 13.0 takes more time to arrive at an optimal solution, so to reduce the complexity and to find the optimal results; Differential Evolution (DE) [13, 14] is employed. To the best of our knowledge, the problem configuration considered in this study is not been considered in the earlier studies in the literature.

2 Proposed Model Formulation

The Objective of the current study is to minimize the total costs discussed in the previous section and maximum reduction in vagueness of the fuzzy environment. The assumptions of this research are essentially the same as those of an EOQ model except for the transportation cost. The section considers a two stage system with finite planning horizon. The demand is uncertain, and no shortages are allowed. Initial inventory of each product is positive at the beginning of the planning horizon and the holding cost is independent of the purchase price and any capital invested in transportation. Inventory deteriorates with constant rate as a percentage of deteriorated products of stored units and inspection cost is also assumed to be constant.

2.1 Sets

Product set with cardinality P and indexed by i, whereas periods set with cardinality T and indexed by t. Price discount break point set with cardinality L and indexed by small l. Freight discount break point set with cardinality K and indexed by small k. And waiting time set at intermediate stoppage with cardinality Γ and indexed by small τ.

2.2 Parameters

$\tilde{C}$ is Fuzzy total cost, C_0 & C_0^* are the aspiration & tolerance level of fuzzy total cost resp.. $\tilde{HS}_{it}$ & $\overline{HS}_{it}$ are the fuzzy & defuzzified holding cost per unit of product i for t^{th} period at source. Φ_{it} is the unit purchase cost for i^{th} item in t^{th} period, and d_{ilt} reflects the fraction of regular price that the buyer pays to purchase products. c_t is the weight freight cost in t^{th} period. f_{kt}, reflects the fraction of regular price that the buyer pays to transport weighted quantity. s, is the Cost/weight of transportation in LTL policy. β_t is the fixed freight cost for each truck load in period t. $\tilde{HD}_{it}$ & $\overline{HD}_{it}$ are the fuzzy & defuzzified holding cost per unit of product i for t^{th} period at destination. λ_{it} is the inspection cost per unit of product i in period t. $\tilde{D}_{it}$ & $\overline{D}_{it}$ are the fuzzy & defuzzified demand for product i in period t. CR_{it} is the Consumption at destination for product i in period t. ISN_{i1} & IDN_{i1} are inventory level at source & destination in beginning of planning horizon for product i. η is deterioration percentage of ith product at destination. a_{ilt}, is the limit beyond which a price break becomes valid in period t for product i for l^{th} price break. w_i is per unit weight of product i. b_{kt} is the limit beyond which a freight break becomes valid in period t for k^{th} price break. O_{1t} & $(\tau\text{-}1)O_{2t}$ are the holding cost at intermediate node for first halting day & second day onwards for $(\tau\text{-}1)$ number of days in period t. ω is weight transported in each full truck.

2.3 Decision Variables

IS_{it} & ID_{it} are the inventory levels at source & destination resp. at the end of period t for product i. X_{it} is the optimum ordered quantity of product i ordered in period t. R_{ilt} is the binary variable, which is 1 if the ordered quantity falls in l^{th} price break, otherwise zero. L_{1t} & L_{2t} are the total weighted quantity transported in stage I & II respectively of period t. Z_{kt} is the binary variable, which is 1 if the weighted quantity transported falls in k^{th} price break, otherwise zero. $v_{\tau t}$, explains that, if the weighted quantity transported waits in t^{th} period for τ number of days then the variable takes value 1 otherwise zero. J_t is the total number of truck loads in period t. y_t is the weighted quantity in excess of truckload capacity. u_t, reflects usage of modes, either TL & LTL mode (value is 1) or only TL mode (value is 0).

2.4 Fuzzy Optimization Model Formulation

Minimize

$$\tilde{C} = \sum_{t=1}^{T}\sum_{i=1}^{P}\tilde{HS}_{it}IS_{it} + \sum_{t=1}^{T}\left\{\left[\sum_{i=1}^{P}\left(\sum_{l=1}^{L}R_{ilt}d_{ilt}\right)\phi_{it}X_{it}\right] + \sum_{k=1}^{K}Z_{kt}f_{kt}C_tL_{1t} + \sum_{\tau=1}^{\Gamma}L_{1t}\left(O_{1t}+(\tau-1)*O_{2t}\right)v_{\tau t}\right\}$$

$$+\sum_{t=2}^{T}\left[\left(sy_t+j_t\beta_t\right)u_t+\left(j_t+1\right)\beta_t\left(1-u_t\right)\right]+\sum_{t=1}^{T}\sum_{i=1}^{P}\tilde{HD}_{it}ID_{it}+\sum_{t=2}^{T}\sum_{i=1}^{P}\lambda_{it}X_{it} \qquad (1)$$

Subject to $\qquad IS_{it} = IS_{it-1} + X_{it} - \tilde{D}_{it}\quad i=1,...,P;\ t=2,...,T \qquad\qquad (2)$

$$IS_{i1} = ISN_{i1} + X_{i1} - \tilde{D}_{i1},\ \ i=1,...,P \qquad\qquad (3)$$

$$\sum_{t=1}^{T}IS_{it} + \sum_{t=1}^{T}X_{it} \geq \sum_{t=1}^{T}\tilde{D}_{it}\ \ i=1,...,P \qquad\qquad (4)$$

$$ID_{it} = ID_{it-1} + \tilde{D}_{it} - CR_{it} - \eta ID_{it}\ \ i=1,...,P;\ t=2,...,T \qquad\qquad (5)$$

$$ID_{i1} = IDN_{i1} + \tilde{D}_{i1} - CR_{i1} - \eta ID_{i1}\ \ i=1,...,P \qquad\qquad (6)$$

$$(1-\eta)\sum_{t=1}^{T} ID_{it} + \sum_{t=1}^{T} \tilde{D}_{it} \geq \sum_{t=1}^{T} CR_{it} \quad i=1,...,P \tag{7}$$

$$X_{it} \geq \sum_{l=1}^{L} a_{ilt} R_{ilt} \quad i=1,...,P; \quad t=1,...,T \tag{8}$$

$$\sum_{l=1}^{L} R_{ilt} = 1 \quad i=1,...,P; \quad t=1,...,T \tag{9}$$

$$L_{1t} = \sum_{i=1}^{P} \left(w_i X_{it} \right) \quad t=1,...,T \tag{10}$$

$$L_{1t} \geq \sum_{k=1}^{K} b_{kt} Z_{kt} \quad t=1,...,T \tag{11}$$

$$\sum_{k=1}^{K} Z_{kt} = 1 \quad t=1,...,T \tag{12}$$

$$\sum_{\tau=1}^{\Gamma} v_{\tau t} = 1 \quad t=1,...,T \tag{13}$$

$$L_{1t} = L_{2\,t+1} \quad t=1,...,T \tag{14}$$

$$L_{2t} \leq \left(y_t + j_t \omega \right) u_t + \left(j_t + 1 \right) \omega \left(1 - u_t \right) \quad t=2,...,T+1 \tag{15}$$

$$L_{2t} = \left(y_t + j_t \omega \right) \quad t=2,...,T+1 \tag{16}$$

$$X_{it}, L_{1t}, L_{2t} \geq 0 \quad R_{ilt} = 0 \text{ or } 1, \quad Z_{kt} = 0 \text{ or } 1 \quad v_{\tau t} = 0 \text{ or } 1 \quad u_t = 0 \text{ or } 1$$

$IS_{it}, ID_{it}, y_t, j_t$ are int *ergers*

In the proposed mathematical model, Eq. (1) is the fuzzy objective function to minimize the sum of cost incurred in holding ending inventory at source, purchasing goods, transportation from source to intermediate stoppage, halting cost at intermediate stoppage, transportation cost from intermediate stoppage to destination, cost of holding at destination and finally inspection cost of the reached quantity at destination.

Constraints (2-7) are the balancing equations, which calculates ending inventory in period t. In eq. (2), ending inventory at source depends upon the inventory left in the last period; the quantity X_{it} ordered in period t and fuzzy demand $\tilde{D}_{it}$. Eq. (3) calculates inventory level at the end of the first period at source for all the products using the inventory level at the beginning of the planning horizon, and the net change at the end of period one. Eq. (4) takes care for shortages i.e. the sum of ending inventory and optimal order quantity is more than the demand of all the periods. Eq. (5) calculates inventory by reducing consumption and fraction of deteriorated inventory from the sum of ending inventory of previous period and fuzzy demand of period t. Eq. (6) calculates inventory by reducing consumption and fraction of deteriorated inventory from the sum of initial inventory and fuzzy demand of period 1. Eq. (7) again explains no shortages at destination i.e. left over inventory after deterioration and demand should be more than the consumption. Eq. (8) shows that the order quantity of all items in period t exceeds the quantity break threshold. Eq. (9) restricts the activation at exactly one level, either discount or no discount situation. The integrator for procurement and distribution is eq. (10), which calculates transported quantity according to product weight. Eq. (11 - 12) work as eq. (8 – 9) with ordered quantity replaced by weighted quantity. Eq. (13) finds out exact number of halting days at intermediate stoppage. Eq. (14) shows the total weighted quantity transported in stage 1 of period t is equal to the total weighted quantity transported in stage 2 of period $t + 1$. In eq. (15), the minimum weighted quantity transported is calculated and further Eq. (16) measures the overhead units from truckload capacity in weights.

2.5 Price Breaks

As discussed above, variable R_{ilt} specifies the fact that when the order size at period t is larger than a_{ilt} it results in discounted prices for the ordered products for which the price breaks are defined as:

Price breaks for ordering quantity are:

$$d_f = \begin{cases} d_{ilt} & a_{ilt} \le X_{it} \le a_{i(l+1)t} \\ d_{iLt} & X_{it} \ge a_{iLt} \end{cases}$$

$$i = 1,...,P; \ t = 1,...,T; \ l = 1,...,L$$

Freight breaks for transporting quantity are:

$$d_f = \begin{cases} f_{kt} & b_{kt} \le L_{1t} \le b_{(k+1)t} \\ f_{Kt} & L_{1t} \ge b_{Kt} \end{cases}$$

$$t = 1,...,T; \ k = 1,...,K$$

Where d_f is only the notation for discount factor and the minimum required quantity is to be transported.

3 Solution Algorithm

3.1 Fuzzy Solution Algorithm

In following algorithm [15] specifies the sequential steps to solve the fuzzy mathematical programming problems.

Step1.   Compute the crisp equivalent of the fuzzy parameters using a defuzzification function. Same defuzzification function is to be used for each of the parameters. Here we use the defuzzification function of the type where   are the triangular fuzzy numbers.

$$F_2(A) = \left(a^1 + 2a^2 + a^3 \right) \Big/ 4$$

Let $\bar{D}_{it}$ be the defuzzified value of $\tilde{D}_{it}$ and $(D_{it}^1, D_{it}^2, D_{it}^3)$ be triangular fuzzy numbers then, $\bar{D}_{it} = D_{it}^1 + 2D_{it}^2 + D_{it}^3 / 4, \ where \ i = 1,...,P; \ t = 1,...,T$

D_{it} and C_0 are defuzzified aspiration levels of model's demand and cost. Similarly, $\overline{HS}_{it}$ and $\overline{HD}_{it}$ are defuzzified aspired holding cost at source and destination.

Step2. Employ extension principle to identify the fuzzy decision, which results in a crisp mathematical programming problem and on substituting the values for

$\tilde{D}_{it}$ as $\overline{D}_{it}$, $\tilde{HS}_{it}$ as $\overline{HS}_{it}$, , $\tilde{HD}_{it}$ as $\overline{HD}_{it}$, the problem becomes given by

$$Min \ \overline{C} = \sum_{t=1}^{T}\sum_{i=1}^{P}\overline{HS}_{it}IS_{it} + \sum_{t=1}^{T}\left\{\left[\sum_{i=1}^{P}\left(\sum_{l=1}^{L}R_{ilt}d_{ilt}\right)\phi_{it}X_{it}\right] + \sum_{k=1}^{K}Z_{kt}f_{kt}C_tL_{1t} + \sum_{\tau=1}^{\Gamma}L_{1t}\left(O_{1t} + (\tau-1)*O_{2t}\right)v_{\tau t}\right\}$$

$$+ \sum_{t=2}^{T}\left[(sy_t + j_t\beta_t)u_t + (j_t+1)\beta_t(1-u_t)\right] + \sum_{t=1}^{T}\sum_{i=1}^{P}\overline{HD}_{it}ID_{it} + \sum_{t=2}^{T}\sum_{i=1}^{P}\lambda_{it}X_{it}$$

Subject to $X \in S = X\{ \quad IS_{it} = IS_{it-1} + X_{it} - \overline{D}_{it} \quad \forall i, t = 2, ..., T; \ IS_{i1} = ISN_{i1} + X_{i1} - \overline{D}_{i1} \quad \forall i$

$$ID_{it} = ID_{it-1} + \overline{D}_{it} - CR_{it} - \eta ID_{it} \quad \forall i, t ; \ ID_{i1} = IDN_{i1} + \overline{D}_{i1} - CR_{i1} - \eta ID_{i1} \quad \forall i$$

$$X_{it} \geq \sum_{l=1}^{L}a_{ilt}R_{ilt} \quad i = 1, ..., P \quad \forall t ; \ \sum_{l=1}^{L}R_{ilt} = 1 \quad \forall i, t$$

$$L_{1t} = \sum_{i=1}^{P}(w_i X_{it}) \quad \forall t ; \ L_{1t} \geq \sum_{k=1}^{K}b_{kt}Z_{kt} \quad \forall t$$

$$\sum_{k=1}^{K}Z_{kt} = 1 \quad \forall t ; \ \sum_{\tau=1}^{\Gamma}v_{\tau t} = 1 \quad \forall t$$

$$L_{1t} = L_{2t+1} \quad \forall t ; \ L_{2t} \leq (y_t + j_t\omega)u_t + (j_t+1)\omega(1-u_t) \quad t = 2, ..., T+1$$

$$L_{2t} = (y_t + j_t\omega) \quad t = 2, ..., T+1 \quad \}$$

$$\sum_{t=1}^{T}IS_{it} + \sum_{t=1}^{T}X_{it} \geq \sum_{t=1}^{T}\overline{D}_{it} \quad \forall i ; \ (1-\eta)\sum_{t=1}^{T}ID_{it} + \sum_{t=1}^{T}\overline{D}_{it} \geq \sum_{t=1}^{T}CR_{it} \quad \forall i$$

$$\tilde{C}(X) \leq C_0$$

$X_{it}, L_{1t}, L_{2t} \geq 0 \ R_{ilt} = 0 \ or \ 1, \ Z_{kt} = 0 \ or \ 1 \ v_{\tau t} = 0 \ or \ 1, \ u_t = 0 \ or \ 1, \ IS_{it}, ID_{it}, y_t, j_t \ are \ intergers \quad \theta \in [0,1]$

$i = 1, ..., P; \ l = 1, ..., L; \ t = 1, ..., T; \ k = 1, ..., K$

Step3. Define appropriate membership functions for each fuzzy inequalities as well as constraint corresponding to the objective function. The membership function for the fuzzy are given as

$$\mu_C(X) = \begin{cases} 1 & ; C(X) \leq C_0 \\ \dfrac{C_0^* - C(X)}{C_0^* - C_0} & ; C_0 \leq C(X) < C_0^* \\ 0 & ; C(X) > C_0^* \end{cases}$$

Where C_0 is the restriction and C_0^* is the tolerance levels to the fuzzy total cost.

$$\mu_{IS_{it}}(X) = \begin{cases} 1 & ; IS_{it}(X) \geq \overline{D}_0^* \\ \dfrac{IS_{it}(X) - \overline{D}_0}{\overline{D}_0^* - \overline{D}_0} & ; \overline{D}_0 \leq IS_{it}(X) < \overline{D}_0^* \\ 0 & ; IS_{it}(X) > \overline{D}_0 \end{cases} ; \quad \mu_{ID_{it}}(X) = \begin{cases} 1 & ; ID_{it}(X) \geq \overline{D}_0^* \\ \dfrac{ID_{it}(X) - \overline{D}_0}{\overline{D}_0^* - \overline{D}_0} & ; \overline{D}_0 \leq ID_{it}(X) < \overline{D}_0^* \\ 0 & ; ID_{it}(X) > \overline{D}_0 \end{cases}$$

Where $\overline{D}_0^* = \sum_{t=1}^{T}\overline{D}_{it}$ is the aspiration and $\overline{D}_0$ is the tolerance level to inventory constraints.

Step4. Employ extension principle to identify the fuzzy decision. While solving the problem its objective function is treated as constraint. Each constraint is considered to be an objective for the decision maker and the problem can be looked as crisp mathematical programming problem

Maximize θ Subject to $\mu_c(X) \leq \theta$, $\mu_{IS_{it}}(X) \geq \theta$, $\mu_{ID_{it}}(X) \geq \theta$, $X \in S$

can be solved by the standard crisp mathematical programming algorithms.

The above described model is coded into Lingo 13.0, but for arriving at solution the code takes more time because of complexities of the model. To reduce the complexity, differential evolution approach is applied, discussed in section 3.2.

3.2 Differential Evolution Algorithm

Differential evolution is an evolutionary algorithm, which is rapidly growing field of artificial intelligence. This class also includes genetic algorithms, evolutionary strategies and evolutionary programming. DE was proposed by Price & Storn [13]. Since then it has earned a reputation as a very powerful and effective global optimizer. The basic steps of DE are as follows

Start   *Step 1*: Generate an initial population of random individuals

Step 2: Create a new population by repeating following steps until the stopping criterion is achieved

　　[*Selection*] Select the random individuals for reproduction

　　[*Reproduction*] Create new individuals from selected ones by mutation and crossover

　　[*Evolution*] Compute the fitness values of the individuals

　　[*Advanced Population*] Select the new generation from target individual and trial individuals

　　End steps

Initialization: Suppose we want to optimize a function of D number of real parameters. We must select a population of size NP. NP parameter vectors have the form $X_{i,G}=(x_{1,i,G}, x_{2,i,G},....,x_{D,i,G})$, where, D is dimension, i is an individual index and G represents the number of generations. First, all the solution vectors in a population are randomly initialized. The initial solution vectors are generated between lower and upper bounds $l=\{l_1,l_2,...,l_D\}$ and $u=\{u_1,u_2,...,u_D\}$ using the equation $x_{j,i,0}=l_j+rand_{i,j}[0,1]*(u_j-l_j)$, Where, i is an individual index, j is component index and $rand_{i,j}[0,1]$ is a uniformly distributed random number lying between 0 and 1. This randomly generated population of vectors $X_{i,0} = (x_{1,i,0}, x_{2,i,0},...,x_{D,i,0})$ is known as target vectors.

Mutation: Each of the NP parameter vectors undergoes mutation, recombination and selection. Mutation expands the search space. For a given parameter vector $X_{i,G}$, three vectors $X_{r_1,G}, X_{r_2,G}, X_{r_3,G}$ are randomly selected such that the indices i, r_1, r_2, r_3 are distinct. The ith perturbed individual, $V_{i,G}$, is therefore generated based on the three chosen individuals as follows: $V_{i,G} = X_{r_1,G} + F*(X_{r_2,G} - X_{r_3,G})$

where, $r_1, r_2, r_3 \in \{1,2,...,NP\}$ are randomly selected, such that $r_1 \neq r_2 \neq r_3 \neq i$, $F \in (0, 1.2]$ and $V_{i,G}$ is called the mutation vector.

Crossover: The perturbed individual, $V_{i,G} = (v_{1,i,G}, v_{2,i,G},..., v_{D,i,G})$ and the current population member, $X_{i,G} = (x_{1,i,G}, x_{2,i,G},..., x_{D,i,G})$ are then subject to the crossover operation, that finally generates the population of candidates, or "trial" vectors, $U_{i,G} = (u_{1,i,G}, u_{2,i,G},...,u_{D,i,G})$, as follows

$$u_{j,i,G} = \begin{cases} v_{j,i,G} \text{ if } rand_{i,j}[0,1] \leq C_r \vee j = j_{rand} \\ x_{j,i,G} \qquad\qquad \text{otherwise} \end{cases}$$

Where, $C_r \in [0,1]$ is a crossover probability, $j_{rand} \in \{1,2,...,D\}$ is a random parameter's index, chosen once for each i.

Selection: The population for the next generation is selected from the individuals in current population and its corresponding trial vector according to the following rule $X_{i,G+1} = \begin{cases} U_{i,G} \text{ if } f(U_{i,G}) \geq f(X_{i,G}) \\ X_{i,G} \qquad \text{otherwise} \end{cases}$ where, $f(.)$ is the objective function value. Each individual of the temporary population is compared with its counterpart in the current population. Mutation, recombination and selection continue until stopping criterion is reached.

Constraint Handling in Differential Evolution: Pareto ranking method is used to handle constraints in DE. The value of constraints is calculated at target and trial vectors. The method is based on the following three rules

1) Between two feasible vectors (target and trial), the one with the best value of the objective function is preferred
2) If out of target and trial vectors, one vector is feasible and the other is infeasible, the one which is feasible is preferred
3) Between two infeasible vectors, the one with the lowest sum of constraint violation is preferred

Stopping Criterion: DE algorithm stops when either
1) Maximum number of generations are reached or
2) Desired accuracy is achieved i.e., $|f_{max} - f_{min}| \leq \varepsilon$

4 Case Study

It is clear that all foods deteriorate, regardless of the classification one may have seen for a particular food product. Some foods become unfit for human consumption more quickly than others. Foods are quite unlike hardware such as nuts and bolts. Yet, modern food markets with their processed and preserved foods tend to give their customers the impression of similar durability for many food products like packed dairy products as milk, cheese, curd, butter. There are several reasons for deterioration of dairy products and their Off-Flavours like Head-Induced Flavours (From Pasteurization, Refrigerator Storage, an Autoclave, High Temperature Processing), Light-Induced Flavour, Oxidized Flavour, Transmitted Flavours, etc.. Same is the problem of BAC Dairy Co. (named changed), who is running a production plants in Delhi area, accumulating milk and milk products from a dairy farmers, process them in packets and sell them off at their outlets. In the case, we are discussing a tiny problem of one dairy farmer (source pt.) and one outlet (destination) with an intermediate stoppage at railway station. Four products as milk (P1), cheese (P2), curd (P3), and butter (P4) are considered for managing procurement & distribution for 3 weeks of summer season. In the case the major objectives of the company is to manage optimal order quantity from farmer of deteriorating type products, as the demand is uncertain, reducing fuzziness in environment so that company will be able to precise demand and cost for future. Company needs to reduce cost of procurement, ending inventory carrying cost, and cost of transportation from dairy farmer to intermediate stoppage (railway station) during stage I, halting cost at intermediate node on the transported weights, and finally in stage II, cost of transportation from intermediate stoppage to dairy outlet, inspection cost of reached quantity at outlet, and ending inventory carrying cost at outlet.

To solve company's problem, a fuzzy optimization model is developed, which is converted into crisp form. Further DE is employed to find the optimal solution with the help of data provided by the company. The data provided by the company is as follows:

The purchase cost of each product for dairy co. from dairy farmer is as, for P1 cost is ₹45, ₹48, ₹50 per 850gms, P2 is ₹64, ₹62, ₹67 per 750gms, P3 is costing ₹53, ₹56, ₹54 per 650gms, and P4 is costing ₹59, ₹54, ₹57 per 550gms. Initial inventory at source in starting of the planning horizon are 70, 80, 40, and 59 packets of all the four products. Similarly initial inventory at destination are 90, 85, 79, and 83 packets. The inspection cost per unit is ₹1 for P1 & P3 and ₹1.5 for P2 & P4. Cost of weight transportation from source to intermediate stoppage is ₹2, ₹3, and ₹3.5 per kg. Halting cost during first halt day is ₹7, ₹8, and ₹9 and second day onwards is ₹2, ₹3, and ₹4 during all the three periods. While transporting weighted quantity from intermediate stoppage to destination, TL & LTL policies may be used. In such case, cost per full truck is ₹1000, ₹1050, and ₹1100, and capacity per truck is 250kg. In the case of LTL policy, per extra unit from full truck capacity is costing ₹6. Ending inventory deterioration fraction is 6% at destination.

Table 1. Holding Cost at Source (Fuzzy – F; Defuzzified - DFZ)

Product	F	DFZ	F	DFZ	F	DFZ
	Period I	Period I	Period II	Period II	Period III	Period III
P1	3, 2, 3	2.5	2.5, 3, 2.7	2.8	2.5, 2, 2.7	2.3
P2	2, 2.6, 2	2.5	2.6, 3.5, 2.4	3	2.3, 2.5, 2.6	2.5
P3	3, 3.4, 3.8	3.4	2.5, 3, 2.3	2.7	2.5, 2, 1.5	2
P4	3, 3.5, 2.8	3.2	2, 2.5, 1.8	2.2	2, 2.4, 1.6	2.1

Table 2. Demand (Fuzzy – F; Defuzzified - DFZ)

Product & Aspiration	F	DFZ	F	DFZ	F	DFZ
	Period I	Period I	Period II	Period II	Period III	Period III
P1, 500	172, 169, 170	170	140, 139, 142	140	189, 191, 189	190
P2, 511	138, 141, 136	139	185, 189, 189	188	183, 186, 181	184
P3, 538	178, 180, 178	179	187, 183, 187	185	175, 173, 175	174
P4, 525	175, 177, 175	176	180, 182, 184	182	165, 169, 165	167

Demand tolerance is always less than the aspired demand, which may vary.

Table 3. Holding Cost at Destination (Fuzzy – F; Defuzzified - DFZ)

Product	F	DFZ	F	DFZ	F	DFZ
	Period I	Period I	Period II	Period II	Period III	Period III
P1	3, 3.5, 2.8	3.2	3.3, 3, 3.1	3.1	2.5, 3.2, 3.1	3
P2	3, 2.6, 3	2.8	2.6, 3, 3	2.9	2.3, 2.9, 2.3	2.6
P3	3, 3.8, 3.4	3.5	3.4, 2.3, 3.6	2.9	2.3, 2.8, 2.1	2.5
P4	3.9, 3.5, 3.5	3.6	2.7, 2.5, 2.7	2.6	2, 2.4, 2.4	2.3

Table 4. Consumption at Destination

Product	Period 1	Period 2	Period 3
P1	169	138	185
P2	135	176	182
P3	167	184	169
P4	174	179	165

Table 5. Quantity Threshold (A) and Discount Factors (B) for P1 & P2 in All Periods

A for (P1)	B	A for (P2)	B
0 – 100	1	0 – 145	1
100 – 200	0.98	145 – 180	0.95
200 – 300	0.95	180 – 220	0.90
300 & above	0.92	220 & above	0.86

Table 6. Quantity Threshold (A) and Discount Factors (B) for P3 & P4 in All Periods

A for (P3)	B	A for (P4)	B
0 – 136	1	0 – 110	1
136 – 170	0.97	110 – 140	0.94
170 – 230	0.95	140 – 190	0.92
230 & above	0.90	190 & above	0.90

Table 7. Weight Threshold and Discount Factors for All Periods

Weight Threshold	Discount Factors
100 – 400	1
400 – 700	0.98
700 – 1000	0.95
1000 & above	0.90

The solution is obtained by employing data in the model. Parameters of DE are given in table 8. A desired accuracy of .001 between maximum and minimum values of fitness function was taken as terminating criteria of the process. The aspiration and tolerance level of total cost are ₹116352 and ₹151352. The total cost of the system is ₹131247.4. Due to page limitation, we are explaining the results for the first period and for remaining period please refer table no. 9, 10, 11, & 12. The optimum ordered quantity & discounts are 645 & 8%, 221 & 14%, 341 & 10%, and 117 & 6% for P1, P2, P3, & P4 resp. The transported quantity in stage I is

1000 with 10% discount. Weighted quantity halts at intermediate stoppage for 1 day. The transported quantity in stage II is 1000 and full TL mode is employed. In this case number of full trucks are 4 and overhead weighted quantity is 0. The Inventory at source is 545, 162, 202, and 0 for all the four products resp. And inventory at destination is 86, 84, 86, and 80. As far as reduction in vagueness is concerned, 57% of vagueness has been reduced from the fuzzy environment of demand, holding cost and total cost.

Table 8. Parameters of Differential Evolution

Parameter	Value	Parameter	Value
Population Size	500	Scaling Factor (F)	.6
Selection Method	Roulette Wheel	Crossover Probability(Cr)	.8

Table 9. Optimum Ordered Quantity/Discounts availed

Product	Period 1	Period 2	Period 3
P1	645/8%	45/0	0/0
P2	221/14%	220/14%	0/0
P3	341/10%	4/0	153/3%
P4	117/6%	353/10%	1/0

Table 10. Ending Inventory at Source//Destination

Product	Period 1	Period 2	Period 3
P1	545//86	450//83	260//83
P2	162//84	194//91	10//87
P3	202//86	21//82	0//82
P4	0//80	171//79	5//76

Table 11. Weighted Transported Quantity in Stage I & II/ Discount

Stage	Period 1	Period 2	Period 3
I	1000/10%	400/2%	100/0

Table 12. Halting days, No. of Trucks, Transportation Mode, Overhead Quantity

	Period 1	Period 2	Period 3
Halting days	1	1	1
No. of Trucks	4	1	0
Transportation Mode	TL	TL & LTL	TL & LTL
Overhead Quantity	0	150	100

5 Conclusion

Research on SCM has been reported for the procurement & distribution coordination with deteriorating natured products incorporating either quantity or freight discounts. A two stage optimization model with the objective of minimizing the total incurred cost and maximizing reduced vagueness in the fuzzy environment (due to fuzzy demand & fuzzy holding cost), during procurement, holding at suppliers, transportation from supply point to intermediate stoppage, halting cost at intermediate stoppage, transportation from intermediate stoppage to destination, inspection cost at destination and finally holding at destinations in a supply chain network with both quantity and freight discounts is not addressed so far. So, the current study proposed a two stage fuzzy mathematical model for the literature gap identified and mentioned in this study. The proposed model was validated by applying to the real case study data.

Acknowledgement

Kanika Gandhi (Lecturer, Quantitative Techniques & Operations) is thankful to her organization "Bharatiya Vidya Bhavan's Usha & Lakshmi Mittal Institute of Management (BULMIM)" to provide her opportunity for carrying research work.

References

[1]   Roy, A. Kar, S., Maiti, M.: A deteriorating multi-item inventory model with fuzzy costs and resources based on two different defuzzification techniques. Applied Mathematical Modelling. 32, 208-223 (2008).

[2]   Minner, S.: Multiple-supplier inventory models in SCM: A review. International Journal of Production Economics. 81/82, 265-279 (2003).

[3]   Ranjan, B., Susmita, B.: A review of the causes of bullwhip effect in a supply chain. The International Journal of Advanced Manufacturing Technology. 54, 1245-1261(2011).

[4]   Alamri, A. A., Balkhi, Z. T.: The effects of learning and forgetting on the optimal production lot size for deteriorating items with time varying demand and deterioraion rates. International Journal of Production Economics. 107, 125-138 (2007).

[5]   Hsu, P. H., Wee, H. M., Teng, H. M.: Preservation technology investment for deteriorating inventory. International Journal of Production Economics. 124, 388-394 (2010).

[6]   Misra, R.B.: Optimum production lot-size model for a system with deteriorating inventory. Int. J. Prod. Res. 13, 495–505 (1975).

[7]   Goyal, S.K., Gunasekaran, A.: An integrated production inventory marketing model for deteriorating item. Comput. Ind. Eng. 28, 755–762 (1995).

[8]   Benkherouf, L.A.: Deterministic order level inventory model for deteriorating items with two storage facilities. Int. J. Prod. Economics. 48, 167–175 (1997).

[9]   Giri, B.C., Chaudhuri, K. S.: Deterministic models of perishable inventory with stock dependent demand rate and non-linear holding cost. Euro. J. Oper. Res. 19, 1267–1274 (1998).

[10] Goyel, S.K., Giri, B.C.: Recent trends in modeling of deteriorating inventory. Euro. J. Oper. Res. 134, 1–16 (2001).

[11] Tu, H.H.J., Lo, M.C., & Yang, M.F.: A two-echelon inventory model for fuzzy demand with mutual beneficial pricing approach in a supply chain. African Journal of Business Management. 5(14), 5500-5508 (2011).

[12] Xu, R., Zhai, X.: Optimal models for single-period supply chain problems with fuzzy demand. Information Sciences. 178(17), 3374–3381 (2008).

[13] Price, K.V., Storn, R.M.: Differential Evolution-A simple and efficient adaptive scheme for global optimization over continuous space (Tech. Rep. No. TR-95-012). ICSI. Available via the Internet: ftp.icsi.berkeley.edu/pub/techreports/1995/tr-95-012.ps.Z. (1995).

[14] Price, K.V.: An introduction to Differential Evolution. In: Corne, D., Marco, D. & Glover, F. (Eds.), New Ideas in Optimization. London, UK: McGraw-Hill. 78-108 (1999).

[15] Zimmermann, H.J.: Description and optimization of fuzzy systems. International Journal of General Systems. 2, 209–215 (1976).

A New Meta-heuristic PSO Algorithm for Resource Constraint Project Scheduling Problem

Tuli Bakshi[1]*,

Arindam Sinharay[c],

Bijan Sarkar[b]

Subir K. Sanyal[b]

*Research Scholar, Jadavpur University, Kolkata,west Bengal,India

[b]Professor , Jadavpur University, Kolkata,west Bengal,India

[c]Assistant Professor , Future Institute of Engineering & Management,

Kolkata,west Bengal,India

Abstract: - In this paper a meta heuristic Particle Swarm Optimization (PSO)-based approach for the solution of the resource-constrained project scheduling problem with the purpose of minimizing project time has been developed. In order to evaluate the performance of the PSO based approach for the resource-constrained project scheduling problem, computational analyses are given. As per the results the application of PSO to project scheduling is achievable.

Keywords: Meta heuristic, Particle Swarm Optimization, Resource-constrained project scheduling, minimizing duration.

1   Introduction

Being a temporary attempt, a project requires to create an unique product, service or result [1]. Temporary emphasizes on definite deadline reaching of which a project may gain its objectives or has already lost its significance of existence. Therefore, a decision maker (DM) has to choose the appropriate alternative among many.

According to Hwang et al.[2], multi-criteria decision making (MCDM)is one of the most widely used decision making methodology.

Many real world projects scheduling problem termed as Resource-Constrained Multi-Project Scheduling Problem (RCMPSP) [3]. To schedule project activities to

---

* Corresponding author. Tel.: +91-943292656583;
E-mail address: tuli.bakshi@gmail.com

J. C. Bansal et al. (eds.), *Proceedings of Seventh International Conference on Bio-Inspired Computing: Theories and Applications (BIC-TA 2012),* Advances in Intelligent Systems and Computing 202, DOI: 10.1007/978-81-322-1041-2_33, © Springer India 2013

complete multiple projects in the minimum possible time under presence of resource constraints is the general objective of this type of problems. Primarily, mathematical models such as linear programming & dynamic programming have been used to solve & to obtain optimal solution.

It was earlier shown that the scheduling problem subject to resource constraints is NP-Hard [4], refraining exact methods time consuming and inefficient for solving large & real world application type problem.

Hence, meta-heuristic algorithm to generate near optimal solution for large problems has drawn special interest. There are many genetic algorithm (GA) applied for RCMPSP [5, 6, 7, 8] with project duration minimization as objective.

Most recently, the particle swarm optimization (PSO) algorithm has been taken into consideration for solving resource constrained project scheduling problem (RCPSP) and multi-criteria resource constrained project scheduling problem (MRCPSP). PSO, developed by Kennedy and Elbe hart [9] is an evolutionary algorithm which simulates the social behaviour of bird flocking to desired place. It is initialized with a population of random solutions. Each individual particle is tagged with a randomized velocity according to its own and its companions flying experiences. In the PSO, the solution is represented as an optimal solution for the RCMPSP.

Compared with GA, PSO has the following advantages:

- It has memory that can be retained by all particles in reference to the knowledge of good solution.

- It has constructive cooperation, between particles, particle in the swarm share information between them.

- It is easy to implement and quickly converges because of its simplicity.

The current researchers have proposed PSO-based approach to resolve the resource-constrained project scheduling problem. It is an evolutionary algorithm. The results are analyzed and described.

2   Particle Swarm Optimization (PSO)

It is a computational intelligence based optimization technique such as genetic algorithm (GA). It is a population based stochastic optimization technique developed by Kennedy and Eberhart in 1995 [10-13] and inspired by the social behaviour of bird flocking in a group looking for food and fish schooling.

Some terms related to PSO

The term PARTICLE refers to a member of population which is mass less and volume less m dimensional quantity. It can fly from one position to other in m dimensional search space with a velocity. POPULATION constitutes a number of such particles. The number of iteration for the solution of the problem is same as the number of generations in GA. The fitness function in PSO is same as the objective function for an optimization problem.

In real number space, each individual possible solution can be represented as a particle that moves through the problem space. The position of each particle is determined by the vector xi and its movement by the velocity of the particle vi represented in (1) and (2) respectively.

$$X_i^{k+1} = X_i^k + V_i^{k+1} \tag{1}$$

The information available for each individual is based on its own experience (the decisions it has made so far, stored in memory) the knowledge of performance of other individuals in its neighbourhood.

Since the relative importance of these two information can vary from one decision to other, a random weight is applied to each part and the velocity is determined as in (2)

$$V_i^{k+1} = V_i^k + c1.rand_1. (p_{best\,i}^{\;k} - X_i^k) + c_2.rand_2. (g_{best}^{\;k} - X_i^k) \tag{2}$$

Where,

- X_i^k = Position vector of a particle $i = [X_{i1}^k, X_{i2}^k, \ldots \ldots X_{im}^k]$ at k^{th} iteration

- V_i^K = Velocity vector of a particle $i = [V_{i1}^K, V_{i2}^K, \ldots \ldots V_{im}^K]$ at k^{th} iteration;

- k = iteration count

- $p_{best\,i}^{\;k}$ = i th particle has a memory of the best position in the search space at k^{th} iteration .

It is computed as $p_{best\,i}^{\;k+1} = X_i^{k+1}$ if the fitness function of ith at k+1 is less then (for minimum) the fitness function at k^{th} iteration otherwise $p_{best\,i}^{\;k+1} = p_{best\,i}^{\;k}$.$g_{best}^{\;k}$ = It is that particle which has the minimum value of fitness function (for minimization) among all the particles in k^{th} iteration.

- c_1 & c_2 = positive acceleration coefficients more then 1.0.

- Normally its value is taken

- $c_1 + c_2 = 4$ or $c_1 = c_2 = 2$.

- rand1 & rand2 are random numbers between 0.0 & 1.0.

  Both the velocity and positions have same units in this case.
  The velocity update equation (2) has three components [14]

1. The first component is referred to "Inertia" or "Momentum". It represents the tendency of the particle to continue in the same direction it has been travelling. This component can be scaled by a constant or dynamically in the case of modified PSO.

2. The second component represents local attraction towards the best position of a given particle (whose corresponding fitness value is called the particles best (p_{best})

scaled by a random weight factor c_1.rand1. This component is referred as "Memory" or "Self knowledge".

3. The third component represents attraction towards the position of any particle (whose corresponding fitness value is called global best ($g_{best)}$, scaled by another random weight c_2.rand2. This component is referred to "cooperation" ,"social knowledge", "group knowledge" or "shared information".

The PSO method is explained as above. The implementation of the algorithm is indicated below:

Initialize the swarm by assigning a random position to each particle in the problem space as evenly as possible.

Evaluate the fitness function of each particle.

For each individual particle, compare the particle's fitness value with its p_{best}. If the current value is better than the p_{best} value , then set this value as the p_{best} and the current particle's position X_i as p_{best} i.

Identify the particle that has the best fitness value and corresponding position of the particle as g_{best}.

Update the velocity and positions of all the particles using equations (1) & (2).

Repeat steps i) to v) until a stopping criterion is met (e.g. maximum number of iterations or a sufficient good fitness value).

On implementation of PSO following considerations must be taken into account to facilitate the convergence and prevent an "explosion" (failure) of the swarm resulting in the variants of PSO.

2.1   Selection of Maximum velocity:

At each iteration step, the algorithm proceeds by adjusting the distance (velocity) that each particle moves in every dimension of problem space. The velocity of a particle is a stochastic variable and it may create an uncontrolled trajectory leading to "explosion". In order to damp these oscillations upper and lower limits of the velocity V_i is defined as

if $V_{id} > V_{max}$ then $V_{id} = V_{max}$

else if $V_{id} < -V_{max}$ then $V_{id} = -V_{max}$

Most of the time, the value V_{max} is selected empirically depending on the characteristic of the problem. It is important it note that if the value of this parameter is too high, then the particle may move erratically, going beyond a good solution , on the other hand, if V_{max} is too small, then the particle movement is limited and it may not reach to optimal solution. The dynamically changing V_{max} can improve the performance given by

$V_{max} = (X_{max} - X_{min})/N$

Where X_{max} and X_{min} are maximum and minimum values of the found so far and N is the number of intervals.

2.2 Selection of Acceleration Constants:

c_1 & c_2 are the acceleration constants; they control the movement of each particle towards its individual and global best positions. Small values limit the movement of the particles, while larger values may cause the particle to diverge. Normally the constants $c_1 + c_2$ limited to 4. If it is taken more than 4 the trajectory may diverge leading to "Explosion". In general a good start is when $c_1 = c_2 = 2$.

2.3 Selection of Constriction Factor or Inertia Constant

Experimental study performed on PSO shows that even the maximum velocity and acceleration constants are correctly chosen, the particles trajectory may diverge leading to infinity, a phenomenon known as "Explosion" of the swarm. Two methods are to control this explosion (a) Inertia control and (b) Constriction factor control, the two variants of PSO.

Inertial Constant.
The velocity improvement represented by equation (2) is modified [15-17] and written as

$Vik+1 = W.Vik + c1.rand1. (pbest ik - Xik) + c2.rand2.(gbestk - Xik)$ (3)

The first right hand side part (velocity of previous iteration) of equation (3) multiplied by a factor W is known as "Inertia Constant". It can be fixed or dynamically changing. It controls the "Explosion" of search space. Initially it is taken as high value (0.9) which finds the global neighborhood fast. Once it is found that it is decreasing gradually to o.4 in order to find narrow search as shown in equation (4)

$W = wmax - (wmax - wmin)*itr/itrmax$ (4)

Where wmax = 0.9, wmin = 0.4,

itrmax= maximum iterations,

itr = current iteration

Since the weighting factor W is changing iteration wise it may be called as Dynamic PSO.

Constriction Factor.
This is another method of control of "Explosion" of the swarm. The velocity in equation (2) is redefined using constriction factor developed by Clark and Kennedy [17], is represented in equation (5) as

$Vik+1 = K*(Vik + c1.rand1.(pbest ik - Xi) + c2.rand2.(gbestk - Xi))$ (5)

Where K is known as constriction factor

$K = 2/(abs(2 - c - sqrt(c^2 - 4*c))$ (6)

Where, c = c1 + c2 > 4.0

Typically when this method is used, c is set to 4.1 and value of K comes out to be 0.729, In general, the constriction factor improves the convergence of the particle by damping oscillations. The main disadvantage is that the particles may follow wider cycles when pbest I is far from gbest (two different regions).A survey is given in reference [18].The present problem is discussed in next section.

3   Problem formulation

In this formulation, the use of a dummy start as well as a dummy end activity is assumed. The problem formulation is as follows:

$$\min \quad f_n \qquad (1)$$

$$s/t.$$

$$f_i \le f_i - d_j, \qquad \forall (i, j) \in A \qquad (2)$$

$$f_1 = 0 \qquad (3)$$

$$\sum_{i \in S_t} r_{it} \le a_k, \qquad \forall k = 1 \ldots m$$

$$\& \quad t = 1 \ldots fn \qquad (4)$$

The decision variable f_i denote the finish times of the different activities, while the d_i denote the duration of each activity, a_k the availability of the kth resource type and r_{ik} the resource requirement of activity i for resource type k. The set S_t that is used in equation (4) denotes the set of activities that are in progress at time t.

The objective function (1) minimises the finish time of the dummy end activity .Equation (2) expresses the precedence relations, while equation (3) forces the dummy start activity to finish at time 0. Finally equation (4) expresses that at no time instant during the project horizon the resource availability may be violated.

4   Case Study

Let us consider the project network with resources and duration as:

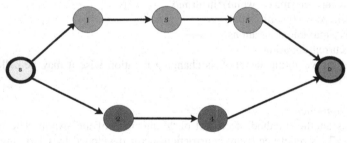

Fig. 1.

So here we consider five activities **1, 3,5,2,4**. The duration of activity **[2, 4, 3, 2, 1]**. The resources of activity are **[2, 3, 4, 1, 2]**. The maximum resource level is 4.

5   Algorithm Of Case Study

5.1   Algorithm 1: To arrange the input elements

Input: elements in B[n][m] form.

Output: arranged elements in $B_{out}[n][m]$ for where m= 1 ...m (column);
n = 1 ...n (row)

```
for row 1 to n
    for column 1 to m
        C (i,j) := B (i,j)
    end
end

for column 1 to m
    for row 1 to n
        if B (i,j) < B (i, j+1)
                then swap (i,j)
        else
                do nothing
        end
end

for  row 1 to n
    for k runs from  1 to m
        for J runs from 1 to m
            if C (i,k) = = B(i,j)
                then sequence ( i,j) = k
            else
                do nothing
end
            end
        end
end

for  row 1 to n
    print sequence (i,j)
end.
```

5.2   Algorithm 2: Time minimization under limited resources

Input: activity sequence matrix S[i][j]
        job duration matrix Du[i][j]
        activity resource matrix Res [i][j]
Output: minimum finish time

```
for  i 1 to n
    for  j 1 to m
        start_time (i,j) = 0
```

```
                        finish_time (i,j) = 0
                        resorce_allocated (i,j) =0
                        resource_available (i,j) =0
            end
    end

    for  i 1 to n
            K1:= 1
            K2:=2
    end

    for k 1 to m
            for  j 1 to n
                    if sequence (i,j) = = K1
                    then finish_time (i,k) = start_time (i,k) + duration_time (i,k)
                                update resource allocation
                                update resource available
                    elseif (k+1) <= m
                                then start_time (i, k+1) = finish_time (i,k)
                                        increment K1 by 2
                                        print total finish_time
                    else if sequence (i,j) = = K2
                                then finish_time (i,k) = start_time (i,k) + duration (i,k)
                                        update resource allocation
                                        update resource available
                                        if (k + 1) <= m
                                        start_time (i, k+1) = finish_time(i, k+1)
                                        increment K2 by 2
                                        print total finish_time
            end
    end.
```

6   Result Set and Graphical Representation of Case Study

Here we are taking the following as input and getting the final duration:

$n =$ 4; $m =$ 5

Sequence:-

$$\begin{bmatrix} 5 & 4 & 2 & 3 & 1 \\ 4 & 1 & 2 & 5 & 3 \\ 3 & 1 & 2 & 4 & 5 \\ 1 & 5 & 2 & 3 & 4 \end{bmatrix}$$

Descending order:-

$$\begin{bmatrix} 9 & 8 & 5 & 4 & 1 \\ 8 & 7 & 6 & 4 & 2 \\ 9 & 6 & 3 & 2 & 1 \\ 8 & 7 & 5 & 3 & 1 \end{bmatrix}$$

Particle No. 1

Activity= 2 ,Start time = 0,Finish time = 4 ,resource = 3 ,Resource available = 1

Activity= 4 ,Start time = 4,Finish time = 6 ,resource = 1 ,Resource available = 3

Activity= 1 ,Start time = 6 ,Finish time = 8 ,resource = 2 ,resource available = 2
Activity= 3 ,Start time = 8 ,Finish time = 11 ,resource = 4 ,resource available = 0
Activity= 5 ,Start time = 11 ,Finish time = 12 ,resource = 2 ,resource available = 2

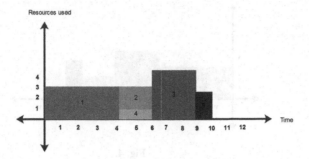

Fig. 2.

Particle No. 2

Activity= 1 ,Start time = 0 ,Finish time = 2 ,resource = 2 ,resource available = 2
Activity= 2 ,Start time = 2,Finish time = 6 ,resource = 3 ,Resource available = 1
Activity= 4 ,Start time = 6,Finish time = 8 ,resource = 1 ,Resource available = 3
Activity= 3 ,Start time = 8 ,Finish time = 11 ,resource = 4 ,resource available = 0
Activity= 5 ,Start time = 11 ,Finish time = 12 ,resource = 2 ,resource available = 2

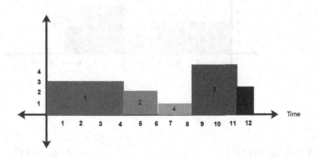

Fig. 3.

Particle No. 3

Activity= 1 ,Start time = 0 ,Finish time = 2 ,resource = 2 ,resource available = 2
Activity= 3 ,Start time = 2 ,Finish time = 5 ,resource = 4 ,resource available = 0
Activity= 2 ,Start time = 5,Finish time = 9 ,resource = 3 ,Resource available = 1
Activity= 4 ,Start time = 9,Finish time = 11 ,resource = 1 ,Resource available = 3
Activity= 5 ,Start time = 11 ,Finish time = 12 ,resource = 2 ,resource available = 2

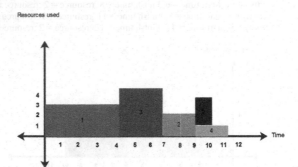

Fig. 4.

Particle No. 4

Activity= 1 ,Start time = 0 ,Finish time = 2 ,resource = 2 ,resource available = 2
Activity= 2 ,Start time = 2,Finish time = 6 ,resource = 3 ,Resource available = 1
Activity= 3 ,Start time = 6 ,Finish time = 9 ,resource = 4 ,resource available = 0
Activity= 5 ,Start time = 9 ,Finish time = 10 ,resource = 2 ,resource available = 2
Activity= 4 ,Start time = 10,Finish time = 12 ,resource = 1 ,Resource available = 3

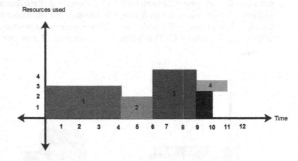

Fig. 5.

F =

$$\begin{bmatrix} 4 & 6 & 8 & 11 & 12 \\ 2 & 6 & 8 & 11 & 12 \\ 2 & 5 & 9 & 11 & 12 \\ 2 & 6 & 9 & 10 & 12 \end{bmatrix}$$

St =

$$\begin{bmatrix} 0 & 4 & 6 & 8 & 11 \\ 0 & 2 & 6 & 8 & 11 \\ 0 & 2 & 5 & 9 & 11 \\ 0 & 2 & 6 & 9 & 10 \end{bmatrix}$$

Final minimum duration = 10

7 Conclusion

This paper presented a new approach to minimize the project duration. The conventional method such as CPM , PERT and others are mainly used for minimizing the duration of project to solve unconstrained project scheduling problem. But it is difficult to use these methods for solving more general scheduling problems. In this paper the current researchers describe the application of PSO which is possible in case of many general problems related to project scheduling without much obstacles. PSO is meta-heuristic approach. So it can be concluded that this meta-heuristic approaches are successful to solve project scheduling problems and the inclusion of much problem specific knowledge is needed for the heuristic. Future research may be included to the development of meta-heuristic algorithms for the RCMPSP and their comparative study with the PSO approach.

8 References

1. Bakshi T.,Sarkar B., MCA Based Performance Evaluation of project selection, International Journal of software engineering & Applications (IJSEA), Vol.2, No2,2011,pp-14-22.
2. C.L. Hwang & K.P.Yoon, Multiple Attribute Decision Making and Introduction, London, Sage publication,1995,pp2.
3. Deng Lin-yi, Wang Yun –long, Lin Yan, A Particle Swarm Optimization Based on Priority Rule for Resource-Constrained Multi-Project Scheduling Problem,978-1-4244-1734-6/08/$25.00@2008 IEEE,pp-1038-1041.
4. M. R. Garey, D. S. Johnson, Computers and intractability: A guide to the theory of NP-completeness, New York, 1979.
5. W. H. Ip, Y Li, K. F. Man, K.S. Tang, Multi-product planning and scheduling using genetic algorithm approach, Computer & Industrial Engineering, Vol.38, No.2, 283-296, 2000.
6. P. Pongcharoen, C Hicks, P M Braiden, The development of genetic algorithm for the capacity scheduling of complex product, with multiple levels of product structure, European Journal of Operational Research, Vol.152, No.l, 215-225, 2004.
7. F. S. C. Lam, B. C. Lin, C. Sriskandarajah, H.Yan, Scheduling to minimize project design time using a genetic algorithm, International Journal of Production Research, Vol.37, No.6, 1369-1386, 1999.
8. M. Zhuang, A. Yassine, Task scheduling of parallel development projects using genetic algorithms, American Society of Mechanical Engineers Design Automation Conference. Salt Lake City, 1-11, 2004.
9. Kennedy J, Eberhart R C, A discrete Binary Version of the Particle Swarm Algorithm,In Proc.1997 Conf. On System, Man and Cybernetics Piscataway, NJ:IEEE Service Center, 1997,4104-4109.
10. Y.Shi and R.C. Eberhart," Particle Swarm Optimization: Developments, Applications And Resources", Proceedings of the 2001 Congress on Evolutionary Computation,Vol. 1, pp. 81-86, 2001.
11. R. C. Eberhart and Y. Shi, "Comparing Inertia Weights and Constriction Factor in Particle Swarm Optimization ", Proceedings of the 2000 Congress on Evolutionary Computation, Vol. 1, pp. 84-88, 2000.

12. J. Kennedy and R. Eberhart, "Particle Swarm Optimization ", Proc. Int. Conf. Neural Networks (ICNN), Nov. 1995, Vol. 4, pp. 1942-1948.
13. R, Eberhart and J. Kennedy, "A New Optimizer Using Particle Swarm Theory", Proc. 6th Int. Symp. Micro Machine and Human Science (MHS), Oct. 1995, pp.39-43.
14. D. Boiringer and D. Werner, "Particle Swarm Optimization versus Genetic Algorithms for Phase Array Synthesis", IEEE Trans. Antennas Propagat. Vol. 52, No. 3, pp. 771-779, Mar. 2004.
15. Y. Shi and R. Eberhart, "A Modified Particle Swarm Optimization", Proc. IEEE World Cong. Comput. Intell., May 1998, pp. 69-73.
16. Y. Shi and R. Eberhart, "Empirical Study of Particle Swarm Optimization", Proc. IEEE Cong. Evol. Comput. July 1999, Vol. 3, pp. 1945-1950.
17. M. Clerc and J. Kennedy, "The Particle Swarm Explosion, Stability and Convergence in a multidimensional Complex Space", IEEE Trans. Evol. Comput. Vol. 6, No. 1, pp. 58-73, Feb. 2002.
18. Yamille del Valle et.al. "Particle Swarm Optimization : Basic Concepts, Variants And Applications in Power Systems", IEEE Trans. on Evolutionary Computation,Vol. 12, No. 2, April 2008.

Single product multi period network design for reverse logistics and remanufacturing using new and old components

Kiran Garg[1]   P C Jha[2]

[1,2]Department of Operational Research, University of Delhi, Delhi
{mittalkiran12@gmail.com; jhapc@yahoo.com}

Abstract. Reverse logistics has received considerable attention due to potentials of value recovery from the used products. Reverse Logistics network contains inputs, processes, and outputs. Inputs refer to used products and recycled materials. Used parts or new parts go through Reverse Logistics processes. Outcomes are remanufactured products, recycled materials and spare parts. In this paper, a mathematical model for the design of a RL network for multi period planning horizon is proposed. It is assumed that returned quantity of a product is collected at collection centers before they are sent to reprocessing centers for inspection and dismantling. Dismantled components are either sent for remanufacturing or to the secondary market as spare parts. Recycling and disposal of these components are also considered in the model. For future modifications in the network structure, we consider multi-period setting. We propose a single product formulation and use a reverse bill of materials. The use of the model is shown through its application in a numerical illustration.

Keywords: Reverse Logistics, Remanufactured Products, Recycle Materials, Reverse Bill Of Material.

1 Introduction

Logistic network design has an important and planned area in an efficient and effective supply chain management. Implementation of government legislation, environmental concern, social responsibility and customer awareness, has forced companies not only to supply environmentally friendly products but also to be responsible for the returned products. The rise of green concerns makes reverse logistics (RL) a time demanding and relevant area of interest. The Original equipment manufacturers (OEM's) are introducing new products in an effort to sustain/increase their market share; hence they are forced to take back their used, end of- lease or end-of-life products through a network for reuse, remanufacture, recycle and disposed of. Hence OEMs have turned to a better design of their products for maximum reuse and recycling so that maximum value can be achieved from their used products. A reverse logistics system comprises a series of activities such as collection, cleaning, disassembly, test and sorting, storage, transport, and recovery operations like reuse, repair, refurbishing, remanufacturing, cannibalization and recycling.

J. C. Bansal et al. (eds.), *Proceedings of Seventh International Conference on Bio-Inspired Computing: Theories and Applications (BIC-TA 2012),* Advances in Intelligent Systems and Computing 202, DOI: 10.1007/978-81-322-1041-2_34, © Springer India 2013

The focus on RL is on waste management, material recovery, parts recovery or product recovery and the cost of recovered products can be reduced by optimal locations and allocations of facilities in RL. It prevents pollution by reducing the environmental burden of End-of-Life (EOL) at its source. Hence, an increasing number of companies now take into account reverse flows, going backwards from customers to recovery centers, within their logistics systems. Remanufacturing is recognized as a main option of recovery in terms of its feasibility and benefits.

2 Literature review

The fundamental studies on reverse logistics network design are driven by an application-oriented approach. Researchers have calculated that for recycling of the returned products, logistics costs account for a large share of the total costs [1], [2]. Transportation of used products is the most challenging issue in RL [3], [4] as smaller return quantities and variability in product types increase the transportation costs [5], [6]. [7] Emphasize on the need for collection centers in a reverse production system to help in maximizing collection of returned products. [8] developed truck sizing models for collection of wastes and transporting them to recovery centers. [9] Review on various quantitative models for RL networks. The location of collection points in a RL system has been examined by [10]. [11] proposed a MILP model by considering the reverse flow of goods.[12] proposed a product-recovery strategy depending on who collects the used products namely the manufacturer; the retailer; or a designated third party. [13] presented a multi-objective and multi- period MILP model for RL network design for modularized products. The authors have not considered the use of new components in remanufactured products.

Majority of the papers focus on recycling-only networks, such as [14] on battery recycling, [15] on tire recycling and [16] on paper recycling. Notable exceptions with a remanufacturing focus are [17] on copiers. [18] proposed a multi-period MILP model for carpet recycling. Their model analyses a set of alternative scenarios identified by the decision maker and provides a near optimal solution for network design. [19] proposed a MILP model to determine the optimal collection and recycling system for end-of-life computers and home appliances. [20] developed a model for the treatment of electrical and electronic wastes in Germany. [21] studied the implications of modular product architecture on RL strategies. Although the models proposed above are realistic representations of the network design problem regarding the specific application, but cannot be generalize to other industries. So, a more solid modeling framework for reverse logistics network design is given by [22] and describe that transportation and other logistics costs are not important factors in designing the RL network rather cost of reprocessing, remanufacturing and the cost of new components are deriving factors in designing of RL network.

Proposed model use a reverse bill of materials (BOM) to fit in component commonality in the product to handle return product for reuse. By using reverse BOM, the model also addresses the possibility of sending certain components to recycling/disposal and the possibility of purchasing new components for remanufacturing. Our modeling framework is applicable when the OEM has fairly reliable estimates of the amount of returns to be collected during the planning horizon as well as the demand at the secondary market for remanufactured products. To consider the possibility of making future adjustments in the network configuration we present a multi-period model of the reverse network design problem.

The remainder of the paper is structured as follows. In Section 3 new model is proposed for multi-period reverse logistics network design. We also state the underlying assumptions and highlight the flexibility of our model in representing a wide variety of possible applications. In section 4, used methodology of differential evolution is described. In Section 5, we present a numerical implementation in order to highlight the features of the proposed model. The paper ends with our concluding remarks.

3 Model Description

Model presented here focused on modular product structure of single product in a RL network and on many features of practical relevance namely, a multi-period setting, reverse BOM, minimum throughput at the facilities, variable operational costs, and finite demands in the secondary market. In multi-period setting all network design decisions are taken over a planning horizon which implemented in the beginning or end of periods. Model considers that used products are collected at collection centers (CC) and send to reprocessing centers (RPC) for inspection and dismantling then inspected components are shipped to spare market, remanufacturing plant (RMP), recycling center (RC) or disposal site (DS) accordingly.

Model considers modular product structure with specifying disposal and recycling fractions. RL network discussed here supplies quality used components for remanufacturing and to the spare markets too. Mismatch of components for remanufacturing is assumed to be tackled by purchasing through pre-qualified suppliers. Larger price and quality differentials between the new and remanufactured product create demand for the remanufactured product. However this factor is not considered explicitly in the model. It is assumed that the spare parts, if any, can fetch a higher unit value compared to the remanufactured products. Also, if the numbers of components are in excess of demand, they are stored in the remanufacturing point till next period. Design of such a network is strategic as it involves a decision on the number of facilities, their locations and allocation of corresponding flow of used products and components at an optimal cost for a given market demand in the network flows. Demands for remanufactured products, spare markets are assumed to be deterministic. Network used for the analysis involves eight echelons: CC, RPC, RMP, RC, DS, spare parts markets, secondary markets (for remanufactured products) and pre-selected new component suppliers. Assumptions are as follows:

1   An infinite source of used product and used products are collected at pre-specify collection centers. Goods collected in each CC transported to the reprocessing centers as soon as possible so that they do not incur any holding costs.

2   Dismantling operations are carried out in the RPC, where the components are disassembled, leaned, tested and sorted for reuse, remanufacture, spare and recycle. As a preference, spare market demands are met due to high value that it fetches from selling spare parts.

3   All the returned products are not suitable for remanufacturing. Therefore some new components may be required for remanufacturing. The final assembly of the product with the used and new components, if any is done inside the RMP.

RMP has inventory carrying cost for used components while it operates on Just-In-Time delivery of new components.

4 CC's, RPC's and RMP's are considered to have a monthly fixed cost. Transport cost is calculated with respect to the distance and overhead costs assuming full truck loads. Cost of new components ordered from pre-selected suppliers includes transportation cost also.

5 Shortages at secondary market are assumed to occur with no loss.

6 If the numbers of components are in excess of demand, then they are either recycled or stored in the RPC till further demand is received.

Notations:

Sets

CC set of collection centers indexed by 'cc'

J set of reprocessing centers (RPC) indexed by 'j'

S set of spare markets indexed by 's'

U set of remanufacturing plants (RMP) indexed by 'u'

X set of recycling plants (RP) indexed by 'x'

V set of disposal centers (DC) indexed by 'v'

Z set of new suppliers indexed by 'z'

H set of secondary markets (SM) indexed by 'h'

C set of components indexed by 'c'

T set of time periods indexed by 't'

Parameters

α No of components in one unit of product

ρ Recycling fraction

σ Disposal fraction

Q_{cc}^t Quantity of used product returned at collection center cc in period $t \in T$

D_{sc}^t Demand of component $c \in C$ in spare market $s \in S$ in period $t \in T$

D_h^t Demand of secondary market $h \in H$ in period $t \in T$

cap_j In bound capacity of RPC $j \in J$

$rcap_j$ Components reprocessing capacity of RPC $j \in J$

$pcap_u$ Production capacity of RMP $u \in U$

$icap_u$ Inventory holding capacity of RMP $u \in U$

$reccap_x$ Recycling capacity of RC $x \in X$

$discap_v$ Capacity of disposal center $v \in V$

$MTCC_{cc}$ Minimum throughput required for collection center $cc \in CC$

f_i^t Set-up cost of facility $i \in CC \cup J \cup U$ in the beginning of period $t \in T$

cdm_j^t Per unit product processing cost in RPC $j \in J$ in period $t \in T$

cr_{jc}^t Per unit component $c \in C$ reprocessing cost in RPC $j \in J$ in $t \in T$

cdp_{vc}^t Unit disposal cost for $c \in C$ in $v \in V$ in period $t \in T$

$casm_u^t$ Assembly cost /product for RMP plant $u \in U$ in period $t \in T$

c_{zuc}^t Unit cost of component c from supplier $z \in Z$ to RMP $u \in U$ in period t

IC_{uc}^t Unit inventory holding cost for $c \in C$ in RMP $u \in U$ in period $t \in T$

Tp_{ccj}^t Unit transportation cost of product from $cc \in CC$ to RPC $j \in J$ in $t \in T$

TS_{js}^t Unit transportation cost of $c \in C$ from RPC $j \in J$ to $s \in S$ in $t \in T$

T_{ju}^t Unit transportation cost of $c \in C$ from RPC $j \in J$ to RMP $u \in U$ in $t \in T$

T_{jx}^t Unit transportation cost of $c \in C$ from RPC $j \in J$ to RP $x \in X$ in $t \in T$

T_{jv}^t Unit transportation cost of $c \in C$ from RPC $j \in J$ to DC $v \in V$ in $t \in T$

Tf_{uh}^t Unit transportation cost of product from $u \in U$ to $h \in H$ in period $t \in T$

Decision Variables

xp_{ccj}^t Amount of used product shipped from $cc \in CC$ to RPC $j \in J$ in $t \in T$

x_{mnc}^t Amount of $c \in C$ shipped from node m to node n in the network in $t \in T$

xfp_{uh}^t Amount of product shipped from RMP $u \in U$ to SM $h \in H$ in $t \in T$

EI_{uc}^t Amount of $c \in C$ hold in inventory in RMP $j \in J$ in the end of $t \in T$

$$v_{cc}^t = \begin{cases} 1, & \text{if collection center } cc \in CC \text{ is operating in period } t \in T \\ 0, & \text{oterwise} \end{cases}$$

$$y_j^t = \begin{cases} 1, & \text{if RPC } j \in J \text{ is operating in period } t \in T \\ 0, & \text{oterwise} \end{cases}$$

$$z_j^t = \begin{cases} 1, & \text{if RMP } u \in U \text{ is operating in period } t \in T \\ 0, & \text{oterwise} \end{cases}$$

MATHEMATICAL MODEL:
Minimize:

$$\sum_t (\sum_{cc} f_{cc}^t (v_{cc}^t - v_{cc}^{t-1}) + \sum_j f_j^t (y_j^t - y_j^{t-1}) + \sum_u f_u^t (z_u^t - z_u^{t-1})) + \sum_t \sum_j (\sum_{cc} xp_{ccj}^t) cdm_j + \sum_t \sum_c \sum_s (\sum_u x_{jsc}^t + \sum_u x_{juc}^t) cr_{jc}^t$$

$$+ \sum_t \sum_c \sum_v (\sum_j x_{jvc}^t) cdp_{vc} + \sum_t \sum_u \sum_h (\sum xfp_{uh}^t) casm_u^t + \sum_t \sum_c \sum_u \sum_z x_{zuc}^t c_{zuc}^t + \sum_t \sum_c \sum_u EI_{uc}^t IC_{uc}^t + \sum_t \sum_c \sum_c xp_{ccj}^t Tp_{ccj}^t$$

$$+ \sum_t \sum_s \sum_j (\sum_c x_{jsc}^t) TS_{js}^t + \sum_t \sum_u \sum_j (\sum_c x_{juc}^t) T_{ju}^t + \sum_t \sum_x \sum_j (\sum_c x_{jxc}^t) T_{jx}^t + \sum_t \sum_v \sum_j (\sum_c x_{jvc}^t) T_{jv}^t + \sum_t \sum_u \sum_h xfp_{uh}^t Tf_{uh}^t$$

Subject to:

$$\sum_j xp_{ccj}^t = Q_c^t \qquad\qquad \forall cc, t \qquad (1)$$

$$\sum_j x_{jsc}^t \le \sum_{cc} \sum_j xp_{ccj}^t - \sum_j \sum_v x_{jvc}^t - \sum_x \sum_j x_{jxc}^t \qquad\qquad \forall s, c, t \qquad (2)$$

$$\sum_j x_{jsc}^t \ge D_{sc}^t \qquad\qquad \forall s, c, t \qquad (3)$$

$$\sum_{cc} xp_{ccj}^t = \sum_s x_{jsc}^t + \sum_u x_{juc}^t + \sum_x x_{jxc}^t + \sum_v x_{jvc}^t \qquad\qquad \forall j, c, t \qquad (4)$$

$$\sum_u x_{juc}^t + EI_{uc}^{t-1} + \sum_z x_{zuc}^t = \sum_h xfp_{uh}^t + EI_{uc}^t \qquad\qquad \forall u, c, t \qquad (5)$$

$$\sum_{u} xfp^t_{uh} \geq D^t_h \qquad\qquad \forall h,t \qquad (6)$$

$$\sum_{cc} xp^t_{ccj} \leq cap_j \qquad\qquad \forall j,t \qquad (7)$$

$$\sum_{c}\sum_{s} x^t_{jsc} + \sum_{c}\sum_{u} x^t_{juc} \leq rcap_j \qquad\qquad \forall j,t \qquad (8)$$

$$\sum_{h} xfp^t_{uh} \leq pcap_u \qquad\qquad \forall u,t \qquad (9)$$

$$\sum_{c} EI^t_{uc} \leq icap_u \qquad\qquad \forall u,t \qquad (10)$$

$$\sum_{x}\sum_{j}\sum_{c} x^t_{jxc} = \rho * \sum_{j}\sum_{cc} xp^t_{ccj} \qquad\qquad \forall t \qquad (11)$$

$$\sum_{c}\sum_{j} x^t_{jxc} \leq reccap_x \qquad\qquad \forall x,t \qquad (12)$$

$$\sum_{v}\sum_{c}\sum_{j} x^t_{jvc} = \sigma * \sum_{j}\sum_{cc} xp^t_{ccj} \qquad\qquad \forall t \qquad (13)$$

$$\sum_{c}\sum_{j} x^t_{jvc} \leq discap_v \qquad\qquad \forall v,t \qquad (14)$$

$$Q^t_c \geq MTCC^t_{cc} * v^t_{cc} \qquad\qquad \forall cc,t \qquad (15)$$

$$\sum_{cc} xp^t_{ccj} \geq MTJ^t_j * y^t_j \qquad\qquad \forall j,t \qquad (16)$$

$$\sum_{h} xfp^t_{uh} \geq MTU^t_u * z^t_u \qquad\qquad \forall u,t \qquad (17)$$

$$v^t_{cc} \leq v^{t+1}_{cc} \qquad\qquad \forall cc,t \qquad (18)$$

$$y^t_j \leq y^{t+1}_j \qquad\qquad \forall j,t \qquad (19)$$

$$z^t_u \leq z^{t+1}_u \qquad\qquad \forall u,t \qquad (20)$$

$$xp^t_{ccj}, x^t_{jsc}, x^t_{juc}, x^t_{jxc}, x^t_{jvc}, x^t_{zuc}, xfp^t_{uh} \geq 0 \qquad\qquad \forall cc,j,s,u,x,v,c \qquad (21)$$

$$EI^t_{uc} \geq 0 \qquad\qquad \forall u,c,t \qquad (22)$$

$$v^t_{cc} \in \{0,1\} \qquad\qquad \forall cc,t \qquad (23)$$

$$y^t_j \in \{0,1\} \qquad\qquad \forall j,t \qquad (24)$$

$$z^t_u \in \{0,1\} \qquad\qquad \forall u,t \qquad (25)$$

Objective function of the above mathematical formulation is cost minimization. Costs incurred are the fixed costs of establishing facilities, operational (dismantling, processing, assembling and disposal) costs, transportation costs, inventory holding costs, and component purchasing costs. Constraints(1) shows that total amount of returned product collected at CC will shipped to the RPC which are to be located. At RPC, dismantled components can be directly sent to recycling or disposal sites and remaining components are shipped to spare market to satisfy the demand at spare market are shown via constraint (2) and (3). After satisfying demand at spare market components are send to RMP to assemble in the product form and if any shortages of the component occur would be purchased from supplier in order to satisfy the demand of secondary market is exposed via constraint (4), (5) and (6). Constraint (5) is the flow balance constraint for RMP. Total inflow, which is composed of components coming from inspection centers, components purchased from suppliers, and components in the inventory, must be equal to the outflow, which is composed of products sold to secondary markets, and the components to be held in inventory.

Constraint (7) ensures that the amount of products that are sent to RPC do not exceed the capacity of the collection center. Constraint (8) satisfy the reprocessing capacity constraint at RPC and (9) for the production capacity of RMP. Inventory to be held in the remanufacturing plants cannot exceed the inventory holding capacity via (10). (11), (12), (13) and (14) are for recycling and disposal capacity at recycling and disposal sites respectively. Constraints (15) - (17) are minimum throughput constraints guaranteeing that a CC, a RMP can only be established if the operation or production amount exceeds the predefined limits. (18) – (20) assure that once a facility is installed it remains operating until the end of the planning horizon. Lastly, Constraints (21)–(25) are domain constraints. The above problem is too complex and big. To get best possible solution we use differential Evolution algorithm.

4. Differential evolution

Differential Evolution (DE) was proposed by Price and Storn in 1995 to solve the polynomial fitting problem. DE is a small and simple mathematical model of a big and naturally complex process of evolution. It optimizes a problem by iteratively trying to improve a solution with regard to a given measure of quality. However, DE does not guarantee an optimal solution is ever found. DE optimizes a problem by maintaining a population of candidate solutions and creating new candidate solutions by combining existing ones, and then keeping whichever candidate solution has the best score or fitness on the optimization problem. The process is repeated and by doing so it is hoped, but not guaranteed, that a satisfactory solution will eventually be discovered. The basic DE Algorithm can be described as:

a) **Initialization**: All solution vectors in a population are randomly initialized. The initial NP, D-dimensional vectors $X_{i,G} = (x_{1,i,G}, x_{2,i,G}, x_{3,i,G}, ..., x_{D,i,G})$ are generated between lower and upper bounds $l = \{l_1, l_2, ..., l_D\}$ and $u = \{u_1, u_2, ..., u_D\}$ using the equation: $x_{j,i,0} = l_j + rand_{i,j}[0,1]*(u_j-l_j)$ where $rand_{i,j}[0,1]$ is uniformly distributed random number lying between 0 and 1.

b) **Mutation:** The mutation process at each generation begins by randomly selecting three individuals in the population. The ith perturbed individual $V_{i,G}$ is therefore generated based on the three choosen individuals as follows: $V_{i,G} = X_{r1,G} + F*(X_{r2,G} - X_{r3,G})$ where $r1, r2, r3 \in \{1, ..., NP\}$ are randomly selected, such that $r1 \neq r2 \neq r3 \neq i$ and $F \in (0,1.2]$

c) **Crossover:** The perturbed individual, $V_{i,G} = (v_{1,i,G}, ...v_{D,i,G})$, and the current population member, $X_{i,G} = (x_{1,i,G}, x_{2,i,G}, x_{3,i,G}, ..., x_{D,i,G})$ are then subject to the crossover operation, that finally generates the population of candidates, or trial vector, $U_{i,G} = (u_{1,i,G}, ..., u_{D,i,G})$ as follows:

$$u_{j,i.G} = \begin{cases} v_{j,i.G} \; if \; rand_{i,j}[0,1] \le C_r \vee j = j_{rand} \\ x_{j,i.G} \qquad\qquad otherwise \end{cases}$$ where $C_r \in [0, 1]$, is crossover

probability and $j_{rand} \in \{1, .., D\}$ is a random parameter's index, for each i

d) **Selection**: The population for the next generation is selected from the individuals in current population and its corresponding trial vector according to

the following rule: $X_{i.G+1} = \begin{cases} U_{i.G} \; if \; f(U_{i.G}) \le f(X_{i.G}) \\ X_{i.G} \qquad\qquad otherwise \end{cases}$ Each individual of the

temporary population is compared with its counterpart in the current population. Trial vector is only compared to one individual, not to all the individuals in the current population. Where f ( ) is objective function.

e) **Constraint handling in differential evolution:** the Pareto ranking method was proposed by Deb, which is based on the following three:

   i.   Feasible solution with the best value of the objective function is preferred.
   ii. Feasible one is preferred over infeasible.
   iii. Infeasible vectors with the lowest sum of constraint violation are preferred.

f) **Stopping criteria:** there are two stopping criteria: Maximum number of generations and accuracy criteria.

5. Numerical illustration

Eight echelon network consisting of 4 CC, 3 RPC's, 3 RMP's, 5 spare markets , 1 RC, 1 DS, 6 new module suppliers and 6 distribution centers has been considered for the model implementation. 50% of returned modules of the returned products are assumed to be disposed. Good modules are either sent to the factory for remanufacturing, or to spare market. 30% of the returned modules are assumed to be sent for recycling. Single returned products with 10 modules are considered. Data used for the analysis are given as: Minimum throughput required for collection center, RPC, and RMP are {4000, 4000, 5000, 5000}, {5000, 3000, 7000} and {1000, 1500, 1000} resp. Apart from that capacities of storage, processing of RPC's are {9350, 6700, 5500} and {70000, 40000, 90000} resp. Production capacities of RMPs are {7000, 6500, 6500}. Capacity of recycling center and disposal site are 50000 and 90000 resp.

Table 1 Data on costs and demand in the network

	Period 1	Period 2	Period 3	Period4	Period 5
Set up cost of CC(cc1,..,cc4)	5750,5500, 4900,6900	5800,5600 4970,7000	5850,5660 5000,7100	5930,6000, 5100,7100	5950,6200, 5100,7300
Set up cost of RPC j1..j3	9350,6700, 5500	9400,6740, 5640	9480,6790, 5700	9500,6820, 5730	9550,6850, 5750
Set up cost of RMP u1,u2,u3	4850,4550, 4600	4900,4600, 4640	4930,4600, 4650	4950,4640, 4670	4970,4650, 4680
Dismantling at RPC(j1j2,j3)	18,16,14	18.5,16.3,14.7	18.9,17,15.2	19.1,17.3,15.2	19.5.17.5,15.5
Assembling cost at u1,u2,u3	34,35.6,34.8	34.3,35.8,35	34.5,35.9,35.3	34.8,36.1,35.2	35,36.4,35.5
Disposal cost of c1,…,c10 at disposal site	.99,1.08,1.64,.0 86,2.27,3.58,2.2 5,1.03,2.09,1.78	1.11,1.23,1.8,0.98, 2.39,3.7,2.37,1.15, 2.21,1.9	1.34,1.46,2.03,1.2 1,2.62,3.93,2.6,1.3 8,2.44,2.13	1.43,1.55,2.12,1.3, 2.71,4.02,2.69,1.4 7,2.53,2.22	1.57,1.69,2.26,1.4 4,2.85,4.16,2.83,1. 61,2.67,2.36
Processing cost of c1,…,c10 at RPC j1	0.64,0.58,0.16,0 .36,0.3,0.48,0.5, 0.58,1.04,1	0.88,0.82,0.4,0.6,0 .54,0.72,0.74,0.82, 1.28,1.24	1.06,1,0.58,0.78,0. 72,0.9,0.92,1,1.46, 1.42	1.30,1.24,0.82,1.0 2,0.96,1.14,1.16,1. 24,1.7,1.66	1.43,1.37,0.95,1.1 5,1.09,1.27,1.29,1. 37,1.83,1.79
Processing cost	0.68,0.64,0.24,0	0.92,0.88,0.48,0.5	1.1,1.06,0.66,0.71,	1.34,1.30,0.9,0.95,	1.47,1.43,1.03,1.0

	Period 1	Period 2	Period 3	Period 4	Period 5
of c1,....,c10 at RPC j2	.29,0.29,0.48,0.6,0.52,0.88,0.76	3,0.53,0.72,0.84,0.76,1.12,1	0.71,0.9,1.02,0.94,1.3,1.18	0.95,1.14,1.26,1.18,1.54,1.42	8,1.08,1.27,1.39,1.31,1.67,1.55
Processing cost of c1,...,c10 at RPC j3	0.66,0.62,0.2,0.31,0.28,0.45,0.55,0.55,0.92,0.84	0.9,0.86,0.44,0.55,0.52,0.69,0.79,0.79,1.16,1.08	1.08,1.04,0.62,0.73,0.7,0.87,0.97,0.97,1.34,1.26	1.32,1.28,.86,0.97,0.94,1.11,1.21,1.21,1.58,1.50	1.45,1.41,0.99,1.10,1.07,1.24,1.34,1.34,1.71
Demand at h1,...,h6	3500,3500,2500 3500,2500,2500	3550,3600,2700 3600,2600,2550	3600,3520,2600, 3580,2600,2600	3600,3550,2600 3600,2700,2700	3610,3520,2700 3620,2650,2700
Demand at spare market s1 for c1,...c10	2400,600,1000 3500,1200,425,1000,0,500,2300	2430,630,1030,3530,1230,4280,1000,0,520,2320	2460,660,1080,3570,1260,4300,1050,0,550,2320	2580,800,1150,3680,1350,4450,1150,1100,600,2400	2550,800,1150,3980,1300,4360,1350,1000,700,2550
Demand at spare market s2 for c1,...c10	0,3500,2500,0,3500,1900,0,2000,600,0	0,3520,2540,0,3530,1950,1010,2040,630,0	0,3510,2520,0,3550,1950,2030,650,0	0,3580,2600,900,3300,1600,0,300,600,0	0,3600,2500,930,3500,1750,0,200,500,0
Demand at spare market s3 for c1,...,c10	2500,2500,2400,1500,2000,800,1500,2500,900,2400	2580,2540,2460,1550,2060,830,1220,2530,980,2430	2570,2550,2460,1550,2060,830,1520,2530,980,2430	2400,2300,2360,1500,2200,800,1400,2300,880,2500	2470,2200,2360,1400,2100,900,1300,2280,880,2200
Demand at spare market s4 for c1,...,c10	2000,550,1800,2700,800,100,1800,1500,4000,1850	2020,570,1850,2730,830,1020,1430,1550,4030,1870	2430,590,1870,2750,830,1020,1830,1550,4030,1900	2100,600,1600,2800,750,1050,1990,1350,3950,1600	2000,550,1630,2700,830,1090,1890,1300,4000,1500
Demand at spare market s5 for c1,...,c10	0,1850,1500,0,800,800,4100,500,1500,950	0,1880,1530,0,820,830,3820,550,1530,970	0,1880,1530,0,850,830,4120,550,1530,970,	0,1800,1400,0,800,700,4100,400,1400,900	0,1900,1400,0,1000,800,4100,400,1400,900

Table 2 data on transportation cost/unit of product or component

	Period 1	Period 2	Period 3	Period4	Period 5
From cc1 to j1,j2,j3	0.95,0.83,0.9	0.97,0.85,0.92	1,0.89,0.95	1.04,0.93,0.99	1.08,0.99,1.02
From cc2 to j1,j2,j3	1.11,1.05,1.02	1.17,1.09,1.08	1.20,1.12,1.12	1.23,1.16,1.18	1.27,1.2,1.22
From cc3 to j1,j2,j3	1.16,1.19,1.23	1.19,1.23,1.27	1.23,1.27,1.31	1.28,1.31,1.34	1.32,1.34,1.37
From cc4 to j1,j2,j3	0.87,0.77,0.83	0.91,0.80,0.87	0.95,0.84,0.90	0.99,0.87,0.95	1.03,0.9,0.99
From j1 to u1,u2,u3	0.24,0.23,0.24	0.26,0.25,0.27	0.28,0.28,0.29	0.3,0.31,0.32	0.33,0.34,0.35
From j2 to u1,u2,u3	0.21,0.24,0.23	0.24,0.27,0.25	0.26,0.29,0.27	0.29,0.33,0.29	0.32,0.35,0.31
From j3 to u1,u2,u3	0.24,0.22,0.23	0.26,0.25,0.25	0.29,0.28,0.26	0.31,0.32,0.28	0.33,0.34,0.3
From j1,j2,j3 to RP	4.65,2.8,3.35	4.95,3,3.55	5.2,3.2,3.8	5.56,3.5,4	5.80,3.75,4.25
From j1,j2,j3 to DS	2.25,2.75,2.2	2.4,2.9,2.55	2.75,3,2.80	2.9,3.25,3	3.1,3.55,3.35
From j1to s1,..,s5	0.9,0.75,0.89,0.78,0.81	0.92,0.78,0.91,0.83,0.88	0.93,0.8,0.94 0.83,0.85	0.95,0.82,0.96,0.85,0.88	0.99,0.84,0.99 0.88,0.9
From j2 to s1,...,s5	0.86,0.84,0.86,0.75,0.78	0.88,0.86,0.89,0.79,0.81	0.9,0.88,0.91,0.80,0.85	0.93,0.90,0.93,0.82,0.86	0.95,0.94,0.96,0.85,0.88
From j3 to s1,...,s5	0.89,0.8,0.87,0.78,0.78	0.92,0.85,0.90' 0.8,0.81	0.94,0.87,0.91,0.82,0.82	0.96,0.9,0.94,0.85,0.84	0.98,0.92,0.97,0.88,0.89
From u1 to h1,...,h6	1.59,3.18,2.85,0.78,1.71,2.43	1.8,3.39,3,0.9,1.99,2.6	2.06,3.7,3.25,1.23,2.23,2.83	2.28,3.89,3.65,1.58,2.73,2.9	2.59,4,3.95,1.88,2.99,3.26
From u2 to h1,..,h6	2.52,2.4,2.28,0.75,1.47,2.73	2.72,2.65,2.56,1,1.77,2.93	2.92,2.88,2.76 1.25,1.96,3.32	3.33,3.12,2.99 2.59,2.26,3.52	3.58,3.46,3.35 2.86,2.45,3.78
From u3 to h1...,h6	2.1,2.07,2.58,0.78,1.89,2.79	2.55,2.48,2.8,1.1,2.99	2.95,2.62,3.05,1.26,1.18,3.3	3.22,2.85,3.35 1.56,1.48,3.75	3.68,3,3.75,1.86,1.68,3.94

Table 3 Data on cost of purchased components/unit

	Period 1	Period 2	Period 3	Period4	Period 5
From z1 to RPC u1	3,6.25,11.25,5.3,6.6,6.85,13.3,7.45,5.75,8.25	3.2,6.45,11.45,5.5,6.8,7.05,13.5,7.65,5.95,8.45	3.39,6.65,11.45,5.65,6.95,7.24,13.95 7.84,6.14,8.64	3.55,6.8,11.81,5.85,7.11,7.4,13.7,8 6.3,8.8	3.91,7.2,12.22,6.25,7.53,7.81,14.26,8.43,6.75,9.22
From z1 to RPC u2	3.3,6.5,11.5,5.6,6.8,7.05,13.5,7.65,6,8.75	3.5,6.7,11.7,5.8,7,7.25,13.7,7.85,6.2,8.95	3.69,6.89,11.85,5.99,7.19,7.44,13.85,8.04,6.39,9.19	3.8,7.05,12.06,6.18,7.33,7.61,14.03,8.2,6.57,9.34	4.26,7.47,12.44,6.54,7.77,8,14.46,8.61,6.95,9.73
From z1 to RPC u3	3.2,6.4,11.35,5.75,6.25,6.5,13.35,7.5 5.6,8.25	3.4,6.6,11.55,5.95,6.45,6.7,13.55,7.7,5.8,8.45	3.59,6.79,11.74,6.14,6.69,6.99,13.74 ,7.89,5.99,8.64	3.77,6.95,11.92,6.35,6.83,7.05,13.9,8.05,6.09,8.81	4.15,7.36,12.3,6.7 3,7.22,7.45,14.35,8.45,6.55,9.26
From z2 to RPC u1	3.2,6.45,11.45,5.5,6.8,7.05,13.5,7.65,5.95,8.45	3.4,6.65,11.65,5.7,7,7.25,13.7,7.85,6.15,8.65	3.95,6.89,11.45,5.8,7.19,7.45,13.95,8.45,6.35,8.85	3.75,7.12.09,6.03,7.3,7.63,14.05,8.2,6.56,9	4.15,7.42,12.45,6.46,7.77,8.03,14.47,8.6,6.9,9.4
From z2 to RPC u2	3.5,6.7,11.7,5.8,7,7.25,13.7,7.85,	3.7,6.9,11.9,6.7.2,7.45,13.9,8.05,	3.85,7.09,12.05,6.19,7.39,7.64,14.05	4.03,7.22,12.23,6.34,7.55,7.8,14.25,	4.45,7.65,12.66,6.77,7.95,8.2,14.63,

	6.2,8.95	6.4,9.15	,8.24,6.59,9.34	8.41,6.7,9.5	8.8,7.15,9.9
From z2 to RPC u3	3.4,6.6,11.55,5.95, 6.45,6.7,13.55,7.7 5.8,8.45	3.6,6.8,11.75,6.15, 6.65,6.9,13.75,7.9' 6,8.65	3.79,6.95,11.95,6. 35,6.85,7.05,13.95 ,8.05,6.15,8.87	3.92,7.13,12.1,6.5 1,7.01,7.22,14.13, 8.22,6.36,9.01	4.36,7.57,12.53,6. 9,7.4,7.656,14.5,8. 65,6.76,9.4
From z3 to RPC u1	3.4,6.65,11.65,5.7, 7,7.25,13.7,7.85,6. 15,8.65,	3.6,6.85,11.85,5.9, 7.2,7.45,13.9,8.05, 6.35,8.85	3.76,7.04,12.05,6. 12,7.39,7.64,14.15 ,8.24,6.63,9.26	3.9,7.2,12.22,6.26, 7.53,7.8,14.26,8.4 1,6.72,9.23	4.35,7.61,12.62,6. 66,7.95,8.2,14.63, 8.8,7.1.9.6
From z3 to RPC u2	3.7,6.9,11.9,6,7.2, 7.45,13.9,9.8,05,6.4 ,9.15	3.9,7.1,12.1,6.2, 7.4,7.65,14.1,8.25, 6.6,9.35	4.09,7.29,12.23,6. 39,7.57,7.84,14.22 ,8.44,6.79,9.54	4.25,7.46,12.4,6.5, 7.7,8,14.4,8.6,6.9, 9.71	4.65,7.86,12.86,6. 96,8.17,8.4,14.83, 9,7.35,10.1
From z3 to RPC u3	3.6,6.85,11.85,5.9, 7.2,7.45,13.9,8.05, 6.35,8.85	3.8,7,11.95,6.35, 6.85,7.1,13.95,8.1 ,6.2,8.85	3.99,7.19,12.13,6. 54, .04,7.29,14.15, 8.26,6.39,9.04	4.15,7.35,12.33,6. 71,7.22,7.43,14.35 ,8.46,6.5,9.2	4.55,7.75,12.7,7.1, 7.6,7.85,14.7,8.85, 6.93,9.6
From z4 to RPC u1	3.6,6.85,11.85,5.9 7.2,7.45,13.9,8.05, 6.35,8.85	3.8,7.05,12.05,6.1, 7.4,7.65,14.1,8.25, 6.55,9.05	3.92,7.23,12.24,6. 25,7.58,7.84,14.27 ,8.46,6.73,9.24	4.15,7.45,12.95,6. 45,7.76,8.01,14.49 ,8.6,6.92,9.43	4.5,7.8,12.3,6.8,8. 1,8.43,14.85,9,7.3, 9.8
From z4 to RPC u2	3.9,7.1,12.1,6,2.7. 4,7.65,14.1,8.25,6. 6,9.35	4.1,7.3,12.3,6.4, 7.6,7.85,14.3,8.45 ,6.8,9.55	4.29,7.49,12.35,6. 95,7.83,8.03,14.48 ,8.64,6.99,9.74	4.49,7.66,12.64,6. 75,7.9,8.2,14.76,8. 8,7.15,9.9	4.8,8.05,13,7.15,8. 35,8.6,15,9.2,7.53, 10.3
From z4 to RPC u3	3.8,7,11.95,6.35,6. 85,7.1,13.95,8.1,6. 2,8.85	4,7.2,12.15,6.55, 7.05,7.3,14.15, 8.3,6.4,9.05	4.19,7.39,12.35,6. 74,7.24,7.49,14.35 ,8.48,6.5,9.24	4.35,7.5,12.68,6.8 7,7.4,7.63,14.52,8. 63,6.78,9.43	4.75,7.96,12.9,7.3, 7.8,8.05,14.9,9.05, 9.05,7.15,9.8
From z5 to RPC u1	3.8,7.05,12.05,6.1, 7.4,7.65,14.1,8.25, 6.55,9.05	4,7.25,12.25,6.3, 7.6,7.85,14.3,8.45, 6.75,9.25	4.20,7.45,12.35,6. 47,7.75,8.05,14.45 8.64,6.93,9.44	4.35,7.6,12.7,6.65, 7.95,8.23,14.67,8. 82,7.1,9.61	3.94,7.19,12.18,6. 24,7.55,7.79,14.24 ,8.39,6.69,9.19
From z5 to RPC u2	4.1,7.3,12.3,6.4,7. 6,7.85,14.3,8.45,6. 8,9.55	4.3,7.5,12.5,6.6 7.8,8.05,14.5,8.65, 7,9.75	4.47,7.66,12.64,6. 78,7.98,8.24,14.62 ,8.83,7.18,9.95	4.63,7.87,12.8,6.9 5,8.1,8.45,14.85,9, 7.35,10.1	4.25,7.46,12.47,6. 54,7.74,7.99,14.5, 8.6,6.95,9.7
From z5 to RPC u3	4,7.2,12.15,6.55, 7.05,7.3,14.15,8.3, 6.4,9.05	4.2,7.4,12.35,6.75, 7.25,7.5,14.35,8.5, 6.6,9.25	4.39,7.53,12.54,6. 93,7.42,7.67,14.54 ,8.69,6.77,9.48	4.55,7.76,12.71,7. 12,7.62,7.85,14.7, 8.85,6.95,9.6	4.14,7.35,12.3,6.6 9,7.2,7.45,14.31,8. 44,6.54,9.19
From z6 to RPC u1	4,7.25,12.25,6.3,7. 6,7.85,14.3,8.45,6. 75,9.25	4.2,7.45,12.45,6.5, 7.8,8.05,14.5,8.65 6.95,9.45	4.31,7.64,12.62,6. 69,7.93,8.24,14.64 ,8.85,7.14,9.64	3.76,7,12.3,6.03,7. 33,7.63,14.14,8.2, 6.5,9	4.15,7.39,12.4,6.4 5,7.76,7.99,14.47, 8.6,6.9,9.39
From z6 to RPC u2	4.3,7.5,12.5,6.6,7. 8,8.05,14.5,8.65,7, 9.75	4.5,7.7,12.7,6.8, 8,8.25,14.7, 8.85,7.2,9.95	4.69,7.86,12.87,6. 99,8.2,8.45,14.8, 9.04,7.39,10.14	4.05,7.24,123.22,6 .33,7.51,7.8,14.25, 8.42,6.76,9.5	4.44,7.66,12.6,6.7 4,7.95,8.2,14.66,8. 8,7.17,9.91
From z6 to RPC u3	4.2,7.4,12.35,6.75, 7.25,7.5,14.35,8.5, 6.6,9.25	4.4,7.6,12.55,6.95 7.45,7.7,14.55, 8.7,6.8,9.45	4.6,7.8,12.74,7.09, 7.63,7.88,14.74 8.89,6,9.64	3.95,7.15,12.1,6.5, 7,7.25,14.13,8.25, 6.3,9	4.35,7.54,12.49,6. 9,7.4,7.64,14.5,8.6 4,6.74,9.56

The proposed model with the above data is solved through differential algorithm with a population size of 3000 and with the value .5 for scaling factor (f) and .6 as crossover probability (C). Resultant values of the variable is listed in the tables below

Table 4   Solution table

	Period 1	Period 2	Period 3	Period4	Period 5
v(cc1...cc4)	1,1,0,1	1,1,0,1	1,1,0,1	1,1,0,1	1,1,0,1
y(j1,j2,j3)	1,1,1	1,1,1	1,1,1	1,1,1	1,1,1
Z(u1,u2,u3)	1,1,1	1,1,1	1,1,1	1,1,1	1,1,1

Table 5 no of units of returned product shipped from CC to RPC

	Period 1	Period 2	Period 3	Period4	Period 5
cc1- j1..j3	1360,4630,2010	1749,4651,1900	7778,5720,6950	0,0,8400	0,0,8470
cc2- j1..j3	6990,0,0	0,0,7100	0,4150,0	6550,0,600	6660,0,530
cc3 - j1..j3	6000,0,0	6300,0,0	0,0,2050	1436,4663,0	1476,4654,0

Table 6 quantity of components shipped from RPCs to spare market in period 1...5

	s1	s2	s3	s4	s5
c1- j1||j2 || j3	0,0,0,0,844||0,0,0,0 ,0||2400,2430,2460 ,2580,1706	0,0,0,0,0||0,0, 0,0,0||0,0,0,0, 0	0,0,0,0,0||0,0,0,0,0||2 500,2580,2570,2400, 2470	2000,0,0,0,200||0,0, 0,0,0||0,2020,2030, 2100,0	0,0,0,0,0|| 0,0,0,0,0|| 0,0,0,0,0

c2- j1 ||j2||j3	0,0,0,0,0|| 600,630,660,800,8 00||0,0,0,0,0	3500,3520,35 10,3580,2068| |0,0,0,0,0||0,0, 0,0,1532	0,0,0,0,0||2490,2491, 2550,2300,2200||10,4 9,0,0,0	0,0,0,0,0||550,570,5 90,600,550||0,0,0,0, 0	0,0,0,0,0||0,0,0,0 ,1104||1850,188 0,1888,1800,796
c3- j1|| j2|| j3	0,0,0,0,0||1000,103 0,1080,115 0,1150||0,0,0,0,0	2500,2540,25 20,2600,2500| |0,0,0,0,0||0,0, 0,0,0	0,0,0,0,0||1040,2460, 1490,1260,474||1360, 0,970,1100,1886	0,0,0,0,0||1800,880, 1870,1600||630,0,9 70,0,0,0	0,0,0,0,0||0,0,0,0 ,1400||1500,153 0,1530,1400,0
c4- j1|| j2|| j3	0,0,1386,1817,0||3 500,3530,2184,186 3,3980||0,0,0,0,0	0,0,0,900,930| |0,0,0,0,0||0,0, 0,0,0	0,659,0,0,0||0,891,0,0 ,0||1500,0,1550,1540 0,1400	2700,0,0,0,2029||0, 230,1800,2800,671| |0,2500,950,0,0	0,0,0,0,0||0,0,0,0,0 ,0||0,0,0,0,0
c5- j1|- j2|| j3	0,0,0,0,0||1200,123 0,1260,1350,1300|| 0,0,0,0,0	3500,3530,35 50,3300,3500| |0,0,0,0,0||0,0, 0,0,0	0,0,0,0,0||0,0,0,0,152 4||2000,2060,2060,22 00,576	0,0,0,0,0||800,0,0,7 50,830|| 0,830,830,0,0	0,0,0,0,0||0,0,0,0 ,1000||800,820,8 50,800,0
c6- j1|| j2|| j3	0,0,0,837,0||3630,4 280,4300,3613,356 4||620,0,0,0,796	1900,1950,19 50,1600,1750| |0,0,0,0,0||0,0, 0,0,0	0,0,0,0,0||0,0,0,0,0||8 00,830,830,800,900	0,0,0,0,0||1000,0,0, 1050,1090||0,1020, 1020,0,0	0,0,0,0,0||0,0,0,0 ,0||800,830,830, 700,800
c7- j1|| j2|| j3	0,0,1150,0||1000, 1000,1050,0,1350|| 0,0,0,0,0	0,1010,0,0,0|| 0,0,0,0,0||0,0, 0,0,0	0,0,0,0,0||240,1220,0, 1400,1300||1260,0,15 20,0,0	0,0,0,0,0|| 1800,1430,1830,19 90,1890|| 0,0,0,0,0	0,319,0,0,1486|| 0,0,0,0,114||410 0,3501,4120,410 0,2500
c8- j1|| j2|| j3	0,0,0,0,0||0,0,0,110 0,1000||0,0,0,0,0,0	2000,2040,20 30,300,200||0, 0,0,0,0||0,0, 0,0	0,909,0,0,0||0,1221,5 80,200,1954|| 2500,400,1950,2100, 326	0,0,0,0||1500,0,15 50,1350,1300||0,15 50,0,0,0	0,0,0,0,0||0,0,0,0 ,400||500,550,55 0,400,0
c9- j1|- j2|| j3	0,0,0,600,0||500,52 0,550,0,700|| 0,0,0,0,0	600,630,650, 600,500||0,0,0 ,0||0,0,0,0,0	0,0,0,0||0,0,0,0,0||9 00,980,980,880,880	4000,0,0,0,1265||0, 0,2412,2696,2735|| 0,4030,1618,1254	0,0,0,0,0||0,0,0,0 ,0||1500,1530,15 30,1400,1400
c10- j1|| j2|| j3	0,0,0,237,0||2300,2 320,2320,2163,255 0||0,0,0,0,0	0,0,0,0,0||0, 0,0,0,0||0,0,0,0, 0	0,439,0,0,0||0,461,50 2,900,0||2400,1530,1 928,1600,2200	1850,0,0,0,60 0||1870,1900,1600, 1440||0,0,0,0,0	0,0,0,0,0||0,0,0,0 ,0||950,970,970, 900,900

Table 7 Units of returned product shipped from RPC to secondary market

	Period 1	Period 2	Period 3	Period4	Period 5
u1 to h1…h6	3500,0,0,500, 500,2500,	3550,0,0,900, 0,2550	3600,0,0,800 0,2600	3600,0,0,700, 0,2700	3610,0,0,690, 0,2700
u2 to h1…h6	0,0,2500,0,2000 0	0,0,2700,2400, 0,0	0,0,2600,2400 0,0	0,2650,2600,0 0,0	0,2600,2700,0 0,0
u3 to h1…h6	0,3500,0,3000 0,0	0,3600,0,300, 2600,0	0,3520,0,380 2600,0	0,900,0,2900, 2700,0	0,920,0,2930,2650 0

Table 8: quantity of components shipped from RPCs to RMPs in the planning horizon

	u1	u2	u3
c1- j1	0,0,0,0,0	400,0,0,0,0	1360,0,0,0,0
c1- j2	0,0,0,0,0	0,0,0,0,0	0,0,0,0,0
c1- j3	0,0,0,0,	4100,1970,1940,1920,0	0,0,0,0,0
c2- j1	0,0,0,0,0	3860,4529,4268,2524,0	0,0,0,0,0
c2- j2	990,960,810,0,0,	0,0,0,0,0	0,0,112,963,0
c2- j3	0,0,0,0,0	640,571,732,2726,0	6500,6500,6388,4474,1372
c3- j1	0,409,258,137,950	0,5100,5000,5250,4686	4860,0,0,0,0
c3- j2	790,281,282,653,0	0,0,0,0,0	0,0,0,0,0
c3- j3	0,0,0,0,0	4500,0,0,0,614	1640,6500,6500,6500,6500
c4- j1	1009,0,1390,20,977	0,4949,5000,5250,4200	3651,0,0,0,0
c4- j2	1127,0,0,0,0	0,0,0,0,0	0,0,0,0,0
c4- j3	0,0,0,1000,0	4651,0,0,0,1100	2849,6500,6500,6500,6500
c5- j1	3860,0,3538,0,4636	0,4519,690,4687,0	0,0,0,0,0
c5- j2	2630,3421,3462,2563,0	0,0,0,0,0	0,0,0,0,0
c5- j3	510,3579,0,4437,2364	4500,581,4310,563,5300	1190,1130,950,1000,760
c6- j1	5460,6099,5828,5550,6386	0,0,0,0,0	0,0,0,0,0
c6- j2	0,371,422,0,0	0,0,0,0,0	0,0,0,0,0
c6- j3	1540,530,750,1450,614	4500,5100,5000,5250,5300	740,690,570,800,590
c7- j1	0,16320,5158,1260,1350	860,5100,2620,5250,5300	6500,0,0,327,0

c7- j2	1590,0,1842,0,0	0,0,0,0,0	0,1001,0,1273,0
c7- j3	0,0,0,0,0	3640,0,2380,0,0	0,5499,980,4900,6500
c8- j1	360,0,748,2437,4810	0,5100,5000,5250,3126	5000,0,0,0,0
c8- j2	3130,3430,2592,2013,0	0,0,0,0,0	0,0,0,0,0
c8- j3	0,0,0,0,0	4500,0,0,0,2174	1500,6500,6500,6500,6500
c9- j1	2760,2319,7000,6787,1071	0,5100,128,0,5300	0,0,0,0,0
c9- j2	2463,2272,0,0,0	0,0,0,0,0	0,0,0,0,0
c9- j3	1777,2409,0,213,5929	4500,0,4872,5250,0	323,51,0,3,791
c10- j1	160,2510,2778,2500,7000	4500,5100,5000,5250,1076	850,0,0,0,0
c10- j2	2330,0,0,0,0	0,0,0,0,0	0,0,0,0,0
c10- j3	0,0,0,0,0	0,0,0,0,4224	5650,6500,6102,6500,6500

Quantity of components sent to recycling center in period 1,2,3,4 and 5 are 6297, 6510, 6450, 6495 and 6537 respectively and to disposal site are 10495, 10850, 10750, 10825 and 10895 respectively. Requirement of new components for remanufacturing to satisfy the demand of secondary market is as follows:

In period 1: 7000, 6010, 6210, 4864, 0, 0, 5410, 3510, 0, 4510 units of c1....c10 are purchased by RMP1 from supplier z1. 5140, 5310, 5760, 6177 units of c1, c5, c6 and c9 are purchased by RMP3 from suppler z1. **In period 2:** RMP1 purchased 7000, 6040, 6310, 7000, 5380, 3570, 4490 units of c1, c2, c3, c4, c7, c8 and c10 resp., RMP2 purchased 3130 units of c1 from z1 and RMP3 purchased 6500,5370,5810and 6449 units of c1, c5, c6 and c9 resp. from z1. **In period 3:** RMP1 purchased 7000, 6190, 5610, 3660, 4222 units of c1, c2, c4, c8 and c10 resp. from supplier1, RMP2 purchased 3060 units of c1, RMP3 purchased 6500,5550,5930,5520, 6500, and 3989 units of c1, c5, c6, c7, c9 and c10 resp. from z1, and RMP1 purchased 6460 units of c3 from z2. **In period 4:** RMP1 purchased 7000, 7000, 6210, 5980, 5740, 2550, and 4500 units of c1, c2, c3, c4, c7, c8, and c10 from z1,. RMP2 purchased 3300 units of c1from z1. RMP3 purchased 6550, 1063, 5500, 5700, 6497 units of c1, c2, c5, c6 and c9 from z1. **In period 5:** RMP1 purchased 7000 units of c1from z1. RMP1 purchased 7000, 6050, 6023, 5650, 2190 units of c2, c3, c4, c7 andc8 resp. from z5. RMP2 purchased 5300 units of c1from z5. RMP3 purchased 6500, 5128, 5740, 5910, 5709 units of c1, c2, c5, c6 and c9 from z5.

6 Conclusion

In this paper, we proposed a mathematical programming framework for multiperiod reverse logistics network design problems of single returned used product. To satisfy the demand of remanufactured products there is a mix and match of old and new components. Therefore, the model incorporates an echelon for suppliers that can provide new components. Model also considers the demand of components in the spare market as it would generally fetch higher value per module for the companies. Decisions to be made regarding the location of the collection centers, RPC and RMP, capacity of the facilities, flow routing through the network, the amount of inventory held and the amount of components to be purchase from the suppliers by RMP's. With advancement in technology and design processes, it is possible to estimate the number and type of components that might have to be disposed. Therefore, we have assumed certain percentages of components going to recycling and disposal centers. Model brings out an important conclusion that, transportation and other logistics costs may not be an important factor in the design of a network.

Rather, the cost of reprocessing, remanufacturing, and the cost of new modules can be the driving factor for the choice of a reverse logistics network. A natural extension to the setting considered in this paper regards the inclusion of uncertainty issues. This is a relevant aspect in many practical reverse logistics planning problems.

References

[1] Beullens, P.: Reverse logistics in effective recovery of products from waste materials. Reviews in Environmental Science and Bio/Technology. 3(4), 283–306 (2004).

[2] Jahre, M.: Household waste collection as a reverse channel – a theoretical perspective. International Journal of Physical Distribution and Logistics Management.25(2), 39–55(1995).

[3] Fleischmann, M.: Quantitative models for reverse logistics. Springer. p. 41(2001).

[4] Krumwiede, D., & Sheu, C.: A model for reverse logistics entry by third-party providers. Omega, 30, 325–333 (2002).

[5] Ferrer, G., & Whybark, C. D.: From garbage to goods: Successful remanufacturing systems and skills. Business Horizons, 43(6), 55–64 (2000).

[6] Tibben-Lembke, R., & Rogers, D. S.: Differences between forward and reverse logistics. Supply Chain Management: An International Journal, 7(5), 271–282 (2002).

[7] Biehl, M., Prater, M., & Realff, M. J.: Assessing performance and uncertainty indeveloping carpet reverse logistics systems. Computers and Operations Research,34, 443–463(2007).

[8] Reimer, B., Sodhi, M., & Jayaraman, V.: Truck sizing models for recyclables pick-up. Computers and Industrial Engineering, 51, 621–636 (2006).

[9] Fleischmann, M., Bloemhof-Ruwaard, J. M., Dekker, R., van der Laan, E. A., van Nunen, J. A. E. E., & van Wassenhove, L. N.: Quantitative models for reverse logistics: A review. European Journal of Operational Research, 103, 1–17 (1997).

[10] Bloemhof-Ruwaard, J., Fleischmann, M., & van Nunen, J.: Reviewing distribution issues in reverse logistics. In M. G. Speranza & P. Stahly (Eds.),New trends in distribution logistics. Springer-Verlag (1999).

[11] Jayaraman, V., Patterson, R., & Rolland, E.: The design of reverse distribution networks: Models and solution procedures. European Journal of Operational Research, 150, 128–149 (2003).

[12] Savaskan, R. C., Bhattacharya, S., & van Wassenhove, L. N. (2004). Closed-loop supply chain models with product remanufacturing. Management Science, 50(2),239–252.

[13] Kusumastuti, R., Piplani, R., & Lim, G.: An approach to design reverse logistics networks for product recovery. In Proceedings of IEEE international engineering management conference, Singapore, pp. 1239–1243 (2004)..

[14] Schultmann, F., Engels, B., Rentz, O.: Closed-loop supply chains for spent batteries. Interfaces 33, 57–71 (2003).

[15] Figueiredo, J., Mayerle, S.: Designing minimum-cost recycling collection networks with required throughput. Transportation Research Part E 44, 731–752(2008).

[16] Pati, R., Vrat, P., Kumar, P.: A goal programming model for paper recycling system. Omega 36, 405–417(2008).

[17] Krikke, H., van Harten, A., Schuur, P.: Business case Oce: Reverse logistics network design for copiers. OR Spectrum 21, 381–409(1999).

[18] Realff, M. J., Ammons, J. C., & Newton, D.: Robust reverse production system design for carpet recycling. IIE Transactions, 36(8), 767–776 (2004).

[19] Shih, L. (2001). Reverse logistics system planning for recycling electrical appliances and computers in Taiwan. Resources, Conservation and Recycling, 32, 55–72.

[20] Walther, G., & Spengler, T. (2005). Impact of WEEE-directive on reverse logistics in Germany. International Journal of Physical Distribution and Logistics Management, 35(5), 337–361.

[21] Fernandez, I., & Kekale, T. (2005). The influence of modularity and clock speed on reverse logistics strategy: Implications for the purchasing function. Journal of Purchasing and Supply Management, 11, 193–205.

[22] Mutha.A.,Pokharel.S.,: Strategic network design for reverse logistics and remanufacturing using new and old product components. Computers & Industrial Engineering56.334-346(2009).

"An Inclusive Survey on Data Preprocessing Methods Used in Web Usage Mining"

Brijesh Bakariya[1], Krishna K. Mohbey[2] G. S. Thakur[3]
Department of Computer Applications
M.A.N.I.T., Bhopal-462051
brijesh_scs@yahoo.co.in[1]
kmohbey@gmail.com[2]
ghanshyamthakur@gmail.com[3]

Abstract: Several data mining techniques applied in Web usage mining applications for discovering user access pattern from web log data. To understand and provide better services it will require Web-based applications. Web usage mining is one of the types of Web mining. Web mining is the technique to extract knowledge from web content, structure and usage. It is the collection of technologies to accomplish the possible of extracting valuable knowledge from the World Wide Web and its usage pattern. Web mining enables to find out relevant result from Web data including web document, hyperlink between documents, usage log of website etc. There are three main areas of web mining research –content, structure and usage. This paper provide an overview of previous and existing work in all three areas, and also define an overview of data preprocessing process like Data Cleaning, User Identification, Session Identification, Transaction Identification, Path Completion used in Web usage mining.

Keywords: data mining, web content mining, web structure mining, web usage mining, data preprocessing.

1 Introduction

Today is the day of Information Technology, accessing information is the most frequent task. Day by day we have to go through various kind of information that we require, we have to just browse the web and get desired information with a single click. Now a day, internet is playing such a crucial role in our daily life that is very difficult to survive without it, because millions of electronic data are included on hundreds of millions data that are previously online today. The amount of data on World Wide Web are huge therefore it is very critical to store all data in an organized way, it also produced problem in data accessing.

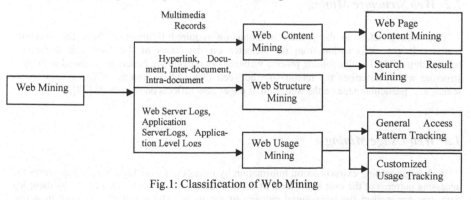

Fig.1: Classification of Web Mining

J. C. Bansal et al. (eds.), *Proceedings of Seventh International Conference on Bio-Inspired Computing: Theories and Applications (BIC-TA 2012),* Advances in Intelligent Systems and Computing 202, DOI: 10.1007/978-81-322-1041-2_35, © Springer India 2013

The World Wide Web has affected a lot to both users as well as the web site owners. The web site owners are able to achieve to all the targeted viewers countrywide and globally [27]. To extract frequent data from huge collection of data [11], data mining techniques can be applied. But the web data is unstructured and semi-structured, so we can not directly apply the technique of data mining .To a certain extent another application is evolved called web mining [10], which is applied on web data. There are several problems like improving web sites, to better understand the visitants behaviour, e-commerce, e-business, advertisements with the help of web mining we can discover interesting patterns in all above problems. Web usage mining [9] is accomplished first by coverage visitors transfer information, which is based on Web server log files and other source of transfer data .Web server log files were used primarily by the webmasters. They may be web architect, web developer, site author, or website administrator and system administrators like Databases, configuring a computer systems, software etc. Web server log files are used to contain the details of the user behaviour [14] and transactional details. These also stores the overall activities of the all users who access the websites.log files may contain the time of session start, details of web page which is access by user, traffic details or error information etc. Web log files plays vital role in web mining process because they provide overall details to the administrator for improving the performance of website in the World Wide Web environment. Web traffic data are handling through Web log file [13] this is the one way. Another way is to find out TCP/IP packets as they cross the network, and to attach to each Web server. After the Web traffic data is obtained, it may joint with other relational databases, over which the data mining techniques are implemented. By different data mining techniques such as association rules mining, path analysis, sequential analysis, clustering and classification, using all these techniques visitors' behaviour patterns are found and interpreted.

1.1 Web Content Mining

It refers to the extraction of useful information from huge data according to the contents. It obtains data from the web pages according to given contents. These contents can either text based or multimedia based. Web content mining generally deal with documents in text or html format and it also get information on the bases of image, audio, video or other contents [28].

1.2 Web Structure Mining

This mining refers to the process of obtaining required information from the structural patterns. It gets information from the websites on the bases of the links and documents relationships [29]. For this mining process websites documents are generally arranged in the tree structure which describes the relationship between different documents. When a user try to search a   particular page on the web, similar pages also reflects on the results [28].

1.3 Web Usage Mining

Web usage mining extracts useful information by using the server logs. Server logs stores the accessing patterns of the user in the form of URL, IP addresses or visiting times etc. by these log data, one can collect the behavioural patterns of the users. These patterns are used to define pattern discovery and associations between documents [28] [29].

2 Web Usage Mining Process

2.1 Data Pre-processing

There are lots of issues for pre-processing like Data Collection, Data Integration, and Transaction Identification. Pre-processing is a methods for converting the Content information, Structure information and usage information enclosed in the different presented data sources into the data abstractions necessary for pattern discovery. Log file pre-processing [18] consists of data cleansing, user identification, session identification. In data cleansing irrelevant records are eliminated. Records with GIF, JPEG, and CSS and so on as suffixes are eliminated. In the second step, we have the task of user and session identification is to find out the diverse user sessions from the original web access log. One way is to branching them based on their IP addresses.

2.2 Pattern Discovery

It is a process to find out patterns in web logs but is frequently approved only on samples of data. The mining process will be unsuccessful if the samples are not a good representation of the larger body of data [19]. According to Literature reviews following methods are used for pattern discovery process:

    A. Statistical Analysis
    B. Association Rules
    C. Clustering
    D. Classification
    E. Sequential Patterns
    F. Dependency Modeling

2.2.1 Statistical Analysis

It is the most general method to take out knowledge about visitors to a web site. We can perform different kinds of expressive statistical analyses like mean, median, mode, frequency etc [33]. On variables such as page visit, the time of visit and navigational path length. There are various web traffic analysis tools produce which generate an intervallic report containing statistical information such as the most commonly accessed pages, average view time of a page or length of navigational path.

2.2.2 Association Rules

It is a procedure for finding frequent patterns, correlations and associations [31] among sets of stuffs and it is used to relate pages that are most frequently located together in a single server session. Association rules [23], [26] are used in order to disclose correlations among pages accessed together throughout a server session. Those types of rules point out the possible relationship between pages that are often viewed together even if they are not directly connected, and can disclose associations between groups of users with specific interests.

2.2.3 Clustering

Clustering is used to group together a set of items that have similar characteristics. In the Web Usage Mining, there are two kinds of interesting clusters to be discovered user clusters and page clusters [20]. User clustering results in groups of users that seem to behave similarly when navigating through a Web site and Page clustering identifies groups of pages that appear to be conceptually related according to the user's perception.

2.2.4 Classification

Classification is the process of mapping a data into one of several predefined classes [6]. In the Web area, one is interested in developing a users profile belonging to a particular category or class. These necessitate selection and extraction of features that best explain the properties of a known class or category. Classification can be done by using supervised inductive learning algorithms [25] such as k-nearest neighbor classifiers, Vector Machines, decision tree classifiers, naive Bayesian classifiers etc.

2.2.5 Sequential Patterns

Sequential patterns indicate the correlation between transactions [32]. The method of sequential pattern discovery challenged to find inter-session patterns such that the presence of a set of objects is followed by another object in a time-ordered set of episodes or session. With the help of this approach, Web marketers can forecast future visit patterns which will be helpful in placing advertisements intended at certain user groups.

2.2.6 Dependency Modeling

The aim to develop this method is to prepare a model which is capable of representing significant dependencies between various variables in web domain. There are different kinds of learning techniques such as Bayesian Belief Networks and Hidden Markov models which can be employed to model the browsing behaviour of users. These models are also useful to analyse the behaviour of the users. Modelling of web usage patterns will not be provide a theoretical framework of users but is useful in forecasting future web resource utilization. By these models, future web resource consumption can be predicted [33].

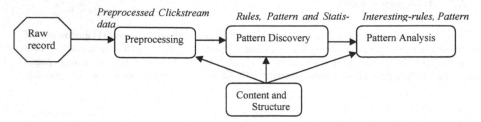

Fig2: Web Usage Mining Process Model

3   Literature Review

Mohd Helmy et al. [1] describes the pre-processing techniques on IIS Web Server Logs ranging from the raw log file until before mining process can be performed. Pre processing is very important and essential for data mining process, pre processing activities can be applied in various ways; it depends on the purpose of algorithm and nature of the applications. Ms. Dipa Dixit et al. [2] discuss two different approaches for data preprocessing one based on XML and other based on text file. But the way and steps involved in pre-processing are considered same for both the approaches. Arshi Shamsi et al. [3] presents, how web server log data is preprocesses, which includes data cleaning, user identification and Sessionization, path completion. If the data is preprocessed by some techniques it is used for discovering some useful patterns. T. Revathi et al. [4] describes an efficient approach for data pre-processing for mining Web based user data in order to speed up the data preparation process. It provides flexibility for data pre-processing and reduce complexity and difficulty of preparation for mining user data. However, we can't directly performed data mining process directly on the Web log data because of the messy and redundant content and other reasons. This paper describes the data pre- processing techniques for Web log data in order to meet the needs of data mining. M. Malarvizhi et al. [5] identifies the problems in existing techniques of preprocessing. It also proposes the possibility of improving the performance of preprocessing with several experiments. The experimental results show that the log error rate, log sizes are reduced and the quality is improved. Suneetha K.R et al. [6] presents algorithm for data cleaning, user identification and session identification. The main new approach of this paper is to access the usage pattern of preprocessed data using snow flake schema for easy retrieval.

4   Web Log Data and its Attributes

Web log file is log file that automatically created and maintained by web server. Every click on the website, include the HTML document, images or other objects are logged. It is essential that every raw web file format on one line of text for each click on the website. This contains information about the users who have already visited the sites. More recent entries are complicated to append at the end of the file. There are various attributes in log files [24] which are mentioned in the table1. This is the statistical analysis of server log which have used to examine traffic pattern on the time of the day, day of week, or user agent. Efficient web site administration adequate hosting resources and the fine turned off sales efforts can be aided by analysis of the web server logs.

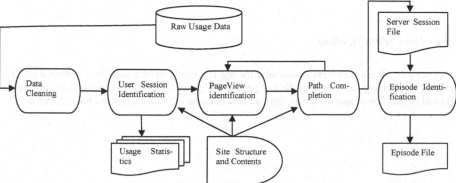

Fig.3: Preprocessing Model for Web Data

The following are the data preprocessing steps-

5.1 Data Cleaning

It is a process in which noise, unused and irrelevant data are removed [11]. It is also useful for web usage mining to clean [15] server log and to eliminate irrelevant information are of importance for any type of web log analysis. The discovery of associations or statistical report are only useful if data represented in the server log and it also gives the accurate picture of the user access to the web site, it is extremely significant, because only this log data that is able to accurately reflect the patterns of user access can be useful to search the correctness of the knowledge, get the model and the results meaningful. In web server log the problem arise when the HTTP protocol requires a separate connection for every file that is requested from the web server. When a user download a particular page then there are different elements are also downloaded with pages like graphics and scripts. In server log entries these all element details are stored. In most cases, only the log entry of the HTML file request is relevant and should be kept for the user session file then the Solution for that problem is to Eliminate some items deemed irrelevant can be reasonably accomplished by checking the suffix of URL name. All log entries with file name suffixes such as gif, jpeg etc. so that the list can be changed according to the site being analysed [33].

Table 1: Sample of web log

IP address	User Name	Timestamp	Access Request	Result Status Code	Bytes Transferred	Referrer URL	User Agent
123.456.78.2	U1	[25/Apr/1998:03: 04:41 -0500]	GET XYZ.html HTTP/1.0	200	1923	XYZ.html	Mozilla/4.7[e n]C-SYMPA (Win95; U)
123.456.78.9	U2	[25/Apr/1998:03: 05:20 -0500]	GET PQR.html HTTP/1.0	200	2828	PQR.html	Mozilla/4.05 (Macintosh; I; PPC)
123.456.78.3	U3	[25/Apr/1998:03: 06:20 -0500]	GET ABC.html HTTP/1.0	200	952	ABC.html	Mozilla/4.05 (Macintosh; I; PPC)

5.2 User Identification

User Identification Process comes after the log file has been cleaned. User Identification [16] means recognizing the user. It is the key part of the process of the server session identification. The identification of users is a very difficult task because of local caches and proxy servers.

5.2.1 User identification by IP address

IP address (computer address) is unique address for each user while browsing the website. So we can just consider that every new IP address represents a new user. But, it is poor user identification method when we are using the user's IP, because of the following problems:

1. Several users can be used the same IP address or computer (i.e. college, internet cafe etc.), so we do not know how many users hidden behind one IP address.
2. One user can have different IP addresses, since a user accesses the Web from different machines will have different IP address.
3. One user can use multiple browsers for the same IP address.

5.2.2 User identification using User registration Data:

Mostly website uses username and password for user identification. When user want to login a website; username and password are essential. These entries are also stored in the web log files; and useful for the next login. But these facilities are not available in every website so that it is not appropriated for the general web browsing [16].

5.2.3 User identification using Cookies:

Cookies are used to store temporary data while WebPages are downloaded on the client, it provides fast accessing if the request come for same data again. They are helpful to solve the problem of user identification. Cookies are HTTP headers in string format. By using Cookies we can extract the details of users and resources which are accessed by the user. If cookies are used for user identification then two problems can be arrived; first if the user lock the use of cookies the server can't store data on local machine. Second, user can delete the cookies. Therefore this technique is not reliable always. [9]

5.3 Session Identification

Session identification process comes after the user identification process. In this process we identify the session of users [17]. If a particular user visited the same site more than one then log entries can be divided in sessions [12].

In other word, if we group the different activities of a single user in the web log files is called session [30]. When a new user starts web page browsing, a new session is created, mostly sites define the time duration of session. Within this session duration that user can visit on multiple pages and these transactions are stored in log files.

Timeout is one of the methods for session identification; it uses assumption for the time duration between two page requests. If this predefines time exceeded then new session is started automatically [7].If proxy servers [8] are uses then log files are creates problems for session identification.

5.4 Transaction Identification

The goal of transaction identification is to create meaningful group of references for each user. By considering the time consumed by users in viewing the page, pages can be categorized as auxiliary or content page. Auxiliary pages are used to navigate from one page to another. Otherwise, Content pages are pages that provide useful contents to the user like information about contents. Based on this consideration two types of transactions [22] are defined. The first type is auxiliary-content transactions, where each transaction including of a single content reference and all of the auxiliary references up to the content reference and another is content based. Mining on these transactions give the common traversal paths to a given content reference.

5.5 Path Completion

Client side caching gives outcomes in accessing references to those pages whose cached are not recorded in the access log. By using heuristic method, the process identifies the missing records of that session. And these are all based on site structure is called path completion [21]. For example the user return the page X which is in its current sessions, if that page is cached at client side then no request is made to server and finally no other request required. On the website we found missing reference by the knowledge of website structure and those reference information available on web server. Fig. 4 shows the missing references. The structure of a site is created by hypertext links. The structure can be obtained and pre-processed in the same way as the site of content. There should be different site structure for every server session [3].

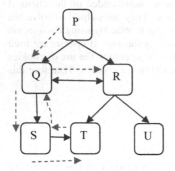
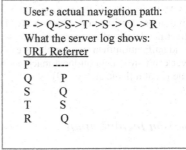

Fig 4: Missing Reference using path completion

6   Conclusion

Web site is considered to be the most important tool for advertisement in wide area. The feature of the website can be evaluated by examine user access of the website by web usage mining. We can identify user behaviour by the log records which is stored when user access the websites. In this paper we survey the research area of web usage mining and processing steps required for web usage mining. We have also discussed one of the processes which is data   pre-processing and its various stages. Data preprocessing stages are mainly Data Cleaning, User Identification, Session Identification, Transaction Identification and Path Completion.

7 Acknowledgment

This work is supported by research grant from MPCST, Bhopal M.P., India, and Endt.No. 2427/CST/R&D/2011 dated 22/09/2011.

References

[1] Mohd Helmy Abd Wahab, Mohd Norzali Haji Mohd, Hafizul Fahri Hanafi, Mohamad Farhan and Mohamad Mohsin "Data Pre-processing on Web Server Logs for Generalized Association Rules Mining Algorithm" World Academy of Science, Engineering and Technology 2008.

[2] Ms.Dipa Dixit and Ms. M. Kiruthika"Preprocessing of Web Logs" (IJCSE) International Journal on Computer Science and Engineering Volume 02, 2010.

[3] Arshi Shamsi, Rahul Nayak, Pankaj Pratap Singh and Mahesh Kumar Tiwari "Web Usage Mining by Data Preprocessing" IJCST Volume 3, Jan. - March 2012.

[4] T. Revathi, M. Mohana Rao and Ch. S. Sasanka "An Enhanced Pre-Processing Research Framework for Web Log Data" IJARCSSE Volume 2, March 2012.

[5] M. Malarvizhi and S. A. Sahaay." Preprocessing of Educational Institution Web Log Data for Finding Frequent Patterns using Weighted Association Rule Mining Technique", 2012.

[6] Suneetha K.R and R. Krishnamoorthi" Data Preprocessing and Easy Access Retrieval of Data through Data Ware House" WCECS, Volume 1, October 2009.

[7] Khasawneh N. And Chan C."Active user-based and ontology-based web log data preprocessing for web usage mining" IEEE/WIC/ACM International Conference on Web Intelligence, December 2006.

[8] Khasawneh N.,Shatnawi M.,Fraiwan M. "Converting Web Applications into Standard XML Web Services" The Tenth International Conference on Intelligent System Design and Applications, Dec 2010.

[9] R. Cooley, B. Mobasher, J. Srivastava,"Grouping web page references into transactions for mining world wide web browsing patterns", University of Minnesota, Dept. of Computer Science, Minneapolis, 1997.

[10] R. Kosala, H. Blockeel. "Web Mining Research: A Survey," In SIGKDD Explorations, ACM press, 2000.

[11] Han, J. and M. Kamber "Data Mining: Concepts and Techniques". A. Stephan. San Francisco, Morgan Kaufmann Publishers is an imprint of Elsevier, 2006.

[12] Raju. G. T. and Satyanarayana. P. S., "Knowledge Discovery from Web Usage Data: Complete Preprocessing Methodology", IJCSNS International Journal of Computer Science and Network Security, Volume8, January 2008.

[13] Suneetha, K. R. and D. R. Krishnamoorthi "Identifying User Behavior by Analyzing Web Server Access Log File" IJCSNS International Journal of Computer Science and Network Security, Volume 9, April 2009.

[14] Etminani, K., Delui, A.R., Yanehsari, N.R. and Rouhani, "Web Usage Mining: Discovery of the Users' Navigational Patterns Using SOM", First International Conference on Networked Digital Technologies, 2009.

[15] Ramya C and Kavitha G, "An Efficient Preprocessing Methodology for Discovering Patterns and Clustering of Web Users using a Dynamic ART1 Neural Network", Fifth International Conference on Information Processing, Springer, 2011.

[16] Renata Ivancsy, and Sandor Juhasz, "Analysis of Web User Identification Methods", World Academy of Science, Engineering and Technology, Volume 34, 2007.

[17] Ling Zheng, Hui Gui and Feng Li, "Optimized Preprocessing Technology for Web Log Mining", International Conference on Computer Design and Applications, Volume1, 2010.

[18] Li Chaofeng," Research and Development of Data Preprocessing in Web Usage Mining", International Journal of computer applications, 2011.

[19] Shaimaa Ezzat Salama, Mohamed I. Marie, "Web Server Logs preprocessing for Web Intrusion Detection",Computer and Information Science, Volume 4, 2011.

[20] Liang Wei and Zhao Shu-hai,"A Hybrid Recommender System Combining Web Page Clustering with Web Usage Mining", International Conference on Computational Intelligence and software Engineering, 2009.

[21] Yan Li, Boqin Feng and Qinjiao Mao, "Research on Path Completion Technique in Web Usage Mining", International Symposium on Computer Science and Computational Technology, Volume 1, 2008.

[22] Jian Chen, Jian Yin, Tung, A.K.H. and Bin Liu, "Discovering Web usage patterns by mining cross-transaction association rules", International Conference on Machine Learning and Cybernetics, Volume 5, 2004.

[23] Yi Dong, Huiying Zhang and Linnan Jiao, "Research on Application of User Navigation Pattern Mining Recommendation", Proceeding of the 6th World Congress on Intelligent Control and Automation, IEEE,China, June 21 – 23, 2006.

[24] Tasawar Hussain, Sohail Asghar and Nayyer Masood "Web Usage Mining: A Survey on Preprocessing of Web Log File" Center of Research in Data Engineering (CORDE) Department of Computer Science, 2010.

[25] Sanjay Bapu Thakare,Sangram and Z. Gawali "A Effective and Complete Preprocessing for Web Usage Mining" (IJCSE) International Journal on Computer Science and engineering, Volume 2, 2010.

[26] D.S. Rajput, R.S. Thakur and G.S. Thakur "Rule Generation from Textual Data by using Graph based Approach" International Journal of Computer Applications New York, USA, Nov. 2011.

[27] Aditi Shrivastava, Nitin Shukla "Extracting Knowledge from User Access Logs" International Journal of Scientific and Research Publications, Volume 2, Issue 4, April 2012.

[28] Liu Wenyun, Bao Lingyun, "Application of Web Mining in E-Commerce Enterprises Knowledge Management", International Conference on E-Business and E-Government, IEEE, 2010.

[29] Zhang Haiyang, "The Research of Web Mining in E-commerce", IEEE, 2011.

[30] Vijayashri Losarwar, Dr. Madhuri Joshi, "Data Preprocessing in Web Usage Mining", International Conference on Artificial Intelligence and Embedded Systems (ICAIES'2012) Singapore, July 15-16, 2012.

[31] R. Agrawal, R. Srikant, "Fast Algorithm for Mining Association Rule", International Conference on Very Large Databases, Santiago, Chile, September 1994.

[32] R. Agrawal, R. Srikant ,"Mining sequential patterns",11th International conference,IEEE Computer Society Press, Taiwan,1995.

[33] Jaideep Srivastav, Robert Cooley, Mukund Deshpande, Pang-Ning Tan," Web Usage Mining: Discovery and Applications of Usage Patterns from Web Data", ACM SIGKDD,Volume 1, Issue 2,Jan 2000.

An Immunity Inspired Anomaly Detection System: A General Framework

Praneet Saurabh[1], Bhupendra Verma[2], Sanjeev Sharma[3]

praneetsaurabh@gmail.com, bk_verma3@rediffmail.com, sanjeev@rgtu.net

[1,2] Department of Computer Science and Engineering, TIT, Bhopal, M.P, India

[3] School of Information Technology, RGPV, M.P, India

Abstract. Exponential growth of internet acted as a centrifugal force in the development of a whole new array of applications and services which drives the e- business/ commerce globally. Now days businesses and the vital services are increasingly dependent on computer networks and the Internet which is vulnerable to the evolving and ever growing threats, due to this the users who are participating in various activities over internet are exposed to many security gaps which can be explored to take advantage. These alarming situations gave rise to the concern about security of computer systems/ networks which resulted in the development of various security concepts and products but unfortunately all these systems somehow fail to provide the desired level of security against ever-increasing threats. Later on it has been observed that there lies a huge analogy between the human immune system (HIS) and computer security systems as the previous protects the body from various external and internal threats very effectively. This paper proposes a general immunity inspired security framework which uses the concepts of HIS in order to overcome the ever growing complex security challenges.

Keywords:-Computer Security, Threat, Anomaly, HIS, Immunity.

1 Introduction

The reliance of the world's infrastructure on computer systems is immense as computers and the services offered by them are used in every domains of life and participation can be through any of the methods such as electronic communication, e-commerce over the Internet [4]. All the information sent or shared on the public network, so the potential chances of its misuse is enormous as it is also exposed to unauthorized "hackers" [5,6]. Pervasiveness of "Network/ Computer Security" makes it a concern of great significance and worth because it has become a prerequisite in the digital age [6, 7].The essence of Network and Computer security is to protect data/ information from unauthorized access, use, disclosure, disruption, modification or destruction [1, 2]. CSI survey 2009 [4] reported big jumps in incidence of financial fraud (19.5 percent); malware infection (64.3 percent); denials of service (29.2 percent) are a few of them.

AIS (Artificial immune System) [14, 15] are modeled after HIS (Human Immune System) are a computational systems inspired by the principles and processes of the biological immune system which enables every organism to survive from the various threats posed by the environment. The way the human body reacts to these different threats and attacks encourages the researchers to model computer security system after Human Immune System as its survives under very demanding circumstances efficiently[9, 19].

Section 2 highlights the background information of computer security and available tools, Section 3 contains perspective of a biologically inspired security system, Section 4 details the proposed work and implementation, discussion about the

J. C. Bansal et al. (eds.), *Proceedings of Seventh International Conference on Bio-Inspired Computing: Theories and Applications (BIC-TA 2012),* Advances in Intelligent Systems and Computing 202, DOI: 10.1007/978-81-322-1041-2_36, © Springer India 2013

results of the experiments performed is covered in Section 5 and Section 6 concludes the discussion.

2 Background Information

The nature of threat has changed over the years and these threats/ attacks needs to be classified in order to determine its severity and origin.

2.1 Types of Attacks

Attacks and intrusions identified at the user level by Harmer et al. [20] are as:

(1) **Misuse/abuse:** unauthorized activities by authorized users.
(2) **Reconnaissance:** findings of systems and services that may be exploitable.
(3) **Penetration:** successful access to computing resources by unauthorized users.
(4) **Denial of service:** an attack that obstructs legitimate access to computing.

2.2 Intrusion Detection System

Intrusion detection was identified as a potential research and valuable area of computer security by Denning [11] in 1987, later on computer professionals started to realize the significance of protecting computer systems. Intrusion detection System can be defined as the tools, methods, and resources to help identify, assess, and report unauthorized or unapproved network activity [12]. It works on the principle that the behavior of the intruder will be different from that of a legitimate user that can be quantified.

On the basis of basic detection techniques IDS are classified into two; the first one is **Misuse Detection** also called Rule-Based Detection or signature detection technique. This technique is based on what's wrong. It contains the attack patterns (or "signatures") [10,12] and match them against the audit data stream, for evidence of known attacks and the second one is **Anomaly based Detection** is also referred as profile-based detection [11] compares desired behavior of users and the applications with the actual ones by creating and maintaining a profile system that flags any events that strays from the normal pattern and passes this information on to output routines [1]. Intrusion detection systems have their own problems, such as false positives, operational issues in high-speed environments, and the difficulty of detecting unknown zero day threats.

All the current solutions for network security are based on static methods that gather, investigate and mine evidences after attacks, these solutions lack in addressing the additional and advanced requirements such as false positive, single point solution, reactive approach to tackle threat and due to this performance suffers heavily on these parameters. An ideal solution will have self-learning and self-adapting abilities which contribute in detection of unknown and new attacks.

3 Biological Perspective in Computer Security

Biologically-inspired computational models offer a wide range of techniques and methods that can be used to develop a computer security solution as the mechanisms of immune responses are self-regulatory in nature and also there is no central organ that controls the functions of the immune system.

3.1 Artificial Immune System

AIS are a computational systems inspired by the principles and processes of the biological immune system [13,15]. The biological immune system is a robust, complex, multilayered and adaptive system which defends the body from foreign pathogens, these excellent features prompts to model in order to develop a security system that is based on BIS as there is a strong correlation between immune system and a computer network security system[23]. BIS is a successful classification system which distinguishes between self and nonself as well as "good" self/ nonself and "not good" self/ nonself [14], also it has a very powerful information processing capabilities, these two qualities are prime necessities of any computer security system[8, 9]. Immune system is constituted by central lymphoid whose purpose is to generate and mature immune cells, bone marrow and thymus to do the task and peripheral lymphoid organ facilitates the interaction between lymphocytes and antigen [13].Thymus produces mature T cells, but it releases only the beneficial T-cells to the blood stream and discards the remaining ones [16]. Immune system can be divided into two categories [16]:

A. **Innate Immune System:** The unchanging defense mechanisms which provide security cover against foreign pathogens and with which an individual is born with. It does not provide complete protection, as it is primarily and static in nature.

B. **Adaptive Immune System:** It is also termed as acquired immunity because it builds a memory over a period of time to achieve a faster response when the same antigen is confronted at a later time. It acts as a supplement to innate immunity.

Table: 1 Correlation Modeling of BIS terms in AIS [27]

BIS Terms	AIS Terms
T cells, B cells, and antibodies	Detectors, clusters, classifiers, and strings
Self-cells, self-molecules, and immune cells	Positive samples, training data, and patterns
Antigens, pathogens, and epitopes	Incoming data, verifying data samples, and test data
String-matching rule Complementary rule and other rules	Distance and similarity measures Affinity measure in the shape–space

AIS is used in developing many computational models, Host intrusion detection (HID) was initially studied by Forrest [15] using sequences of system calls as the detectors. Aickelin [26] discussed various Immune system approaches to intrusion detection, Kim and Bentley [18] studied Network intrusion detection (NID) and employed a variety of AIS techniques including static and dynamic clonal selection with immune memory. Harmar et. al.[20] in various experiments used the traits of human immune system in their system which detects foreign substances and responds appropriately to the foreign substances.

3.2 Models for AIS: State of the art.

Based on Human Immune System following are the paradigms for A.I.S.[15, 21] which can be used for modeling computer security system.

3.2.1 Negative Selection Paradigm

Negative selection paradigm is the immune system's ability to detect unknown antigens while not reacting to the self-cells [13, 20]. A lot of variations of negative selection algorithms have been frequently proposed, but the crux remains the same [15, 16] is to build self profile of the normal network patterns as self and other patterns are as non-self. Once this profile has been built, the non self patterns are can be easily filtered out and termed as anomalous. If the random incoming pattern matches a self pattern then it is removed, otherwise it becomes a detector pattern and monitors newly arriving patterns. Forrest [17] applied the idea of negative selection along with the concept of detectors to detect anomaly inspired by T-cells, represented in binary strings. A detector string match an antigen string if the two strings shared the same characters in an uninterrupted stretch of r-bits; this is known as the r-contiguous bits matching rule. If the detector matches any new pattern, then it has detected an anomaly and becomes a memory cell which is stored for future attacks. Zhang et. al. [24] proposed a novel distributed intrusion detection model based on immune agents (IA-DIDM) which is inspired by Negative Selection paradigm lymphocyte's working in the BIS. The detectors represent non-self and mature through negative selection algorithm. Chung et. al.[22] discussed Host-based intrusion detection systems adapted from artificial immune systems. The system by Ishida [25] introduced a mechanism of diversity generation in immunity-based anomaly detection system inspired by positive and negative selection in the thymus.

3.2.2 Clonal Selection Theory

Clonal selection theory is population-based search and optimization algorithm which generates a memory pool of suitable antibodies in order to solve a problem. It is based on proliferation of immune cells [14] and can be viewed as a random event because a particular lymphocyte which is assigned to produce clones from a large pool of lymphocytes. These activated lymphocytes flourish through the process called cloning [16].

All these models and the current state of the art embodies that biological immune mechanism is in process of developing new computational intelligence applications through the computational models (Negative selection algorithm (NSA), Clonal Selection). There lies a huge potential in AIS which can be used in the development of a computer security solution which will look after the increasing threat perception in a more structured and organized manner despite the heterogeneity of the networks.

4 Immunity Inspired Anomaly Detection Framework

This section presents an Immunity Inspired Anomaly Detection Framework (IIADF), which embodies desirable immune attributes, such as adaptation, learning capability, and self-configuration. Biologically-inspired approaches for anomaly detection systems have proven to be very interesting; it has the ability to perform anomaly detection as it detects pathogens that has been never encountered.

4.1 Scheme

The proposed framework uses the concepts of natural immune system for anomaly detection as it finds it apt for the purpose. IIADF involves concept of random detector generation, preprocessing of the incoming traffic which will be used as input in the selection of active and fine-tuned detectors, matching with the test dataset to identify the anomaly.

4.2 Model Narration

The block diagram the proposed paradigm Immunity Inspired Anomaly Detection Framework **(IIADF)** shown in **Fig: 1** consists of two modules: **Detector Generation and Selection Module (DGSM)** and **Anomaly Detection Module (ADM).** DGSM is all about generating detectors; then creating finetuned detector sets by utilizing the preprocessed training set (containing network traffic features) and the Real-Valued Negative Selection Algorithm (RV-NSA). ADM is then applied on the test set (unseen network traffic) to detect anomalies by comparing it with the detectors of the detector set generated by **DGSM.**

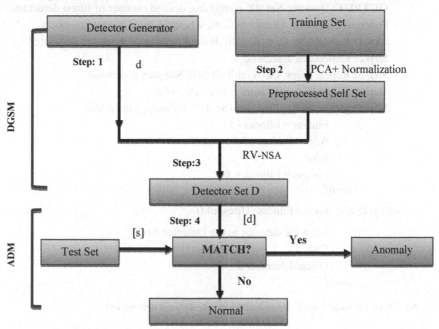

Fig: 1 Block Diagram of Immunity Inspired Anomaly Detection System

(i) Detector Generation and Selection Module (DGSM)

This module forms the basis for the whole framework. DGSM uses a real-value detector generator engine to generate random detectors; once the detectors are generated, its uniqueness is verified using preprocessed training set (Self Set). RV-NSA is used to select unique detectors and are kept in the detector-set forming the input for second module ADM. In DGSM, the *Self-Set S* is first preprocessed by applying *Principle Component Analysis (PCA)* and *Min-Max Normalization*. The most significant Principal Components (Network Features) are extracted using PCA

Min-Max Normalization is applied on these Principal Components to reduce the detector search space and increase detector generality.

Algorithm for DGSM

INPUTS:

Preprocessed Self Set (PCA + Normalization): S = {S1,S2,...,Si}, where S is a feature vector containing different network features (S1, S2,.., Si)extracted from the network traffic; i belongs to the set of positive integer numbers.

Detector 'd': generated by the Detector Generator Engine.

Affinity_Threshold: any real value in the range [0,1]

Fitness_Threshold = any real value in the range [0,1]

Radius = any random value in the range [0,1]

OUTPUT: Detector-Set 'D' containing desired number of fittest detectors.

Function Euclidean_Distance (Si,d), where Si and d are real numbers.

Step 1: Repeat Steps 2 to 6, until desired number of detectors generated

Step 2: Generate a detector d

Step 3: Repeat Steps 4 to 6 until all Self-Samples processed.

Step 4: Select feature Si from S and detector di.

Step 5: if (Euclidean_Distance (Si, d) < Affinity_Threshold)

        Fitness = Fitness - 1

        Adjust Radius of detector

        Else

        Fitness = Fitness + 1

    endif

Step 6: if (Fitness >Fitness_Threshold)

        Keep the detector in the Detector-Set D

        Else

        Discard detector d

    endif

Note: Steps 1 through 6 could be repeated to generate different detector sets.

PCA (Principal Component Analysis) is used to extract the most significant fields in KDD Cup Dataset [3] which is used as *Self Set, S* for training of detectors. The KDD Cup 99 dataset, is derived from the DARPA IDS evaluation dataset, The complete dataset has almost 5 million input patterns and each record represents a TCP/IP connection that is composed of 41 features, both qualitative and quantitative in nature. The dataset used in this work is a smaller subset (10% of the original training set), that contains 494,021 instances. 38 features are selected for the classification about data to quantify it as normal or anomaly. Classical Signature based R-Contiguous matching algorithm for 38 features would require string (signatures) of length 38 bits for matching, and it would have taken large amount of memory, disk space and processing time. Principal Component Analysis (PCA) is applied on all vectors each containing

38 features yields the most significant Principal Components called the network features. The output of PCA is spread in a wide range (e.g. -32322 to 54334), the Min-Max Normalization is used to normalize these principal components in the range [0,1], as in the current state it would have taken much time and processing to converge the training curve towards the goal, also the detectors generated by using unnormalized values would have failed to detect anomalies. PCA with Min Max Normalization make the system stable and helps in selecting better trained detectors. Detectors are generated with an objective of detecting the unknown or even a zero day attack. Euclidean distance between each *Self Sample 'S'* and the generated detector *'d'* is obtained using formula:

$$d(x,y) = \sqrt{\sum_i (x_i - y_i)^2} = \| x - y \|$$

In immunological terms affinity of an antibody is the power it boasts to bind with an antigen; it is based on a distance measure between points in the shape–space. Small distance between an antibody and an antigen represents high affinity between them. In the proposed framework **IIADF** a detector is matched with the normalized incoming traffic, to achieve this task affinity of detectors is calculated because it is considered as a criterion for keeping or rejecting the detector. More affinity represents higher chances of the selection of detectors.

Euclidean Distance < *affinity threshold* reflects that the detector *'d'* is close to the *"self"* sample and NSA discards such a detector in the normal scenario after just one comparison; this limits the selection of detector for the detector set but in IIADF concept of "Fitness" is introduced in the existing RV-NSA as an additional selection and rejection criteria for the detectors generated. A fitness-threshold of detector having value 0.7 means that the detector would be selected only when it fails to match at least 70 self samples out of 100. To increase the fitness of a detector dynamically it's value is modified slightly everytime it matches with a self sample. It is done by introducing a parameter called *"radius"*. These different values of *r* also enable the detectors to cover up the unfilled space in state space representation. This whole concept of fitness actually fine tunes the detectors which enable them to detect an anomaly with a higher precision.

(ii) Anomaly Detection module (ADM)

ADM is the second module of **IIADF** which performs anomaly detection in a preprocessed test set, using detector-sets generated by DGSM. *Euclidean Distance*'d' is used as a similarity measure and when this distance between a test sample and a detector *'d'* is found to be below a predefined threshold, the test sample is classified as an anomaly. For the experimental purpose, three custom datasets are generated with KDD CUP 1999 10% dataset by random selection of feature vectors. The KDD Cup dataset contain 41 features of a network packet/connection and 1 additional classification field to show the category of the packet – attack type/normal. There are total 4 categorical fields those could be converted into numeric values but we decide to not to use them because 38 features are sufficient enough to classify the data as normal or anomaly.

5 Experimental Results & Analysis

Experiment-1 Detector Selection/ Rejection vs. Strength of Detector

Detectors form the basis of anomaly detection, as the prime task of detectors is to identify the anomaly. This experiment is performed with a view to find how many detectors are generated and then selected from them to detect anomalies.

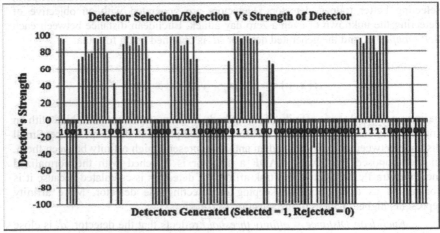

Fig: 2 Detector Selection/ Rejection vs. Strength of Detector

In the first selection criteria NSA selects the detectors; this selection signifies that the particular detector was not previously kept in the detector-set, as a second criteria Detectors having strength > 70% are selected to form the detector set. Rest of the detectors which fails to match either of the set criterions is rejected. As depicted in figure 5.1, selected detectors are labeled as '1' on the X-axis and rejected detectors are depicted with a '0' on the X-axis. Strength of detectors is depicted on Y-axis.

Experiment- 2 Matching Threshold vs. Generation/Rejection Rate

This experiment is carried out in order to identify the fittest detectors that will be put into the decector set which will identify the anomalies.

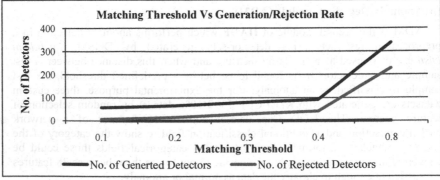

Fig: 3 Number of Detectors vs. Detection Rate

Matching threshold is the metric which signifies degree of matching or power of matching, a detector will have in its quest to identify anomaly. The more the value of matching threshold the more likely is its chances to identify an anomaly.

The results depicted in Figure: 3 also justifies this theory, result illustrates that as the matching threshold increases from 0.1 to 0.8 the number of detectors generated increases, in the same manner the rate of rejection of the detectors which do not pass the set of conditions are rejected. Detector set with detectors having high threshold and higher rejection ratio facilitates and enhances the chnace of identifying the anomalies.

Experiment-3 Generalization of Detectors (Fine Tuning)

Anomaly based detection approach requires consistent and constant modeling of normal traffic and fine tuning of the detectors achieves the task of identification of unknown patterns or detection. This experiment is done to with a motive to find how the selected and finetuned detectors perform the task of detecting anomalies and what the percentage of false positive is when a completely new data set is used for the testing purpose.

Test Set Used: Unseen Self Set Containing 33014 Normal Packets.
Details about Training File: normal.csv
No. of Self Samples: 58000; No of Detectors to be Generated: 500; Euclidean Threshold: 0.7; Detector's Fitness Threshold: 0.7; Detector's Radius Manipulator: 0.1; Alert Threshold: 1.

Experimental Result: Number of Detectors Generated: 1672; No. of Detectors Selected: 500; Number of Detectors Rejected: 1172; Test File used: .\normal_Unseen.csv; Number of Normal Packets in Test File: 33014; Number of Packets to be processed: 33014 No. of Anomalous Packets in Test File: 0; Number of Anomaly Detected: 1; Number of Normal: 33013

Anomaly Detection Rate for Unseen Normal Packet: 0.0030%

The anomaly detection rate for all of these normal packets is 0.003%; this reflects the fact that fine tuned detectors detect anomalies in a more efficient manner and the results reflect that they are not detecting the self-set, the lesser percentage of false positive just indicates that theory. Unseen self set is used for training of detectors, this test set is used for validation of detectors and to fine tune their ability to detect anomaly, hence reducing the false alarm rate. Lower false alarm is very good reflector that the proposed system is working fine and it can even detect a 0-day attack which still is very much one of the open issues in the field of ID.

Experiment-4 Number of Detectors vs. Detection Rate

The results of experiments with different test sets show and point that higher number of detectors can achieve higher detection rates when applied to different test sets.
Test Set 1.
Total Packets = 3269, Normal Packets = 2000, Anomalous Packets = 1269
Test Set 2.
Total Packets = 730, Normal Packets = 300, Anomalous Packets = 430
Test Set 3.
Total Packets = 1000, Normal Packets = 450, Anomalous Packets = 550

This testing approach based on different detector sets detects a group of different attack types more efficiently with lesser false positives. The experimental results also indicate that the detection rate of anomalies is higher when there is more number of detectors to do the task. The mechanism of polling fine tunes the detectors that results in lowering the rejection of effective detectors so that the number of detectors must be sufficient enough so that detection ratio remains high.

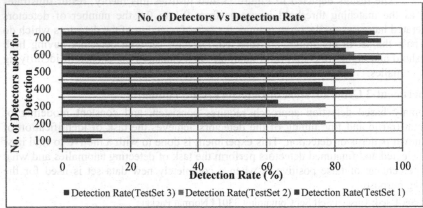

Fig: 4 Number of Detectors vs. Detection Rate

Experiment-5 Attack Type vs. Detection Rate

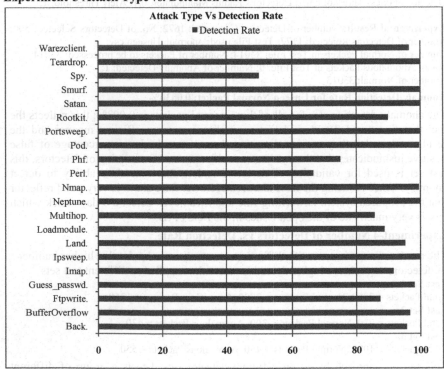

Fig- 5 Attack Type vs. Detection Rate

Training File: normal.csv;
No. of Self Samples: 58000; No of Detectors to be Generated: 500
Euclidean Threshold: 0.5; Detector's Fitness Threshold: 0.7;
Alert Threshold: 1; Detector's Radius Manipulator: 0.1

The objective of this experiment to summarize the detection of different anomalies present in the given traffic set by using the proposed technique IIADF which uses the concepts of fitness, strength and finetuning of detectors.

The system uses fine tuned detectors to identify the anomalies in the given traffic. The criteria of fitness in the selection of detectors and then fine-tuning it helps in achieving high detection rate when tested against all the different types of attack patterns present in the self samples. Detector is only selected when it fails to match at least 70 self samples out of 100. These features in IIADS results in higher detection rate with low false positive which is encouraging, as false positive plays a vital role in determining the effectiveness of the method.

6 Conclusion

The framework IIADF discussed in this paper uses the mechanism of NSA along with fitness, strength and fine tuning of detectors. This paper introduces the concepts of fitness and strength of detectors which help in covering the open unaddressed space in the problem state space modeling, the mechanism of polling fine tunes the detectors that results in lowering the rejection of effective detectors. All these features of IIADF make the detector set very robust which is reflected in the results which are encouraging as the detection rate falls between 93-99%, and lower false positive when tested against different set for unseen data. IIADF forms a base for host based real time anomaly detection system. This framework is still in its nascent phase of its development and can pave way in growth of self configuring prevention system with lesser false positive and more intelligence.

References-

[1.] Marin A G., Network Security Basics, Security and Privacy, PP 68- 72, IEEE, 2005.
[2.] Bishop M, Computer Security Art and Science, Pearson Education, 2003.
[3.] http://kdd.ics.uci.edu/databases/kddcup99/kddcup99.html, Ocotber 2007, KDD Cup 1999.
[4.] Richardson Robert, "CSI Computer Crime & Security Survey 2010", 2011.
[5.] Zafar F. M., Naheed F., Ahmad Z. and Anwar M.M., Network Security: A Survey of Modern Approaches, The Nucleus, 45 (1-2), pp 11-31, 2008.
[6.] Bishop M, An Overview of Computer Viruses in a Research Environment, 4th DPMA, IEEE, ACM Computer virus and Security Conference, pp 154-163, 1997.
[7.] Guillen E, Weakness and Strength Analysis over Network-Based Intrusion Prevention & Prevention Systems" Communications, pp 1-5, 2009.
[8.] Chia-Mei Chen, Ya-Lin Chen, Hsiao-Chung Lin, An efficient network intrusion detection, Computer Communications 33, pp 477–484, 2010.
[9.] Xiaonan S, Banzhaf W, The use of computational intelligence in intrusion detection systems: A review, App Soft Computing,1–35,2010.
[10.] Endorf C., Schultz E. and Mellander J.; Intrusion Detection & Prevention, Published by McGraw Hill, 2004.
[11.] Denning E..D., An Intrusion-Detection Model, IEEE Transactions on Software Engineering, Vol. SE-13.No. 2, pp 222-232, 1987.

[12.] Mukerjee B., Heberlein T. L., Levitt N. K. Network Intrusion Detection, IEEE Network, pp.26-41, 1994.

[13.] Castro L., Timmis J., Artificial Immune Systems as a Novel Soft Computing Paradigm. Soft Computing, Journal, vol. 7, Issue 7, pp 526-544, 2003.

[14.] Dasgupta D., Forrest S. An Anomaly Detection Algorithm Inspired by the Immune System. Chapter 14 in the book entitled Artificial Immune Systems and their Applications, Publisher: Springer-Verlag, Inc., pp. 262–277, January 1999.

[15.] Forrest S., Hofmeyr S., Somayaji A., "Computer Immunology," In Communications of the ACM, vol. 40, no. 10, pp. 88–96, 1997

[16.] Dasgupta D., Yu S., Nino F, Recent Advances in Artificial Immune Systems: Models and Applications, Applied Soft Computing, 11, 1574–1587, 2011.

[17.] Forrest S., Hofmeyr A. S., Somayaji A., Longstaff , A sense of self for Unix processes, in Proceedings of the 1996, IEEE Symposium on Security and Privacy, pp 120-128, 1996.

[18.] Jungwon K., Bentley P, "The Human Immune System and Network Intrusion Detection", EUFIT 99, pp 1244-1252, September 1999.

[19.] Nishiyama H, Mizoguchi F, "Design of Security System Based on Immune System", pp 138-143, IEEE 2001.

[20.] Harmer K.. P., Williams D. P., Gunsch H.G., Lamont B. G., An artificial immune system architecture for computer security applications. IEEE transactions on evolutionary computation 6(3), pp 252–280, 2002.

[21.] Dasgupta D., Immunity-based Intrusion Detection System: A General Framework, In: Proceedings of 22nd national information systems security conference, Arlington, Virginia, USA, pp 147–160, 1999.

[22.] Ou M Chung, Host-based intrusion detection systems adapted from agent-based artificial immune systems, Neurocomputing, Elsevier, pp 78-86, 2012

[23.] Timmis J., Hone A., Stibor T., Clark E., Theoretical advances in artificial immune systems, TCS, Elsevier, 403, pp 11-32, 2008.

[24.] Zhang Z., Luo W., Wang X., Designing abstract immune mobile agents for distributed intrusion detection. In: Proceedings of the international conference on neural networks and brain, ICNN&B '05, Beijing, (2), pp 748–753, 2005.

[25.] Ishida Y., The next generation of immunity-based systems: from specific recognition to computational intelligence, Studies in computational intelligence (SCI), vol 115. Springer, Berlin, pp 1091–1121, 2008.

[26.] Aickelin U., Greensmith J., Twycross J., Immune system approaches to intrusion detection-a review. ICARIS 2004, LNCS 3239.Springer, pp 316–329, 2004.

[27.] Overil E R. "Computational immunology and anomaly detection", Information Security Technical Report, Science Direct, Vol.12, pp 188-191,2007.

Promotional Resource Allocation for a Product Incorporating Segment Specific and Spectrum Effect of Promotion

Sugandha Aggarwal[1], Yogender Singh[2], Anshu Gupta[3], P.C. Jha[4]

[1,2,4] Department of Operational Research, University of Delhi, Delhi

[3] SBPPSE, Dr B.R. Ambedkar University, Delhi

{sugandha_or@yahoo.com; aeiou.yogi@gmail.com; anshu@aud.ac.in; jhapc@yahoo.com}

Abstract. Firms invest a huge proportion of their resources in promotion to acquire higher adoption of their products. To get maximum possible returns from promotional resources judicious and effective spending is essential, especially in a segmented market. Mass and differentiated market promotion are typically two different techniques of promotion used in the segmented market. With mass promotion product is promoted in the entire market using a common strategy, thereby creating a spectrum effect in all segments. Differentiated market promotion targets the segments specifically. In this paper we formulate a mathematical programming problem to optimally allocate the limited promotional resources for mass market promotion and differentiated market promotion among various segments of the market, maximizing the total sales measured through product adoption under budgetary constraint and minimum target sales level constraints in each segment. A recent innovation diffusion model is used to measure the adoption in each segment which describes the sales through the combined effect of mass and differentiated promotion. The solution procedure has been discussed using NLPP methods. The optimization model is extended incorporating aspiration constraint on total sales from all the segments. Such a constraint can result in an infeasible problem. To obtain best compromised solution, differential evolution approach is used. Results are demonstrated through a numerical illustration.

Keywords: Innovation Diffusion, Mass Market Promotion, Differentiated Market Promotion, Spectrum Effect, Promotional Effort Allocation, Differential Evolution.

1 Introduction

Promotion is a vital part of the firm's marketing mix which by informing, educating, persuading, and reminding the target market about the benefits of the product, positions the product in the market place. A major reason of failure of most of unsuccessful products in marketplace is ineffective promotion, irrespective of their product design, pricing and distribution. Thus, firms develop promotional campaigns for its products in a way that it effectively and efficiently communicates message to the target market. The target market for almost all products is heterogeneous, consisting of buyers with different interests, preferences and perceptions. To capture maximum from the target potential market, marketers stresses on segmenting the market among

J. C. Bansal et al. (eds.), *Proceedings of Seventh International Conference on Bio-Inspired Computing: Theories and Applications (BIC-TA 2012)*, Advances in Intelligent Systems and Computing 202, DOI: 10.1007/978-81-322-1041-2_37, © Springer India 2013

relatively homogeneous groups with similar characteristics. These characteristics can be defined either geographically (region, state, countries, cities, neighbourhoods); demographically (age, gender, income, family size, occupation, education); psychographically (social class, life style, personality, value) or behaviorally (user states, usage rate, purchase occasion, attitude towards product). A product is usually promoted in such markets using two types of promotional approaches: *Mass market promotion* and *differentiated market promotion*. Mass promotion addresses all segments of the market alike which results in promoting a product using a single promotional plan in the entire market. In this way the firm appeals to a wide variety of potential customers by ignoring differences that exists between them [1, 2]. Being visible to audience in the entire market such type of promotion creates an effective segment-spectrum, which is distributed across all the segments of the market [3]. Mass promotion focuses on the average behavior of the entire population of potential customers. The potential customers in one segment have some distinguished characteristics which differentiate them from the other segment. The influence of mass promotion on any segment is only partial. In order to target a specific customer segment, differentiated promotion is carried by targeting individual segments through distinct promotional strategies [1, 2, 4]. This kind of promotion can also be called segment-specific promotion in which each segment is tailored separately through unique promotional strategies.

In a lure to capture maximum adopter population, firms carry out both mass market as well as differentiated market promotion. In this way it is made sure that they do not miss out any potential consumers. Mass promotion also allows them to reach average adopter population who may not fall in the supposed target market but would still be interested in buying the product in future. Whereas differentiated promotion targets the specific segments. Product acceptance arises due to promotional influence of these two strategies. This concept holds significantly in a country like India which has a large culture and language base. Indian market can be segmented geographically into various regional segments as the difference in the marketing environments of various regions of the country suggests that each market is different and requires a distinct marketing and promotional programs [5]. Companies cater diversified customer base of India by promoting products in each region independently keeping in mind their geographic, psychographic and behavioral aspects such as promotional messages are given in their native regional language, promoting the product through local event, regional TV channels etc. Along with this the product is promoted using the techniques of mass promotion such as promotion through national TV channels in national language, promotion through national events etc. which reaches several regions and influences the product awareness as well as product acceptance in each of the geographical segments. For instance firms in Indian automobile industry such as Maruti Suzuki India Limited, Hyundai, Honda hits several platforms and connects them efficiently to produce sales. They use multitude of national and regional promotional vehicles for designing their mass and differentiated promotional strategies. Targeting the potential customers with mass as well as differentiated promotion is not only seen in the durable technology product segment but also in the consumer good section. For example, Hindustan Unilever Limited promotes most of its products through these two strategies viz. Fair & Lovely, a skincare cream for women is promoted using TV commercials on national TV channels like Doordarshan, Zee, Star, Sony etc. These commercials are dubbed in various state dominant regional languages and are also telecasted in regional TV channels. In this way, higher adoption rates are yielded among different regions.

Every organization sets an upper bound on resources to be spent on promotion. Moreover promotion takes off a large amount of company's investment, so measuring its effectiveness and utilizing it efficiently in different market segments becomes important. As each segment responds uniquely to the promotional activities done for it, the problem faced by a firm is to determine the amount of resources it should allocate for mass and differentiated promotion so that the sales generated is maximum. Numerous promotional allocation problems exist in literature [6-11], but none addresses combined effect of mass promotion and differentiated promotion on sales explicitly. This study deals with this issue by proposing a promotional effort allocation model to optimally allocate the company's limited promotional resources for mass and differentiated promotion in different segments. The objective is to maximize the total sales generated through the impact of both the promotional strategies under budgetary constraints and minimum desired level of sales constraint from each segment. Sales measured through adoption growth in each market segment is described using a recent innovation diffusion model developed due to Jha *et al* [12] which enables to capture the product adoption due to the combined effect of mass as well as differentiated segmented promotion for durable technology products.

The problem formulated is a nonlinear programming problem (NLPP). Solution of problem is found by making suitable transformations using NLPP methods in LINGO Software [24]. A constraint corresponding to the minimum aspiration of total sales to be achieved from all the segments collectively is also incorporated in the problem. Such a constraint is usually imposed by the management to make sure that certain minimum percentage of total adopters, adopt the product. In such situations it is possible that the aspired sales from each segment or from all segments collectively is high, as a result, the problem becomes infeasible due to limited availability of promotional resources. To get the best possible compromised solution Differential Evolution (DE) [13, 14] is applied. DE stands as a powerful tool for global optimization. It is a population based optimizer that generates new points that are perturbations of existing points using the scaled difference of two randomly selected population vectors. There is no particular requirement on the problem before using DE, as it can be applied to solve any kind of problem. Here we use DE to solve our problem to get best possible allocations of promotional resources in case of infeasibility condition.

Rest of the paper is structured as follows. Section 2 reviews literature of innovation diffusion modeling and promotional allocation problems in marketing. An optimization problem is formulated in Section 3 for allocating a fixed promotional budget among the different consumer segments for mass market and differentiated market promotion, maximizing the total sales of the product measured through the adoption growth. The section also discusses the mathematical model of durable technology adoption [12]. Some extensions to the problem and their solution methodology have also been discussed in this section. Numerical illustration has been given in section 4 to validate the results. Section 5 concludes the paper and provides direction for future research.

2 Literature Review

Diffusion theory is most widely known and accepted theory for measuring adoption over the life cycle of new products. Theory of innovation diffusion relates how a new idea, a new product and/or a new service are accepted among the members of its potential consumer population over time. Since years, models of innovation diffusion are used successfully by various practitioners to evaluate the market response over the product life cycle and make valuable decisions related to product modifications, price differentiation, optimization of resource allocation etc. Earliest and most famous first purchase model of new product diffusion in marketing are by Fourt & Woodlock [15], and Bass [16] that attempted to describe the penetration and saturation aspects of the diffusion process. The main impetus underlying diffusion research is the Bass model for durable technology products. It assumes that a potential customer either makes the purchase decision independently or is influenced by a previous purchaser. The first category of consumer is called innovators whereas the second category is known as imitators. The model describes the growth in purchaser population w.r.t. time. Bass *et al* [17] also developed a generalized Bass model based on the basic Bass model which primarily reflects the current effect of dynamic marketing variables on the conditional probability of adoption at time *t* as various marketing mix variable such as price, promotion which has a considerable effect on adoption of the product. An alternative formulation of GBM was proposed by Jha *et al* [18], wherein the current effect of dynamic marketing variables is represented by the promotional effort intensity function. Promotion is the most dominant marketing mix variable in the product category and therefore other marketing mix elements are assumed at constant level. Recently Jha *et al* [12] developed a sales growth model for a segmented market in which the sales are assumed to be evolved through a combination of differentiated promotion done exclusively for each segment and the spectrum effect of mass promotion. Spectrum effect was discussed by Burrato *et al* [3] assuming that an advertising channel has an effectiveness segment-spectrum, which is distributed over potential market segments and solutions to deterministic optimal control problems were obtained.

Main contribution in the area of promotional effort allocation problems was done by Kapur *et al* [6] and Jha *et al* [7, 8]. They formulated optimization problems for marketing single product and multi products respectively in segmented market using Bass model of innovation diffusion to describe the adoption and proposed solution methods based on dynamic programming, goal programming and multi-objective programming approach. Further Jha *et al* [10, 11] formulated promotional effort allocation problem for single product and multi products respectively in segmented market where the market is subject to dynamic potential adopter population and repeat purchasing. These problems aimed to maximize the total sales of the products subject to the budget and technical constraints imposed by the management. Multi-criteria optimization and goal programming approaches were used to solve the problems. Kapur *et al* [9] formulated an optimal promotional effort allocation problem of a single product in a segmented market using innovation diffusion model with consumer balking and repeat purchasing and solved it using genetic algorithm. The research done so far somehow developed promotional allocation problems considering the impact of differentiated promotion on product adoption but none considered the proportional effect of mass market promotion explicitly. Though many other factors may contribute in generating the sales of the product but the effect of mass market promotion cannot be ignored. To fill this gap, we formulate an optimization problem that optimally allocates promotional resources to mass market promotion and differentiated market promotion in various segments of the market. Optimization model formulation requires an appropriate model that can describe sales growth of a product in respect to promotional efforts where promotion effort in each segment is a combination of mass and differentiated market promotion employed in all the segments. Here we use the diffusion model proposed by Jha *et al* [12].

3 Model Formulation

3.1 Notations

i Segments in the market; $i = 1, 2, \ldots K$.

$\bar{N}_i$ Expected potential adopter population in i^{th} segment of the market.

p_i Coefficient of external influence in the ith segment.

q_i Coefficient of internal influence in the ith segment.

$x_i(t)$ Instantaneous rate of promotional effort at time t in i^{th} segment; $X_i(t) = \int_0^t x_i(u)\,du$

X_i Amount of promotional resources allocated for differentiated market promotion in the ith segment.

X Amount of promotional resources allocated for mass market promotion.

Z Total promotional budget available.

$N_i(t)$ Expected number of adopters of the product in the ith segment.

X_i^* Optimal value of X_i (Amount of promotional resources allocated for differentiated market promotion in the ith segment).

X^* Optimal value of X (Amount of promotional resources allocated for mass market promotion in all the K segments collectively).

3.2 Marketing Model

The relationship between time and adoption for developing the optimization problem is described using diffusion model proposed due to Jha *et al* [12]. The model describes the cumulative adoption of a product considering promotional expenditure in mass and differentiated market promotion. The model is based on the following assumptions

1. The market for a new product is divided in to K disjoint segments.
2. Each purchaser buys a single unit of the product.
3. The consumer decision process is binary (adopt or not adopt).
4. The potential consumer population for the product in each segment is finite and remains constant during the promotional campaign.
5. In each segment buyers can be categorized into two groups: Innovators and Imitators. Innovators make their purchase decisions independently, whereas imitators buy the product through the world of mouth influence.
6. The consumer behavior of segments is independent of each other and independent promotion done in one segment has no impact on others.

7. The parameters of external and internal influence are fixed over the diffu-
    sion process of the innovation in each segment.
8. The rate of purchase with respect to promotional effort intensity is propor-
    tional to the number of non-purchasers of the product.

Following the model given by Jha $et\ al$ [12], growth in sales of a product in the ith market segment with respect to the promotional effort is given as

$$N_i\left(X_i(t), X(t)\right) = \frac{\overline{N_i}\left(1 - e^{-(p_i+q_i)(X_i(t)+\alpha_i X(t))}\right)}{\left(1 + \left(\dfrac{q_i}{p_i}e^{-(p_i+q_i)(X_i(t)+\alpha_i X(t))}\right)\right)}, i = 1, 2, ..., K \qquad (1)$$

Cumulative adoption of any product is a function of promotional efforts spent by that time; resources are spent continuously in the market and sales of the product increases. At the same time, the planning period for the product promotion is almost fixed. Therefore without loss of generality the cumulative adoption can be assumed to be a function of promotional effort spent for both differentiated market and mass market promotion in all the segments, explicitly in the above equations

$$N_i(X_i, X) = \frac{\overline{N_i}\left(1 - e^{-(p_i+q_i)(X_i+\alpha_i X)}\right)}{\left(1 + \left(\dfrac{q_i}{p_i}e^{-(p_i+q_i)(X_i+\alpha_i X)}\right)\right)}, i = 1, 2, ..., K \qquad (2)$$

Equation (2) describes the expected number of adopters of product in i^{th} segment. The parameters of model (2) can be estimated statistically after observing market for certain duration. Alternatively data on similar product launches can be used. Note that the model is applicable to category of durable technology product [16].

3.3 The Optimization Problem

Promotion of a product is carried out for fixed time period and also there is an upper bound on the resources available for promotion. Judicious allocation of limited promotional resources is essential (especially in a segmented market where adoption grows through differentiated and mass market promotion).

3.3.1 Optimization model under budget constraint

The problem for finding the optimal amount of promotional resources to be allocated for differentiated promotion in each of the ith segment (X_i) and mass promotion for all the K segments (X) that would maximize the total sales in each of the K segments under the budget constraint can be formulated as follows

$$\textbf{Maximize } \sum_{i=1}^{K} N_i(X_i, X) = \sum_{i=1}^{K} \frac{\overline{N_i}\left(1 - e^{-(p_i+q_i)(X_i+\alpha_i X)}\right)}{\left(1 + \dfrac{q_i}{p_i}e^{-(p_i+q_i)(X_i+\alpha_i X)}\right)}$$

$$\textbf{Subject to } \quad N_i(X_i, X) \geq r_i \overline{N_i} = N_i^0 \qquad i = 1, 2, ..., K$$

$$\sum_{i=1}^{K} X_i + X \leq Z$$

$$X >= m * Z$$

$$X, X_i \geq 0 \qquad\qquad i = 1, 2, ..., K \qquad\qquad (P1)$$

The objective function maximizes the cumulative number of adopters for the product. First constraint guarantees that firm fetches a minimum proportion (r_i) of market share in each of the ith market segment by doing some promotional activities in it. Through this constraint it is made sure that none of the segment gets a very low or no allocation of resources and the product is promoted in all the segments. Second constraint guarantees that the total amount of resources allocated does not exceed the total resources available. Third constraint ensures that some minimum proportion (m) of the total promotional resources is allocated for mass market promotion. This constraint is usually imposed by management as without it, the solution may suggest a zero or some minimal allocation for mass promotion. Reason being, mostly the adoption of the product is more due to differentiated promotion as compared to mass promotion. This situation may not be advisable and non-acceptable to the management as the main aim of the mass market promotion is to reach everyone in the potential market, create image for the product and influence the population who is expected to be in the potential segment in near future. The proportion m is set by the management on the basis of experience from the past product behavior. Fourth constraint ensures that X_i, X are non-negative.

Let $f_i(X_i, X) = \overline{N}_i(1 - e^{-(p_i + q_i)(X_i + \alpha_i X)})$, $g_i(X_i, X) = (1 + \frac{q_i}{p_i} e^{-(p_i + q_i)(X_i + \alpha_i X)})$

and $F_i(X_i, X) = f_i(X_i, X) / g_i(X_i, X)$ $i = 1, 2, ..., K$. Hence resulting problem becomes maximization of a sum of ratios (fractional functions) under specified promotional effort expenditure which is again a fractional function

Maximize $\sum_{i=1}^{K} F_i(X_i, X)$

Subject to $F_i(X_i, X) \geq r_i \overline{N}_i = N_i^0 \qquad i = 1, 2, ..., K$

$$\sum_{i=1}^{K} X_i + X \leq Z \qquad\qquad\qquad (P2)$$

$$X >= m * Z$$

$$X, X_i \geq 0 \qquad\qquad i = 1, 2, ..., K$$

The Hessian matrices $H_i = \begin{bmatrix} \frac{\delta^2 f_i}{\delta X_i^2} & \frac{\delta^2 f_i}{\delta X_i X} \\ \frac{\delta^2 f_i}{\delta X X_i} & \frac{\delta^2 f_i}{\delta X^2} \end{bmatrix}$ and $G_i = \begin{bmatrix} \frac{\delta^2 g_i}{\delta X_i^2} & \frac{\delta^2 g_i}{\delta X_i X} \\ \frac{\delta^2 g_i}{\delta X X_i} & \frac{\delta^2 g_i}{\delta X^2} \end{bmatrix}$ are negative semi definite

and positive semi definite respectively, therefore, functions $f_i(X_i, X)$ and $g_i(X_i, X)$, $i = 1, 2...K$ are concave and convex respectively. The ratio of concave and convex functions is a pseudo-concave function and sum of pseudo-concave functions is not necessarily a pseudo-concave function. There does not exist any direct method to obtain an optimal solution for such class of problems. Dur *et al* [19] proposed a method to solve such class of problems converting sum of ratio functions of the objective to a multiple objective fractional programming problem. It has been established that every optimal solution of original problem is an efficient solution of the equivalent multiple objective fractional programming problem. Dur's equivalent of problem (P2) can be written as

Maximize
$$F_i(X_i,X)=\left(f_1(X_1,X)/g_1(X_1,X),f_2(X_2,X)/g_2(X_2,X),...,f_K(X_K,X)/g_K(X_K,X)\right)^T$$
or
$$F_i(X_i,X)=\left(F_1(X_1,X),F_2(X_2,X),...,F_K(X_K,X)\right)^T \qquad (P3)$$

Subject to $X_i,X\in S=\{X_i,X\in R/\,F_i(X_i,X)\ge r_i\bar{N}_i=N_i^0,X_i\ge0,X\ge0,i=1,2,...,K,\sum_{i=1}^{K}X_i+X\le Z,X\ge=m*Z\}$

Let $y_i=F_i(X_i,X)=f_i(X_i,X)/g_i(X_i,X)$, $i=1,2...K$, then the equivalent parametric problem for multiple objective fractional programming problem (P3) is given as

Maximize $y=(y_1,y_2,...,y_K)^T$

Subject to $f_i(X_i,X)-y_ig_i(X_i,X)\ge0 \qquad i=1,2,...,K \qquad (P4)$

$y_i\ge N_i^0 \qquad\qquad\qquad\qquad i=1,2,...,K$

$\sum_{i=1}^{K}X_i+X\le Z$

$X>=m*Z$

$X,X_i\ge0 \qquad\qquad\qquad\qquad i=1,2,...,K$

The Geoffrion's [20] equivalent scalarization for the problem (P4) for fixed weights for the objective functions is as follows

Maximize $\sum_{i=1}^{K}\lambda_iy_i$

Subject to $f_i(X_i,X)-y_ig_i(X_i,X)\ge0 \qquad i=1,2,...,K$

$y_i\ge N_i^0 \qquad\qquad\qquad\qquad i=1,2,...,K$

$\sum_{i=1}^{K}X_i+X\le Z$

$X>=m*Z \qquad\qquad\qquad\qquad\qquad\qquad\qquad (P5)$

$X,X_i\ge0 \qquad\qquad\qquad\qquad i=1,2,...,K$

$$\lambda\in\Omega=(\lambda\in R^K/\sum_{i=1}^{K}\lambda_i=1,\lambda_i\ge0,i=1,2,...,K)$$

where λ_i's are the weights assigned to the objective function for segment i.

Here we define some Lemmas using some definition from the theory of convexity and multiple optimization programming.

Definition 1 [21]: A function $F(X)$ is said to be pseudo-concave if for any two feasible points X_1,X_2; $F(X_1)\ge F(X_2)$ implies $F'(X_1)(X_2-X_1)\le0$.

Definition 2[22]: A feasible solution $X^*\epsilon S$ is said to be an efficient solution for the problem (P3) if there exists no $X\epsilon S$ such that $F(X)\ge F(X^*)$ and $F(X)\ne F(X^*)$.

Definition 3 [20]: An efficient solution $X^*\epsilon S$ is said to be a properly efficient solution for the problem (P3) if there exists $\alpha>0$ such that for each r, $F_r(X)-F_r(X^*))/F_j(X^*)-F_j(X)<\alpha$ for some j with $F_j(X)<F_j(X^*)$ and $F_r(X)>F_r(X^*)$ for $X\epsilon S$.

Lemma 1 [19]: The optimal solution X^* of the problem (P1) is an efficient solution of the problem (P3).

Lemma 2 [23]: A properly efficient solution $(X_i^*,X^*,y_i^*$ for $i=1,2,...,K)$ of the problem (P4) is also a properly efficient solution $(X_i^*,X^*$ for $i=1,2,...,K)$ for the problem (P3).

Lemma 3 [20]: The optimal solution $(X_i^*,X^*,y_i^*$ for $i=1,2,...,K)$ of the problem (P5) is a properly efficient solution $(X_i^*,X^*,y_i^*$ for $i=1,2,...,K)$ for the problem (P4).

Theorem 1: If relative importance is attached to each of the objective of the problem (P5) and $(X_i^*, X^*, y_i^*$ for $i = 1,2,...,K)$ is an optimal solution of the problem (P5) then $(X_i^*, X^*$ for $i=1,2,...,K)$ is an optimal solution of the original problem (P1).

By Lemma 1, 2 and 3 an optimal solution of the problem (P5) is also an optimal solution of original problem (P1). Optimal solution of problem (P5) can be found out by assigning different weights to different segments. These weights are decided either by top management or expert's judgment etc. Mathematical approach to facilitate decision making process could be Analytical Hierarchy Process (AHP) which prioritizes goals or alternatives based on pair wise comparison or judgment. The purpose of assigning different weights is to prioritize the segments according to which resources can be allocated to these segments. It remains to obtain an optimal solution of the problem (P5) by assigning these weights or one can assign equal weight to all the segments. Problem (P5) can be solved by standard mathematical programming approach in LINGO [24].

3.3.2 Incorporating minimum total market potential aspiration constraint

The diffusion rate varies in each segment. In some segments people adopt the product with a higher rate and in some segments they adopt slowly. The optimization model developed in the previous section imposes constraints that promotion allocation should be made in such a way that some minimum proportion of adoption should be observed from each segment. As the segments having low diffusion rates also has this restriction it may lead to large proportion of promotional resource allocation to hard segments. This results in lower resource allocation in segments having high propensity to adopt the product. Consequently, total market share of the product reduces. This situation is not advisable/desirable by the management and it imposes a constraint in order to fetch a certain minimum proportion of total sales for product collectively from all segments.

Hence the problem (P1) needs to be suitably modified. The problem is redefined as

$$\text{Maximize} \quad \sum_{i=1}^{K} N_i(X_i, X) = \sum_{i=1}^{K} \frac{\overline{N}_i \left(1 - e^{-(p_i + q_i)(X_i + \alpha_i X)} \right)}{\left(1 + \frac{q_i}{p_i} e^{-(p_i + q_i)(X_i + \alpha_i X)} \right)}$$

$$\text{Subject to} \quad \sum_{i=1}^{K} N_i(X_i, X) \ge r_0 \sum_{i=1}^{K} \overline{N}_i = N_0$$

$$N_i(X_i, X) \ge r_i \overline{N}_i = N_i^0 \qquad i = 1,2,...,K \qquad \text{(P6)}$$

$$\sum_{i=1}^{K} X_i + X \le Z$$

$$X >= m * Z$$

$$X, X_i \ge 0 \qquad\qquad i = 1,2,...,K$$

where r_0 = minimum proportion of the total market potential desired to capture.

This problem can be written equivalently with suitable transformations as,

$$\text{Maximize} \quad \sum_{i=1}^{K} \lambda_i y_i$$

$$\text{Subject to} \quad f_i(X_i, X) - y_i g_i(X_i, X) \ge 0 \qquad i = 1,2,...,K$$

$$\sum_{i=1}^{K} y_i \geq N_0 \qquad\qquad\qquad\qquad\qquad\qquad (P7)$$

$$y_i \geq N_i^0 \qquad\qquad\qquad i = 1, 2, ..., K$$

$$X >= m * Z$$

$$\sum_{i=1}^{K} X_i + X \leq Z$$

$$X, X_i \geq 0$$

$$\lambda \in \Omega = (\lambda \in R^K / \sum_{i=1}^{K} \lambda_i = 1, \lambda_i \geq 0, i = 1, 2, ..., K)$$

If minimum level of target adoption of product in each segment to be obtained is very high, also the management aim to obtain a minimum return on investment measured through target level of market share to be obtained from the total sales in all the segments, (P7) may lead to an infeasible solution. Infeasibility suggests either to increase the level of promotional resources or to obtain compromised solution. In most of the situations a compromised solution is obtained as it gives a better decision making capability to the management as to what needs to be compromised and by how much. In order to incorporate such aspirations in the problem formulation, differential evolution algorithm can be used to obtain the best possible solution. It should be noted here that transformation (P7) is required for solving the problem using NLPP technique, however DE being a numerical optimizer can be directly applied to solve problem (P6). Procedure for applying DE is presented in the following subsection.

3.3.2.1 Differential Evolution Algorithm

Differential evolution is an evolutionary algorithm, which is rapidly growing field of artificial intelligence. This class also includes genetic algorithms, evolutionary strategies and evolutionary programming. DE was proposed by Price & Storn [13]. Since then it has earned a reputation as a very powerful and effective global optimizer. The basic steps of DE are as follows

Start
Step 1: Generate an initial population of random individuals
Step 2: Create a new population by repeating following steps until the stopping criterion is achieved
    [*Selection*] Select the random individuals for reproduction
    [*Reproduction*] Create new individuals from selected ones by mutation and crossover
    [*Evolution*] Compute the fitness values of the individuals
    [*Advanced Population*] Select the new generation from target individual and trial individuals
End steps

Initialization - Suppose we want to optimize a function of D number of real parameters. We must select a population of size NP. NP parameter vectors have the form
$$X_{i,G} = (x_{1,i,G}, x_{2,i,G}, ..., x_{D,i,G})$$

where D is dimension, i is an individual index and G represents the number of generations.

First, all the solution vectors in a population are randomly initialized. The initial solution vectors are generated between lower and upper bounds $l = \{l_1, l_2, ..., l_D\}$ and $u = \{u_1, u_2, ..., u_D\}$ using the equation

$$x_{j,i,0} = l_j + rand_{i,j}[0,1] \times (u_j - l_j)$$

where, i is an individual index, j is component index and $rand_{i,j}[0,1]$ is a uniformly distributed random number lying between 0 and 1. This randomly generated population of vectors $X_{i,0} = (x_{1,i,0}, x_{2,i,0}, ..., x_{D,i,0})$ is known as target vectors.

Mutation - Each of the NP parameter vectors undergoes mutation, recombination and selection. Mutation expands the search space. For a given parameter vector $X_{i,G}$, three vectors $X_{r_1,G}, X_{r_2,G}, X_{r_3,G}$ are randomly selected such that the indices i, r_1, r_2, r_3 are distinct. The ith perturbed individual, $V_{i,G}$, is therefore generated based on the three chosen individuals as follows

$$V_{i,G} = X_{r_1,G} + F * (X_{r_2,G} - X_{r_3,G})$$

where, $r_1, r_2, r_3 \in \{1, 2, ..., NP\}$ are randomly selected, such that $r_1 \neq r_2 \neq r_3 \neq i$, $F \in (0, 1.2]$ and $V_{i,G}$ is called the mutation vector.

Crossover - The perturbed individual, $V_{i,G} = (v_{1,i,G}, v_{2,i,G}, ..., v_{D,i,G})$ and the current population member, $X_{i,G} = (x_{1,i,G}, x_{2,i,G}, ..., x_{D,i,G})$ are then subject to the crossover operation, that finally generates the population of candidates, or "trial" vectors, $U_{i,G} = (u_{1,i,G}, u_{2,i,G}, ..., u_{D,i,G})$, as follows

$$u_{j,i,G} = \begin{cases} v_{j,i,G} & \text{if } rand_{i,j}[0,1] \leq C_r \vee j = j_{rand} \\ x_{j,i,G} & \text{otherwise} \end{cases}$$

where, $C_r \in [0,1]$ is a crossover probability, $j_{rand} \in \{1, 2, ..., D\}$ is a random parameter's index, chosen once for each i.

Selection - The population for the next generation is selected from the individuals in current population and its corresponding trial vector according to the following rule

$$X_{i,G+1} = \begin{cases} U_{i,G} & \text{if } f(U_{i,G}) \geq f(X_{i,G}) \\ X_{i,G} & \text{otherwise} \end{cases}$$

where, $f(.)$ is the objective function value. Each individual of the temporary population is compared with its counterpart in the current population. Mutation, recombination and selection continue until stopping criterion is reached.

Constraint Handling in Differential Evolution - Pareto ranking method is used to handle constraints in DE. The value of constraints is calculated at target and trial vectors. The method is based on the following three rules

1) Between two feasible vectors (target and trial), the one with the best value of the objective function is preferred

2) If out of target and trial vectors, one vector is feasible and the other is infeasible, the one which is feasible is preferred

3) Between two infeasible vectors, the one with the lowest sum of constraint violation is preferred

Stopping Criterion - DE algorithm stops when either
1) Maximum number of generations are reached or
2) Desired accuracy is achieved i.e., $|f_{max}-f_{min}| \leq \varepsilon$.

4 Numerical Illustration

Here we show the practical application of the optimization problem formulated in the paper. The product which has been taken under consideration is a new hatchback car of an ABC automobile company. The company name and data has not been disclosed for the confidentiality reasons. Data is available for the mass and differentiated promotion in four geographic segments in the country. The estimates of the parameters $\bar{N}_i$, p_i, q_i, and α_i for all four segments are taken from Jha *et al* [12] as shown in table 1. It is given that maximum promotional resource available are 88 units where cost per unit promotional effort is □2,50,00,000. The management sets a target that at least 50% of the potential adopters in each segment buy the product during the promotional campaign. Also the firm wants to allocate a minimum of 30% of the total promotional resources for the mass market promotion. With this data, the problem of stage 1 (P5) is solved and results obtained are given in table 3.

In case 80% of the total market share is required, the resulting problem (P7) has no feasible solution and hence it is imperative to use DE to obtain a compromise solution. Parameters of DE are given in table 2. A desired accuracy of .001 between maximum and minimum values of fitness function was taken as terminating criteria of the algorithm. The solution obtained on solving the problem of stage 2 (P6) is given in table 3. Table 3 also shows the cumulative adoption of the product and the percentage of market potential captured in each segment along with the total market potential captured corresponding to the solution obtained in both stages.

Table 1. Parameters of the sales growth model

Segment	Estimated parameters			
	$\bar{N}_i$	p_i	q_i	α_i
S1	287962	0.000671	0.132113	0.372663
S2	156601	0.001128	0.470658	0.197823
S3	106977	0.001344	0.566035	0.165732
S4	223291	0.000621	0.331664	0.263569

Table 2. Parameters of Differential Evolution

Parameter	Value	Parameter	Value
Population Size	200	Scaling Factor (F)	0.7
Selection Method	Roulette Wheel	CrossoverProbability(C_r)	0.9

Table 3. Results of Numerical illustration

Segment	Xi^*	X^*	$Ni^*(Xi, X)$	% of Captured Market Size
		Stage 1		
S1	30.0215	26.4	143981.02	50.01
S2	9.6633		114050.78	72.83
S3	6.9709		63804.32	59.65
S4	14.9443		163009.43	73.01
Total	87.9999		484845.54	62.57
		Stage 2		
S1	35.9969	26.4	198451.89	68.92
S2	11.6530		136668.55	87.27
S3	9.1877		89735.19	83.88
S4	17.7600		195009.13	87.33
Total	100.9976		619864.77	80.00

It can be seen that in stage 1, 62.5% of the total market potential is covered. When in stage 2 an aspiration of 80% is set on the total sales to be achieved from all the segments, the total market potential covered shows an increment. However differential evolution compromised on the promotional resources utilized during allocation in this stage. The solution suggests increasing the promotional resources limit to 101 units (approx.). In case promotional resources can't be increased the management must decrease the aspiration on total sales.

5 Conclusion

In this paper we have formulated and solved an optimization problem allocating the limited promotional resources for mass market promotion and differentiated market promotion of a product maximizing the sales measured through adoption. A recent innovation diffusion model which captures the adoption of a product in segmented market due to the joint effect of two types of promotion is used to measure the adoption process. This is for the first time that a promotion allocation optimization problem is formulated for a situation when the effect of mass and dif-ferentiated promotion is explicitly captured. The optimization model is developed under upper bound constraint on promotional resources and minimum target sales level constraints to be attained from distinct segments individually and in total. The solution procedures to the problems are also discussed. Infeasibilities are han-dled using differential evolution approach using which we get the best possible so-lution to the problem. To show the applicability of the proposed problem a numer-ical illustration is presented. This paper shows a big scope of future research such as allocation of promotional resources for mass market promotion and differen-tiated market promotion can be done for multiple products. Also dynamic promo-tional resource allocation for a single product and multi product in multiple time periods over a planning horizon can be worked upon.

References

[1]   Rao, K. R.: Services Marketing. Second edition, Dorling Kindersley (India) Pvt. Ltd., Noida (2011)

[2]   Egan, J.: Marketing Communications. Thomson Publishing, London(UK) (2007)

[3]   Buratto, A., Grosset, L., Viscolani, B.: Advertising a new product in segmented market. EJOR. 175,1262–1267 (2006)

[4]   Kotler, P., Armstrong G.: Principles of Marketing. 12th edition, Prentice Hall, India (2009)

[5]   Ramaswamy V.S., Namakumari S.: Marketing Management: Global Perspective - Indian Context. Fourth edition, Macmillan India Ltd. (2009)

[6]   Kapur, P.K., Jha, P.C., Bardhan, A.K. & Singh O.: Allocation of promotional resource for a new product in a segmented market. In V. K. Kapoor (Ed.), Mathematics and Information Theory: Recent topics and applications. New Delhi, India: Anamaya Publishers., 52-61 (2004)

[7]   Jha, P.C., Mittal, R., Singh, O., Kapur, P.K.: Allocation of Promotional Resource for Multi-Products in A Segmented Market. Presented in ICORAID-2005-ORSI and published in Proceedings (2005)

[8]   Jha, P.C., Gupta, A., Kapur, P.K.: On Maximizing the Adoption of Multi-Products in Multiple Market Segments. Journal of Information and Optimization Sciences. 28(2), 219-245 (2007)

[9]   Kapur, P.K., Aggarwal, A.G., Kaur, G. & Basirzadeh, M.: Optimising adoption of a single product in multi-segmented market using innovation diffusion model with consumer balking. International Journal of Technology Marketing. 5(3), 234–249 (2007)

[10] Jha, P. C., Aggarwal, R., Gupta, A.: Optimal allocation of promotional resource under dynamic market size for single product in segmented market. Paper presented at the International Conference on Development & Applications of Statistics in Emerging areas of Science & Technology, Jammu, India (2010)

[11] Jha, P. C., Aggarwal, R., Gupta, A., Kapur, P.K.: Optimal allocation of promotional resource for multi-product in segmented market for dynamic potential adopter and repeat purchasing diffusion models. International Journal of Advanced Operations Management, Inderscience. 3(3/4), 257-270 (2011)

[12] Jha, P.C., Aggarwal, S., Gupta, A., Dinesh, U.: Innovation Diffusion Model for a Product Incorporating Segment-specific Strategy and the Spectrum Effect of Promotion. Communicated Paper. (2012)

[13] Price, K.V., Storn, R.M.: Differential Evolution-A simple and efficient adaptive scheme for global optimization over continuous space. Technical Report TR-95-012, ICSI, March 1995. Available via the Internet: ftp.icsi.berkeley.edu/pub/techreports/1995/tr-95-012.ps.Z. (1995)

[14] Price, K.V.: An introduction to Differential Evolution. In: Corne, D., Marco, D. & Glover, F. (eds.), New Ideas in Optimization, McGraw-Hill, London(UK), 78-108 (1999)

[15] Fourt, L.A., Woodlock J.W.: Early prediction of Market Success for grocery products. Journal of Marketing. 25, 31-38 (1960)

[16] Bass, F.M.: A new product growth model for consumer durables. Management Science. 15, 215-227 (1969)

[17] Bass, F.M., Krishnan, T.V., Jain, D.C.: Why the Bass model fits without decision variables. Marketing Science. 13 (3), 203-223 (1994)

[18] Jha, P.C., Gupta, A., Kapur, P.K.: Bass Model Revisited. Journal of Statistics and Management Systems. Taru Publications. 11 (3), 413-437 (2006)

[19] Dur, M., Horst, R., Thoai, N.V.: Solving Sum-of-Ratios Fractional Programs Using Efficient Points. Optimization. 41: 447-466 (2001)

[20] Geoffrion, A.M.: Proper Efficiency and Theory of Vector Maximization. Journal of Mathematical Analysis and Application. 22: 613-630 (1968)

[21] Bazaraa, S.M., Setty, C.M.: Nonlinear Programming: Theory and Algorithm. John Wiley and Sons (1979)

[22] Steuer, R.E.: Multiple Criteria Optimization: Theory, Computation and Application. Wiley, New York (1986)

[23] Bector, C.R., Chandra, S., Bector, M.K.: Generalized Fractional Programming, Duality: A Parametric Approach. Journal of Optimization Theory and Applications. 60: 243-260 (1989)

[24] Thiriez, H.: OR Software LINGO. EJOR. 124: 655-656 (2000)

Gray level Image Enhancement by Improved Differential Evolution Algorithm

Pravesh Kumar[1], Sushil Kumar[1], Millie Pant[1]

[1] Department of Applied Sciences and Engineering, IIT Roorkee, India

{praveshtomariitr@gmail.com; kumarsushiliitr@gmail.com; millidma@gmail.com}

Abstract. In this paper, an enhanced version of DE named MRLDE is used to solve the problem of image enhancement. The parameterized transformation function is used for image enhancement which uses both local and global information of image. For image enhancement, an objective criterion is considered which use the entropy and edge information of image. The objective of the DE is to maximize the objective fitness criterion in order to improve the contrast. Results of MRLDE are compared with basic DE, PSO, GA and with histogram equalization (HE) which is another popular enhancement technique. The obtained results indicate that proposed MRLDE yield better performance in the comparison of other techniques.

Keywords: Image enhancement, Differential evolution, Mutation, Parameter optimization.

1 Introduction

Image enhancement, is one of the important image processing techniques, where an image can be transform into other image in order to improve the explanation or observation of information for human viewers, or to provide better input for other automated image processing techniques. According to [1], image enhancement techniques can be divided into four main categories: point operation, spatial operation, transformation, and pseudo coloring. The work done in this paper is based on spatial operation. .

There are several method available for image enhancement out of which some are Histogram equalizer for contrast enhancement for gray images [2], Linear contrast stretching [2] which employs a linear transformation that maps the gray-levels in a given image to fill the full range of values, and Pseudo-coloring which is an enhancement technique that artificially "color" the gray-scale image based on a color mapping [1], [3].

J. C. Bansal et al. (eds.), *Proceedings of Seventh International Conference on Bio-Inspired Computing: Theories and Applications (BIC-TA 2012)*, Advances in Intelligent Systems and Computing 202, DOI: 10.1007/978-81-322-1041-2_38, © Springer India 2013

In recent years evolutionary algorithms like GA and PSO, have been successfully applied to image enhancement [3]-[9]. In this paper Differential Evolution (DE) and its improved version (MRLDE) has been used for contrast enhancement for gray image.

DE is a stochastic population based, direct search technique for global optimization. In Comparison to Particle Swarm Optimization (PSO) and Genetic Algorithm (GA), DE has many advantages, such as faster convergence speed, stronger stability, easy to understand [10]. It has been successfully applied to many real life problems of science and engineering field [11]. Some recent survey of DE can be found in [12] and [13].

In the present study, Modified Random Localization based DE (MRLDE) an improve variant of DE is proposed for image enhancement. In MRLDE, a novel selection procedure is proposed for choosing three randomly vectors to perform mutation operation.

The rest of paper is organized as follows: In Section 2 the model of enhancement problem is discussed. In Section 3 theory of DE and proposed MRLDE are discussed. Results and discussion are given in Section 4, and finally the conclusions derived from the present study are drawn in Section 5

2 Problem Formulation

In this section we have described the transformation function that is required for image enhancement and the criteria which is required to evaluate the quality of enhanced image

2.1 Transformation Function

The transformation function T generates the new intensity value for each pixel of size $M \times N$ input image to produce the enhanced output image. The enhancement process can be denoted by,

$$g(i,j) = T(f(i,j)) \tag{1}$$

where $f(i, j)$ is the gray value of the $(i, j)^{th}$ pixel of the input image and $g(i, j)$ the gray value of the $(i,j)^{th}$ pixel of output image. The function used here is designed in such a way that takes both global as well as local information to produce the enhanced image.

Considering literature [3] and [9], the transformation function can be defined as:

$$g(i,j) = \frac{kM}{\sigma(i,j)+b}[f(i,j)-c\times m(i,j)] + m(i,j)^a \tag{2}$$

where $m(i, j)$ is local mean and $\sigma(i, j)$ is local standard deviation of the $(i,j)^{\text{th}}$ pixel of input image over $n \times n$ window; M is global mean of image; a, b, c and k are the parameters defined over real positive and same for whole image.

DE's task is to solve the image enhancement problem by tuning the four parameters a, b, c and k, in order to find the best combination according to an objective criterion that describes the contrast in the image.

2.2 Evaluation Criterion

To evaluate the quality of an enhanced image without human intervention, we need an objective function which will say all about the image quality [3]. According to [3], enhanced images should have the following characteristics – sufficient information, higher contrast and clear texture. Thus, the objective evaluation function is the mixture of three performance measures, namely entropy value, mean value of edge intensities and the number of edges (in pixels).

According to literature [3] and [9], objective evaluation function can be written as:

$$F(I_e) = \log(\log(E(I_s))) \times \frac{n\_edgels(I_s)}{M \times N} \times H(I_e) \tag{3}$$

where I_e is enhance image of input image I_o by transform function given in Equation-2. To obtained edge image I_s, Sobel edge operator has been used in the study. $n\_edgels(I_s)$ is sum of non-zero pixels and $E(I_s)$ is sum of gray level pixels in the image I_s.

In (3), entropy's calculation is based on the histogram of the enhanced image, so it can be expressed as:

$$H(I_e) = -\sum_{i=0}^{255} e_i \tag{4}$$

where $e_i = h_i \log_2(h_i)$, if $h_i \neq 0$, otherwise $e_i = 0$ and h_i is the probability of occurrence i^{th} intensity value of enhance image.

3 Basic DE and MRLDE Algorithms

In this section, the brief overview of basic DE and proposed MRLDE is discussed.

3.1 Basic DE

Basic DE algorithm is a kind of evolutionary algorithm, was proposed by Storn and Price [14]. The structure of DE is similar to the GA with both using the operators' *selection, mutation and crossover* to guide the search process. The main difference between standard GA and DE is mutation operation. Mutation is a main operation of DE, and it revises each individual's value according to the difference vectors of the population. The algorithm uses mutation operation as a search mechanism; crossover operation is applied to induce diversity and selection operation is to direct the search toward the potential regions in the search space.

The working of DE is as follows: First, all individuals are initialized with uniformly distributed random numbers and evaluated using the fitness function provided.

The following are executed until maximum number of generation has been reached or an optimum solution is found.

i. **Mutation**: For a D-dimensional search space, for each target vector $X_{i,G}$ at the generation G, its associated mutant vector is generated via certain mutation strategy. The most often used mutation strategy implemented in the DE is given by equation-1.

$$V_{i,G} = X_{r_1,G} + F * (X_{r_2,G} - X_{r_3,G}) \tag{5}$$

where $r_1, r_2, r_3 \in \{1, 2, ..., NP\}$ are randomly chosen integers, different from each other and also different from the running index i. F (>0) is a scaling facor which controls the amplification of the difference vectors.

ii. **Crossover**: Once the mutation phase is over, crossover is performed between the target vector and the mutated vector to generate a trial point for the next generation.

The mutated individual, $V_{i,G} = (v_{1,i,G}, \ldots, v_{D,i,G})$, and the current population member (target vector), $X_{i,G} = (x_{1,i,G}, \ldots, x_{D,i,G})$, are then subject to the crossover operation, that finally generates the population of candidates, or "trial" vectors, $U_{i,G} = (u_{1,i,G}, \ldots, u_{D,i,G})$, as follows:

$$u_{j,i,G} = \begin{cases} v_{j,i,G} & \text{if } rand_j \leq Cr \vee j = k \\ x_{j,i,G} & \text{otherwise} \end{cases} \tag{6}$$

iii. **Selection**: The final step in the DE algorithm is the selection process. Each individual of the temporary (trial) population is compared with the corresponding target vector in the current population. The one with the lower objective function value survives the tournament selection and goes to the next generation. As a result, all the individuals of the next generation are as good as or better than their counterparts in the current generation.

$$X_{i,G+1} = \begin{cases} U_{i,G} & if \quad f(U_{i,G}) \le f(X_{i,G}) \\ X_{i,G} & otherwise \end{cases} \qquad (7)$$

3.2 Proposed MRLDE

During mutation operation, generally, we do not follow any rule of selection of three random vectors X_{r1}, X_{r2} and X_{r3} from population, except for the fact that these should be mutually different from each other and also from the target vector X_i. By this procedure we are not sure about the position of these vectors. These vectors may be selected from either a small cluster or may be selected as very far from each others. This procedure may lead to the loss of some important information about the search space.

To avoid this shortcoming, a new mutation scheme is presented in which instead of having a random selection we make use of localized selection where each solution vector represents a particular region of the search space.

The proposed strategy is very simple. After sorting the initial population according to the fitness function value, we divide it into three regions say R-I, R-II and R-III.

➤ R-I represent the region having the fittest individuals or the elite individuals.

➤ R-II represents the set of next best individuals.

➤ R-III represents the remaining.

➤ Now, we select the three candidates for mutation; X_{r1}, X_{r2} and X_{r3} from R-I, R-II and R-III respectively.

We can easily see that this scheme, tries to cover the maximum of the search space making it more exploratory in nature.

The size of R-I R-II and R-III are taken as $NP*\alpha\%$, $NP*\beta\%$ and $NP*\gamma\%$ respectively, where α, β and γ are integers to be decided by the user. The proposed strategy is named Modified Random localization (MRL) and the corresponding DE variant is called 'MRLDE'.

More details of MRLDE can be found in [15], where the effectiveness of MRLDE has been proved on some benchmark functions, in term of accuracy and fast convergence speed.

The main steps of MRLDE for image enhancement are given;

MRLDE for Image enhancement

1	*Begin*
2	Input original image I_o
3	Transform I_o into enhance image I_e by equation -2.

4	Start with $G=0$ // generation
5	Generate uniformly distribution random population $P_G= \{X_{i,G},$ $i=1,2,...NP\}$
	$X_i^G = X_{lower} +(X_{upper} -X_{lower})*rand(0,1),$
	where X_{lower} and X_{upper} are lower and upper bound.
6	Evaluate the fitness as $F(X_{i,G})$ given by equation-3
7	*Sort ($F(X_{i,G})$) /* put whole population in increasing order on the basis of fitness value*/*
8	*while* (termination criteria is nor met) *do*
9	*for i=1:NP*
10	Divide P into 3 regions say $R-I, R-II$ ans $R-III$ where size of $R-I, R-II$ and $R-III$ are $(NP * \alpha\%), (NP * \beta\%)$ *and* $(NP * \gamma\%)$ respectively
11	Select r_1, r_2 and r_3 as;
	$do\{r_1 =(int)(rand(0,1)*\alpha)\}while(r_1 ==i)$
	$do\{r_2 =(int)(\alpha+rand(0,1)*\beta)\}while(r_2 ==i)$
	$do\{r_3 =(int)(\beta+rand(0,1)*\gamma)\}while(r_3 ==i)$
12	Perform mutation operation by equation-5
13	Perform crossover operation by equation-6
14	Evaluate the fitness $F(U_{i,G})$ by equation -3
15	Select best vector among $X_{i,G}$ and $U_{i,G}$ for next generation by equation-7
16	*End for*
17	*Sort ($F(X_{i,G+1})$)*
18	*End while*
19	*END*

4   Experimental Result and Discussion

4.1   Experimental Settings

Following parameter settings have been taken during experiments for DE and proposed MRLDE:

➤ The population size NP is taken as 30 for DE and MRLDE as suggest in [9].

➤ Scale factor F and Crossover probability is taken as 0.5 and 0.9 respectively [15]

➤ For MRLDE, the value of α, β, γ is taken as 20, 80 and 80 respectively [15].

➤ Maximum number of iteration is set to 300 [9].

➤ The range of parameter a, b, c, k have been taken similar as [3] i.e.

$a \in [0,1.5], b \in [0,0.5], c \in [0,1], k \in [0.5,1.5]$

➢ All experiments are simulated using by Matlab-7.7.0

4.2 Results

3 images of different size have been selected for implementation. The detail of these images is given in Table-1 as below:

Table 1. Detail of input images

Image	Size	No of edge	Fitness value
Cameraman	256*256	2485	99.32
Tire	205*232	1823	124.43
Pout	291*240	1492	2.01

The results of enhance images are given in Table-2. The results are given for number of edges and fitness value. Here the results of GA and PSO are taken from [9].

The results show that MRLDE gives maximum number of edge pixels and fitness value in the comparison of all other techniques, while DE perform better than GA and PSO but not perform better in the comparison of histogram equalizer (HE) in all the cases.

Table 2. Results and comparison of enhanced images by different techniques

Image		HE[9]	GA [9]	PSO [9]	DE	MRLDE
Cameraman	Edge	2604	2575	2674	2688	**2765**
	Fitness	128.66	102.98	128.82	129.21	**137.33**
Tire	Edge	2194	1917	2020	2034	**2455**
	Fitness	142.29	130.03	136.39	139.73	**154.48**
Pout	Edge	2044	2040	2048	2056	**2088**
	Fitness	8.96	2.97	10.45	10.74	**12.65**

The above results ensure the better quality of enhanced image by MRLDE comparative to other techniques. It can be seen from Figure 1, that the brightness and contrast of the enhance images using MRLDE, appear visible and is more than the brightness and contrast of the original images. Also, it can be shown clearly, that the brightness of the enhanced images using MRLDE is better than the brightness of the enhanced images using HE and DE.

Image: Cameraman

(a)   (b)

(c)   (d)

Image: Pout

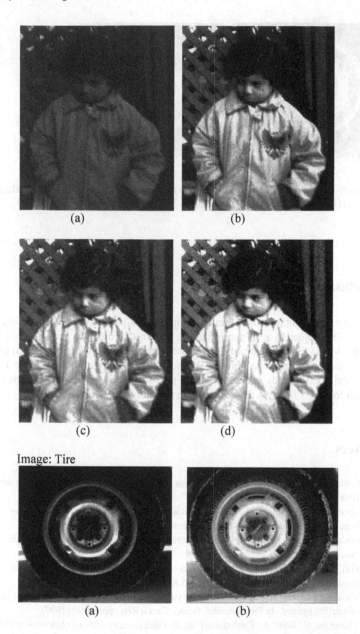

(a)   (b)

(c)   (d)

Image: Tire

(a)   (b)

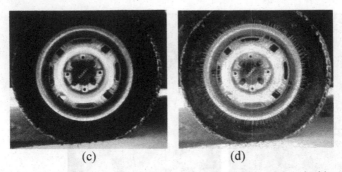

(c)   (d)

Fig. 1. Results of all image (a) Original image, (b) Enhance image by histogram equalizer (HE) , (c) Enhance image by DE, (d) Enhance image by MRLDE

5 Conclusion

In the present study, image enhancement an important part of image processing is considered. The objective is to maximize the contrast in order to get a better picture quality. MRLDE, an enhanced DE variant is employed as an optimization tool for solving the problems. The numerical results when compared with basic DE, GA, PSO and HE methods indicate that the performance of MRLDE is better in comparison to others in terms of fitness value and edge detection.

References

[1] Gonzales R C, Woods, R. E.: Digital Image Processing. New York: Addison-Wesley (1987).

[2] Gonzalez, R.C., Fittes, B.A.: Gray-level transformations for interactive image enhancement. Mechanism and Machine Theory, 12, 111-122 (1977)

[3] Gorai, A., Ghosh, A.: Gray level image enhancement by particle swarm optimization. Proceeding of IEEE (2009)

[4] Poli, R., Cagnoni, S.: Evolution of pseudo-coloring algorithms for image enhancement. Univ. Birmingham, Birmingham, U.K., Tech. Rep. CSRP-97-5 (1997).

[5] Munteanu, C., Lazarescu, V.: Evolutionary contrast stretching and detail enhancement of satellite images. In Proc. Mendel, Berno, Czech Rep., pp. 94-99 (1999)

[6] Munteanu, C., Rosa, A.: Evolutionary image enhancement with user behavior modeling. ACM SIGAPP Applied Computing Review,9(1), 8-14 (2001).

[7] Saitoh, F.: Image contrast enhancement using genetic algorithm. In Proc. IEEE SMC, Tokyo, Japan, pp. 899-904 (1993)

[8] Pal, S.K., Bhandari, D., Kundu, M.K.: Genetic algorithms for optimal image enhancement. Pattern Recognition Letter, 15, 261-271 (1994)

[9] Braik, M., Sheta, A., Ayesh, A.: Image enhancement using particle swarm optimization. In Proc of the World Congress on Engineering (WCE-2007), London UK (2007)

[10] Vesterstrom, J., Thomsen, R.: A comparative study of differential evolution, particle swarm optimization and evolutionary algorithms on numerical benchmark problems. Congress on Evolutionary Computation, pp. 980-987 (2004).

[11] Plagianakos, V., Tasoulis, D., Vrahatis M.,: A review of major application areas of differential evolution. In: Advances in differential evolution, Springer, Berlin, vol. 143, pp 197–238 (2008)

[12] Neri, F., Tirronen, V.: Recent advances in differential evolution: a survey and experimental analysis. Artif Intell Rev. 33 (1–2), 61–106 (2010)

[13] Das, S., Suganthan, P.N.: Differential evolution: a survey of the state-of-the-art. IEEE Transaction of Evolutionary Computing. 15(1), 4-13 (2011)

[14] Storn, R., Price, K.: Differential evolution—a simple and efficient adaptive scheme for global optimization over continuous. Spaces. Berkeley, CA, Tech. Rep. TR-95-012 (1995)

[15] Kumar, P., Pant, M.: Enhanced mutation strategy for differential evolution. In: Proc of IEEE Congress on Evolutionary Computation (CEC 12) (2012)

Honey Bee Based Vehicular Traffic Optimization and Management

Prasun Ghosal[1], Arijit Chakraborty[2], Sabyasachee Banerjee[3]

[1] Bengal Engineering and Science University, Shibpur, Howrah, WB, India

[2] Heritage Institute of Technology, Kolkata, WB, India

{prasun@ieee.org; arijitchakraborty.besu@gmail.com; sabyasachee.banerjee@gmail.com}

Abstract. Traffic densities in highly populated areas are more prone to various types of congestion problems. Due to the highly dynamic and random character of congestion forming and dissolving, no static and pre deterministic approaches like shortest path first (SPF) etc. can be applied to car navigators. Sensors are adequate here. Keeping view in all the above mentioned factors, our contributions in this paper include the development of a novel Bio Inspired algorithm on multiple layers to solve this optimization problem, where, car routing is handled through algorithms inspired by nature [Honeybee behavior]. The experimental results obtained from the implementation of the proposed algorithm are quite encouraging.

Keywords: Vehicular traffic management; Honey bee; Traffic optimization

1 Introduction

Traffic densities in highly populated areas are more prone to congestion problems, to some extent. Due to highly dynamic and random character of congestion forming and dissolving, no static and pre deterministic approaches like shortest path first (SPF) etc. can be applied to car navigators.

In this paper we tried to emphasize on the progress of a highly adaptive and innovative algorithm to deal with this problem. We got inspired primarily by the ideas of Swarm Intelligence technique, as the same is being applied rationally to address similar kinds of problems since decades back and also that have been detected in the honeybee communication. As a major development towards this field we present an idea based on honeybee based self-organizing vehicle routing algorithm termed as *honey jam* (ideally aimed at traffic congestion).

J. C. Bansal et al. (eds.), *Proceedings of Seventh International Conference on Bio-Inspired Computing: Theories and Applications (BIC-TA 2012)*, Advances in Intelligent Systems and Computing 202, DOI: 10.1007/978-81-322-1041-2_39, © Springer India 2013

2 Historical Background

Honey bees live in a colony and play two types of functional roles. In one type, they discover new food sources and are termed as *scouts*, and the second one be-ing *foragers*, that transport nectar from an already discovered flower site by following the dances of other scouts or foragers. Foragers use a special kind of mechanism called waggle dance to specify information about the quality, direction and distance to a distant food source. The intensity of the dance (reflecting the quality of the food source) determines the number of additional foragers required to be recruited to exploit the source. These foragers fly in the rough direction of the food source.

Once they have arrived at the approximate location, the foragers use their senses to precisely find their destination. These recruited foragers arrive in greater numbers at more profitable food sources because the dances for richer sources are more conspicuous and hence likely to be encountered by more number of unem-ployed foragers.

3 Previous Works

The paper proposed by Jake Kononov, Barbara Bailey, and Bryan K. Allery [1], first explores the relationship between safety and congestion and then examines the relationship between safety and the number of lanes on urban freeways. The relationship between safety and congestion on urban free-ways was explored with the use of safety performance functions [SPF] calibrated for multi-lane free-ways in Colorado, California, Texas. The Focus of most SPF modeling efforts to date has been on the statistical technique and the underlying probability distributions. The modelling process was informed by the consideration of the traffic operations parameters described by the Highway Capacity Manual [1].

In 2006, H Ludvigsen et al., has published Differentiated speed limits allowing higher speed at certain road sections whilst maintaining the safety standards are presently being applied in Denmark [2]. The typical odds that higher speed limits will increase the number of accidents must thus be beaten by the project [2].

In another important work, C.J. Messer et al. [3] presented a new critical lane analysis as a guide for designing signalized intersections to serve rush-hour traffic demands. Physical design and signalization alternatives are identified, and me-thods for evaluation are provided. The procedures used to convert traffic volume data for the design year into equivalent turning movement volumes are described, and all volumes are then converted into equivalent through auto-mobile volumes. The critical lane analysis technique is applied to the proposed design and signali-zation plan. The resulting sum of the critical lane volumes is then checked against established maximum values for each level of service (A, B, C, D, E) to determine the acceptability of the design. In this work, the authors have provided guidelines, a sample problem, and operation performance characteristics to assist the engineer in determining satisfactory design alternatives for an intersection [3].

There is one more design called Design of a Speed Optimization Technique in an Unplanned Traffic (DSOTU) [4] finding methods in other literature are a family of optimization algorithms, which incorporate level of traffic services in the algorithms. There are two major issues, in the first part; we have analyzed the major issues residing in the latest practice of the accidental lane; and, in the last part, we have discussed the possible applications of this new technique and new algorithm [4]. Other works in this area are also reported in [5-8].

4 Proposed Work

In our previous work we specially concentrated on maximum speed utilization of any vehicle as well as planning lanes for an unplanned traffic, but in this work we are also considering the speed of each lanes and their speed difference. As too much speed difference drives vehicles to be biased to only one lane though other lanes are not utilized properly. So our algorithm ensures maximum utilization of the lanes present in traffic without affecting the optimum speed of the vehicle too much, because vehicles can transit after the lanes threshold value is reached. But as a trade of, this eventually increases the number of transitions required to give a vehicle its optimum high speed.

4.1 Assumption

To implement this algorithm as a simulation of the real life scenario under consideration, certain assumptions are made without loss of generality of the problem. During the execution of the algorithm it is assumed that there will be no change in the current speed of the vehicle, if accidentally any vehicle's speed becomes '0' then totally discard the vehicle from the corresponding lane. Our algorithm runs periodically and continuously tries to optimize the speed of the lanes by reducing the speed difference of present lanes, but to achieve this we might have to increase the number of transitions of vehicles entered in to the lanes.

4.2 Description of the proposed algorithm

The primary sections of the proposed algorithm and their major functionalities are described below.

Step 1. During this step, inputs are taken from sensors, e.g. current speeds of vehicles, arrival time etc., and numbers of vehicles are counted input by the
sensor, and numbers of lanes present in the traffic with their corresponding
threshold values are input, too.

Step 2. In this step, lanes are assigned to different vehicles having different current
speeds. The way is first fill up the first lane up to its threshold value then
when the first lane's threshold value is filled up the vehicles which are allocated to the first lane is moved to the next lane until its population
reaches to the threshold value and the population of the first lane get de-
cremented as vehicles moved from the first lane to next lanes. Then the
first lane is also get populated with remaining vehicle's speed simultaneously. This process is continued until all the lanes got filled up to its
threshold value.

Step 3. Categorizing them depending on their assigned lane.

Step 4. This step finds the lane for remaining number of vehicles, where, the
difference between the vehicle's current speed and lane's speed buffer's
average speed is minimum and takes the vehicle to the lane, categorizes it
same as the lane's other vehicles, increases the population of the lane,
and stores the vehicle's current speed in the speed buffer of the lane.

Step 5. This step is used for checking total numbers of transitions, i.e. at which
point of the lane and from which lane to where the transition will occur
for all vehicles, thereby calculating the average speed of the lanes also.

Table 1. Table of Notations used in algorithm HoneyJam

Notations used	Meaning
V_i	Speed of the ith vehicle
Li	Lane of the ith vehicle
type(i)	Type of the vehicle
T	Arrival time difference between a high and low speed vehicles
t1	Time interval to overtake a vehicles at lower speed
D	Distance covered by low speed Vehicle
d1	Distance covered by high speeding Vehicle
Bn	Buffer of Lane n or population of Lane n
Count	Total no. Vehicle in traffic
t_Count	Total no. of transitions while assigning vehicles to a particular lane up to it's threshold value
trans_l	Total number Of transitions
L_th(i)	Threshold value of lane i
L_V a(i)	Speed of the vehicle a present in the lane i
L_avg(i)	Average speed of the lane i
Lane	Number of lanes
X	Marking after all the lanes filled up by its threshold value

4.3 Pseudo Code of Algorithm HoneyJam

INPUT: Vehicle's name, current speed, arrival time, number of lanes, threshold value of each lane.
OUTPUT: Total number of transitions and the average speed of the each lane present in the traffic.

Step 1.1: Set count=1; /*used to count the number of vehicles*/
Step 1.2: get_input ()/*Enter the inputs when speed of the vehicle is non-zero. */
Step 1.3: Continue Step 1.1 until sensor stops to give feedback.
Step 2: for b=1 to Lane
   Set x=x+L_th(b);
   Set Bb=0 and i=1;
   While (Bb! =L_th (b))
   if(b==0)
      Enter Vi into b lane's speed buffer,
      Bb++;
      i++;
   else
      Enter V_i into 1st lane's speed buffer and transfer 1st lane's populated vehicle to next   unpopulated lane.
      t_Count= t_Count+(b-1);
      Bb++;
      i++;
   End Loop
   End Loop
   return x;
Step 3: for b=1 to Lane
   for a=1 to L_th(b)
      for c=1 to Count
        if(Vc==L_V a(b))
          type(c)=b
      End Loop
   End Loop
   End Loop
Step 4: for a=x to Count
   set min=|Va-L_avg(1) |;
   set buf=1;
   set type(a)=1;
      for b=1 to Lane
        d=|Va-L_avg(b)|;/*Taking only magnitude of the difference*/
        if(d==0)
          set type(a)=b;
          set buf=b;

break;

if(d<min)

                  set type(a)=b

                  set buf=b;

        End Loop

        update Bbuf++;

        update L_V Bbuf(buf);

    End Loop

Step 5: for 1≤i≤Count

    for 2≤j≤Count

        If type(i)= type(j) and Vi<Vj and (i) vehicle's arrival time≤ (j) vehicle's arrival time

        Set t=(j) vehicle's arrival time - (i) vehicle's arrival time

        Set t1=0

        Begin loop

            Set t1=t1+1

            Set d=Vi*(t+t1)

            Set d1=Vj*t1

            If d1≤d set trans_l = trans_l+1;

        End loop

        trans_l= trans_l+t_Count;

    End loop

    End loop

    For 1≤m≤Lane

    Calculate each lane's average speed from its speed buffer.

Step 6: Return Number of transitions required= trans_l

Step 7: End.

5 Results and Analysis

5.1 Analysis of Algorithm HoneyJam

- The algorithm stated above is implemented on a planned traffic area where number of lanes and their population's threshold value is known beforehand.

- The objective will follow linear queue as long as speed/value/cost of proceeding to greater than the immediate next.

- Transitions/Crossovers are calculated and appropriate data structures are used in order to maintain the uniformity.

- We assume that the lanes are narrow enough to limit the bi-directional approach.

- We tried to implement logic in the algorithm *HoneyJam* in order to maintain optimized speed for each lane by reducing the average speed difference amongst lanes.

- The algorithm also calculates the transition points if speed/value/cost of a vehicle whenever found unable to maintain the normal movement and failed to transit in all possible calculated lanes.

- Transition points are recorded with their positions and numbers using appropriate data structures to maintain the record.

5.2   Time Complexity Analysis

Time complexity of the proposed algorithm and its subsections has been analyzed in the following Table 2.

Table 2. Time complexity analysis

Section	Time complexity	Final Time complexity
Step 1	O(Count)	
Step 2	O(Lane*L_th(b))	O ((Count-1)*(Count-1))
Step 3	O(Lane*L_th(b)*Count)	Since Count>>Lane and
Step 4	O(Count-(Lane*L_th(b))*Lane)	L_th(b)
Step 5	O((count-1)* (count-1))	

5.3   Graphical Analysis

The proposed algorithm has been implemented using C++ under GNU GCC compiler environment running Linux operating system with an Intel 3 GHz chip and 1 GB of physical memory. Experimental datasets have been plotted in bar chart form to study the variations. Some sample bar charts are shown in the following Figure 1. In the graphical analysis we can easily find that the speed difference between the lanes are far much decreased in the bar chart (Figure 1a) we have fixed the sample size and varied the number of lanes present in the traffic and in next bar chart (Figure 1b) we have fixed the number of lanes present in the traffic and varied the number of samples.

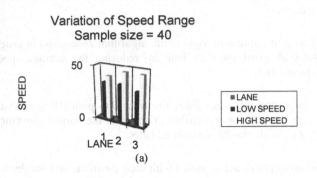

Variation of Speed Range
Sample size = 40

(a)

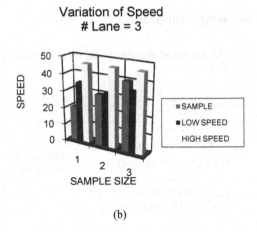

Variation of Speed
Lane = 3

(b)

Fig. 1 Graphical plot of experimental results

6 Conclusions and Future Scopes

The article presented through this paper mainly emphasize on optimal usage of
lanes using threshold information as knowledge base, but at the cost of transitions,
because in real life scenario transitions may be too high, hence our future effort
will be certainly in this direction.

In this article amount of time taken to transit between lanes has been considered cannot be ignored. The cumulative sum of transition time between lanes in real world problem contributed much in optimality of the proposed solution.

Bio inspired algorithms (like swarm intelligence) has been used with population information as knowledge base in our previous works, but partial modification of the stated concept taking threshold level information of the respective lanes will certainly be taken into consideration but implementation and formulation of algorithms along with optimality in transition, there by optimizing various aspects of traffic movement in real world will be considered in our future effort.

References

[1] Jake Kononov, Barbara Bailey, and Bryan K. Allery, *"The relationship between safety and congestion"*, Journal of the Transportation Research Board, No. 2083.

[2] *"Differentiated speed limits"*, European Transport Conference Differentiated speed limits, 2007.

[3] C.J. Messer and D.B. Fambro, *"Critical lane analysis for intersection design"*, Transportation Research Record No. 644; 1977, pp, 26-35.

[4] Prasun Ghosal, Arijit Chakraborty, Amitava Das, Tai-Hoon Kim, Debnath Bhattacharyya, *"Design of Non-accidental Lane"*, In Advances in Computational Intelligence, Man-Machine Systems and Cybernetics, pp. 188-192, WSEAS Press, 2010.

[5] Prasun Ghosal, Arijit Chakraborty, Sabyasachee Banerjee, *"Design of Efficient Knowledge Based Speed Optimization Algorithm in Unplanned Traffic"*, IUP Journal of Computer Sciences.

[6] Prasun Ghosal, Arijit Chakraborty, and Sabyasachee Banerjee, *"Particle Swarm Optimization of Speed in Unplanned Lane Traffic"*, International Journal of Artificial Intelligence & Applications (IJAIA), Vol.3, No.4, July 2012, pp. 51-63. DOI : 10.5121/ijaia.2012.3404

[7] Prasun Ghosal, Arijit Chakraborty, Sabyasachee Banerjee, and Satabdi Barman, *"Speed Optimization in Unplanned Traffic Using Bio-inspired Computing And Population Knowledge Base"*, Computer Science & Engineering: An International Journal (CSEIJ), Vol. 2, No. 3, June 2012, pp. 79-97. DOI : 10.5121/cseij.2012.2307

[8] Prasun Ghosal, Arijit Chakraborty, and Sabyasachee Banerjee, *"Computational Optimization of Speed in an Unplanned Lane Traffic"*, IEM International Journal of Management & Technology (IEMITMT) [ISSN: 2296-6611], pp. 160-163.

[9] P Ghosal, A Chakraborty, S Banerjee, "Bio-inspired Computational Optimization of Speed in an Unplanned Traffic and Comparative Analysis Using Population Knowledge Base Factor", Springer, Advances in Computer Science, Engineering & Applications, 2012, pp. 977-987

[10] P Ghosal, A Chakraborty, S Banerjee, "Speed Optimization in an Unplanned Lane Traffic Using Swarm Intelligence and Population Knowledge Base Oriented Performance Analysis", Springer, Advances in Computer Science, Engineering & Applications, 2012, pp. 471-480

Recent trends in supply chain management: A soft computing approach

Sunil Kumar Jauhar[1], Millie Pant[2]

[1] Research scholar, Indian Institute of Technology, Roorkee, India
[2] Associate Professor, Indian Institute of Technology, Roorkee, India
{jauhar.sunil@gamil.com; millidma@gmail.com}

Abstract. Increasing globalization, diversity of the product range and increasing customer awareness are making the market highly competitive thereby forcing different supply chains to adapt to different stimuli on a continuous basis. It is also well recognized that overall supply chain focus should be given an overriding priority over the individual goals of the players, if one were to improve overall supply chain surplus. Therefore supply chain performance has attracted researcher's attention. A variety of soft computing techniques have been employed to improve effectiveness and efficiency in various aspects of supply chain management. The aim of this paper is to summarize the findings of existing research concerning the application of soft computing techniques to supply chain management.

Keywords: computing; Supply chain management; Genetic algorithm, Fuzzy logic, Neural network.

1 Introduction

This research aims at reviewing the common soft computing techniques applied to supply chain management, exploring the current research trends and identifying opportunities for further research. The main issues to address include: what are the main problems within supply chain that have been investigated using soft computing techniques? What techniques have been employed? What are the main findings and achievements up to date? This paper is organized in five sections. Subsequent to the introduction in Section 1, the supply chain management and soft computing techniques are briefed in Sections 2 and 3. Section 4 describes the research methodology used in this paper. Finally, a summary of existing studies and a discussion on the future research directions are provided.

2 Supply chain management

Supply chain management as the management of upstream and downstream relationships with suppliers and customers to deliver superior customer value at less

J. C. Bansal et al. (eds.), *Proceedings of Seventh International Conference on Bio-Inspired Computing: Theories and Applications (BIC-TA 2012),* Advances in Intelligent Systems and Computing 202, DOI: 10.1007/978-81-322-1041-2_40, © Springer India 2013

cost to the supply chain as a whole [1]. Harrison described the supply chain management as a plan and controls all of the processes that link partners in a supply chain together in order to meet end-customers requirements [2]. As the sub-process of supply chain management, logistics deals with planning, handling, and control of the storage of goods between manufacturer and consumer [3]. Rushton described another well-known definition of logistics as the strategic management of movement, storage, and information relating to materials, parts, and finished products in supply chains, through the stages of procurement, work-in-progress and final distribution. A pictorial classification of supply chain linkage is shown in Fig. 1.

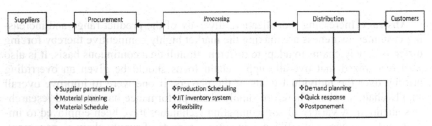

Fig. 1. Supply chain linkages

3 Soft computing

According to Prof. Zadeh, in contrast to traditional hard computing, soft computing exploits the tolerance for imprecision, uncertainty, and partial truth to achieve tractability, robustness, low solution-cost, and better rapport with reality. In other words, soft computing provides the opportunity to represent ambiguity in human thinking with the uncertainty in real life [5].Soft computing is a group of unique methodologies, contributed mainly by Fuzzy Logic (FL), Neural Networks (NN), and Genetic Algorithms (GA), which provide flexible information processing capabilities to solve real-life problems. The major soft computing techniques are briefed as following.

3.1 Genetic algorithms

The genetic algorithm is a probabilistic search algorithm that iteratively transforms a set (called a population) of mathematical objects (typically fixed-length binary character strings), each with an associated fitness value, into a new population of offspring objects using the Darwinian principle of natural selection and using operations that are patterned after naturally occurring genetic operations, such as crossover (sexual recombination) and mutation [6]. Genetic algorithms (GA) are a special subclass of a wider set of EA techniques. In resolving difficult problems where little is known, their pioneered work stimulated the development of a broad class of optimisation methods [7]. Based on the principles of natural evolution, genetic algorithms are robust and adaptive methods to solve search and optimisation problems [3]. Because of the robustness of genetic algorithms, a vast interest had been attracted among the researchers all over the world [8].In addition, by simulating some features of biological evolution; genetic algorithms can solve problems where traditional search and optimisation methods are less effective. Therefore, genetic algorithms have been demonstrated to be promising techniques which have been applied to a broad range of application areas [9].

3.2 Neural network

DARPA Neural Network Study (1988): Defines a neural network as a system composed of many simple processing elements operating in parallel whose function is determined by network structure, connection strengths, and the processing performed at computing elements or nodes. A neural network is a parallel distributed information processing structure consisting of a number of nonlinear processing units called neurons. The neuron operates as a mathematical processor performing specific mathematical operations on its inputs to generate an output [10]. It can be trained to recognize patterns and to identify incomplete patterns by duplicating the human-brain processes of recognizing information, burying noise literally and retrieving information correctly. In terms of modeling, remarkable progress has been made in the last few decades to improve artificial neural networks (ANN).Artificial neural networks are strongly interconnected systems of so called neurons which have simple behavior, but when connected they can solve complex problems. Changes may be made further to enhance its performance [11].

3.3. Fuzzy logic

Fuzzy logic is a mathematical formal multi-valued logic concept which uses fuzzy set theory. Its goal is to formalize the mechanisms of approximate reasoning [12]. It provides a mathematical framework to treat and represent uncertainty in the perception of vagueness, imprecision, partial truth, and lack of information [7]. As the basic theory of soft computing, fuzzy logic supplies mathematical power for the emulation of the thought and perception processes [9]. To deal with qualitative, inexact, uncertain and complicated processes, the fuzzy logic system can be well-adopted since it exhibits a human-like thinking process [13].One of the reasons for the success of fuzzy logic is that the linguistic variables, values and rules enable the engineer to translate human knowledge into computer evaluable representations seamlessly [7]. Fuzzy logic is one of the techniques of soft computing which can deal with impreciseness of input data and domain knowledge and giving quick, simple and often sufficiently good approximations of the desired solutions.

4 Methodology

The research methodology involves reviewing papers for soft computing techniques applied to the related processes in supply chain management. Initially two groups of keywords were used to cross-search related papers in specific databases. The first group of key words includes soft computing, neural network, fuzzy logic and genetic algorithm while the second group includes supply chain, transportation, logistics, forecasting, and inventory. The framework applied in this research is defined and developed by the Global Supply Chain Forum (GSCF) sponsored by the Council of Logistics Management (since 2005 it is called the Council of Supply Chain Management Professionals). The following eight processes of supply chain management have been categorized by the GSCF:

1. Demand management
2. Manufacturing flow management
3. Order fulfillment
4. Product development and commercialization
5. Returns management
6. Supplier relationship management
7. Customer service management
8. Customer relationship management

To refer to the eight processes of supply chain management categorized by the GSCF, the review of existing papers is classified into the following sections,

4.1. Demand management

Selen and Soliman have defined Demand Cycle Management as a set of practices aimed at managing and coordinating the whole demand chain, starting from the end customer and working backward to raw material supplier. Demand management plays a critical role within supply chain management. A reliable demand forecast can improve the quality of organizational strategy [15].The domain of demand management has been a major interest in soft computing since 1990s. A pictorial classification framework on demand chain is shown in Fig. 2.

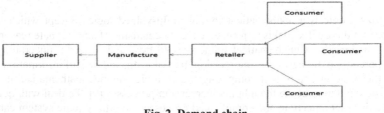

Fig. 2. Demand chain

4.1.1. Sales and demand forecasting

Accurate forecasting is an essential tool for many management decisions, for both strategic and tactical business planning. Advances in data analysis and software capabilities have the potential to offer effective forecasting to anticipate future demands, schedule productions and reduce inventories [15]. Artificial neural networks have been recognized as a valuable tool for forecasting. The major advantages to employ artificial neural networks in forecasting include its self-adaptive capability to learn from experience as well as to generalize results from sample data with noise. In addition, to compare with conventional statistical methods, artificial neural networks can model continuous functions to any desired accuracy [17]. Furthermore, as opposed to the traditional linear and nonlinear time series models, artificial neural networks are nonlinear data-driven approaches with more flexibility and effectiveness in modeling for forecasting [18]. Besides, a prototype supply planning system to enhance short-term demand forecast [19]. Ansuj et al. and Luxhoj et al. presented a neural network-based model to achieve more accurate sales forecasting results [20-21]. In addition, Kimbrough et al. and Strozzi et al. analysed the famous beer game for order policy optimisation [22-23]. Liang and Huang developed a multi-agent system for agents in supply chain to share information and minimise total cost [24].

4.1.2. Bullwhip effect

An effective supply chain management means efficient flow of quality and timely information between customer and suppliers which shall enable the supplier to uninterrupted and timely delivery of material to the customer but in practical life, there are situations which are never planned, and create oscillations in demand resulting in distortions in the supply chain. There can be a single cause or combinations of many factors. Suppliers, manufacturers, sales people, and customers have their own, often incomplete, understanding of what real demand is. Each group has control over only a part of the supply chain, but each group can influence the entire chain by ordering too much or too little. This lack of coordination coupled with the ability to influence while being influence by others lead. Drivers of bull whip effect can be from Customers, suppliers, systems, processes, sales, manufacturing, external factors etc [25].The bullwhip effect is one of the most popular research problems in supply chain management. It describes the distortion on demand forecasting throughout supply chain partners. Soft computing techniques proved to be effective to reduce bullwhip effect in supply chains [26].

4.2. Manufacturing flow management

Manufacturing flow management is the supply chain management process that includes all activities necessary to move products through the plants and to obtain, implement and manage manufacturing flexibility in the supply chain. Manufacturing flexibility reflects the ability to make a wide variety of products in a timely manner at the lowest possible cost.

Fig.3. Manufacturing flow in SCM

To achieve the desired level of manufacturing flexibility, planning and execution must extend beyond the four walls of the manufacturer in the supply chain[27].The work flow of the Manufacturing division encompasses sections devoted to Parts Management, Assembly, and Inspection. A framework on manufacturing flow in SCM is shown in Fig. 3. The initial paper with respect to application of soft computing in manufacturing flow management was accepted in 1990. There were only a few works in this area before 2001. Nevertheless, it demonstrates a steady rise in the number of papers since 2003 and reaches a peak in 2008. The challenge to improve manufacturing performance has drawn the attention of researchers to employ diverse soft computing techniques. The evidence seems to be strong that more studies can be anticipated in the near future.

4.2.1. Supply chain planning

In most organizations, supply chain planning is the administration of supply-facing and demand-facing activities to minimize mismatches, and thus create and capture value requires a cross-functional effort [28]. A framework on supply chain planning in SCM is shown in Fig.4.

Fig.4. Supply chain planning steps

Supply chain planning is focused on synchronizing and optimizing multiple activities involved in the enterprise from procurement of raw materials to the delivery of finished products to end customers [29].Genetic algorithms and artificial neural networks have been applied to derive optimal solutions for collaborative supply chain planning [30]. Moon et al. integrated process planning and scheduling model for resource allocation in multi-plant supply chain and Huin et al. presented a knowledge-based model for resource planning [31-32]. Subsequently Huang et al. designed a supply chain model to integrate production and supply sourcing decisions [33].

4.2.2. Production planning

Production planning involves looking ahead, anticipating bottlenecks and identifying the steps necessary to ensure smooth and uninterrupted flow of production.

Fig. 5. Production Planning in SCM

Production planning is such a key issue that both directly and indirectly influences on the performance of the facility. Different approaches are proposed in the literature for production planning, each of them has its own characteristics [34]. A classification framework on production planning in SCM is shown in Fig. 5.Genetic algorithms have been applied to solve production planning problems. The general capacitated lot-sizing problem was studied by Xie and Dong initially [35]. Ossipovthen proposed a heuristic algorithm to optimise the sequence of customer orders in production line [36]. Moreover, Kampf and Kochel focused on simulation-based sequencing and lot size optimisation while Bjork and Carlsson analysed the effect of flexible lead times by developing a combined production and inventory model [37-38].

4.2.3. Materials planning/inventory management

Supply chain inventory management is an integrated approach to the planning and control of inventory, throughout the entire network of cooperating organizations from the source of supply to the end user. SCIM is focused on the end-customer demand and aims at improving customer service, increasing product variety, and lowering costs [39]. For a business to be successful it requires a lot of hard work and a well thought out mind that will plan wise methods and useful ones to manage inventory and keep stocks low The economic lot-size scheduling problems were solved by a GA-based heuristic approach as well [40].There were also a few studies concentrated on fuzzy order and production quantity with or without backorder problems [41]. Recently the typical inventory problems such as the order quantity and reorder-point problem or the two storage inventory problem have been solved by the development of multi-objective inventory model [42]. Shelf space allocation problems [43], determination of base-stock levels in a serial supply chain [44].

4.3. Order fulfillment

Order fulfillment process is viewed as a key business process for achieving and maintaining competitiveness and is frequently the subject of re-engineering initiatives. Developing more responsive order fulfillment processes is generally recognised as being desirable [45]. The key components to grade actual order fulfillment are whether orders were delivered on time, in full, damage free, with accurate and complete documentation. A pictorial cycle on order fulfillment in SCM is shown in Fig. 6.

Fig.6.Order Fulfillment Cycle

Genetic algorithms have been applied to some challenging tasks successfully, such as logistics network design, vehicle routing, and vehicle scheduling problems. In addition to that, there are other interesting works that develop genetic algorithm approaches for customer allocation and shipping alternatives selection.

4.3.1. Vehicle routing

Consider the situation shown below where we have a depot surrounded by a number of customers who are to be supplied from the depot. The depot manager faces the task of designing routes (such as those shown below) for his delivery vehicles and this problem of route design is known as the vehicle routing or vehicle scheduling problem. A pictorial route of vehicle in a depot is shown in Fig. 7.

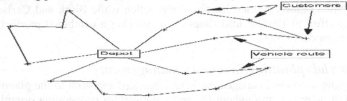

Fig. 7. Vehicle route

Vehicle routing is the problem of designing routes for delivery vehicles (of known capacities) which are to operate from a single depot to supply a set of customers with known locations and known demands for a certain commodity. Routes for the vehicles are designed to minimise some objective such as the total distance travelled [46]. In order to pick up and deliver within specific time window, Slater used expert system and artificial intelligence to predict e-commerce customer orders [47]. Also, Pankratz justified that a GA-based approach is able to find quality solution to meet the increasing demands on flexible and prompt transportation services [48].Torabi et al. found that a hybrid genetic algorithm is more promising in minimizing transportation cost in a simple supply chain [49]. A survey of different heuristic shortest path algorithms for demand-responsive transportation applications was presented [50]. In terms of vehicle assignment; Vukadinovic et al. concluded that neural networks can refine the fuzzy system to achieve better performance [51]. In addition, Potvin et al. reported an experimental result with data provided by a courier service company and proved that the neural network outperform the linear programming model in vehicle dispatching [52].

4.3.2. Logistics network design

A supply chain distribution network's physical structure can substantially affect its performance and profit margin. Most existing research on supply chain network design pursues a cost-minimization objective and tries to satisfy all the demands. However, the additional revenue generated from serving some retailers could be much lower than the cost associated with serving them. Thus, trying to satisfy all the retailers' demands might not give us the highest profit [53]. Teodorovic proved that fuzzy logic could be a very promising mathematical approach to solve complex traffic and transportation problems [54]. Sheu first presented a hybrid fuzzy-based methodology to identify global logistics strategies then achieved a remarkable cost saving and customer service enhancement by allocating logistics resources dynamically [55]. Genetic algorithms have been employed to solve dynamic logistics network design and planning problems, such as multistage logistic network design and optimisation [56], freight transportation planning [57], multi-time period production and distribution planning [58], logistic process optimisation, and vehicle transshipment planning in seaport terminal [59].

4.4. Product development and commercialization

The product development and commercialization process requires effective planning and execution throughout the supply chain, and if managed correctly can provide a sustainable competitive advantage. Developing product rapidly and moving them into the market place efficiently is important for the long term corporate success [60]. A pictorial classification framework on product development and commercialization in SCM is shown in Fig. 8.

Fig.8. Product Development and commercialization steps

The soft computing techniques that have been applied to the sub processes of product development include product quality enhancement and cost reduction [61], the relationship between the shelf space assigned to various brands and the market share [62], the optimal variable selections of R&D and quality design [63], and evaluation of supply chain performance for new product [64].

4.5. Returns management

Returns management is the supply chain management process by which activities associated with returns, reverse logistics, gate keeping, and avoidance are managed within the firm and across key members of the supply chain [65].Min et al. proposed a GA-based approach to solve reverse logistics problem of managing returned products [66]. Furthermore, Lieckens and Vandaele developed an optimal solution to solve the reverse logistics network design problem while Min and Ko addressed the similar problem from 3PL service providers' perspective [67-68].

4.6. Supplier relationship management

Herrmann and Hodgson defined SRM as a process involved in managing preferred suppliers and finding new ones whilst reducing costs, making procurement predictable and repeatable, pooling buyer experience and extracting the benefits of supplier partnerships. It is focused on maximizing the value of a manufacturer's supply base by providing an integrated and holistic set of management tools focused on the interaction of the manufacturer with its suppliers. [69].

Fig.9. Supplier Relationship management in SCM

A pictorial classification of supplier relationship management different steps in SCM is shown in Fig. 9 Several papers used fuzzy logic approach to monitor and measure suppliers' performance based on different criteria [70]. For example, Lau et al. analysed suppliers' product quality and delivery time while Shore and Venkatachalam (2003) focused on the information sharing capability of potential partners [71-72]. Deshpande et al. achieved an outstanding performance in assigning tasks to suppliers [73]. Furthermore, decision support models were proposed to enable a more effective selection of suppliers, vendors, and 3PL service providers [74-76]. Choy et al. used artificial neural network to design an intelligent supplier relationship management system in order to benchmark suppliers 'performance and shorten the cycle time of outsourcing [77].

4.7. Customer service management

Customer service management (CSM) offers a service oriented management interface between customer and service provider .CSM includes a wide range of activities, ranging from the time that there is a customer need for a product such as, requisition of a quotation to eventually providing ongoing support to customers, who have purchased the product. Since customer service processes are becoming more complex and a large number of decisions have to be made within a short period of time, the conventional way of customer services based on fax, e-mail and telephone might not satisfy customer needs in electronic business. [78]. Bottani and Rizzi presented a fuzzy quality function deployment approach to address customer needs, improve logistics performance, and ensure customer satisfaction [79].

4.8. Customer relationship management

Customer Relationship Management (CRM) is a process by which a company maximizes customer information in an effort to increase loyalty and retain customers' business over their lifetimes. It involves using technology to organize, automate, and synchronize business processes—principally sales activities, but also those for marketing, customer service and technical support [80]. It seems that there is a lack of papers addressing related issues in this area.

5 Discussion, conclusions and future research

The numerous and complex data sources are always needed to solve most of the problems in supply chain management. Soft computing tools seem promising and useful to analyse this data and to support manager's decision making in a complex environment. Both genetic algorithms and fuzzy logic approach are the most popular techniques adopted to solve supply chain management problems, particularly in the manufacturing management and order fulfillment issues. By examining the number of papers in manufacturing flow management, order fulfillment and demand management, the evidence seems to be strong that the issues in supply chain management have attracted a growing attention. It could be identified that there has been a significant upward trend of applying soft computing techniques to solve diverse supply chain management problems. The reasons may not only be that more researches have been involved in traditional supply chain domain, but also far more studies have been developed in new areas such as supplier relationship management and product development and commercialization

Some of the main problems in supply chain management have been addressed by soft computing techniques; there are still some areas of possible application which have not yet been well explored. This is particularly true in the field of customer service management. The qualitative issues dominate customer service management research. The qualitative nature of this domain also implies that it is difficult to frame problems in this area in a way that soft computing techniques can be readily applied. This may have resulted in the limited number of studies in this area. It is therefore expected that this paper can stimulate more research in the field of supply chain management.

References

[1] Christopher, M.: Logistics and Supply Chain Management, 2nd ed., Prentice Hall, Norfolk, 2004.

[2] Harrison, A., Hoek, R.: Logistics Management and Strategy, 2nd ed., Prentice Hall, Essex, 2005.

[3] Silva, C.A., Sousa, J.M.C., Runkler, T., Palm, R.: Soft computing optimization methods applied to logistic processes, International Journal of Approximate Reasoning 40 (3) (2005) 280–301.

[4] Rushton, A., Oxley, J., Croucher, P.: The Handbook of Logistics and Distribution Management, 2nd ed., Kogan Page, London, 2000.

[5] Roy, R., Furuhashi, T., Chawdhry, P.K.: Advances in Soft Computing: Engineering Design and Manufacturing, Springer, London, 1999.

[6] Sivanandam, S.N., Deepa, S.N.:Principles of Soft Computing, 2nd Edition copyright , 2011 Wiley India Pvt. Ltd.

[7] Tettamanzi, A., Tomassini, M.: Soft computing: Integrating Evolutionary, Neural, and Fuzzy Systems, Springer, Heidelberg, 2001.

[8] Goldberg, D.E.: Genetic Algorithms in Search, Optimization and Machine Learning, Addison-Wesley Publishing Company, Massachusetts, 1989.

[9] Sinha, N.K., Gupta, M.M.: Soft Computing and Intelligent Systems: Theory and Applications, Academic Press, San Diego, 2000.

[10] Musilek, P., Gupta, M.M.: Neural networks and fuzzy systems, in: Sinha, N.K, Gupta M.M. (Eds.), Soft Computing and Intelligent Systems: Theory and Applications, Academic Press, San Diego, 2000.

[11] Kartalopoulos, S.V.: Understanding Neural Networks and Fuzzy Logic, IEEE Press, New York, 1996.

[12] Ross, T.J., Fuzzy Logic with Engineering Applications, 2nd ed., John Wiley and Sons, West Sussex, 2004.

[13] Du, T.C., Wolfe, P.M.: Implementation of fuzzy logic systems and neural networks in industry, Computers in Industry 32 (3) (1997) 261–272.

[14] Selen, W., & Soliman, F. (Operations in today's demand chain management framework. Journal of Operations Management, (2002), 20. 667–673. [15] Kuo, R.J., Xue, K.C.: Fuzzy neural networks with application to sales forecasting, Fuzzy Sets and Systems 108 (2) (1999) 123–143.

[16] Bermúdeza, J.D., Segurab, J.V., Verchera, E. .: A decision support system methodology for forecasting of time series based on soft computing Computational Statistics & Data Analysis 51 (2006) 177 – 191.

[17] Sprecher, D.A.: On the structure of continuous functions of several variables, Transactions of the American Mathematical Society 115 (1965) 340–355.

[18] Zhang, G., Patuwo, B.E., Hu, M.Y.: Forecasting with artificial neural networks: the state of the art, International Journal of Forecasting 14 (1) (1998) 35–62.

[19] Wall, B., Higgins, P., Browne, J., Lyons, G.: A prototype system for short-term supply planning, Computers in Industry 19 (1) (1992) 1–19.

[20] Ansuj, A.P., Camargo, M.E., Radharamanan, R., Petry, D.G.: Sales forecasting using time series and neural networks, Computers & Industrial Engineering 31 (1–2) (1996) 421–424.

[21] Luxhoj, J.T., Riis, J.O., Stensballe, B.: A hybrid econometric—neural network modeling approach for sales forecasting, International Journal of Production Economics 43 (2–3) (1996) 175–192.

[22] Kimbrough, S.O., Wu, D.J.,Zhong, F.: Computers play the beer game: can artificial agents manage supply chains? Decision Support Systems 33 (3) (2002) 323–333.

[23] Strozzi, F., Bosch, J., Zaldivar, J.M.: Beer game order policy optimization under changing customer demand, Decision Support Systems 42 (4) (2007) 2153– 2163.

[24] Liang, W., Huang, C.: Agent-based demand forecast in multi-echelon supply chain, Decision Support Systems 42 (1) (2006) 390–407.

[25] http://knowscm.blogspot.in/2008/02/bullwhip-effect-in-supply-chain.html/25/07/(2012

[26] Carbonneau, R., Laframboise, K., Vahidov, R.: Application of machine learning techniques for supply chain demand forecasting, European Journal of Operational Research 184 (3) (2008) 1140–1154.

[27] Douglas M. Lambert and Terrance L. Pohlen, Supply Chain Metrics, The International Journal of Logistics Management, Vol. 12, No. 1, (2001), pp. 1-19.

[28] Braunscheidel, M.J., Suresh, N.C., 2009. The organizational antecedents of a firm's supply chain agility for risk mitigation and response. Journal of Operations Management 27, 119–140.

[29] Roghanian, E., Sadjadi, S.J., Aryanezhad, M.B.: A probabilistic bi-level linear multiobjective programming problem to supply chain planning, Applied Mathematics and Computation 188 (1) (2007) 786–800.

[30] Chiu, M., Lin, G.: Collaborative supply chain planning using the artificial neural network approach, Journal of Manufacturing Technology Management 15 (8) (2004) 787–796.

[31] Moon, C., Kim, J., Hur, S.: Integrated process planning and scheduling with minimizing total tardiness in multi-plants supply chain, Computers & Industrial Engineering 43 (1–2) (2002) 331–349.

[32] Huin, S.F., Luong, L.H.S., Abhary, K.: Knowledge-based tool for planning of enterprise resources in ASEAN SMEs, Robotics and Computer-Integrated Manufacturing 19 (5) (2003) 409–414.

[33] Huang, G.Q., Zhang, X.Y., Liang, L.: Towards integrated optimal configuration of platform products, manufacturing processes, and supply chains, Journal of Operations Management 23 (3–4) (2005) 267–290.

[34] Nasab , M.K., Konstantaras, I.: A random search heuristic for a multi-objective production planning Computers & Industrial Engineering 62 (2012) 479–490.

[35] Xie, J., Dong, J.: Heuristic genetic algorithms for general capacitated lot-sizing problems, Computers & Mathematics with Applications 44 (1–2) (2002) 263– 276.

[36] Ossipov, P.: Heuristic optimization of sequence of customer orders, Applied Mathematics and Computation 162 (3) (2005) 1303–1313.

[37] Kampf, M., Kochel, P.: Simulation-based sequencing and lot size optimisation for a production-and-inventory system with multiple items, International Journal of Production Economics 104 (1) (2006) 191–200.

[38] Bjork, K., Carlsson, C.: The effect of flexible lead times on a paper producer, International Journal of Production Economics 107 (1) (2007) 139–150.

[39] Verwijmeren, M., Vlist, P., Donselaar, K.: Networked inventory management I formation systems: Materializing supply chain management, International Journal of Physical Distribution and Logistics Management 26 (6) (1996) 16–31.

[40] Chang, P., Yao, M., Huang, S., Chen, C.: A genetic algorithm for solving a fuzzy economic lot-size scheduling problem, International Journal of Production Economics 102 (2) (2006) 265–288.

[41] Wang, X., Tang, W., Zhao, R.: Fuzzy economic order quantity inventory models without backordering, Tsinghua Science & Technology 12 (1) (2007) 91–96.

[42] Maiti, M.K.: Fuzzy inventory model with two warehouses under possibility measure on fuzzy goal, European Journal of Operational Research 188 (3) (2008) 746–774.

[43] Hwang, H., Choi, B., Lee, M.: A model for shelf space allocation and inventory control considering location and inventory level effects on demand, International Journal of Production Economics 97 (2) (2005) 185–195.

[44] Daniel, J.S.R., Rajendran, C.: Heuristic approaches to determine base-stock levels in a serial supply chain with a single objective and with multiple objectives, European Journal of Operational Research 175 (1) (2006) 566–592.

[45] Kritchanchai, D., MacCarthy, B.L.: Responsiveness of the order fulfillment process, International Journal of Operations & Production Management, Vol. 19 Iss: 8 pp. 812 - 833(1999)

[46] Thanh P., Bostel N., Pe´ton, O.: A DC programming heuristic applied to the logistics network design problem Int. J. Production Economics 135 (2012) 94–105)

[47] Slater, A.: Specification for a dynamic vehicle routing and scheduling system, International Journal of Transport Management 1 (1) (2002) 29–40.

[48] Pankratz, G.: Dynamic vehicle routing by means of a genetic algorithm, International Journal of Physical Distribution & Logistics Management 35 (5) (2005) 362–383

[49] Torabi, S.A., Ghomi, S.M.T.F., Karimi, B.: A hybrid genetic algorithm for the finite horizon economic lot and delivery scheduling in supply chains, European Journal of Operational Research 173 (1) (2006) 173–189.

[50] Fu, L., Sun, D., Rilett, L.R.: Heuristic shortest path algorithms for transportation applications: state of the art, Computers & Operations Research 33 (11) (2006) 3324–3343.

[51] Vukadinovic, K., Teodorovic, D.A., Pavkovic, G.: An application of neurofuzzy modeling: the vehicle assignment problem, European Journal of Operational Research 114 (3) (1999) 474–488.

[52] Potvin, J., Shen, Y., Dufour, G. , Rousseau, J.: Learning techniques for an expert vehicle dispatching system, Expert Systems with Applications 8 (1) (1995) 101–109.

[53] Jia Shu a, ZhengyiLi a, HoucaiShen b, TingWuc,n, WeijunZhong a A logistics network design model with vendor managed inventory Int. J. Production Economics 135 (2012) 754–761

[54] Teodorovic, D.: Fuzzy logic systems for transportation engineering: the state of the art, Transportation Research Part A: Policy and Practice 33 (5) (1999) 337– 364.

[55] Sheu, J.: A hybrid fuzzy-based approach for identifying global logistics strategies, Transportation Research Part E: Logistics and Transportation Review 40 (1) (2004) 39–61.

[56] Altiparmak, F., Gen, M., Lin, L., Paksoy, T.: A genetic algorithm approach for multiobjective optimization of supply chain networks, Computers & Industrial Engineering 51 (1) (2006) 196–215.

[57] Ma, H., Davidrajuh, R.: An iterative approach for distribution chain design in agile virtual environment, Industrial Management & Data Systems 105 (6) (2005) 815–834.

[58] Gen, M., Syarif, A.: Hybrid genetic algorithm for multi-time period production/ distribution planning, Computers & Industrial Engineering 48 (4) (2005) 799– 809.

[59] Fischer, T., Gehring, H.: Planning vehicle transhipment in a seaport automobile terminal using a multi-agent system, European Journal of Operational Research 166 (3) (2005) 726–740.

[60] Cooper, Robert G., Scott J. Edgett, and Elko J. kleinschmidt, Protfolio management for New products, Reading, MA; Perseus Books ,1998.

[61] Tong, L., Liang, Y.: Forecasting field failure data for repairable systems using neural networks and SARIMA model, International Journal of Quality & Reliability Management 22 (4) (2005) 410–420.

[62] Suarez, M.G.: Shelf space assigned to store and national brands: a neural networks analysis, International Journal of Retail & Distribution Management 33 (11) (2005) 858–878.

[63] Tsai, C.: An intelligent adaptive systemfor the optimal variable selections of R&Dand quality supply chains, Expert Systems with Applications 31 (4) (2006) 808–825.

[64] Wang, J., Shu, Y.: A possibilistic decision model for new product supply chain design, European Journal of Operational Research 177 (2) (2007) 1044–1061.

[65] Dale S. Rogers, Douglas M. Lambert, Keely L. Croxton and Sebastián J. García-Dastugue, The Returns Management Process, The International Journal of Logistics Management, Vol. 13, No. 2 (2002), pp. 1-18.

[66] Min, H., Ko, C.S., Ko, H.J.: The spatial and temporal consolidation of returned products in a closed-loop supply chain network, Computers & Industrial Engineering 51 (2) (2006) 309–320.

[67] Lieckens, K., Vandaele, N.: Reverse logistics network design with stochastic lead times, Computers & Operations Research 34 (2) (2007) 395–416.

[68] Min, H., Ko, H.: The dynamic design of a reverse logistics network from the perspective of third-party logistics service providers, International Journal of Production Economics 113 (1) (2008) 176–192.

[69] Herrmann, J. W., & Hodgson, B (2001). SRM: Leveraging the supply base for competitive advantage. Proceedings of the SMTA International Conference, Chicago, Illinois, 1 October, 2001.

[70] Gunasekaran, N., Rathesh, S., Arunachalam, S., Koh, S.C.L.: Optimizing supply chain management using fuzzy approach, Journal of Manufacturing Technology Management 17 (6) (2006) 737–749.

[71] Lau, H.C.W., Pang, W.K., Wong, C.W.Y.: Methodology for monitoring supply chain performance: a fuzzy logic approach, Logistics Information Management 15 (4) (2002) 271–280.

[72] Shore, B., Venkatachalam, A.R.: Evaluating the information sharing capabilities of supply chain partners: a fuzzy logic model, International Journal of Physical Distribution & Logistics Management 33 (9) (2003) 804–824.

[73] Deshpande, U., Gupta, A., Basu, A.: Task assignment with imprecise information for real-time operation in a supply chain, Applied Soft Computing 5 (1) (2004) 101– 117.

[74] Moghadam, M.R.S., Afsar, A., Sohrabi, B.: Inventory lot-sizing with supplier selection using hybrid intelligent algorithm, Applied Soft Computing 8 (4) (2008) 1523–1529.

[75] Wang, J., Zhao, R., Tang, W.: Fuzzy programming models for vendor selection problem in a supply chain, Tsinghua Science & Technology 13 (1) (2008) 106– 111.

[76] Isiklar, G., Alptekin, E., Buyukozkan, G.: Application of a hybrid intelligent decision support model in logistics outsourcing, Computers & Operations Research 34 (12) (2007) 3701–3714.

[77] Choy, K.L., Lee, W.B., Lo, V.: An intelligent supplier management tool for benchmarking suppliers in outsource manufacturing, Expert Systems with Applications 22 (3) (2002) 213–224.

[78] Langer, M., Loidl, S., & Nerb, M. (1999). Customer service management: towards a management information base for an IP connectivity service. The Fourth IEEE Symposium on Computers and Communications, Red Sea, Egypt, pp. 149–155.

[79] Bottani, E., Rizzi, A.: Strategic management of logistics service: a fuzzy QFD approach, International Journal of Production Economics 103 (2) (2006) 585–599.

[80] Robert, S.: Computer Aided Marketing & Selling (1991) Butterworth Heinemann ISBN 978-0-7506-1707-9.

Modified Foraging Process of Onlooker Bees in Artificial Bee Colony

Tarun Kumar Sharma[1],Millie Pant[1] and [2]Aakash Deep

[1] Indian Institute of Technology Roorkee, India
[2]Jaypee University of Engineering and Technology Guna, India
{taruniitr1, millidma, aakash791}@gmail.com

Abstract. Artificial Bee colony (ABC), a recently developed optimization algorithm has gained the attraction of many researchers. The foraging behavior of bees is used to search the optimum solution to the problem. In this study the foraging process for food sources by onlooker bees is being modified, which combines the information of the best food sources (based on fitness/nectar value) and also the information of the location of current food source to find new search directions. The proposed variant is named as MF-ABC and is tested in a set of 5 well known benchmark functions. The simulated results demonstrate the performance and efficiency of the proposal over basic ABC.

Keywords: Artificial Bee Colony, ABC, Optimization, Metaheuristic.

1. Introduction

Artificial Bee Colony (ABC), proposed by Karaboga [1][2] is the recent addition to the group of evolutionary algorithms. Like genetic algorithm (GA) [3], particle swarm optimization (PSO) [4], differential evaluation (DE) [5], and ant colony optimization (ACO) [6] ABC has also proved its performance when applied to the various complex real world problems. Karaboga and Basturk have compared the performance of the ABC algorithm with the performance of other well-known modern heuristic algorithms such as genetic algorithm (GA), differential evolution (DE), particle swarm optimization on unconstrained and constrained problems [7][8].

ABC, inspired by the intelligent foraging behavior of honey bees is characterized by three groups of bees (1) employed (2) onlooker and (3) scout bees. The bees intelligently organize themselves and divide the labor to perform the tasks, like searching for the nectar, sharing the information about the food source etc. The position of a food source represents a possible solution to the optimization problem and the nectar amount of a food source corresponds to the quality (fitness) of the associated solution. An epigrammatic mathematical narration of ABC is presented in section 3.

ABC has gained a vibrant attraction of researchers to solve the many real world problems arising in various fields of engineering, biology, and finance etc. ABC has been successfully applied for solving a variety of real life and benchmark problems, the latest applications and survey can be found in [9]-[18] and [19] respectively.

J. C. Bansal et al. (eds.), *Proceedings of Seventh International Conference on Bio-Inspired Computing: Theories and Applications (BIC-TA 2012)*, Advances in Intelligent Systems and Computing 202, DOI: 10.1007/978-81-322-1041-2_41, © Springer India 2013

In the present study, onlooker bees search mechanism is modified by allowing each location of food sources (solution) to generate better food source (offsprings). The search mechanism is improved by using the collective information of the best food source as well as the information of current food source. This search process tries to balance exploration and exploitation in ABC. The proposed MF-ABC is tested in a small test bed of 5 benchmark functions with different dimensions which shows its performance over basic ABC.

The paper is structured as follows: Section 3 presents brief explanation and algorithm of the proposed work. The experimental settings and results are given in section 4. Findings and future work are presented as conclusion in section 5.

2 Artificial Bee Colony: An Outline

Define SN as the total number of bees, N_e as the colony size of the employed bees and N_o as the size of onlooker bees, which satisfy the equation $SN = N_e + N_o$. The number of food sources is equal to the number of employed bees because each food source is exploited by only one employed bee around the hive. The standard ABC algorithm can be expressed as follows:

1. Randomly initialize a set of feasible food sources $(x_1; \ldots; x_{SN})$, and the specific solution x_i can be generated by:

$$x_i^j = x_L^j + rand(0,1)(x_U^j - x_L^j) \tag{1}$$

   where $j \in \{1,2,\ldots,D\}$ is the jth dimension of the solution vector. Calculate the fitness value of each solution vector respectively.

2. For an employed bee in the nth iteration $x_i(n)$, search new solutions in the neighborhood of the current position vector according to the following equation:

$$\hat{x}_{i,G}^j = x_{i,G}^j + \phi_{i,G}^j(x_{i,G}^j - x_{k,G}^j) \tag{2}$$

where $\hat{x} \in SN, j \in \{1,2,\ldots,D\}, k \in \{1,2,\ldots,N_e\}, k \neq i$. ϕ_i^j is a random number between -1 and 1.

3. Apply the greedy selection operator to choose the better solution between searched new vector $\hat{x}_i^j$ and the original vector x_i^j into the next generation. The greedy selection operator ensures that the population is able to retain the elite individual, and accordingly the evolution will not retreat.

4. Each onlooker bee selects an employed bee from the colony according to their fitness values. The probability distribution (p_i) of the selection operator can be described as follows.

$$p_i = \frac{fit_i}{\sum_{i=1}^{N_e} fit} \tag{3}$$

where fit_i is the fitness value of the solution i which is proportional to the nectar amount of the food source in the position i.

5. The onlooker bee searches in the neighborhood of the selected employed bee's position to find new solutions using Eq. (2). The updated best fitness value can be denoted with f best, and the best solution parameters

6. If the searching times surrounding an employed bee exceeds a certain threshold *limit*, but still could not find better solutions, then the location vector can be re-initialized randomly according to the Eq. (1).

7. If the iteration value is larger than the maximum number of the iteration then stop, else, go to 2.

3 Proposed Variant: MF-ABC

The success of any evolutionary algorithm simply depends upon two antagonist i.e. exploration and exploitation. Like other evolutionary algorithms ABC also has some drawbacks which impede its performance. ABC is good at exploration while poor at exploitation. In this study an attempt is made to balance exploration and exploitation by improving the search mechanism of the onlooker bees. In the proposal the probability of each parent to generate a better food source is increased by allowing each location of food source (solution) to generate more than one food source. This is done by using different mutation operator that incorporate information of the best food source in the current population and as well as the information of the current parent to define the new search directions. This mutation operator allows each parent to generate more than one food sources in the same generation. The pseudocode of the proposed variant to generate multiple food sources is given in Fig. 1.

```
For k=1 to q
    Select randomly r₁≠r₂≠r₃≠i
    For j=1 to D
        If (U(0,1))>0.5 Then
```
$$v_{i,G}^j = x_{r3,G}^j + \alpha(x_{best,G}^j - x_{r2,G}^j) + \beta(x_{i,G}^j - x_{r1,G}^j)$$
```
        Else
```
$$v_{i,G}^j = x_{i,G}^j$$
```
        End if
    End For
    If k>1 Then
        If ((f(v_{i,G}^j)<f(x̂_{i,G}^j))Then
```
$$\hat{x}_{i,G}^j = v_{i,G}^j$$
```
        End If
    Else
```
$$\hat{x}_{i,G}^j = v_{i,G}^j$$
```
    End If
End For
```

Fig.1. Multiple food sources generation, where q is user defined parameter to keep the best offspring generated. r_1, r_2 & r_3 are randomly chosen random numbers such that $r_1 \neq r_2 \neq r_3 \neq i$. $x_{i,G}^j$ and $x_{best,G}^j$ is the current parent and best individual in the current population G respectively.

The α and β factors indicate the influence of the best and parent food sources (solutions), respectively, in the search direction of the offspring

4 Experimental Settings, Simulated results and Discussions

Benchmark Functions
A test bed of 5 benchmark functions is taken to test the performance of the MF-ABC.
➤ The first function is the Sphere function described by

$$f_1(x) = \sum_{i=1}^{D} x_i^2$$

where the initial range of x is $[-100, 100]^D$. The minimum solution of the Sphere function is $x^* = [0,0,\dots, 0]$ and $f_1(x^*)=0$.
➤ The second function is the Ackley function described by

$$f_2(x) = 20 + e - 20\exp\left(-0.2\sqrt{\frac{1}{D}\sum_{i=1}^{D} x_i^2}\right) - \exp\left(\frac{1}{D}\sum\cos(2\pi x_i)\right)$$

where the initial range of x is $[-32.768, 32.768]^D$. The minimum of the Ackley function is $x^* = [0, 0,\dots,0]$ and $f_2(x^*)=0$.
➤ The third function is the Griewank function described by

$$f_6(x) = \frac{1}{4000}\left(\sum_{i=1}^{D}(x_i - 100)^2\right) - \left(\prod_{i=1}^{D}\cos\left(\frac{x_i - 100}{\sqrt{i}}\right)\right) + 1$$

where the initial range of x is $[-600, 600]^D$. The minimum of the Griewank function is $x^* = [100, 100,\dots,100]$ and $f_6(x^*)=0$.

➤ The fourth function is the Rastrigin function described by

$$f_3(x) = \sum_{i=1}^{D}(x_i^2 - 10\cos(2\pi x_i) + 10)$$

where the initial range of x is $[-5.12, 5.12]^D$. The minimum of the Rastrigin function is $x^* = [0, 0,\dots,0]$ and $f_3(x^*)=0$.
➤ The fifth function is the Rosenbrock function described by

$$f_5(x) = \sum_{i=1}^{D-1}100(x_{i+1} - x_i^2)^2 + (x_i - 1)^2$$

where the initial range of x is $[-50, 50]^D$. The minimum solution of the Rosenbrock function is $x^* = [1,1,\dots,1]$ and $f_5(x^*)=0$.

Parameter Settings
The population size, MCN (maximum cycle numbers) and limit are fixed to 80, 10000 and 100 respectively. q is taken as 6 where as α and β are fine tuned at 0.8 and 0.2 respectively, such that $\alpha+\beta=1$. All the considered test functions are tested for 10, 20 &

30 dimensions. And maximum iteration are fixed to 500, 750 & 1000 for 10, 20, & 30 dimensions respectively. Each of the experiments was repeated 25 times independently. And the reported results are the means and standard deviations of the statistical experimental data. All the algorithms have been executed on dual core processor with 1GB RAM. The programming language used is DEV C++. The random numbers are generated using inbuilt *rand* () function with same seed for every algorithm.

Performance Analysis
In order to analyze the relative performance of MF-ABC and ABC algorithms, a comparison is done on the basis of a performance index (PI) [20]. This index gives specified weighted importance to the rate of success to observe the **reliability**, and the number of function evaluations to observe the **efficiency** of the algorithm. For the computational algorithms under comparison the value of performance index PI_j for the jth algorithm is computed as under:

$$PI_j = \frac{1}{N}\sum_{i=1}^{N}(k_1\alpha_1^i + k_2\alpha_2^i)$$

$$where\ \alpha_1^i = \frac{Sr_i}{Tr^i},\ \alpha_2^i = \begin{cases} \dfrac{Mf^i}{Af^i}, & if\ Sr^i \geq 0, \\ 0, & if\ Sr^i = 0 \end{cases}$$

where $i = 1, 2..., N.$

Here, Tr^j is total number of times the ith problem solved, and Sr^j is the number of times i^{th} problem solved successfully. Af^i is the average number of function evaluations used by the jth algorithm in obtaining the optimal solution of ith problem in the case of the successful runs, and Mf^i the minimum of the average number of function evaluations of successful runs used of the algorithms under comparison in obtaining the optimal solution of ith problem. N is total number of problems on which the performance of algorithms has been tested. Further, k_1; and k_2 are nonnegative constants such that $k_1 + k_2 = 1$ (these are in fact the weights assigned by the user to the percentage of success, and the average number of function evaluations used in successful run, respectively). The larger the value of PI_j, better is the performance of the algorithm. In order to analyze the relative performance of MF-ABC and ABC algorithms equal weights are assigned to two of the terms (k_1; and k_2) at a time. Therefore, PI_j becomes a function of a single variable. The case considered is $k1 = w$; $k2 = (1 - w)$; $0 \leq w \leq 1$.

Result Discussion
The optimization results of 5 benchmark functions are presented in Table 1, which demonstrates the performance of the proposed MF-ABC over ABC. The performance curves of sphere, ackley, griekwank and restrign's benchmark functions are shown in Fig. 2.
The average error values of 25 runs for each algorithm are shown in Table 1.The obtained results show that our algorithm performed well against four benchmark functions in terms of average fitness value as well as standard deviation else for rosenbrock where ABC performed well. But for the same function when dimension is

20 and iterations are 750, MF-ABC performed better than basic ABC. In Table 2 the performance of the proposed variant MF-ABC is compared with GABC [21] when C=1.5. It can be analysed from the Table 2 that MF-ABC performed well for the sphere, Ackley, griekwank and rastrigin functions in terms of average error, where as GABC out performed for the rosenbrock benchmark function. Best results are highlighted in bold in Table 1 & Table 2.

Further when the performance is compared on the basis of PI, shown in Fig. 3, PI values of both the algorithms, at each value of w between 0 to 1, show that MF-ABC is better than ABC.

Table 1. Optimization results of benchmark problems in terms of mean and standard deviation (SD)

Function	D	Max. Iteration	ABC		MF-ABC	
			Mean	SD	Mean	SD
Sphere	10	500	1.426E−16	8.212E−17	**3.960e-17**	3.802e-18
	20	750	2.353E−12	2.198E−12	**1.563e-16**	4.298e-17
	30	1000	2.649E−10	2.136E−10	**3.145e-16**	9.922e-17
Ackley	10	500	1.400E−10	7.657E−11	**1.038e-13**	5.244e-14
	20	750	3.681E−07	1.302E−07	**4.347e-08**	4.441e-08
	30	1000	5.246E−06	1.702E−06	**4.462E-06**	2.987e-06
Griewank	10	500	1.040E−03	2.741E−03	**4.590e-17**	1.162e-17
	20	750	4.734E−09	2.177E−08	**1.518e-16**	4.325e-17
	30	1000	5.369E−09	2.664E−08	**2.994e-16**	9.683e-17
Rastrigin	10	500	4.911E−16	1.073E−15	**2.688e-17**	1.286e-18
	20	750	2.933E−11	5.728E−11	**1.119e-17**	2.937e-18
	30	1000	3.593E−07	3.565E−06	**9.642e-15**	2.132e-14
Rosenbrock	10	500	**0.07022**	0.09035	0.66038	0.35901
	20	750	0.427111	0.611375	0.41741	0.39873
	30	1000	**0.681802**	0.774655	0.71943	0.69139

Table 2. Comparative simulated results with the state-of-art algorithm. Max. generation = 5000; Function evaluation=400,000.

Function	D	ABC		MF-ABC		GABC	
		Mean	SD	Mean	SD	Mean	SD
Sphere	30	6.3791e-16	±1.203e-16	**7.391e-18**	±6.673e-19	4.1761e-16	±7.365e-17
Ackley	30	4.6955e-13	±5.954e-15	**2.736e-15**	±2.847e-16	3.2152e-14	±3.252e-15
Griewank	30	1.2730e-14	±1.464e-15	**1.374e-17**	±1.692e-17	2.9606e-17	±4.993e-17
Rastrigin	30	1.3453e-13	±7.966e-14	**9.642e-15**	±2.953e-14	1.3263e-14	±2.445e-14
Rosenbrock	3	6.44947e-02	±4.852e-02	1.3747e-02	±2.962e-02	**2.6591e-03**	±2.220e-03

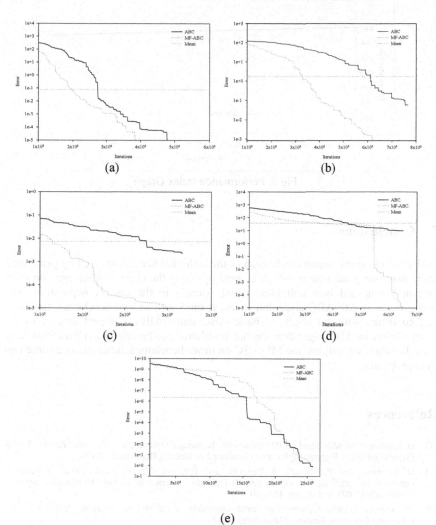

Fig.2 Convergence graphs of (a) Sphere (b) Ackley (c) Griekwank (d) Rastrign's (e) Rosenbrock benchmark functions for D=30 & Iteration = 1000

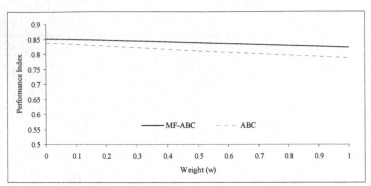

Fig.3. Performance Index Graph

5   Conclusions

In the present study search mechanism of the onlooker bee is improved by generating more than one good source, which is done by using the combined information from current parent and best individual (food source) in the current population. The performance of MF-ABC algorithm is tested on 5 well known benchmark functions and compared with the results of basic ABC and GABC. The performance of the modification in ABC algorithm for the problems can handle tested functions very well. In future we will test the MF-ABC on more benchmark functions and some real time problems.

References

[1]   D. Karaboga, An Idea based on Bee Swarm for Numerical Optimization, Technical Report, TR-06, Erciyes University Engineering Faculty, Computer Engineering Department (2005).

[2]   D. Karaboga and B. Basturk, A Powerful and Efficient Algorithm for Numerical Function Optimization: Artificial Bee Colony (ABC) algorithm, Journal of Global Optimization, Springer Netherlands (2007), Vol. 39, pp. 459–471.

[3]   D. Goldberg, Genetic Algorithms in Search Optimization and Machine Learning, Addison Wesley Publishing Company, Reading, Massachutes (1986).

[4]   J. Kennedy and R. C. Eberhart, Particle Swarm Optimization, *Proceeding of IEEE International Conference on Neural Networks*, Perth, Australia, IEEE Service Center, Piscataway, NJ (1995), pp. 1942–1948.

[5]   K. Price and R. Storn, Differential Evolution – a Simple and Efficient Adaptive Scheme for Global Optimization Over Continuous Spaces, Technical Report, International Computer Science Institute, Berkley (1995).

[6]   M. Dorigo, V. Maniezzo, A. Colorni, Positive feedback as a search strategy, Technical Report 91-016, Politecnico di Milano, Italy, 1991.

[7]   Karaboga, D., Basturk B.: On the performance of artificial bee colony (ABC) algorithm, *Applied Soft Computing*, Vol. 8, pp. 687-697, (2008)

[8] D. Karaboga and B. Basturk, Artificial Bee Colony (ABC) Optimization Algorithm for Solving Constrained Optimization Problems, LNCS: Advances in Soft Computing: Foundations of Fuzzy Logic and Soft Computing, Springer-Verlag, IFSA (2007), pp. 789–798.

[9] Karaboga D et al., Artificial bee colony programming for symbolic regression, Information Sciences (2012), http:// dx.doi.org/10.1016/j.ins.2012.05.002.

[10] Kashan MH, Nahavandi N, Kashan AH (2012) DisABC: A new artificial bee colony algorithm for binary optimization, Applied Soft Computing 12:342–352.

[11] Ma M, Liang J, Guo M, Fan Y, Yin Y (In Press) SAR image segmentation based on Artificial Bee Colony algorithm, Applied Soft Computing, doi:10.1016/j.asoc.2011.05.039, in press.

[12] Yeh WC, Hsieh TJ (2012) Artificial bee colony algorithm-neural networks for s-system models of biochemical networks approximation. Neural Comput Appl. doi:10.1007/s00521-010-0435-z.

[13] Li G, Niu P and Xiao X (2012) Development and investigation of efficient artificial bee colony algorithm for numerical function optimization. Applied Soft Computing 12:320– 332.

[14] Bahriye A and Karaboga D (2012). A modified Artificial Bee Colony algorithm for real-parameter optimization. Information Sciences 192: 120-142.

[15] Gao WF, Liu S, Huang L (2012) A global best artificial bee colony algorithm for global optimization. Journal of Computational and Applied Mathematics 236:2741-2753.

[16] F. Gao, Feng-xia Fei, Qian Xu, Yan-fang Deng, Yi-bo Qi, Ilangko Balasingham. A novel artificial bee colony algorithm with space contraction for unknown parameters identification and time-delays of chaotic systems, Appl. Math. Comput. (2012), http://dx.doi.org/10.1016/j.amc.2012.06.040

[17] T.K. Sharma, M. Pant, Enhancing scout bee movements in artificial bee colony algorithm, in: International Conference on Soft Computing for Problem Solving, SocProS 2011, AISC of Advances in Intelligent and Soft Computing, Vol. 130, Springer Verlag, 2011, pp. 601–610. December 20, 2011 – December 22, 2011.

[18] T.K. Sharma, M. Pant, Enhancing different phases of artificial bee colony for continuous global optimization problems, in: International Conference on Soft Computing for Problem Solving, SocProS 2011, AISC of Advances in Intelligent and Soft Computing, Vol. 130, 2011, pp. 715–724. December 20, 2011 – December 22, 2011.

[19] Dervis Karaboga, Beyza Gorkemli, Celal Ozturk, Nurhan Karaboga: A comprehensive survey: artificial bee colony (ABC) algorithm and applications, Artif Intell Rev 2011, DOI 10.1007/s10462-012-9328-0.

[20] Bharti (1994), Controlled random search technique and their applications. Ph.D. Thesis, Department of Mathematics, University of Roorkee, Roorkee, India, 1994.

[21] Zhu GP, Kwong S. Gbest-guided artificial bee colony algorithm for numerical function optimization. Applied Mathematics and Computation 2010, doi:10.1016/j.amc.2010.08.049.

A performance study of Chemo-inspired Genetic Algorithm on Benchmark Functions

Kedar Nath Das[1], Rajashree Mishra[2]

[1] NIT Silchar, Assam

[2] KIIT University, Orissa

{kedar.iitr@gmail.com; rajashreemishra011@gmail.com}

Abstract. In solving non-linear optimization problems, Bacterial Foraging Optimization (BFO) is a novel heuristic algorithm inspired from foraging behavior of E. Coli bacterium. In the other hand, Genetic algorithm (GA) has attracted increased attention from the academic and industrial communities to deal with such problems. In recent literature, it is discovered that the hybrid techniques provides the better solution with faster convergence. In this paper, a novel approach of hybridization is presented. The Chemotactic step (from BFO) is only hybridized with GA, namely CGA. The better performance of the proposed CGA than Quadratic Approximation hybridized GA, is experimentally verified through a set of 22 benchmark problems taken from recent literature.

Keywords: Genetic Algorithm, Quadratic Approximation, Bacterial Foraging Optimization, Hybridization, Benchmark Problems

1 Introduction

Genetic algorithm (GA) is an evolutionary optimization approach which is an alternative paradigm to the traditional optimization methods. It gains increased popularity for complex non-linear models where the location of the global optimum is a difficult task. It may be possible to use GA techniques to consider problems which may not be modeled as accurately using other approaches. Therefore, GA appears to be a potentially useful approach. It follows the concept of solution evolution by stochastically developing generations of solution populations using a given fitness statistic (for example, the objective function in mathematical programs). They are particularly applicable to problems with high degree of complexity, highly non-linear and even possibly discrete in nature. Due to the probabilistic development of the solution, GA does not guarantee optimality even when it may be reached. However, they are likely to be close to the global optimum.

J. C. Bansal et al. (eds.), *Proceedings of Seventh International Conference on Bio-Inspired Computing: Theories and Applications (BIC-TA 2012)*, Advances in Intelligent Systems and Computing 202, DOI: 10.1007/978-81-322-1041-2_42, © Springer India 2013

Current literature provides an idea of the increased research interest of researchers that focuses on hybrid genetic algorithm approaches for optimization problems. Fan et al. [1] integrate the Nelder–Mead simplex search method with genetic algorithm and particle swarm optimization in an attempt to locate the global optimal solution of nonlinear continuous variable functions. It focuses mainly on response surface methodology. Comparative performance on ten test problems is demonstrated in their paper. Hwang et al. [2] presented a novel adaptive real-parameter simulated annealing genetic algorithm which maintain the merits of genetic algorithms and simulated annealing. Comparative result is presented on 16 benchmark problems and two engineering design problems. Zhang and Lu [3] define a new real valued mutation operator and use it to design a hybrid real coded GA with quasi-simplex technique. A nitche hybrid genetic algorithm is proposed by Wei and Zhao [4] and results are reported on 3 benchmark functions. Chen, Tsai and Pan [5] proposed a novel optimization approach Bacterial–GA foraging. Kim, Abraham, Cho [6] presented an efficient genetic algorithm hybridized with bacterial foraging (namely GA-BF) approach for Global optimization. GA-BF is tested on 4 benchmark functions. Deep and Das [15] have introduced a hybrid GA incorporating the Quadratic approximation method, in which Quadratic approximation is used as an additional operator at the completion of each GA cycle. An attempt has been worked out in this paper in order to hybridize GA with chemotactic step of BFO to increase the searching ability in a precise manner. The same set of 22 problems from [15] has been picked up for a fair comparison.

This paper is organized as follows. The different components of the hybrid mechanism along with proposed algorithm are presented in the next section. Section 3 describes the result and discussion. The conclusion of the paper is drawn in section 4.

2 The hybrid system

2.1 Genetic Algorithm

A possible population of solutions at a particular generation undergoes the following steps to explore the search space and approaches towards the optimal solution. The Pseudo-code for Genetic Algorithm is given as follows:

Begin
t=1;
initialize $p(t)$;
evaluate fitness of $p(t)$;
while (termination criterion is not satisfied) do
t=t+1 ;

select from $p(t-1)$;

apply crossover on $p(t)$;

apply mutation on $p(t)$;

end while;

end begin.

2.2 Overview of Bacterial Foraging Optimization

Bacterial foraging optimization (BFO) is a novel heuristic approach based on swarm intelligence, to solve optimization problems [8, 9, 10]. In BFO, each individual string in the population is treated as an *E. Coli* bacterium that moves alternatively through running and tumbling, using its rigid set of spinning flagella. BFO is a bio-heuristic technique proposed by K.M Passino for finding the minimum of the cost function $P(\theta), \theta \in R^n$, where θ is the position of the bacterium. $P(\theta) < 0$, $P(\theta) = 0$ and $P(\theta) > 0$ represent the presence of nutrients, a neutral medium and the presence of noxious substances, respectively. The position of each member in the population of N bacteria at the jth chemotactic step, k^{th} reproduction step and l^{th} elimination-dispersal event is given by

$$P(j,k,l) = \{(\theta^i(j,k,l))/i = 1,2.....s\} \tag{1}$$

BFO mainly simulates four principal mechanism of E.Coli bacterium namely Chemotaxis, Swarming, Reproduction, and Elimination-Dispersal which are described as follows.

Chemo taxis: This step is simulation of the movement of E. Coli bacterium via flagella through run followed by tumble or tumble followed by tumble. The computational step is as follows.

$$\theta^i(j+1,k,l) = \theta^i(j,k,l) + C(i)\frac{\Delta(i)}{\sqrt{\Delta^T(i)\Delta(i)}} \tag{2}$$

where $\theta^i(j,k,l)$ is the position of the ith bacterium at j^{th} chemotactic step, l^{th} reproduction step, $C(i)$ is the step size. $\Delta(i) \in R^n$ is the random vector in the direction of tumble, with each element $\Delta_m(i)$, $m = 1,2...p$, a random number on [-1, 1].

Swarming: Group behavior of bacteria through cell-to-cell attraction and repulsion is defined as follows

$$J_{cc}\left(\theta, P(j,k,l)\right) = \sum_{i=1}^{s} J_{cc}^{i}\left(\theta, \theta^{i}(j,k,l)\right)$$

$$= \sum_{i=1}^{s}\left[-d_{aattract}\ \exp\left(-w_{attract}\ \sum_{m=1}^{p}\left(\theta_{m} - \theta_{m}^{i}\right)^{2}\right)\right]$$

$$+ \sum_{i=1}^{s}\left[h_{repellant}\ \exp\left(-w_{repellant}\ \sum_{m=1}^{p}\left(\theta_{m} - \theta_{m}^{i}\right)^{2}\right)\right] \qquad (3)$$

Here, J_{cc} denotes the swarming effect and θ_{m}^{i} is the mth component of the ith bacterium position. $d_{aattract}$ and $w_{attract}$ are depth and width of attractant released by bacteria while $h_{repellant}$ and $w_{repellant}$ are the height and width of the repellants. $\theta = \left(\theta_{1}, \theta_{2}, ... \theta_{p}\right)$ is a random point in the search space.

Reproduction: The least healthy bacteria die off and the healthier bacteria replicate themselves so that the size of the population remains constant.

Elimination –Dispersal: Some bacteria are liquidated randomly and reinitialized with very small probability.

2.3 The proposed Chemo-inspired Genetic Algorithm

In the recent past the popular BFO has been presented in many incarnations to find the optimal solution of a non-linear optimization problem with high complexity. Interestingly, researchers attempted in hybridizing the BFO with PSO, DE [11, 12, 13, 14]. Moreover, to solve the same, GA also itself empire the evolutionary world since last few decades and become user-friendly for the researchers in order to find the near optimal solutions. However, this paper proposes a different kind of hybridization keeping in view of both the mechanism of GA and BFO. The keys of motivation behind this paper are listed below.

a) Basically GA consists of 4 fundamental operators viz. Selection, Crossover, Mutation, Elitism. BFO is based on 3 principal mechanisms viz. Chemotaxis, Reproduction, Elimination-Dispersal. In fact Selection mechanism in GA is based on principle of retaining multiple copies of the best strings in the population which is similar in case of reproduction step followed in classical BFO, in which the least healthy bacteria dies and the healthier bacteria reproduces asexually.

b) Elitism in GA cycle does the work of the elimination of worst half strings and stores the best half of the string at the end. This mechanism is almost similar to that followed in Elimination-Dispersal step in classical BFO, where elimination of strings and simultaneously insertion of new strings takes place with a very small probability at the the end of BFO cycle.

Therefore, keep in view the above, it is experienced that while hybridizing to BFO with GA, probably the reproduction and elimination-dispersal steps of BFO become inefficient or it behave just like a repetition of few of the existing operators in GA. Hence, only the Chemotaxis step of BFO has been picked up from BFO to hybridize with GA as an additional operator. The hybridized algorithm thus proposed is named as **Chemo-inspired Genetic Algorithm (CGA)**. Therefore CGA has 5 major steps viz. Selection, Crossover, Mutation, Elitism and Chemotaxis. To improve the solution quality further, three productive properties have been employed in the mechanism as illustrated in the improved BFO [7] as given below.

i. Adaptive step size:

The adaptive step-size C_{step} is taken in this algorithm as follows.

$$C_{step} = C_{max}(i) - \frac{C_{max}(i) - C_{min}(i)}{N_c} j \qquad (4)$$

where $C_{max}(i), C_{min}(i)$ are maximum and minimum step size. N_c is the no of chemotactic step. The adaptive step size can exploit and explore the search space in a better way.

ii. Squeezed search space:

In order to improve the search performance in subsequent step the search space has been reduced. The range of the search space is as follows.

$$Range = (Min (j), Max (j))$$

$$Min (j) = X_{best} - \frac{R}{2^j} \qquad (5)$$

$$Max (j) = X_{best} + \frac{R}{2^j} \qquad (6)$$

where X_{best} is the current best position in the search space R, where R is the sphere of activity of the bacteria swarm.

iii. Fitness function criterion:

Let F_{best} represents the fitness of the best bacterium in the swarm so far. It will guide the swarm around the local optima. This method ensures both the Swarming effect and less computing time and contributes to the further improvement in the search procedure.

After the GA operators have been applied for the first 10 generations, the further refinement is inspired by the chemotactic step and again the operators of

GA have been applied. Another step has been proposed for the improvement of the search precision within the chemotactic step defined as follows.

2.4 Proposed Modified Chemotactic Step Size

In order to avoid the premature convergence, care has been taken in the chemotactic step size. If the solution remains unchanged for continuously 10 steps then a new and modified step size C_{step} is proposed as

$$C_{step} = i \cdot k + C_{step} \qquad (7)$$

where i represents the ith bacterium and k is a constant. This is acting as a mutation operator. If the optimum is trapped in the local optima a sudden change in step size is required to get out of that. The stepwise algorithm is defined as follows. The step wise algorithm is defined as follows.

Consider the following initial parameters.

P : Dimension of the variable.

S : No of bacteria

N_c : Number of chemotactic steps

N_S : Swim steps.

C_{step} : Step size of the bacterium.

X^i : Initial position of the ith bacterium.

$X^i(j)$: position of the ith bacterium at j^{th} the chemotactic step.

F_{best} : The best fitness value of the bacteria swarm.

X_{best} : The corresponding position of the F_{best}.

$Min(j), Max(j)$: Search scope of the bacteria.

i : Index of bacteria number, $i = 1, 2, \ldots \ldots S$.

j : Index of chemotactic step, $j = 1, 2, \ldots \ldots N_c$.

m : Index of the swimming steps, $m = 1, 2, \ldots \ldots N_s$.

Pseudo Code for proposed CGA:
[Step 1] Initialize the population. Calculate the fitness function.
While (termination criterion is not satisfied) do
[Step 2] Apply binary tournament selection.
[Step 3] Apply crossover operator.
[Step 4] Apply mutation operator.
[Step 5] Use Elitism Operator.
[Step 6] Chemotaxis loop:
        The detailed steps are

[substep a] *Re-initialization of the search space:*
For $j = 1 : N_c$
For $i = 1 : S$
Compute the fitness function $fit(X^i(j))$.
Update F_{best} and X_{best}
$Min(j)$ and $Max(j)$ are computed according to eqn. (5) and (6) and
X^i will be reinitialized in the range of $Min(j)$ to $Max(j)$
[substep b] *Tumble and Move:*
For $i = 1 : S$

Calculate the new position: $X^i(j+1) = X^i(j) + C_{step}\Delta(i)/\sqrt{\Delta(i)\Delta^T(i)}$ (8)

where $\Delta(i) \in R^n$ with each element $\Delta_m(i)$, $m = 1,2...p$, a random number on
[-1, 1]. This results in a step size C_{step} in the direction of the tumble for
bacterium i.
Calculate the C_{step} according to eqn. (4) or eqn.(7) as the case may be.
[substep c] *Swim Step:*
Let $m = 0$ (counter for swim length).
While $m < N_S$ (If have not climbed down too long)
Let $m = m + 1$.
If $fit(X^i(j+1)) < F_{best}$
Update F_{best} and X_{best}, and calculate the new position according to the
following equation:

$$X^i(j+1) = X^i(j+1) + C_{step}\Delta(i)/\sqrt{\Delta(i)\Delta^T(i)} \qquad (9)$$

Then compute the $fit(X^i(j+1))$ again. Else keep the bacterium stay still.
$X^i(j+1) = X^i(j)$. Calculate the fitness value of $X^i(j+1)$, let m = N_S
.This is the end of the while statement.
End for $(i = 1 : S)$
End for $(j = 1 : N_c)$
End do
End begin
The detail flow diagram of the proposed algorithm is shown in fig. 1.

3. Result and Discussion

In this section the impact of Chemo-inspired GA is explained through some
experimental results.

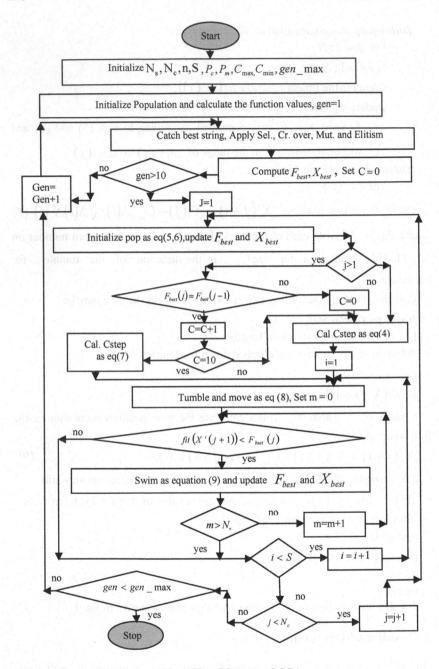

Fig.1. Flow Diagram of CGA

3.1. Experimental Set up

The proposed CGA program code is designed in C++ and the experiment is carried out on a P-IV, 2.8 GHz machine with 512 MB RAM under WINXP platform. In order to realize the efficiency of CGA, it is compared with Quadratic approximation based hybrid genetic algorithm, where simple GA has been hybridized with the Quadratic approximation [15]. They generated 4 versions of GAs and 4 versions of Quadratic approximation Hybrid GAs, depending on all possible combination of selection (Roulette Wheel and Tournament) and crossover (One-point and Uniform) operators of GA . The outcomes of CGA are again compared with best reported version in [15] for a set of 22 scalable test functions given in Table -1. The problems are of variable levels of difficulties.

For a fair comparison, the ranges of the decision variables remain unchanged. Problem size for all problems is fixed at 10 and the population size is at 60. A total of 100 runs are conducted for each of CGA and HGAs. A run is said to be a success if the value obtained by the algorithm is within 1% accuracy of the known optimal solution. But for Rosenbrock,Schwefel,Neumier-3 and Salomon it is taken to be 3, 0.4, 2, 0.1 respectively .The stopping criteria is a maximum of 1000 generations or if no improvement is observed in consecutive100 generations. But for Rosenbrock function the considered maximum generation is 10000. The chemotactic loop is allowed after 100 generations for 6 functions i.e Ackley, Sinusoidal, Levymontalvo-1, Levymontalvo-11, Griewank and Cigar function and for the remaining functions after 10 generations. In chemotactic loop, out of 60 bacteria only 4 bacteria are considered for re-initialization and exploitation of the search operation in a better way.

After a series of hand tuning experiments, the recommended values of the parameters are as follows. Probability of crossover (P_c=0.9), probability of mutation (P_m=0.01), maximum step size (C_{max} =0.1), minimum step size (C_{min} =0.008), chemotactic step size=40. But for Neumier-3 function 1500 chemotactic step has been considered. Recommending these parameters, further study is carried out.

3.2. Analysis of Result

The average minimum function value, success rate, function evaluation, standard deviation, CPU time of 100 independent runs both for CGA and HGAs/ GAs are reported in table-2. The efficacy of CGA is evaluated in terms of either average function value or success rate.

If the function values are found same for a particular function, then the next priority is given to success rate, and so on. The reported result assures that CGA achieved either much better optimal solution or success rate than HGAs/GAs in 17 problems and tie occurs in 2 cases. It also achieved better standard deviation for those problems. The CPU time is better in 14 problems as compared to HGAs/GAs. The best function value, success rate, standard deviation, time and number of function evaluation are mentioned in bold face letters in Table 2.

Table-1: List of test functions with the bounds of the decision variables

Sl. No.	Funtion Name	Function	Bounds
1	Ackley	$-20\exp(-0.02\sqrt{\frac{1}{n}\sum_{i=1}^{n}x_i^2})-\exp(\frac{1}{n}\sum_{i=1}^{n}\cos(2\pi x_i))+20+e$	[-30, 30]
2	Cosine Mixture	$0.1n+\sum_{i=1}^{n}x_i^2-0.1\sum_{i=1}^{n}\cos(5\pi x_i)$	[-1, 1]
3	Exponential	$1-\left(\exp\left(-0.5\sum_{i=1}^{n}x_i^2\right)\right)$	[-1, 1]
4	Griewank	$1+\frac{1}{4000}\sum_{i=1}^{n}x_i^2-\prod_{i=1}^{n}\cos\left(\frac{x_i}{\sqrt{i}}\right)$	[-600, 600]
5	Levy and Mantalvo-1	$\frac{\pi}{n}\left(10\sin^2(\pi y_1)+\sum_{i=1}^{n-1}(y_i-1)^2[1+10\sin^2(\pi y_{i+1})]+(y_n-1)^2\right),$ $y_i=1+\frac{1}{4}(x_i+1)$	[-10, 10]
6	Levy and Montalvo-2	$0.1*\left(\left(\sin^2(3\pi x_1)+\sum_{i=1}^{n-1}\dfrac{(x_i-1)^2[1+\sin^2(3\pi x_{i+1})]}{+(x_n-1)^2[1+\sin^2(2\pi x_n)]}\right)\right)$	[-5, 5]
7	Paviani	$45.778+\sum_{i=1}^{10}[(\ln(x_i-2))^2+(\ln(10-x_i))^2]-\left(\prod_{i=1}^{10}x_i\right)^{0.2}$	[2, 10]
8	Rastrigin	$10n+\sum_{i=1}^{n}[x_i^2-10\cos(2\pi x_i)]$	[-5.12, 5.12]
9	Rosenbrock	$\sum_{i=1}^{n-1}[100(x_{i+1}-x_i^2)^2+(x_i-1)^2]$	[-30, 30]
10	Schwefel	$418.9829n-\sum_{i=1}^{n}x_i\sin(\sqrt{|x_i|})$	[-500, 500]
11	Sinusoidal	$3.5-[2.5\prod_{i=1}^{n}\sin(x_i-\pi/6)+\prod_{i=1}^{n}\sin(5(x_i-\pi/6))]$	[0, π]
12	Zakharovs	$\sum_{i=1}^{n}x_i^2+\left(\sum_{i=1}^{n}\frac{i}{2}x_i\right)^2+\left(\sum_{i=1}^{n}\frac{i}{2}x_i\right)^4$	[-5.12, 5.12]
13	Sphere	$\sum_{i=1}^{n}x_i^2$	[-5.12, 5.12]
14	Axisparallel hyperellipsoid	$\sum_{i=1}^{n}ix_i^2$	[-5.12, 5.12]
15	Schewefel-3	$\sum_{i=1}^{n}|x_i|+\prod_{i=1}^{n}|x_i|$	[-10, 10]

16	Neumaier-3	$\dfrac{n(n+4)(n-1)}{6}+\sum_{i=1}^{n}(x_i-1)^2-\sum_{i=2}^{n}x_i x_{i-1}$	$[n^2, n^2]$
17	Salomon	$1-\cos\left(2\pi\|x\|\right)+0.1\|x\|,\ \|x\|=\sqrt{\sum_{i=1}^{n}x_i^2}$	$[-100, 100]$
18	Ellipsoidal	$\sum_{i=1}^{n}(x_i-i)^2$	$[-n, n]$
19	Schaffer-1	$0.5+\left(\left(\sin\sqrt{\sum_{i=1}^{n}x_i^2}\right)^2-0.5\right)\Big/\left(1+0.001\left(\sum_{i=1}^{n}x_i^2\right)\right)^2$	$[-100, 100]$
20	Brown-3	$\sum_{i=1}^{n-1}[(x_i^2)^{(x_{i+1}^2+1)}+(x_{i+1}^2)^{(x_i^2+1)}]$	$[-1, 4]$
21	New function	$\sum_{i=1}^{n}\left(0.2x_i^2+0.1x_i^2\sin 2x_i\right)$	$[-10, 10]$
22	Cigar	$x_1^2+100{,}000\sum_{i=2}^{n}x_i^2$	$[-10, 10]$

Table-2: Comparison of CGA with best reported method for each function in [15]

Sl. No.	Function Name	Function Value	Success Rate	S.D.	Time	No. of Fun. Evaluations
1	HGA3	1.67E-08	100	1.32E-08	4.06	28757
	CGA	**7.70e-15**	100	**9.85e-15**	**2.62**	112064
2	HGA3	2.00E-015	100	1.63E-15	2.44	16749
	CGA	**2.22E-18**	100	**1.56E-17**	**1.8**	109477
3	HGA3	2.49E-16	100	1.33E-16	2.00	15849
	CGA	**4.88E-17**	100	**5.51E-17**	**1.33**	71885
4	GA4	3.53E-03	23	4.43E-03	8.04	26204
	CGA	**1.23E-03**	8	**3.26E-03**	16.2	197492
5	HGA3	**1.18E-12**	100	9.17E-16	**3.40**	16023
	CGA	5.23E-09	98	**1.16E-21**	4.48	219182
6	HGA3	**1.89E-13**	100	4.62E-017	**2.02**	13418
	CGA	1.67E-09	83	**9.52E-22**	3.56	155836
7	HGA3	4.70E-04	100	8.41E-14	3.00	24872
	CGA-	**2.64E-14**	100	**2.67E-14**	**0.15**	18802
8	HGA3	4.73E-13	35	2.64E-13	4.80	34276
	CGA	**2.72E-15**	100	**1.67E-15**	**3.25**	276187
9	HGA3	6.92E-01	100	1.75E+00	**88.4**	309516
	CGA	**4.42E-05**	78	**4.78E-05**	181	3.82E+06
10	HGA4	**1.27E-04**	75	3.45E-12	4.61	28134
	CGA	**1.27E-04**	99	**3.55E-13**	**4.2**	75782

11	HGA3	3.02E-15	100	1.23E-15	2.33	**19580**
	CGA	**5.86E-16**	91	**2.70E-16**	**1.37**	58993
12	HGA3	2.04E-14	100	1.01E-14	3.00	**17088**
	CGA	**1.33E-22**	100	**4.27E-23**	**1.17**	156966
13	HGA3	2.69E-15	100	1.49E-15	2.27	**16773**
	CGA	**6.36E-41**	100	**1.49E-41**	**1.06**	169683
14	HGA3	2.89e-15	100	1.59e-15	2.16	**16703**
	CGA	**3.495E-40**	100	**1.32E-40**	**1.72**	321818
15	HGA3	3.83E-07	100	1.21E-07	**3.00**	**19684**
	CGA	**1.71E-08**	100	**1.84E-08**	3.42	197641
16	HGA4	**7.17E-04**	100	**2.98E-04**	14.3	**98728**
	CGA	5.007E-01	92	1.44E-01	**6.01**	1.03E+06
17	GA4	**9.99E-02**	35	3.370E-07	7.71	**28657**
	CGA	**9.99E-02**	44	**1.39E-017**	**3.83**	216620
18	HGA3	1.06E-14	100	6.74E-15	2.94	**17311**
	CGA	**1.80E-31**	100	**2.86E-031**	**0.92**	71123
19	HGA3	**9.72E-03**	16	5.89E-017	**2.81**	**25799**
	CGA	**9.72E-03**	34	**2.62E-17**	3.06	350332
20	HGA4	5.96E-16	100	2.12E-16	3.00	**16466**
	CGA	**3.05E-29**	100	**7.99E-30**	**1.72**	146612
21	HGA3	1.85E-15	100	9.42E-16	2.47	**16781**
	CGA	**4.69E-041**	100	**1.12E-41**	**1.18**	187579
22	HGA3	6.39E-10	100	3.46E-10	**3.00**	**18113**
	CGA	**2.03E-12**	100	**2.94E-12**	11	1.17E+08

5 Conclusion

Inspired by the mechanism of chemotaxis in bacterial foraging optimization, an attempt has been made in this paper to hybridize only the chemotactic step of BFO with GA cycle. It is observed from the result and discussion that the chemo-inspired GA gives efficient and challenging result as compared to Quadratic approximation based hybrid genetic algorithm as well as simple GA, not only in terms of better objective function value but also in terms of less computational time for most of the functions. Though the function evaluation is more, it consumes less computational time in return for most of the cases.

References

[1]   Fan, Shu-Kai. , Liang, Y.C., Zahara, E. : A genetic algorithm and a particle swarm optimizer hybridized with Nelder–Mead simplex search, Computers and Industrial Engineering 50 , 401–425 (2006).

[2]   Hwang, Shun-Fa., He, Rong Song. : A hybrid real parameter genetic algorithm for function optimization, Advance Engineering Informatics 20 ,7–21(2006).

[3]   Zhang, G., Lu, H. : Hybrid real coded genetic algorithm with quasi-simplex technique, International Journal of Computer Science and Network Security 6(10) 246–255(2006).

[4]   Wei, L., Zhao, M. : A nitche hybrid genetic algorithm for global optimization of continuous multi modal functions, Applied Mathematics and Computations 160 ,649–661 (2005).

[5]   Chen, T.C. ,Tsai, P.W., Chu, S.C., Pan, J.S.:A novel optimization approach- bacterial GA foraging, IEEE, Proceedings of Second International Conference on Innovative Computing, Information and Control, ICICIC'07,PP.31-1391,(2007).

[6]   Dong, Hwa .Kim., Abraham, Dong . Hwa .Ajith., Cho, Jae. Hoon. : A Hybrid Genetic Algorithm and Bacterial Foraging Approach for Global Optimization, Information Sciences,vol.177,no.18,pp.3918-3937,(2007).

[7]   Chen, Yanhai., Lin ,Weixing. : An Improved Bacterial Foraging Optimization , Proceedings of the IEEE International Conference on Robotics and Biomimetics, December 19-23,Guilin,China. (2009).

[8]   Sastry, V.R.S.Gollapudia. , Shyam, S.Pattnaika., Bajpaib ,O.P . , Devi ,Swana., Bakwada, K.M. : Velocity Modulated Bacterial Foraging Optimization Technique (VMBFO), Applied Soft Computing, www.elsevier.com.

[9]   Chen, Hanning., Zhu, Yunlong., Hu, Kunyuan.:Research Article, Adaptive Bacterial Foraging Optimization, Hindawi Publishing Corporation, Abstract and Applied Analysis, Volume 2011, Article ID 108269,27 pages doi:10.1155/2011/10826.

[10]   Passino, K.M.: Biomimicry of bacterial foraging for distributed optimization and control, IEEE Control Systems Magazine (2002) 52-67, doi:10.1109/MCS.2002.1004010.

[11]   Long, Liu Xiao., Jun, Li. Rong., Ping, Yang., : A Bacterial Foraging Optimization Algorithm Based On the Particle Swarm Optimization, IEEE proceedings of International Conference on Intelligent Computing and Intelligent System held in SCUT Guangzhou, China, vol.2, pg.22-27 (2010).

[12]   Biswas, Arijit., Dasgupta. , Sambarta., Das, Swagatam., Abraham, Ajith.: A synergy of PSO and Bacterial Foraging Optimization- A Comparative Study on Numerical Benchmarks, in proceedings of the 2nd international symposium on Hybrid Artificial Intelligent System(HAIS) , Advances Soft Computing, vol.44, pp.253-263,(2007).

[13]   Biswas, Arijit., Dasgupta, Sambarta., Das, Swagatam., Abraham, Ajith.,: A synergy of differential evolution and bacterial Foraging optimization for global optimization, International Journal on Neural and Mass parallel computing and Information Systems, Neural Network world, vol.17, No.6, pp.607-626, (2007).

[14]   Biswas, Arijit., Dasgupta, Sambarta., Das, Swagatam., Abraham, Ajith.,: A synergy of PSO and bacterial Foraging optimization-A Comparative Study on Numerical Benchmarks: Innovations in Hybrid Intelligent Systems, ASC 44, springerlink.com, pp.255-263, (2007).

[15]   Deep, K., Das, Kedar. Nath. : Quadratic approximation based Hybrid Genetic Algorithm for Function Optimization, AMC, Elsevier, Vol. 203, pp. 86-98, (2008).

SELF ADAPTIVE HYBRIDIZATION OF QUADRATIC APPROXIMATION WITH REAL CODED GENETIC ALGORITHM

Kedar Nath Das [1], **Tapan Kumar Singh** [2]

Department of Mathematics, National Institute of Technology, Silchar-788010

{kedar.iitr@gmail.com[1]; tksingh1977.nits@rediffmail.com[2]}

Abstract: The real coded Genetic Algorithm (LX-PM) that uses Laplace Crossover and Power mutation became popular to find optimal solution. In recent past, the LX-PM is being hybridized with a local search called Quadratic Approximation (QA) to improve the solution quality. However, there are some instances to improve it further, just by checking the frequency of hybridization. In this paper, a self adaptive strategy of hybridization is incorporated in the cycle of LX-PM. The improved efficiency and efficacy of the adaptive hybridization of QA over the simple hybridization of QA with LX-PM, is being realized through a set of 22 unconstrained benchmark problems. Comparative result is being analyzed through the numerical results, in terms of five different aspects.

Keywords: Hybrid real coded genetic algorithm, Laplace Crossover, Power Mutation and Quadratic Approximation

1 Introduction

The genetic algorithm (GA) is a search technique based on the mechanics of natural genetics and survival of the fittest. GA is an attractive and alternative tool for solving complex non-linear optimization problems [4,5,6]. Any carefully designated GA is only able to balance the exploration of the search effort, which mean that an increase in the accuracy of a solution can only come at the sacrifice of the convergent speed, and vice versa. Despite their superior search ability, GA still fails to meet the high expectations that theory predicts for the quality and efficiency of the solution. In order to solve the optimal control problem of a class of hybrid system M S Arumugam et. al. [3] introduced two new hybrid operators in real coded genetic algorithm. They named them as hybrid selection and hybrid crossover operators. The designed algorithm has been applied to solve the optimal

J. C. Bansal et al. (eds.), *Proceedings of Seventh International Conference on Bio-Inspired Computing: Theories and Applications (BIC-TA 2012),* Advances in Intelligent Systems and Computing 202, DOI: 10.1007/978-81-322-1041-2_43, © Springer India 2013

control problem with the number of jobs varying from 5 to 25. Deep and Thakur [2,7] introduce a real coded GA with Laplace Crossover (LX) and Power Mutation (PM). They name it as LX-PM. For maintaining the efficiency, reliability and stability Deep and Das [1] hybridized LX-PM with a local search called Quadratic Approximation (QA). They name it as H-LXPM. The performance is evaluated by using a set of 22 benchmark test problems. Further It is proposed that H-LXPM outperforms LX-PM in most of the cases. However, it is learnt that the process of hybridization and the frequency of appearance in the GA cycle play an important role in improving the solution quality. Therefore, an attempt is made in this paper to hybridize QA with GA with a self adaptive mechanism. In the later part of this paper, the superior quality of the same is observed through the same set of 22 benchmark problems.

Y.Yun et.al. [8] developed a adaptive local search to implement in GA cycle for solving multistage base supply chain problems. Later in 2010, S.Junjie and Z.Qiuhai[9] mixed the internal mechanism of Radio Forecast Network (RFN) ,self adaptive GA and gradient descent method to overcome the short coming of learning rule of RFN Network. In the same year J.Li and H.R.fung [10] proposed an adaptive adjustment (self adaptive) of the control parameters to improve the excellence of the rate of convergence. A mixed strategies of adapting operation parameter with hybridization of GA and SA (Simulated Annealing) is proposed as a mixed excellence algorithm in [11]. S.L.Xiu and HE Y.Yao [12] tried to improve two drawbacks of quantum evolution algorithms. They are (a) slow convergence rate and (b) poor robustness. They introduce a novel self adaptive genetic algorithm.

The rest part of the paper is organized as follows. In section 2, the Hybrid Real Coded GA (H-LXPM) is described. In Section 3, the proposed hybridization of self adaptive QA with LX-PM is discussed. In Section 4, parameter settings and the numerical results and their analysis are presented. Section 5 contains the conclusion of this paper.

2 H-LXPM Algorithm

In solving unconstrained optimization problems, Deep and Thakur [2, 7] proposed a real coded GA. They used a new crossover operator namely Laplace Crossover (LX) [2] and a new mutation operator called Power Mutation (PM) [7]. The GA thus designed is named as LX-PM. The working mechanism is presented briefly in the next subsection.

2.1 Pseudo Code of LX-PM

begin
    Gen=1
    Generate initial population randomly using real coding
    Evaluate fitness of each individual in the population
    While (termination criterion is not satisfied) **do**

Gen=Gen+1
Apply Tournament Selection operator
Apply Laplace Crossover operator
Apply Power Mutation operator
Apply Complete Elitism
Apply Quadratic Approximation Operator
end do
end begin.

2.2 Quadratic Approximation

Quadratic Approximation (QA) is a method to generate one 'child' from three parents from the population. The working principle of QA is as follows.

1. Select the individuals R_1, with the best fitness value. Choose two random individuals R_2 and R_3 such that out of R_1, R_2 and R_3, at least two are distinct.

2. Find the point of minima (child) of the quadratic surface passing through R_1, R_2 and R_3 defined as:

$$\text{Child} = 0.5\left(\frac{(R_2^2 - R_3^2)f(R_1) + (R_3^2 - R_1^2)f(R_2) + (R_1^2 - R_2^2)f(R_3)}{(R_2 - R_3)f(R_1) + (R_3 - R_1)f(R_2) + (R_1 - R_2)f(R_3)}\right) \quad (1)$$

Where $f(R_1)$, $f(R_2)$ and $f(R_3)$ are the fitness function values at R_1, R_2 and R_3, respectively.

2.3 H-LXPM

Deep and Das [1] proposed a hybrid real coded GA, where the real coded GA (LX-PM) undergoes the hybridization of QA as a local search. In the cycle of LX-PM as shown above, they applied QA just after the completion of the process of elitism. This hybrid version is named as H-LXPM. As a conclusion, authors presented that the H-LXPM outperforms LX-PM, though a set of 22 scalable benchmark problems of varying difficulty levels.

3 The proposed self adaptive hybridization of real coded GA

In general, GA cycle plays with a population of individuals. While the generation goes on, initially much diversity is maintained in the population. Hence, the hybridization is essential to implement for faster convergence. But later, the individuals of the population come closer to each other. Then the hybridization does not appear to be much effective. Rather it increases not only the number of function calls but also the computational time. Keeping in view the above, a self adaptive hybridization of QA is proposed as follows.

3.1 Self Adaptive QA

To maintain the diversity in the population, in each LXPM cycle, the number of children need to be generated, should be decided. The frequency of QA is proposed as a function of generation number. Hence it is called self adaptive QA and is defined as follows.

$$\text{The frequency of QA in a GA cycle} = F_{QA} = PS - \left\lceil \frac{Generation\ Number}{Population\ Size} \right\rceil \quad (2)$$

3.2 Pseudo Code for S-LXMP

The proposed self adaptive QA is hybridized with LX-PM and is named as S-LXPM. The hybridization idea is reflected below.

```
begin
    Gen=1
            Generate initial population randomly using real coding
            Evaluate fitness of each individual in the population
            While (termination criterion is not satisfied) do
                    Gen=Gen+1
                    Apply Tournament Selection operator
                    Apply Laplace Crossover operator
                    Apply Power Mutation operator
                    Apply Complete Elitism
                    Apply Self-adaptive Quadratic Approximation Operator
        end do
end begin.
```

The flow mechanism of S-LXPM is presented in Fig. 1.

4. Test Problems, Experimental Setup and Result Analysis

4.1 Test Problems

In order to compare the performance of self adaptive hybridized GA (S-LXPM) with H-LXPM, a set of 22 benchmark scalable problems are picked from Deep and Das [1]. The problems are listed in Table-1. The reason of considering such a set is that the problems carry 4 main characteristics as follows.

i.   They are scalable (where the problem size can be varied with the user's choice).
ii.   They are of various difficulty levels.

iii. They contain uni-modal as well as multi-modal functions.
iv. Each function value has an optimal solution '0'.

4.2 Experimental Setup

The H-LXPM and S-LXPM are implemented in C++ and the experiments are carried out on a Pentium Dual Core, 2.00 GHz machine with 1GB RAM under Windows 7 platform. A total of 100 runs are conducted for each hybridized GA, a different seed for the generation of random numbers is considered to start a run. A run is considered to be success if the value obtained by the algorithm is within 1% accuracy of the known optimal solution. But for Rosenbrock and Schwefel problem it is taken as 12 and 0.4 respectively.

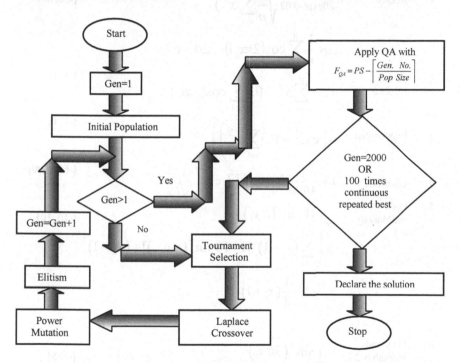

Fig. 1: Schematic representation of S-LXPM

An extensive series of experiment is carried out to fine tune the different parameters in S-LXPM. Finally they fixed as follows.

    i. For Laplace Crossover
        a) The location parameter a=0
        b) The scale parameter b=0.35
        c) The probability of crossover Pc=0.70

ii. For Power Mutation
  a) The probability of mutation Pm=0.007
  b) The index of mutation p=0.25

To stop a run, the criterion is either the objective function value is within 1% accuracy for 100 consecutive generations or a maximum of 2000 generation is attained. For a fair comparison problem size is fixed at 10 as in Deep and Das [1] and the population size (PS) is taken to be ten times the problem size.

Table 1. A set of Unconstrained Benchmark Problems

Sl. No.	Name of the Function	Function	Bounds
1	Ackley	$-20\exp(-0.02\sqrt{\dfrac{1}{n}\sum_{i=1}^{n}x_i^2})-$ $\exp(\dfrac{1}{n}\sum_{i=1}^{n}\cos(2\pi x_i))+20+e$	[-30, 30]
2	Cosine Mixture	$0.1n+\sum_{i=1}^{n}x_i^2-0.1\sum_{i=1}^{n}\cos(5\pi x_i)$	[-1, 1]
3	Exponential	$1-\left(\exp\left(-0.5\sum_{i=1}^{n}x_i^2\right)\right)$	[-1, 1]
4	Griewank	$1+\dfrac{1}{4000}\sum_{i=1}^{n}x_i^2-\prod_{i=1}^{n}\cos\left(\dfrac{x_i}{\sqrt{i}}\right)$	[-600,600]
5	Levy andMantalvo-1	$\dfrac{\pi}{n}\left(\begin{array}{l}10\sin^2(\pi y_1)+\\ \sum_{i=1}^{n-1}(y_i-1)^2[1+10\sin^2(\pi y_{i+1})]+(y_n-1)^2\end{array}\right)$ $y_i=1+\dfrac{1}{4}(x_i+1)$	[-10, 10]
6	Levy and Montalvo-2	$0.1\left(\begin{array}{l}\sin^2(3\pi x_1)\\ +\sum_{i=1}^{n-1}(x_i-1)^2[1+\sin^2(3\pi x_{i+1})]+\\ (x_n-1)^2[1+\sin^2(2\pi x_n)]\end{array}\right)$	[-5, 5]

7	Paviani	$45.778 + \sum_{i=1}^{10} [(\ln(x_i - 2))^2 + (\ln(10 - x_i))^2]$	[2, 10]
		$-\left(\prod_{i=1}^{10} x_i\right)^{0.2}$	

| 8 | Rastrigin | $10n + \sum_{i=1}^{n} [x_i^2 - 10\cos(2\pi x_i)]$ | [-5.12,5.12] |

| 9 | Rosenbrock | $\sum_{i=1}^{n-1} [100(x_{i+1} - x_i^2)^2 + (x_i - 1)^2]$ | [-30, 30] |

| 10 | Schwefel | $418.9829n - \sum_{i=1}^{n} x_i \sin\left(\sqrt{|x_i|}\right)$ | [-500,500] |

| 11 | Sinusoidal | $3.5 - [2.5 \prod_{i=1}^{n} \sin(x_i - \pi/6)$ | $[0, \pi]$ |
| | | $+ \prod_{i=1}^{n} \sin(5(x_i - \pi/6))]$ | |

| 12 | Zakharovs | $\sum_{i=1}^{n} x_i^2 + \left(\sum_{i=1}^{n} \frac{i}{2} x_i\right)^2 + \left(\sum_{i=1}^{n} \frac{i}{2} x_i\right)^4$ | [-5.12, 5.12] |

| 13 | Sphere | $\sum_{i=1}^{n} x_i^2$ | [-5.12, 5.12] |

| 14 | Axisparallel hyperellipso id | $\sum_{i=1}^{n} i x_i^2$ | [-5.12, 5.12] |

| 15 | Schewefel-3 | $\sum_{i=1}^{n} |x_i| + \prod_{i=1}^{n} |x_i|$ | [-10, 10] |

| 16 | Neumaier-3 | $\frac{n(n+4)(n-1)}{6} + \sum_{i=1}^{n} (x_i - 1)^2 - \sum_{i=2}^{n} x_i x_{i-1}$ | $[-n^2, n^2]$ |

| 17 | Salomon | $1 - \cos\left(2\pi\|x\|\right) + 0.1\|x\|, \|x\| = \sqrt{\sum_{i=1}^{n} x_i^2}$ | [-100,100] |

| 18 | Ellipsoidal | $\sum_{i=1}^{n} (x_i - i)^2$ | [-n, n] |
| | | $0.5 +$ | |

19	Schaffer-1	$$\left(\left(\sin\sqrt{\sum_{i=1}^{n}x_i^2}\right)^2-0.5\right)\bigg/\left(1+0.001\left(\sum_{i=1}^{n}x_i^2\right)\right)$$	[-100, 100]
20	Brown-3	$$\sum_{i=1}^{n-1}[(x_i^2)^{(x_{i+1}^2+1)}+(x_{i+1}^2)^{(x_i^2+1)}]$$	[-1, 4]
21	New function	$$\sum_{i=1}^{n}(0.2x_i^2+0.1x_i^2\sin 2x_i)$$	[-10, 10]
22	Cigar	$$x_1^2+100{,}000\sum_{i=2}^{n}x_i^2$$	[-10, 10]

4.3 Results and Analysis

In this section, the comparative computational results of H-LXPM and S-LXPM on 22 benchmark problems are recorded. Table 2 contains results for comparative success rate and average number of function evaluations, Table 3 contains computational time and Table 4 contains the mean and standard deviations of optimal fitness (objective function) values. All the values are reported for successful runs only.

Table 2. Comparative success rate and average number of function evaluation, using H-LX PM & S-LXPM

Problem No.	Success Rate		Average Function Evaluations	
	H-LXPM	S-LXPM	H-LXPM	S-LXPM
1	1	5	35700	22245
2	100	100	22526	13097
3	100	100	21214	12552
4	24	5	32833	12123
5	100	100	22242	16634
6	92	93	21593	20105
7	100	100	26936	14740
8	7	12	34557	17757
9	100	100	370352	145530
10	87	2	61831	20352

11	98	100	23640	16856
12	100	100	23690	13413
13	100	100	22618	16556
14	100	100	22492	17056
15	100	100	26272	18059
16	100	100	162066	67504
17	*	*	*	*
18	100	100	24058	17168
19	2	2	25600	12528
20	100	100	23286	16887
21	100	100	22470	16358
22	100	100	27404	16266

Table 3. Comparative computational time (in seconds) using H-LXPM and S-LXPM

Problem No.	H-LXPM	S-LXPM	Problem No.	H-LXPM	S-LXPM
1	**0.531**	0.557	12	0.492	**0.328**
2	0.343	**0.25**	13	0.347	**0.343**
3	0.27	**0.207**	14	**0.346**	0.359
4	0.517	**0.312**	15	0.375	**0.332**
5	**0.46**	0.469	16	2.356	**1.847**
6	0.546	**0.325**	17	*	*
7	0.49	**0.405**	18	0.362	**0.354**
8	**0.462**	0.499	19	0.399	**0.257**
9	7.301	**5.039**	20	0.575	**0.572**
10	0.875	**0.6**	21	0.481	**0.437**
11	0.401	**0.37**	22	0.395	**0.37**

Table 4. Comparative Mean and Standard Deviation of the optimal objective
function value   using H- LXPM & S-LXPM

Problem No.	Mean		Standard Deviation	
	H-LXPM	S-LXPM	H-LXPM	S-LXPM
1	6.10E-14	**9.80E-15**	0.00E 00	**6.11E-05**
2	1.57E-15	**7.69E-20**	1.50E-15	**5.79E-20**
3	1.32E-15	**6.61E-20**	2.68E-15	**5.91E-20**
4	4.01E-03	**3.35E-03**	4.43E-03	**4.14E-03**
5	1.18E-12	**5.55E-13**	3.07E-18	2.73E-13
6	**1.89E-13**	1.98E-13	**2.64E-18**	1.87E-13
7	4.70E-04	**4.69E-04**	7.23E-14	4.92E-10
8	1.65E-14	**5.78E-18**	1.67E-14	**6.22E-18**
9	1.29E+00	**1.08E+00**	1.81E+00	**1.40E+00**
10	**1.27E-04**	**1.27E-04**	6.82E-13	8.96E-11
11	3.46E-15	**8.21E-16**	1.39E-15	**1.37E-16**
12	7.41E-21	**1.63E-21**	6.04E-20	**5.55E-21**
13	1.94E-22	**7.46E-31**	5.32E-22	**2.99E-30**
14	1.73E-21	**1.03E-30**	1.17E-20	**7.05E-30**
15	1.43E-13	**1.07E-17**	2.79E-13	**4.40E-17**
16	2.98E-03	**9.53E-04**	1.05E-03	**4.67E-04**
17	*	*	*	*
18	1.78E-15	**5.47E-16**	1.19E-15	**8.31E-16**
19	9.72E-03	**9.71E-03**	2.78E-17	3.14E-15
20	3.81E-21	**6.48E-30**	2.25E-20	**4.65E-29**
21	1.76E-21	**2.46E-29**	1.01E-20	**1.93E-28**
22	3.93E-21	**8.01E-25**	2.76E-20	**5.21E-24**

The computational results clearly shows that the proposed S-LXPM
outperforms from H-LXPM on problem no 1-3, 8, 13-15, 20-22 and slight
improved values on problem no 4-5, 7, 9, 11-12, 16, 18, 19 but for problem
number 6, 10 the performance of S-LXPM is not as good as H-LXPM. The
problem number 17 could not be solved by either of the two methods. Out of 21
solved out problems, it is worth here to note that the success rate for S-LXPM is
more than that of H-LXPM. In terms of computational time and average number
of functions calls S-LXPM performs better. Hence from above discussion of
results we observed that the proposed variants produce good results on considered
test problems and could be recommended to solve other test problems and real life
based applications.

5 Conclusion

This paper presents a self adaptive Real Coded Genetic Algorithm called S-LXPM by incorporating Quadratic Approximation with its adaptive mechanism. The performance of both algorithms is evaluated with a wide variety of 22 benchmark test problems taken from the literature. The experimental results show that the new variant S-LXPM is efficient for solving uni-modal as well as multi-modal functions. This proposed algorithm seems to avoid the premature convergence up to some greater extend. It also manage to maintain the diversity in the population while solving non-linear optimization problems, with a higher efficiency and reliability.

References

[1] Deep, K., Das, K.N.: Performance improvement of real coded genetic algorithm with Quadratic Approximation based hybridization , Int J Intelligent Defence Support Systems. 2(4),319-334(2009).
[2] Deep, K., Thakur, M.: A new crossover operator for real coded genetic algorithm .Applied Mathematics and Computations, Elsevier. 188(1), 895-911(2007a).
[3] Rao,M.V.C., Palaniappan, R., Arumugam, M. S. : New Hybrid genetic operators for real coded genetic algorithm to compute optimal control of class of hybrid systems, Applied Soft computing ,Elsevier. 6, 38-52(2005).
[4] Goldberg, D. : Genetic algorithm in search, in Optimization and Machine Learning , Addison Wesley, Masschusetts USA(1989).
[5] Michalewicz ,Z.: Genetic Algorithm+Data Structures= Evoluation Programs, Springer-Verlag (1994).
[6] Mohan ,C., Nguyen, H.T. : A random search technique for global optimization based on quadratic approximation,   Asia Pacific Journal of Operations Research.11,93-101(1994).
[7] Deep, K., Thakur, M.: A new mutation operator for real coded genetic algorithm", Applied Mathematics and Computations, Elsevier doi:10.1016/j.amc.2007.03.04 (2007b)
[8] YoungSu, Yun., Chiung, Moon., Daeho, Kim. : Hybrid genetic algorithm with adaptive local search scheme for solving multistage-based supply chain problems, Computers & Industrial Engineering , Elsevier. 56, 821-838(2009).
[9] Su, Junjie., Zhong, Qiuhai. : Research on Prediction of Breath Period Signal based on RFN Network of Self adaptive Genetic Algorithm, International Conference on Electrical and Control Engineering ,IEEE.1798-1801(2010).
[10] Jing, Li., Han, Rui-feng. : A self adaptive genetic algorithm based on real coded. Biomedical Engineering and Computer Science (ICBECS), IEEE.1-4 (2010).
[11] LI, Xiaoquan., ZHANG, Xing., RENJianbo. : The study of SA with Improved GA, Procedia Engineering , Elsevier.15 ,168-172(2011).
[12] SHA, Lin-xiu., HE, Yu-yao. : A Novel Self- Adaptive Quantum Genetic Algorithm, 8th International Conference on Natural Computation, IEEE. 618-621(2012).

Partitional Algorithms for Hard Clustering Using Evolutionary and Swarm Intelligence Methods: A Survey

Jay Prakash, Pramod Kumar Singh

Computational Intelligence and Data Mining Research Lab
ABV-Indian Institute of Information Technology and Management, Gwalior
{jayprakash.iiitm@gmail.com; pksingh@iiitm.ac.in}

Abstract. Evolutionary and swarm intelligence methods attracted attention and gained popularity among the data mining researchers due to their expedient implementation, parallel nature, ability to search global optima and other advantages over conventional techniques. These methods along with their variants and hybrid approaches have emerged as worthwhile class of methods for clustering. Clustering is an unsupervised classification method. The partitional clustering algorithms look for hard clustering; they decompose the dataset into a set of disjoint clusters. This paper describes a brief review of evolutionary and swarm intelligence methods with their variants and hybrid approaches designed for partitional clustering algorithms for hard clustering of datasets.

Keywords: Evolutionary algorithm, Swarm intelligence, Clustering, Unsupervised classification.

1 Introduction

Clustering has been approached by many disciplines in the past few decades because of its wide applications. The clustering classifies or groups the objects of an unlabeled dataset on the basis of their similarity. Each group known as cluster consists of objects such that the objects belonging to same cluster have more similarity than objects belonging to the other clusters. Though several similarity measures, e.g., Manhattan distance, Euclidean distance, cosine distance, Mahalanabis distance, are available for different applications of clustering, the most widely used distance measure is the Euclidean distance [2]. Clustering problem is unsupervised in nature as the clusters are unknown in number, volumes, densities, shapes and orientation, if any [1].

J. C. Bansal et al. (eds.), *Proceedings of Seventh International Conference on Bio-Inspired Computing: Theories and Applications (BIC-TA 2012)*, Advances in Intelligent Systems and Computing 202, DOI: 10.1007/978-81-322-1041-2_44, © Springer India 2013

Clustering methods can be broadly categorized as partitional or hierarchical [25]. Hierarchical algorithms can be further categorized as agglomerative (bottom-up) or divisive (top-down). The agglomerative approach starts with each data point in its own cluster and merges them successively based on some proximity criterion until all data points are finally in a single cluster or some termination criterion meets. Divisive approach starts with all the data points in single cluster and splits them into different sub clusters based on some splitting criterion until each data point make a cluster of its own or some termination criterion meets. On the other hand, the partitional methods decompose the datasets directly into a set of disjoint clusters based on some optimization criterion.

Clustering can be carried out in two different modes fuzzy or Hard. In fuzzy clustering, each object may belong to each cluster with a certain fuzzy membership grade [7]. In hard clustering, the clusters are disjoint and each object belongs to exactly one cluster. We can mathematically represent hard partitioning of a data set X, based on the descriptions in [5] as follows. Consider $X = \{x_1, x_2, ..., x_N\}$, where every data point x_i in X corresponds to a n-dimensional feature vector. Then, hard partitioning of X is a collection $C = \{C_1, C_2, ..., C_k\}$ of K number of nonempty and non-overlapping groups of data such that

$$C_i \neq \phi \qquad\qquad i=1......K \qquad\qquad\qquad (1)$$
$$C_i \cap C_j = \phi \qquad\quad i, j=1,......,K \text{ and } i \neq j \qquad (2)$$

Clustering involves several issues and challenges, e.g., no standard guideline is available for feature selection/extraction, no standard design of clustering algorithm or selection criteria is available, no widely accepted validity and interpretation criteria is available for measuring the quality of results, no standard guideline is available for choosing clustering validity index for a particular application or dataset, no standardized solutions are available for clustering tasks having wide applications in many areas. However, many clustering algorithms, e.g., K-means [4], k-Medoid [15], self-organizing map (SOM) [9] are available in the literature for solving partitional clustering problems. Nonetheless, these prototype-based conventional algorithms have many problems, e.g., they are quite sensitive to the initialization of prototypes, they do not provide any guarantee to the global optimality rather they easily stuck into local minima [20], they are not computationally feasible especially for the large datasets and large number of clusters. Specifically, for these reasons, powerful metaheuristics such as evolutionary algorithms and swarm intelligence based algorithms offer to be more effective methods to overcome the deficiencies of the conventional clustering methods as they posses several desired key features like up gradation of the candidate solutions iteratively based on objective function (fitness function), decentralization, parallel nature, flexibility, robustness, no need of prior information about domain knowledge, and self organizing behavior [14].

In this paper, we present a brief survey of evolutionary algorithms genetic algorithm [12] and differential algorithm [10], a swarm intelligence based method Particle swarm optimization [11] along with their variants and hybrids of these algorithms within and with conventional algorithms for partitional hard clustering. The key aspects of the comparative study are (i) the algorithm and the operators in the backdrop of their design, (ii) representation of clusters – medoid-based, centroid-based, label-based, tree-based, or graph-based, (iii) encoding scheme of solutions - binary, integer, or real [6], (iv) type of dataset - real or synthetic, and (v) clustering validity indexes or objective function. In addition, we take note of an important issue that whether the number of clusters is known *apriori* or is determined at run time. However, Cura [3] presents a method, which may be employed for both known and unknown number of clusters by defining separate fitness function for each.

Rest of this paper is organized as follows. Section 2 presents a brief introduction of the evolutionary algorithms Genetic algorithm and Differential Evolution, and a most widely used swarm intelligence method Particle swarm optimization. Section 3 provides a review of solution methodologies based on evolutionary algorithms and swarm intelligence methods for the hard partitional clustering. Conclusions and few important issues for future research are presented in Section 4.

2 Evolutionary algorithms and swarm intelligence Techniques

In this section, we present a brief introduction to the particle swarm optimization, differential evolution and genetic algorithm.

2.1 Particle swarm optimization (PSO)

PSO is a swarm intelligence method modeled on social behavior of birds within a flock, where each particle represents a potential solution [16]. A change in the particle's position is influenced by its won best position in the history and best particle's position in the swarm; it is mathematically represented in equation (4). The performance of each particle is evaluated based on a predefined fitness function. In this way, the particles move towards the optimal solution within a defined search area. Based on neighborhood size of each particle within the swarm, it is categorized as global best PSO (gbest PSO) and local best PSO (lbest PSO). In gbest PSO, entire swarm is neighborhood of a particle whereas in lbest PSO a part of the swarm is neighborhood. Evaluation of current velocity of a particle (refer, equation (3)) consists of three parts (i) the previous velocity, which serves as a momentum of the previous flight direction; it is also referred to as inertia component, (ii) the cognitive component, which quantifies how much a particle is influenced by its own best position, and (iii) the social component, which quantifies how much a particle is influenced by the best particle in the neighborhood. In each generation, personal best (y) of a particle and global best particle (ŷ) in the neighborhood is evaluated using an objective function. Change of velocity (v) and position (x) of a particle in the swarm is expressed in equations (3) and (4).

$$v_{ij}(t+1) = v_{ij}(t) + c_1.r_{1j}(y_{ij}(t)-x_{ij}(t)) + c_2.r_{2j}(\hat{y}_j(t)-x_{ij}(t)) \qquad (3)$$
$$x_{ij}(t+1) = x_{ij}(t) + v_{ij}(t) \qquad (4)$$

Where j is the dimension of a particle, t is the iteration, r_1 and r_2 are random numbers in the interval $[0, 1]$, and c_1 and c_2 are positive acceleration constants; commonly, $c_1 = c_2 = 2$. The PSO is sensitive to number of parameters, e.g., number of particles in the swarm, dimension of the search space, inertia weight, acceleration coefficients, and neighborhood size. Inertia weight controls the exploration and exploitation abilities of the swarm [17]. The key strengths of PSO are its fast convergence towards global optimal solution, self organizing behavior, and few parameters to tune. Therefore, PSO has been widely applied to solve many diverse optimization problems in various areas including clustering. However, the main downside of the algorithm is premature convergence. Though it is originally proposed for continuous valued search space, Kennedy and Albert [19] proposed discrete PSO, known as binary PSO, to operate in discrete search space.

2.2 Differential Evolution (DE)

Storn and Price [21] propose DE, a stochastic population-based search strategy for continues –valued problem. The basic DE strategy can be described as the notation DE/x/y/z, where x is the vector to be mutated (a random vector or the best vector), y is the number of difference vectors, and z is the crossover scheme (binomial or exponential) [21]. Some of the more frequently used strategies comprise DE/rand/2/bin, DE/best/1/bin, and DE/best/2/bin. In each generation, for each individual X_1 in the current population, a mutation operator generates a mutant parameter vector T_1 by mutating a target vector X_2 with difference vector of two individuals X_3 and X_4, making sure that all these four individuals are different as shown in equation (5).

$$T_{1j} = X_{2j} + \beta(X_{3j}-X_{4j}) \qquad (5)$$

Where j represents dimension of an individual and β is a scaling factor, which manages the magnification of difference vector. For smaller values of β, the algorithm converges slowly but it can be used to explore search in local area. However, larger value of β assists greater diversity in search space but it may cause the algorithm to escape good solution. Therefore, the value of β should be balanced enough. Empirical results suggest that for large values of β and population size, solutions often converge prematurely [23]. Therefore, empirically decided value of $\beta = 0.5$ generally exhibits good performance [21] [13]. A final trial vector is then generated by crossover operator through implementing a discrete recombination of the mutant vector T_1 and the parent vector X_1. Further, fitness value of the trial vector is compared with the corresponding parent vector and the one with the higher fitness value is selected for the next generation. It is gaining attention of researchers, for solving optimization problems, as it is easy to implement and requires tuning of few parameters.

2.3 Genetic Algorithms (GAs)

GA is a stochastic optimization method that mimics the process of natural evolution [12]. In GA, a solution to the problem is represented by a chromosome; a set of chromosomes is called a population. Typically, solutions of initial population are generated randomly and then three genetic operators crossover, mutation, and selection are applied on solutions of current generation to produce new candidate solutions. The candidate solutions are selected for the next generation based on their fitness value. The crossover, which usually operates on two parents and produces two offsprings with some crossover probability, is primarily responsible for diversification of solutions in the entire search space. The mutation alters the gene of a chromosome to find solutions near it with a very small probability. Selection operator selects solutions from current generation for next generation based on some selection strategy. This process of applying genetic operators to produce candidate solutions for next generation continues till termination criteria meets. GAs provides optimal or near optimal solutions to a vast range of optimization problems as they are efficient, robust, self-adaptive, and parallel in nature. As GA is more suitable for discrete optimization problems and the clustering requires evaluating optimal number of groups in a multidimensional search space, it is suitable for the problem. Raghavan and Birchand [26] first introduced GA to the clustering.

3 Literature survey

In this section, we present a brief survey of recent ideas proposed for the clustering based on particle swarm optimization, differential evolution, and genetic algorithm along with their variants. In recent years, many researchers found that hybridization of metaheuristics within and with conventional algorithms increases the efficiency and accuracy of clustering. Therefore, hybrid methods are also an integral part of this study.

3.1 PSO variants for clustering

Das et al. [32] propose a multi-elitist particle swarm optimization based algorithm, which is able to obtain proper number of clusters and proper clusters automatically for complex and non-separable datasets. As the canonical PSO converges prematurely especially when swarm uses a small inertia weight or constriction coefficient, the authors bring a concept of growth rate β as an additional parameter to each particle. Its value is increased if fitness function of the particle improves in successive iteration. If local best of a particle has higher fitness value than the global best of the swarm, it is moved to the candidate area, and finally, a particle in the candidate area that has the highest growth rate becomes the global best of the swarm. The authors use real-coded encoding scheme, Gaussian kernelized distance measure, and Gaussian kernelized CS index as cluster validity index. The main downside of the algorithm is that the classification accuracy and mean number of classes obtained are considerably unsatisfactory when number of dimensions in the dataset exceeds 40. Chuang et.al [18] propose accelerated chaotic particle swarm optimization (ACPSO), which combines chaotic particle swarm optimization (CPSO) with an acceleration strategy; it works well when the number of clusters are known *apriori* and clusters are well defined. The CPSO maintains population diversity in PSO and saves it in entrapping in local convergence. It finds comparatively better solutions to its competitors, e.g., K-means, PSO, NM-PSO, K-PSO, on variety of datasets with reference to the sum of the intra-cluster distance and error rate.

3.2   PSO with conventional algorithms for clustering

Omran et al. [29] propose a method, which uses lbest-PSO to avoid the local optima. It initially selects a group of centroids, say X, randomly chosen from the dataset, say D, and cluster centroids of global best particle, say Y, using binary-PSO. Here, role of binary-PSO to choose active cluster centroids is similar as that of the activation threshold in [32]. The cluster centroids of global best particle are refined by K-means. The fitness of particles is evaluated on the basis of fitness function proposed by Turi [36]. To provide diversity in search space, new (X-Y) centroids are randomly selected from D to make group of centroids X for next iteration. The primary advantage of this algorithm is that the user can select any validity index suitable for his/her dataset. Qian and Li [30] present an effective data clustering method where preprocessing of dataset is done by principal component analysis (PCA). PCA is not only used to reduce the dimension of features but also to overcome multicollinearity between features. The authors use K-means to initialize particles in the PSO in order to improve its convergence rate. The authors use quantization error as fitness function to achieve quality of clustering. Some more aspects of this paper are described in Table 1. The main drawback of the algorithm is that it does not deal with the issue of incomplete attribute sets in the dataset.

For fast convergence and avoiding premature convergence, Tsai et al. [22] present selective regeneration particle swarm optimization (SRPSO), which improves original PSO with two new features as follows. First, setting of cognitive and social parameters C_1 and C_2; they enable fast convergence of algorithm. It is suggested that C_2 should be assigned greater value as compared to C_1 so that the new position of particle will be closer to the global best position. Second, selective particle regeneration is designed to avoid premature convergence; a likely trapped particle is re-assigned to a new position with certain degree of randomness in an attempt to escape from the local optimum. Taking advantages of K-means and SRPSO, the authors also develop a hybrid method KSRPSO. Here, K-means is used to generate the initial solutions for SRPSO to improve current best solution. Performance of algorithms is compared on the basis of sum of intra-cluster distances and error rate. The authors experiment on two synthetic and seven real datasets to show that the SRPSO and KSRPSO are efficient, accurate and robust over original PSO algorithm and K-means algorithm.

3.3 DE variants for clustering

Das et al. [25] propose a method that determines number of clusters at run time. The authors suggest that it may yield proper number of clusters of previously unhandled dataset in a reasonable time if a suitable choice of validity index is considered. It modifies classical DE by tuning parameters in two different ways to improve its slow conversance property. First, the scale factor is changed in random manner in the range (0.5, 1); it helps to retain diversity as the search progresses. Second, the crossover rate (Cr) linearly decreases from Cr_{max} to Cr_{min} with iterations during the run; it helps to initially explore a large search space and then search for the solutions in a relatively small local search space near the good solutions where the suspected global optimal solution resides. The chromosome representation and encoding scheme are similar to [32]. The authors compare the performance of their algorithm with the algorithms proposed in [28], [29], and [21] focusing on three important parameters (i) quality of solutions measured by DB index and CS index, (ii) ability to find proper number of clusters, and (iii) computational time required to find the solution. The major drawback of the method is that its performance heavily depends on the choice of a suitable clustering validity index.

3.4 DE with conventional algorithms for clustering

Kwedlo [20] proposes a method (DE-KM) that combines DE and K-means to obtain clusters based on sum of square errors (SSE) criterion when number of cluster is known. Here, DE is a global search algorithm with slow convergence and K-means is local search algorithm with fast convergence. The purpose of combining these two is to make them compliment of each other or to take advantages of both. Here, K-means algorithm is used in two ways, first, K-means is used to generate initial solutions in DE population, and second, K-means is applied to refine offsprings obtained by the mutation and crossover operators in DE. In both the cases, K-means is run till convergence. Tvrd'ık and K˘riv'y [37] empirically find that DE hybridized with K-means performs comparatively superior to the non-hybrid DE based on optimizing two basic criteria, trace of within scatter matrix and variance ratio criterion. Tian et.al [38] use K-Harmonic mean in the place of K-means with DE when number of clusters is fixed.

3.5 GAs variants for clustering

He et al. [35] propose a two-stage genetic clustering algorithm (TGCA) that can obtain proper number of clusters and proper clusters automatically. It employs real-coded encoding scheme, CH-value index as fitness function, and two-stage selection and genetic operators. Initially, it focuses on finding best number of clusters, and then gradually moves finding global optimal cluster centers. The authors show its performance in comparison to other automatic clustering algorithms, e.g., hierarchical agglomerative K-means, automatic spectral algorithm, standard genetic K-means clustering algorithm on four artificial and seven real-life datasets. Downside of the algorithm is that its search ability and grouping of data reduces in high dimensional datasets. Chang et al. [39] develop a GA for automatic clustering based on dynamic niching with niche migration to overcome the downsides of the fitness sharing approach. It works comparatively well even with datasets having very noisy background and high dimensions.

3.6 GAs with conventional algorithms for clustering

Kwedlo et al. [27] use GA to generate initial cluster centers for K-means as suggested in [29]. They use binary encoding for centroid based representation of clusters; a chromosome consists of N bits for a dataset having N objects. As the number of clusters K is fixed in the algorithm, a chromosome with exactly K bits set encodes a feasible solution. Therefore, two chromosome repairing methods random repair and distance based repair are used to modify each infeasible chromosome created by crossover and mutation operator before fitness function (SSE) evaluation. In reproduction, the authors use modified two-point crossover and bit-flip mutation to reduce the infeasible solutions. The Primary contribution of a method proposed by Wang et al. [40] is to create a new hybrid crossover operator in which string-coded crossover operator plays a role to retain the genetic characteristics of the parent generation while real-coded crossover carry out to maintain population diversity.

3.7 Hybrid PSO/DE/GA for clustering

As PSO suffers from premature convergence, Xu et al. [31] propose a hybrid DEPSO algorithm by using DE operator into the PSO procedure to add diversity to the PSO. It follows a two-step process; it carries canonical PSO in every odd iteration and DEPSO in every even iteration. In DEPSO, for every particle z_i with its personal best P_i, four other particles with their personal best are randomly selected from canonical PSO. An average weight is calculated by taking difference of personal best of two pairs selected randomly. This weighted difference is added to the P_i to create trial vector. Fitness of offspring created in trial vector is calculated against that of the parent; one with higher fitness is selected for the next generation. The authors use real encoding for centroid based representation of clusters as suggested in [32] and compare the performance of DEPSO with DE and PSO with regard to the number of iterations required for reaching a *cutoff* value of CH index in maximum of 100 runs. As the real partitions of the experimental datasets are already known, they use Rand index and Adjusted rand index to compare the performance of partitions to the real partitions. The same authors compare the performance of eight cluster validity indices on same hybrid algorithm [8].

Rehab [34] proposes a hybrid algorithm that works in two phases. In first phase, a genetically improved particle swarm optimization (GAI-PSO), a metaheuristic, which combines the standard velocity and position update rules of PSO with the idea of selection, mutation, and crossover from GA, performs a global search to produce optimal initial seed as cluster centroids for next phase. In second phase, K-means locally improves the seed of cluster centroids of first phase to optimal centroids of clusters. In this way, a combination of global search ability of nature-inspired algorithms and fast convergence of local optimal algorithm K-means avoid the drawback of both and produces good solution. Some more aspects of this paper are described in Table 1. Kuo et al. [24] present an approach, which combines binary-PSO and GA. A chromosome generated through binary-PSO goes through the crossover and mutation operators in GA to generate new chromosomes. It uses elitist selection to send chromosomes to the next iteration. This process is continued until a termination criterion is met. Finally, the best chromosome is refined using K-means. However, their work is highly influenced by the work in [29] as encoding scheme, cluster representation, fitness function, and diversity maintenance are similar to [29].

Table 1. A brief review of the proposed methods

Pub	Purpose/Feature	Encod	Rep	K	FF	DS
[25]	Clustering using improved DE	RC	CB	Auto	DB Index and CS measure	R
[31]	Clustering with DEPSO	RC	CB	Auto	CH index and SIL index	R & S

[32]	Automatic kernel clustering with a Multi-Elitist PSO	RC	CB	Auto	Gaussian kernelized CS measure	R & S
[29]	Dynamic Clustering using PSO	RC	CB	Auto	Proposed by Turi [36]	R
[35]	Clustering using two-stage GA	RC	CB	Auto	CH-index	R & S
[27]	GA is used to generate Initial cluster centres for the K-means	BC	CB	Fixed	Sum of square errors (SSE)	R & S
[30]	Data clustering using hybrid K-means and PSO where PCA serves to preprocess the dataset	RC	CB	Fixed	Quantization errors described in [30]	R
[34]	Data clustering using genetically improved PSO followed by K-means to refine the solutions	RC	CB	Fixed	Sum of clustering Error	R
[22]	Clustering using selective regeneration particle swarm optimization (SRPSO) method	RC	CB	Fixed	Sum of intra-cluster distances	R & S
[20]	Integration of DE and K-means	RC	CB	Fixed	Sum of square errors criteria (SSE)	R
[24]	Integration of PSO and GA for dynamic clustering	BC	CB	Auto	As proposed by Turi [36]	R

Descriptions of abbreviations used are as follows:
Pub: Publications; Encod: Encoding scheme of chromosome /particle; Rep: Representation of clusters; k: Whether number of clusters are fixed or determined automatically; FF: Cluster validity index /fitness function; DS: Type of Data set used; RC: Real-coded; BC: Binary-coded; CB: Centroid-based; Auto: Automatically; R: Real; S: synthetic

4 Conclusion

This paper presents a survey of recent literature for the hard partitional nature-inspired algorithms PSO, DE, and GA for data clustering. It describes key issues in the design of the algorithms for data clustering for known or unknown number of clusters such as kinds of algorithms with different operators and parameters, encoding scheme of solutions, objective function, and type of data set used. As algorithms have their own pros and cons, in some cases, authors describe some advantages and limitations too. As nature-inspired methods are global search heuristics with relatively slow convergence and the traditional algorithms, e.g., K-means, are local search methods with relatively fast convergence, researchers usually hybridize the two to take advantages of both. The literature further suggests that evolutionary algorithms and swarm intelligence methods are better alternative for clustering in different domains. In addition, hybridization of evolutionary and swarm intelligence algorithms themselves produces improved results by combing their supporting features as well as by taking care of exploration and exploitation capabilities of algorithms in the search space.

The authors are of the opinion that as the clustering algorithms are designed based on certain assumptions, under certain constraints, and for some specific applications, it is impossible to solve all the problems by a single clustering algorithm. Therefore, there are several issues, which require attention of the researchers.

a) Computational analysis of the algorithms is an important issue as clustering is a NP-Hard problem [33] that usually involves large datasets.

b) Develop algorithms that can handle variety of data types, e.g., numerical, categorical, binary, and varying number of clusters with equal ease and efficiency.

c) Develop algorithms that are capable to identify and rectify possible outliers and noise.

d) Design efficient clustering algorithms that work for fixed numbers of clusters as well as determine optimal number of clusters automatically based on the decision made by user.

e) Develop computationally efficient (hybrid) algorithms for automatic clustering when the dataset is unknown.

f) Developing an appropriate fitness function is still a challenging task for many clustering problems.

g) The most frequently used K-means to hybrid nature-inspired computing methods may be replaced with a better conventional local search method to increase the efficiency of hybrid algorithms.

References

[1] Fraley, C., Raftery, A. E.: How many clusters? Which clustering method? Answer via model-based cluster analysis. The Computer Jounal. 41, 578–588 (1998).

[2] Xu, R., Wunsch II, D.: Survey of clustering algorithms. IEEE Transaction Neural Networks. vol. 16, no. 3, pp. 645–678 May (2005).

[3] Cura, T.: A Particle Swarm Optimization approach to clustering. Expert Systems with Applications.39,1582-1588(2012)

[4] Jain, A. K., Dubes, R. C. :Algorithms for clustering data. Engle-wood Cliffs, NJ: Prentice-Hall, 1988.

[5] Hansen, P., Jaumard, B.:Cluster analysis and mathematical programming. Mathematical Programming. 79, 191–215 (1997).

[6] Hruschka, E. R., Campello , R. J. G. B., Alex, A. F., de Carvalho, A. C. P. L. F.,:A survey of evolutionary algorithms for clustering" IEEE Transactions on Systems, Man, and Cybernetics—Part C:Applications and Reviews. vol. 39, no. 2, pp. 133–155 March (2009).

[7] Jain, A. K., Murty, M. N., Flynn, P. J.: Data clustering: A review .ACM Comput. Surv., vol. 31, no. 3, pp. 264–323, Sep. (1999).

[8] Xu, R., Xu, J., Wunsch II,D.C: A comparison study of validity indices on swarm-intelligence-based clustering. IEEE Transactions on systems, man, and cybernatics-part B: Cybernetics. vol. 42, no. 4, pp. 1243–1256, Aug.(2012).

[9] Bigus, J. P.: Data mining with neural networks. McGraw-Hill, New York (1996).

[10] Price, K.V., Storn, R.M., Lampinen, J. A.: Differential evolution: A practical approach to global optimization. Berlin: Springer. (2005).

[11] Kennedy, J., Eberhart, R.: Morgan Kaufmann Publishers Inc. San Francisco, CA, USA (2001).

[12] Goldberg, D.E.: Genetic Algorithms-in Search, optimization and machine learning. Addison- Wesley Publishing Company Inc., London (1989).

[13] Ali, M.M., T˙orn, A.: Population set-based global optimization algorithms: Some Modifications and Numerical Studies. Computers & Operations Research. 31(10),1703–1725 (2004).

[14] Velmurugan, T., Santhanam, T.:A Survey of partition basedclustering algorithms on data mining: An experimental approach:,International Technology Journal. 10, 478-484(2011).

[15] Kaufman, L., Rousseeuw, P.J.:Clustering by means of Medoids, in Statistical Data Analysis Based on the L–Norm and Related Methods, edited by Y. Dodge, North-Holland, 405–416(1987).

[16] Kennedy, J., Eberhart, R.C.: Particle Swarm Optimization. In Proceedings of the IEEE International Joint Conference on Neural Networks, 1942–1948 (1995).

[17] Eberhart, R.C., Shi, Y.: Particle Swarm Optimization: Developments, applications and re-sources. In Proceedings of the IEEE Congress on Evolutionary Computation. 1, 27–30, May (2001).

[18] Chuang, L., Hsiao, C., Yang, C.: Chaotic Particle Swarm Optimization for data cluster-ing. Expert Systems with Applications. 38, 14555–14563 (2011).

[19] Kennedy, J., Eberhart, R.C.: A Discrete binary version of the Particle Swarm algorithm. In Proceedings of the World Multiconference on Systemics, Cybernetics and Informatics, 4104–4109 (1997).

[20] Kwedlo, W.: A clustering method combining differential evolution with the K-means algorithm. Pattern Recognition Letters. 32, 1613-1621 (2011).

[21] Storn, R., Price, K.: Differential evolution—A simple and efficient heuristic for global optimization over continuous spaces. J. Global Optim. Vol 11, no. 4, 341–359, Dec. (1997).

[22] Tsai, C.-Y., Kao, I.-W.: Particle swarm optimization with selective particle regeneration for data clustering. Expert Systems with Applications .38, 6565–6576 (2011).

[23] Chiou, J-P., Wang, F-S.: A Hybrid method of Differential Evolution with application to optimal control problems of A bioprocess system. In IEEE World Congress on Computational Intelligence, Proceedings of the IEEE International Conference on Evolutionary Computation, 627–632 (1998).

[24] Kuo, R.J., Syu, Y.J., Chen, Zhen-Yao , Tien,F.C.: Integration of Particle Swarm Optimization and Genetic Algorithm for dynamic clustering . Information Sciences.195,124-140 (2012).

[25] Das , S., Abraham, A., Konar, A.: Automatic clustering using an improved Differential Evolution algorithm. IEEE Transactions on System, Man, and Cybernetics-Part A: Systems and Humans, VOL. 38, NO. 1, JAN. (2008).

[26] Raghavan , V.V., Birchand, K.:A clustering strategy based on a formalism of the reproductive process in a natural system," in Proc. 2nd Int. Conf. Inf. Storage Retrieval. 10–22 (1979).

[27] Kwedlo,W., Iwanowicz, P. : Using Genetic Algorithm for Selection of initial cluster centers for the K-Means method . ICAISC, Part II, LNAI 6114, 165–172 (2010).

[28] Bandyopadhyay, S., Maulik, U.:Genetic clustering for automatic evolution of clusters and application to image classification, Pattern Recognition., vol. 35, no. 6, 1197–1208, Jun.(2002).

[29] Omran, M., Engelbrecht,A., Salman, A.:Dynamic clustering using Particle Swarm Optimization with application in unsupervised image classification. Proceedings of world academy of science, engineering and technology. Vol. 9, 199-204, Nov. (2005).

[30] Qian, X.-D., Li-Wie.: Date Clustering using principal component analysis and Particle Swarm Optimization. In: Proceedings of the 5th International Conference on Computer Science & Education Hefei, China.493-497 (2010).

[31] Xu, R., Xu, J., Wunsch II,D.C: Clustering with Differential Evolution Particle Swarm Optimization. In: IEEE Congress on Evolutionary Computation CEC (2010).

[32] Das, S., Abraham, A., Konar, A.: Automatic kernel clustering with a multi-elitist Particle Swarm Optimization algorithm. Pattern Recognition Letters 29, 688–699 (2008).

[33] Aloise, D., Deshpande, A., Hansen, P., Popat, P.: NP-hardness of euclidean sum-of-squares clustering. Machine Learning. 75, 245–248 (2009).

[34] Abdel-Kader, R.F.: Genetically Improved PSO algorithm for efficient data clustering. In: Proceedings of the IEEE Second International Conference on Machine Learning and Computing.71-75 (2010).

[35] He, H., Tan, Y.: A Two-stage Genetic Algorithm for automatic clustering. Neurocomputing. 81, 49-59 (2012).

[36] Turi, R.H.: Clustering-based colour image segmentation, PhD Thesis, Monash University, Australia (2001).

[37] Tvrd'ık, J., K ̆riv ́y, I.: Differential Evolution with competing strategies Applied to Partitional Clustering. LNCS 7269. 136-144 (2012).

[38] Tian, Y., Liu, D.,Qi,H.: K-Harmonic means data clustering with Differential Evolution. International Conference on Future BioMedical Information Engineering. 369-372(2009).

[39] Chang, D., Zhang, X., Zheng, C., Zhang, D.: A robust dynamic niching Genetic Algorithm with niche migration for automatic clustering problem. Pattern Recognition. 43, 1346–1360 (2010).

[40] Wang, J., Zhang, H., Dong, X., Xu, B., Mei, B.: An effective hybrid crossover operator for Genetic Algorithms to solve K-means clustering problem. Sixth International Conference on Natural Computation (ICNC). 2271-2275(2010).

A Two-Stage Unsupervised Dimension Reduction Method for Text Clustering

Kusum kumari bharti, Pramod kumar singh

Computational Intelligence and DataMining Research Lab
ABV-Indian Institute of Information Technology and Management Gwalior
Morena Link Road, Gwalior, MP, India
{kkusum.bharti@gmail.com; pksingh@iiitm.ac.in}

Abstract. Feature selection is widely used in text clustering to reduce dimensions in the feature space. In this paper, we study and propose two-stage unsupervised feature selection methods to determine a subset of relevant features to improve accuracy of the underlying algorithm. We experiment with hybrid approach of feature selection – feature selection (FS-FS) and feature selection – feature extraction (FS-FE) methods. Initially, each feature in the document is scored on the basis of its importance for the clustering using two different feature selection methods individually Mean-Median (MM) and Mean Absolute Difference (MAD).In the second stage, in two different experiments, we hybridize them with a feature selection method absolute cosine (AC) and a feature extraction method principal component analysis (PCA) to further reduce the dimensions in the feature space. We perform comprehensive experiments to compare FS, FS-FS and FS-FE using k-mean clustering on Reuters-21578 dataset. The experimental results show that the two-stage feature selection methods are more effective to obtain good quality results by the underlying clustering algorithm. Additionally, we observe that FS-FE approach is superior to FS-FS approach.

Keywords: Feature selection, feature extraction, relevant, redundant, text clustering.

1 Introduction

With rapid development of the Internet technology, the amount of digital information is drastically increasing day by day, thus making it difficult to extract relevant information in time from the huge corpus. Text clustering is one of the efficient ways to organize digital documents in a well-structured format to facilitate quick and efficient retrieval of relevant information in time. Traditionally, documents

J. C. Bansal et al. (eds.), *Proceedings of Seventh International Conference on Bio-Inspired Computing: Theories and Applications (BIC-TA 2012),* Advances in Intelligent Systems and Computing 202, DOI: 10.1007/978-81-322-1041-2_45, © Springer India 2013

are represented in a form of vector space model (VSM) [1], also known as feature vector model (FVM), where each dimension corresponds to a single term. It increases the number of dimensions in the representation model unmanageably even for a modest corpus of documents. It necessitates a requirement of relevant feature selection as a lot of features are irrelevant, redundant and noisy which adversely affect the efficacy and efficiency of the clustering algorithm and sometimes even misguide them. Moreover, high dimensional feature space also makes it difficult to apply computationally intensive algorithm on all datasets. In this context, dimension reduction plays a key role in reducing the number of features.

The primary aim of the dimension reduction methods is to generate low dimensional subspace from high dimensional feature space without scarifying the performance of the underlying algorithm. These methods not only improve the performance of the algorithm but also improve the understandability of the datasets, and reduce the storage requirements and computational complexity of the modelling process. These methods are broadly categorized as feature selection (FS) methods and feature extraction (FE) methods.

Feature selection (FS) is a process of selecting discriminative set of features based on documents intrinsic characteristics. A number of FS methods, e.g., information gain (IG) [2], mutual information (MI) [4], chi-square (χ^2) [5], document frequency (DF) [6], term variance (TV) [6], term strength (TS) [7], feature dispersion (FD) [8], are available in the literature for relevant features identification. Filter and wrapper are two subcategories of FS methods. Filter methods [2] quantify the relevancy of features based on statistical feature set, while wrapper methods [3] use classifier for selecting discriminative features subspace. These methods can be further subcategorized into two categories based on whether they used class label information or not for quantifying the relevancy of features. The FS methods, which use class label information for quantifying the relevancy of features, are known as supervised FS methods. The IG, MI, Chi-Square are all examples of supervised FS methods. The primary limitation of these methods is that they can only be used when class label information is known apriori. The other methods, which quantify relevancy of terms in data driven way, are known as unsupervised FS methods. The DF, TV, TS and FD are examples of unsupervised FS methods. Recently, a new concept of dispersion measures [9] has been proposed for quantifying the relevancy of terms. In this study, we use dispersion measures mean-median (MM) and mean absolute difference (MAD) for identifying relevant terms.

Feature extraction (FE) methods transform high dimensional feature space into a low dimensional subspace without losing much information. The Principal component analysis (PCA) [10], latent semantic indexing (LSI) [11], and independent component analysis (ICA) [12] are examples of FE methods. Among these, PCA has received alot of attention for dimension reduction [13, 14].

In this study, we experiment with two-stage unsupervised dimension reduction methods to reduce high dimensional feature space into low dimensional subspace by removing irrelevant and noisy features. The experimental methods hybridize the FS methods by integrating with FS and FE methods. Such methods embed the advantages of both of these methods for dimension reduction. In the first stage, each term is ranked by the discriminative power of features using MM [9] and MAD [9] individually. In the second stage, we transform thus obtained relevant feature subspace into low dimensional feature subspace using absolute cosine (AC) [9] measure to make a hybrid FS-FS method and using PCA to make a hybrid FS-FE method. It transforms high dimensional irrelevant and noisy feature space into low dimensional informative feature subspace. To evaluate the effectiveness of the two-stage models, experiments are conducted using k-mean clustering algorithm on Reuters-21578 dataset[1]. Experimental results explicitly indicate that the two-stage models are more effective as comparatively the reduction in the dimension of feature space is more as well as they improve the performance of the K-mean clustering algorithm measured by sum-of-square error (SSE).

The reminder of this paper is organized as follows. Section 2 discusses the related work. Section 3 describes the mean-median, mean absolute difference, absolute cosine, principal component analysis for dimension reduction followed by k-mean clustering for creating clusters of documents. Section 4 outlines the two-stage unsupervised dimension reduction methods. The effectiveness of the two-stage methods is demonstrated through experimental results for the clustering of text documents in Section 5. Finally, we conclude the paper in section 6.

2 Literature Review

Various studies [5, 6, 8, 9, 13, 14, 15, 16, 17, 18] are available in the literature in the area of feature selection for effectively and efficiently reducing the dimension of feature space without scarifying the performance of an algorithm. Recently, several authors propose hybrid approaches, which take advantages of both the FS and the FE methods [13, 14], filter and wrapper methods [17, 18] and filter-filter methods [9,20] for dimension reduction. Some of the examples of the hybrid algorithms are IG-PCA [13, 14], feature contribution degree and latent semantic indexing (FCD-LSI) [15], and maximum relevance minimum redundancy PSO (mr^2PSO) [25]. More recently, some authors [9, 18, 20] introduce the concept of redundant terms removal along with relevant term identification for creating discriminative feature subspace. In this section, we describe some works following this approach.

Uguz [13] uses hybrid methods for selecting discriminative set of features. He uses filter-wrapper (IG-GA) and FS-FE method (IG-PCA) to convert high dimensional feature space into low dimensional subspace. First, each term in the document is ranked on the basis of its importance for classification using the IG. In the

---

[1]http://kdd.ics.uci.edu/databases/reuters21578/reuters21578.html

second stage, a FS method (GA) and a FE method (PCA) have been applied separately to reduce the dimension of feature space. To evaluate the effectiveness of his proposed dimension reduction methods, he uses k-nearest neighbour (KNN) and C4.5 decision tree algorithm on Reuters-21,578 and Classic3 datasets collection for text categorization. The experimental results show that the proposed model is very effective in terms of precision, recall and F-measure. Similar approach has been used by Uguz [14] for Doppler signal selection.

Usually strict term matching is used to represent a document in the vector space model. It does not take into account the semantic correlation between features(terms)hence, leads to a poor classification/clustering accuracy. To overcome this problem, Meng et al. [15] use LSI to replace the individual terms with statistically derived conceptual indices. Initially, they use a feature selection method FCD to select discriminative features set and then construct a new semantic space between features based on the LSI. The results show effectiveness of the proposed model on spam database categorization. Song et al. [16] also use LSI to propose a GA based method GAL to reduce the dimension of feature space. They show the superiority of their approach GAL over conventional GA applied in VSM model for Reuter-21578 document clustering results.

Hsu et al. [17] introduce a hybrid feature selection method, which combines filer and wrapper methods. Initially, they select candidate feature subset from original feature space using computationally efficient filter methods F-Score and IG, and then refine the subspace using wrapper method inverse sequential floating search. Thus, the hybrid mechanism takes advantage of both the filters and the wrappers. They experiment with two bioinformatics problems, namely, protein disordered region prediction and gene selection in microarray cancer data and show that equal or better prediction accuracy can be achieved with a smaller feature set also. Foithong et al. [28] continue this concept of hybrid filer-wrapper method for feature selection. They select feature subspace using mutual information without requiring a user-defined factor for the selection of the candidate feature subset. Additionally, they reduce the computational cost of the algorithm and avoid local maxima of wrapper search method. Experimental results on 10 UCI datasets show the effectiveness of proposed model.

Akadi et al. [18] propose a two-stage selection process for gene selection. They combine MRMR (Minimum Redundancy–Maximum Relevance) and GA to select informative genes subset. Initially, they use MRMR to filter out noisy and redundant genes from the high dimensional space. Then, they use GA to select discriminative subset of features and use classifier accuracy as a fitness function to select the highly discriminating genes on five different datasets. The comparison of the MRMR-GA with MRMR and GA show that the proposed model is able to find the smallest gene subset that gives the most classification accuracy in leave-one-out cross-validation (LOOCV). Zhang et al. [20] and Unler et al. [25] also use MRMR to select discriminative features subspace. Zhang et al. [20], combine MRMR with ReliefF [27], which is an extension of Relief [26]. Unler et al. [25]

integrate MRMR with discrete particle swarm optimization (PSO) to bring the efficiency and accuracy of filters and wrappers for selecting discriminative feature subspace.

Many methods have been proposed in the literature [9, 18, 20] for relevancy and redundancy assessments of the terms. These methods are computationally expensive in nature. To weed out this problem, Ferreira et al. [9] propose an efficient unsupervised and supervised feature selection filter method for high dimension feature space. Their findings show the efficiency and effectiveness of proposed model with lower generalization error than state-of -the-art techniques.

3 Background

In this section, we briefly describe feature selection methods MAD, MM and AC, feature extraction method PCA, and the K-means algorithm used for creating clusters of documents.

3.1 Mean absolute difference (MAD)

This measure [9] is simplified form of term variance [6]. It quantifies the relevancy of feature based on their sample variance. In other words, it quantifies the relevancy by computing the difference of sample from the mean value. It is shown in equation 1.

$$MAD_i = \frac{1}{n} \sum_{j=1}^{n} |X_{ij} - \overline{X}_i| \qquad (1)$$

Here, X_{ij} is the value of feature i with respect to document j and $\overline{X}_i$ is the mean of the features, given by equation 2.

$$\overline{X}_i = \left(\frac{1}{n}\right) \sum_{j=1}^{n} X_{ij} \qquad (2)$$

3.2 Mean-median (MM)

This measure [9] is simplified form of skewness. It quantifies the relevancy of term based on absolute difference between the mean and median of $\overline{X}_i$. It is shown in equation 3.

$$MM_i = |\overline{X}_t - median(X_i)| \tag{3}$$

3.2 Absolute cosine (AC)

This measure [9] quantifies the similarity between two vectors by measuring the cosine of the angle between them.It is shown in equation 4.

$$\cos(\theta_{w_i,w_t}) = \left| \frac{\langle w_i. w_t \rangle}{\|w_i\|\|w_t\|} \right| \tag{4}$$

Here w_i and w_t are observed features, $\langle . \rangle$ denotes the dot product, and$\|.\|$is the Euclidean norm. It measures the angle between given vectors. As the term frequencies (tf-idf weights) cannot be negative, its value ranges between 0 and 1; 0means both the vectors are orthogonal and 1 means both the vectors are collinear.

3.3 Principal component analysis

Principal component analysis (PCA) [10] is introduced by Karl Pearson in 1901. It is a mathematical technique that uses an orthogonal linear transformation to convert the set of correlated variables into a set of new uncorrelated variables called principal components. The number of transformed principal components may be less than or equal to the original variables. Transformation of data to new space is carried out in such a way that the greatest variance of the data by any projection lies on the first component (called the first principal component), the second greatest variance lies on the second component, and so on. A detailed description of mathematical process of PCA is available in [22]. The most important step in computation of PCA is the determination of the appropriate principal components. The [21, 24] present some methods for automatically determining the number of suitable principal components in data driven ways. In this study, we use cumulative percentage of variance for determining the number of principal components as used by [21]. It automatically selects informative features based on defined threshold. In this study, the threshold value used in all experiments is <95%.

3.4 Text clustering method

In this study, we use k-mean [19] clustering algorithm for creating clusters of documents represented through various feature selection methods. It is the most widely used clustering algorithm in the literature, for document clustering too, because of its simplicity and efficiency. It is shown below.

Algorithm:
1. *Select k points randomly form the feature vector space. These points represent initial centres of the clusters.*
2. *Assign each vector (document) to the nearest cluster centre.*
3. *Recompute the cluster centres when all vectors are assigned to the nearest cluster centres.*
4. *Repeat Steps 2 and 3 until the stopping criteria (i.e., number of iteration, improvement over centre positions) is not satisfied.*

Here, we use cosine measure for computing difference between vectors and cluster centres. Cosine similarity of two vectors (d_i, c_j) is defined as:

$$cos(d_i, c_j) = \frac{dot\,(d_i, c_j)}{\|d_i\|\|c_j\|} \qquad (5)$$

In this study, we use k=5, where k is the required number of clusters. The k-mean is susceptible to the selection of initial cluster centre; apoor selection of initial cluster centre may lead to the local optimum solution. Therefore, we run the k-mean clustering algorithm with 10 different sets of initial cluster centres and finally select a set of partitions where the SSE is minimum.

4 Experimental model for two-stage unsupervised dimension reduction methods

Here, we discuss the two-stage experimental model for reducing dimensions in the representation model to improve the performance of the underlying algorithm. The first stage uses feature selection methods MM, MAD individually to remove irrelevant terms from high dimensional feature space, whereas the second stage uses feature selection method AC in one set of experiments and feature extraction method PCA in other set of experiments to transform the high dimensional feature subspace into a low dimensional subspace. The schema of the proposed model is shown in Fig. 1.

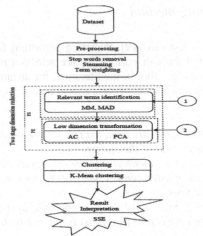

Fig 1: The general scheme of the three steps model for dimension reduction

The first stage uses the feature selection measures MM and MAD individually to compute relevancy of the features. Then, first p and q features are selected respectively to form a new feature subspace. Instead of manually selecting the optimal number of features we use cumulative relevance criteria as used by [16].

The second stage, for transformation of the high dimensional feature space into a low dimensional subspace, uses the feature selection method AC in one set of experiments as it does not depend upon the length and captures a scale invariant understanding of similarity and uses feature extraction method PCA in other set of experiments as it transforms a high dimensional subspace into a low dimensional subspace without losing much information. The MAD_AC and MM_AC are proposed by Ferreira et al. [9]. In this article, we use these models only for the comparison purpose. Next, the k-mean clustering algorithm has been used to evaluate the effectiveness of the transformed features space.

5 Experimental setup

The experiments are conducted on Retures-21578 dataset for text clustering using K-means to examine performance of the proposed model. All the pre-processing steps and feature selection methods are implemented in java whereas the feature extraction and clustering methods are implemented in MATLAB. The experiments are run on machine with 2.80 GHz CPU and 2 GB of RAM under windows 7 environment.

5.1 Performance measure

To evaluate the quality of various feature selection methods, we use the sum of square error (SSE) in K-means. It is an internal criterion and attempts to measure well a given partitioning is corresponding to the natural cluster structure of the data. It is formally defined as shown in equation 6:

$$SSE = \sum_{i-1}^{k} \sum_{x \in C_i} dist(c_i, x)^2 \qquad (6)$$

Here, $dist(c_i, x)^2$ is square of the Euclidean distance between the ith centroid and the data points belonging to that cluster. The centroid c_i is defined as:

$$c_i = \frac{1}{m_i} \sum_{x \in C_i} x \qquad (7)$$

5.2 Results& discussion

The pre-processing of the documents is performed in four stages. Initially, we re-move the stop words[2] -the common words that carry less weight and unnecessarily increase complexity of the underlying algorithm. In the second stage, we use por-ter stemmer[3]for reducing derivationally related terms into common root form. In third stage, we remove duplicate terms and finally in stage four we construct the document vector using tf-idf (term frequency inverse document frequency) term weighting scheme [23]. It results in a feature vector model with dimension 1339*7164, where 1339 is the number of documents and 7164 is the number of terms extracted after pre-processing steps.

We conduct three sets of experiments. The first set of experiments consists of ob-taining the relevant sets of features using FS methods, MAD and MM, individu-ally (refer, rows 1 and 2 of Table 1). The second set of experiments consists of ob-taining the relevant sets of features using FS-FS methods, MAD_AC and MM_AC, individually (refer, rows 3 and 4 of Table 1). The third set of experi-ments consists of obtaining the relevant sets of features using FS-FE methods, MAD_PCA and MM_PCA, individually (refer, rows 5 and 6 of Table 1).

---

[2]http://www.unine.ch/Info/clef/
[3]http://tartarus.org/~martin/PorterStemmer/

Table 1: Summary of experimental models

Model Name	Description	Idea	# of features selected
MAD	Mean absolute difference	MAD is used for relevant terms identification.	4522
MM	Mean-Median	MM is used for relevant terms identification.	4394
MAD_AC	Mean absolute difference relevant and redundant	Initially MAD is used for relevant terms identification, then AC is used for removing redundant terms.	4468
MM_AC	Mean-Median relevant and redundant	Initially MM is used for relevant terms identification, then AC is used for removing redundant terms.	3108
MAD_PC A	Mean absolute difference principal component analysis	Initially MAD is used for relevant terms identification, then PCA is used for transforming high dimensional feature space into low dimensional subspace.	658
MM_PCA	Mean-Median principal component analysis	Initially MM is used for relevant terms identification, then PCA is used for transforming high dimensional feature space into low dimensional subspace.	360

Let the first set of experiments, where the dimension reduction is done by relevant terms identification through FS methods MM and MAD individually, be called as one stage feature selection as the dimension reduction of feature space is done once only. The obtained results are shown in Fig. 2.The MM performs better than MAD. It suggests that the features selected by MM measure are more informative than features selected by MAD.

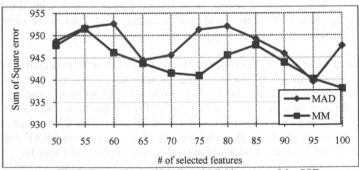

Fig 2: Comparison of MAD and MM in terms of the SSE

Our next sets of experiments are based on two-stage procedure for dimension reduction. We call them two-stage dimension reduction methods asthe dimension reduction of the feature space is done two times in succession. Primarily, two-stage dimension reduction can be performed in two ways FS-FS and FS-FE. We perform both sets of experiments and present the obtained results in Fig. 3 and Fig. 4 respectively. It shows that the two-stage dimension reduction methods, whether FS-FS or FS-FE, perform better than a one stage methods as performance of the K-means clustering algorithm improves with the two-stage dimension reduction methods in comparison to the one stage dimension reduction methods. The performance of FS-FS methods MAD_AC and MM_AC are comparatively better than one stage methods MAD and MM (refer, Fig. 3). Similarly, FS-FE methods MAD_PCA and MM_PCA perform better than MAD and MM (refer, Fig. 4).

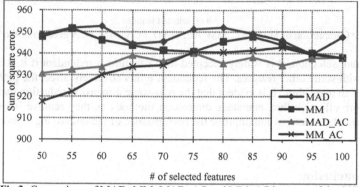

Fig.3: Comparison of MAD, MM, MAD_AC and MM_AC in terms of the SSE

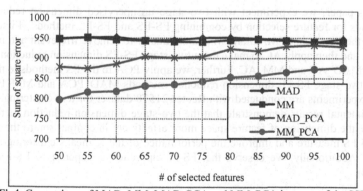

Fig4: Comparison of MAD, MM, MAD_PCA and MM_PCA in terms of the SSE

Further, we compare the performance of FS-FS methods and FS-FE methods. It is observed that FS-FE methods MAD_PCA and MM_PCA perform comparatively better than FS-FS methods MAD_AC, MM_AC (refer, Fig. 5).

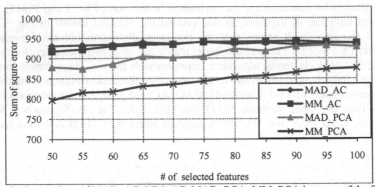

Fig. 5: Comparison of MAD_AC, MM_AC, MAD_PCA, MM_PCA in terms of the SSE

In feature selection, it is widely accepted that a feature is redundant if it is correlated with some other features. The redundant terms adversely affect the performance of an algorithm. It is evident that the two-stage dimension reduction methods
are more effective than one stage dimension methods as they remove irrelevant
and redundant features and improve the performance of the underlying algorithm.

6 Conclusion

In this paper, we experiment with two-stage unsupervised dimension reduction
methods for feature selection by combining FS-FS and FS-FE methods. First, we
apply the FS methods MAD and MM individually to remove irrelevant and noisy
terms, and then apply AC to make them two-stage FS-FS dimension reduction methods MAD_AC and MM_AC. Further, we apply PCA to the FS methods to make
them two-stage FS-FE dimension reduction methods MAD_PCA and MM_PCA.
The experiments are conducted with Reuter-21578 dataset for text clustering. The
experimental results demonstrate that the two-stage dimension reduction methods
reduce the dimension of feature space more effectively in comparison to the individual FS measure and improve the performance of the K-means clustering algorithm. Additionally, we observe that FS-FE approach is superior to FS-FS approach.

References

[1] Salton, G.: Wong, A.: Yang, C.S.: A Vector Space Model for Automatic Indexing. Communications of the ACM 18(11), 613–620 (1975)

[2] Quinlan, J.R.: Induction of decision tree. Machine learning 1(1), 81-106 (1986)

[3] Maldonado, S.: Weber, R.: A wrapper method for feature selection using Support Vector Machines. Information Sciences179(13), 2208-2217 (2009)

[4] Church, K.W.: Hanks, P.: word association norm, mutual information and lexicography. In proceeding of ACL 27, 76-83, Vancouver, Canada (1989)

[5] Li, Y.: Luo, C.: Chung, S.M.: Text Clustering with Feature Selection by Using Statistical Data. IEEE Transactions On Knowledge And Data Engineering, 20(5), 641-652 (2008)

[6] Liu, L.: Kang, J.: Yu, J.: Wang, Z.: A comparative study on unsupervised feature selection methods for text clustering. In: IEEE International Conference on Natural Language Processing and Knowledge Engineering 597–601 (2005)

[7] Yang, Y.: Noise reduction in a statistical approach to text categorization. In proceedings of the ACM SIGIR Conference on Research and Development in Information Retrieval 256-263 (1995)

[8] Ferreira, A.: Figueiredo, M.: Unsupervised Feature Selection for Sparse Data. In proceedings, European Symposium on Artificial Neural Networks, Computational Intelligence and Machine Learning, 339-344 (2011)

[9] Ferreira, A.J.: Figueired, M.A.T.: Efficient Feature Selection Filters for High-Dimensional Data. Pattern Recognition Letters 33(13), 1794-1804 (2012)

[10] Pearson, K..On Lines and Planes of Closest filt to Systems of Points in Space. Philosophical Magazine 1(6), 559-572 (1901)

[11] Deerwester, S.: Improving Information Retrieval with Latent Semantic Indexing. In proceedings of the 51st Annual Meeting of the American Society for Information Science 25, 36–40 (1988)

[12] Hyvärinen, A.: Oja, E.: Independent component analysis: a tutorial. In Helsinki University of Technology, Laboratory of computer and Information Science (1999)

[13] Uguz, H.: A two-stage feature selection method for text categorization by using information gain, principal component analysis and genetic algorithm. Knowledge-Based Systems 24(7), 1024-1032 (2011)

[14] Uguz,H.:A hybrid system based on information gain and principal component analysis for the classification of transcranial Doppler signals. Computer Methods and Programs in Biomedicine 107(3), 598-609, 2012.

[15] Meng, J.: Lin, H.: Yu, Y.: A two-stage feature selection method for text categorization. Knowledge-Based Systems 62(7), 2793-2800 (2011)

[16] Song, W.: Park, S.C.: Genetic algorithm for text clustering based on latent semantic indexing. Computers and Mathematics with Applications 57(11-12), 1901-1907 (2009)

[17] Hsu, H.H.: Hsieh, C.W.: Lu, M.D.: Hybrid feature selection by combining filters and wrappers. Expert Systems with Applications 38(7), 8144–8150 (2011)

[18] Akadi, A.E.: Amine, A.: Ouardighi, A.E.: Aboutajdine, D.: A two-stage gene selection scheme utilizing MRMR filter and GA wrapper. KnowlInfSyst26(3), 487–500 (2011)

[19] MacQueen, J. B.: Some Methods for classification and Analysis of Multivariate Observations". 1. Proceedings of 5th Berkeley Symposium on Mathematical Statistics and Probability. University of California Press. 281–297 (1967)

[20] Zhang, Y.: Ding, C.: Li, T.: Gene selection algorithm by combining reliefF and mRMR. IEEE 7th International Conference on Bioinformatics and Bioengineering. 1-10 (2008)

[21] Valle, S.: Li, W.: Qin, S.J.: Selection of the number of principal components: the variance of the reconstruction error criterion with a comparison to other methods. Ind, Engineering Chemistry Research 38(11), 4389–4401 (1999)

[22] Jilliffe, T.: Principal component analysis. ACM Computing Survey, Springer, Verlag, 1-47 (1986)

[23] Singh, P.K.: Machavolu, M.: Bharti, K.: Suda, R.: Analysis of Text Cluster Visualization in Emergent Self Organizing Maps Using Unigrams and Its Variations after Introducing Bigrams. In proce. of international conference on soft computing for problem solving, 967-978 (2011)

[24] Ferr, L.: Selection of components in principal component analysis: a comparison of methods, Computing and Statistical Data Analysis 19(6), 669–682 (1995)

[25] Unler, A.: Murat, A.: Chinnam, R.B.: mr^{2}PSO: A maximum relevance minimum redundancy feature selection method based on swarm intelligence for support vector machine classification. Information Sciences 181(20), 4625–4641 (2011)

[26] Kira, K.: Rendell, L.: The feature selection problem: Traditional methods and a new algorithm. In: Association for the Advancement of Artificial Intelligence. AAAI Press and MIT Press, Cambridge, MA, USA. 129–134 (1992)

[27] Kononenko, I.: Estimating attributes: Analysis and extensions of RELIEF. In: Proc. of the European Conference on Machine Learning. Springer, Verlag, 171–182 (1994)

[28] Foithong, S.: Pinngern, O.: Attachoo, B.: Feature subset selection wrapper based on mutual information and rough sets. Expert Systems with Applications 39(1), 574-584, (2012)

Author Index

A
Aggarwal, S., 429
Akashe, S., 139, 151, 189, 245, 265
Ali, S. S., 367
Annal Deva Priya Darshini, C., 359
Arya, A., 211

B
Babu, S., 139, 189
Bakariya, B., 407
Bakshi, T., 381
Banerjee, S., 455
Bharathi, M. A., 29
Bharti, K. K., 529
Bhateja, A. K., 101
Bhattacharyya, S., 127
Bhushan, S., 89

C
Chakraborty, A., 455
Chaudhary, K., 347
Cyriac, R., 41

D
Das, K. N., 489, 503
Das, V., 65
Deep, A., 479
Deep, K., 229
Din, M., 101
Dubey, M., 211

F
Fouzia Sayeedunnissa, S., 299

G
Gandhi, K., 367
Garg, K., 393
Ghosal, P., 455
Gupta, A., 429
Gupta, R., 211

H
Hameed, M. A., 299
Hans Raj, K., 323
Hussain, A. R., 299

J
Jain, A., 15
Jain, M. B., 15
Jammu, B. R., 77
Janarthanan, R., 127
Jauha, S. K., 465
Jha, P. C., 347, 367, 393, 429
Johri, R., 245

K
Khandelwal, S., 89
Khasnobish, A., 127
Konar, A., 127
Krishnaiah, J., 221